Springer Natural Hazards

The Springer Natural Hazards series seeks to publish a broad portfolio of scientific books, aiming at researchers, students, and everyone interested in Natural Hazard research. The series includes peer-reviewed monographs, edited volumes, textbooks, and conference proceedings. It covers all categories of hazards such as atmospheric/climatological/oceanographic hazards, storms, tsunamis, floods, avalanches, landslides, erosion, earthquakes, volcanoes, and welcomes book proposals on topics like risk assessment, risk management, and mitigation of hazards, and related subjects.

More information about this series at http://www.springer.com/series/10179

Radu Vacareanu · Constantin Ionescu
Editors

Seismic Hazard and Risk Assessment

Updated Overview with Emphasis
on Romania

 Springer

Editors
Radu Vacareanu
Seismic Risk Assessment Research Center
Technical University of Civil Engineering
  of Bucharest
Bucharest
Romania

Constantin Ionescu
National Institute for Earth Physics
Magurele
Romania

ISSN 2365-0656 ISSN 2365-0664 (electronic)
Springer Natural Hazards
ISBN 978-3-319-74723-1 ISBN 978-3-319-74724-8 (eBook)
https://doi.org/10.1007/978-3-319-74724-8

Library of Congress Control Number: 2018931926

Printed on acid-free paper

This Springer imprint is published by the registered company Springer International Publishing AG part of Springer Nature
The registered company address is: Gewerbestrasse 11, 6330 Cham, Switzerland

Preface

The Sixth National Conference on Earthquake Engineering and the Second National Conference on Earthquake Engineering and Seismology—6CNIS & 2CNISS—took place during June 14–17, 2017, in Bucharest, Romania, at the Technical University of Civil Engineering. The opening ceremony of the 6CNIS & 2CNISS was hosted by the Romanian Academy. The event was jointly organized by Technical University of Civil Engineering of Bucharest (*UTCB*), National Institute for Earth Physics (*INCDFP*), National Institute for Research and Development in Construction, Urban Planning and Sustainable Spatial Development (*URBAN-INCERC*) and General Inspectorate for Emergency Situations (*IGSU*), with the support of Romanian Academy Institute of Geography (*IGAR*), State Inspectorate for Construction (*ISC*), Romanian Association of Civil Engineers (*AICR*), and Romanian Association for Earthquake Engineering (*ARIS*). The peak audience to the conference amounted at 200 participants.

The 6CNIS & 2CNISS offered a stimulating and challenging environment to scientists, engineers, contractors, urban planners, and policy officials for the exchange of ideas, knowledge, lessons, and experience. The main topics of the 6CNIS & 2CNISS, mirrored in the conference sessions, were:

- Seismicity and hazard analysis;
- Geotechnical earthquake engineering;
- Seismic design and evaluation of buildings and structures;
- Innovative solutions for seismic protection of building structures;
- Seismic risk assessment and management of emergency situations.

During the three-day conference, a workshop devoted to the recently completed national project "RO-RISK—Disaster risk assessment at national", as well as three roundtables were organized. The latter addressed the issues of resilience-based assessment of structures, revision of Romanian seismic evaluation code, and quick post-earthquake evaluation of buildings.

The papers accepted by the International Advisory Committee and Scientific Committee were published by CONSPRESS (UTCB publishing house) in the Conference Proceedings. The authors of the accepted papers presented their contribution in the conference. The detailed program of the symposium is presented in Annex 1. The Conference Proceedings includes 14 keynote and invited papers, 15 papers in Seismology and Engineering Seismology, and 40 papers in Earthquake Engineering.

Renowned international scholars from the International Advisory Committee provided 11 keynote lectures in the plenary sessions. Moreover, ten invited lectures were delivered to the participants, in parallel sessions, by prominent international and national researchers. In addition, 53 contributions were presented in parallel sessions.

The most valuable papers selected by the members of International Advisory and Scientific Committees are published in this contributed volume, given the permission of CONSPRESS Publishing House. The papers selected from the Conference Proceedings were further extended by the authors and rereviewed before the final submission to Springer. The book benefits from the input of renowned researchers and professionals from Germany, Japan, Netherlands, Portugal, Romania, Turkey, and United Kingdom.

The book puts forward an updated overview of seismic hazard and risk assessment activities, with an emphasis on recent developments in Romania, a very challenging case study because of its peculiar intermediate-depth seismicity and evolutive code-compliant building stock. The content of the book focuses on seismicity of Romania, geotechnical earthquake engineering, structural analysis and seismic design regulations, innovative solutions for seismic protection of building structures, seismic risk evaluation, resilience-based assessment of structures, and management of emergency situations.

The book provides:

- Contributions of top researchers from seven countries;
- An integrated view on seismic hazard, risk, and resilience, with a perspective from civil protection, as well;
- Reliable and updated information on seismic hazard and risk of Romania based on the outcome of several recent research project: BIGSEES (BrIdging the Gap between Seismology and Earthquake Engineering: From the Seismicity of Romania towards a refined implementation of Seismic Action EN 1998-1 in earthquake resistant design of buildings), COBPEE (Community Based Performance Earthquake Engineering), and RO-RISK (Disaster risk assessment at national level);
- Comprehensive information on a scientifically challenging seismic source (Vrancea intermediate-depth) and a building stock designed according to compulsory seismic codes since 1963, constantly upgraded, and spanning all the progresses and paradigm shifts in engineering seismology and earthquake engineering.

This contributed volume aims at addressing some major challenges faced by Romanian researchers, educators, building officials, and decision-makers in disaster risk management and industry:

- The seismic evaluation and retrofitting of a large building stock; the national program for seismic retrofitting of residential buildings is very hard to implement because of social and institutional issues; meanwhile, the national programs for seismic retrofitting of public buildings, albeit its important achievements, need more focus and visibility;
- The highest seismic risk in Romania concentrated, by far, mostly in Bucharest; the expected social and economic impacts of destructive earthquakes are very high but mitigation is possible through a comprehensive and dedicated approach;
- The seismic design of buildings and structures for very large displacement demands in Romanian plain, and especially in Bucharest area;
- The rather weak public awareness; on average, there are two major earthquakes per century in Romania; thus, the education of population and the increase of public awareness are daunting tasks;
- The dormant shallow crustal seismic sources that, besides Vrancea intermediate-depth seismic source, endanger the territory of Romania;
- The quest for seismic resilience—a paradigm shift absolutely needed in Romania;
- The insurance premiums, versatile tools for enabling performance based design and boosting seismic rehabilitation, are not used so far up to their potential; moreover, the involvement of reinsurance companies and industry is rather scarce, so far.

The international cooperation is a major opportunity to address these challenges. Meanwhile, an approach similar to National Earthquake Hazards Reduction Program (NEHRP)—A research and implementation partnership—is definitely needed for focusing the activities aiming at seismic risk reduction in Romania.

The general readership of this book consists of researchers, engineers, decision-makers, and professionals working in the fields of seismic hazard and risk, seismic design, evaluation and rehabilitation of buildings and structures, building officials, insurers and reinsurers, and decision-makers for emergency situations, preparedness and recovery activities.

Acknowledgements

The Sixth National Conference on Earthquake Engineering and the Second National Conference on Earthquake Engineering and Seismology—6CNIS & 2CNISS—greatly benefited from the comprehensive and professional support of the International Advisory, Scientific and Technical Committees. The full list of the members is given in Annex 2.

The editors wish to extend their gratitude to the members of the International Advisory, Scientific and Organizing Committees, to the reviewers, and to all the contributing authors for their full involvement and cooperation in developing this book. The generosity of the 6CNIS & 2CNISS sponsors and the kind permission of CONSPRESS Publishing House of Technical University of Civil Engineering of Bucharest are gratefully acknowledged.

The valuable constructive comments and suggestions of the reviewers consistently enhanced the quality of the manuscripts. The editors deeply acknowledge the dutiful and careful checking of all the manuscripts performed, before the final submission to Springer, by our colleagues from Technical University of Civil Engineering of Bucharest (UTCB), Veronica Colibă and Ionuţ Crăciun.

A final word of gratitude is conveyed to Johanna Schwarz, Dörthe Mennecke-Bühler, Claudia Mannsperger, Ashok Arumairaj, Sooryadeepth Jayakrishnan and all the editorial staff from Springer International Publishing AG for their professional coordination and support in preparing this contributed volume.

Bucharest, Romania   Radu Vacareanu
March 2018   Constantin Ionescu

Contents

Part I
Seismicity Analysis

Earthquake Hazard Modelling and Forecasting for Disaster Risk Reduction

Alik Ismail-Zadeh

Abstract Understanding of lithosphere dynamics, tectonic stress localization, earthquake occurrences, and seismic hazards has significantly advanced during the last decades. Meanwhile despite the major advancements in geophysical sciences, yet we do not see a decline in earthquake disaster impacts and losses. Although earthquake disasters are mainly associated with significant vulnerability of society, comprehensive seismic hazards assessments and earthquake forecasting could contribute to preventive measures aimed to reduce impacts of earthquakes. Modelling of lithosphere dynamics and earthquake simulations coupled with a seismic hazard analysis can provide a better assessment of potential ground shaking due to earthquakes. This chapter discusses a quantitative approach for simulation of earthquakes due to lithosphere dynamics that allows for studying the influence of fault network properties and regional movements on seismic patterns. Results of earthquake simulations in several seismic-prone regions, such as the Vrancea region in the southeaster Carpathians, the Caucasian region, and the Tibet-Himalayan, are overviewed. A use of modelled seismicity in a probabilistic seismic hazard analysis is then discussed.

Keywords Lithospheric dynamics · Faults · Earthquake simulation
Earthquake disasters

1 Introduction

Challenges posed by disasters due to earthquakes or related natural hazards can result in negative impacts for the sustainable development. Since the beginning of the 21st century the impacts of earthquake-related disasters have risen rapidly,

A. Ismail-Zadeh (✉)
Institute of Applied Geosciences, Karlsruhe Institute of Technology,
Adenauerring 20b, 76137 Karlsruhe, Germany
e-mail: alik.ismail-zadeh@kit.edu

© Springer International Publishing AG, part of Springer Nature 2018
R. Vacareanu and C. Ionescu (eds.), *Seismic Hazard and Risk Assessment*,
Springer Natural Hazards, https://doi.org/10.1007/978-3-319-74724-8_1

e.g., the 2004 Sumatra-Andaman earthquake and induced tsunamis, the 2005 Kashmir earthquake, the 2008 Wenchuan earthquake and induced landslides, the 2011 Tohoku earthquake and induced tsunamis and flooding, and the 2015 Nepal earthquake and landslides. The disasters affect developed and developing countries and almost all sectors of economy at local, national, and regional levels. The vulnerability of our civilization to seismic events is still growing in part because of the increase in the number of high-risk objects and clustering of populations and infrastructure in the areas prone to earthquakes. Today an earthquake may affect several hundred thousand lives and cause significant damage up to hundred billion dollars. A large earthquake may trigger an ecological catastrophe, if it occurs in close vicinity to a nuclear power plant; the 2011 Tohoku earthquake and subsequent tsunamis damaged the cooling system of the Fukushima Dai-ichi nuclear power plant and resulted in nuclear radiation leaks (Ismail-Zadeh 2014).

Earthquake disasters continue to grow in number and impact, although the number of strong earthquakes a year is not growing with time (strictly speaking, the logarithm of the cumulative number of earthquakes shows a linear dependence on the earthquake magnitudes; Gutenberg and Richter 1944). Reducing disaster risk using scientific knowledge is a foundation for sustainable development (Cutter et al. 2015). Our knowledge on seismic hazard and other geohazards and their interaction with human systems is lacking in some important areas and is being challenged by the unforeseen or unknown repercussions of a rapidly changing and increasingly interdependent world. In such a tightly coupled world a disaster not only affects the immediate area where it occurs, but also may have cascading impacts that can affect other nations near and far.

Understanding of disasters associated with earthquakes (and/or other geohazards) comes from recent advances in basic sciences, engineering, and applied research including advances in: (1) geophysics, Earth's lithosphere dynamics, and understanding of hazardous event occurrences, all gained from Earth observations, analysis, and modelling; (2) comprehensive hazard assessments combining knowledge on seismology, geology, geodesy, geodynamics, electro-magnetism, hydrology, and soil properties with modelling tools and forecasting; (3) engineering related to development of earthquake resistant constructions; and (4) analysis of physical and social vulnerabilities and exposed values as well as studies of resilience of the society that help to prepare, respond and adapt to possible disruptions due to disasters. The understanding may become full, if it is based on co-designed and co-productive transdisciplinary work of all stakeholders involved in disaster risk reduction, including natural, social and behavioural scientists, engineers, insurance industry, media, emergency management and legislation authorities, and policymakers (Ismail-Zadeh et al. 2017a).

2 Lithospheric Dynamics, Tectonic Stresses and Earthquakes

According to plate tectonic theory (e.g. Turcotte and Schubert 2014), lithospheric plates are continually created and consumed. At ocean ridges, adjacent plates diverge and move away from the ridges cooling, densifying, and thickening. Once the lithosphere becomes sufficiently dense compared to the underlying mantle rocks, it bends, founders, and subducts into the mantle due to gravitational instability. The downward buoyancy forces, which are generated due to the excess density of the rocks of the subducting lithosphere, promote the lithosphere descent, but elastic, viscous and frictional forces resist it. The combination of these forces produces shear stresses high enough to cause earthquakes. Earthquakes occur as a sudden release of stresses. At ocean trench zones, they occur along the subducting lithosphere to the depths of about 660 km depending on the thermal state in the mantle. When an earthquake occurs, part of the released energy generates elastic waves propagating through the Earth. These waves generate sudden ground motions and shaking, which may result in building damage or collapse, landslides, tsunami wave generation, etc.

Ocean trenches are the sites of the world greatest earthquakes, which produce significant ruptures every century or rarely in the same ocean trench. According to Lay and Kanamori (2011), great earthquake occurrences can be understood from plate-boundary frictional characteristics. A slip may generate an earthquake at some patches of a fault surface, whereas the slip may occur without an earthquake. Conditionally stable patches normally slip continuously, but can slip seismically, when loaded abruptly during the failure of neighbouring seismic patches. "A failure of one seismic patch may produce a large earthquake. But when two or more patches fail in a cascade that also prompts conditionally stable regions between them to slip seismically, the result is a much larger earthquake than one would otherwise expect from just the seismic patches alone" (Lay and Kanamori 2011).

Although the majority of large earthquakes occur in subduction zones, some of them happen inside of continents (so called 'intraplate earthquakes'), especially in the regions of continental collisions, rifts, and grabens. For example, the Vrancea intermediate-depth strong earthquakes occur far away from lithosphere plate boundaries in the southern Carpathians (Romania) at the depths of about 70–180 km (e.g. Ismail-Zadeh et al. 2012). These earthquakes are considered to be associated with the relic slab sinking beneath the old Carpathian continental collision zone (e.g. Ismail-Zadeh et al. 2005, 2008). Other examples are the 2001 Bhuj M7.7 earthquake, which occurred in the Kutch rift zone, India, and caused widespread damage and death toll of over 20,000 people (Gupta et al. 2001); and large earthquakes, which took place in 1811 and 1812 in the New Madrid Rift complex, USA (Braile et al. 1986). According to the global risk analysis (Dilley et al. 2005), an area of about 10 million km^2 is estimated to undergo significant shaking by earthquakes or more precisely, peak ground acceleration of at least 2 m s^{-2} are expected in the area for 50 years with probability 0.1. This area is inhabited by more than one billion people.

Tectonic stress generation and its localization due to lithosphere plate motions is an important component in studies of earthquake-prone regions (e.g. Aoudia et al. 2007; Ismail-Zadeh et al. 2005, 2010). For example, Ismail-Zadeh et al. (2005) analysed stress localization in and around a descending lithospheric slab in the Vrancea region using a three-dimensional numerical model of mantle flow induced by the slab. The numerical model, which was based on temperatures derived from seismic P-wave velocity anomalies (Martin et al. 2006) and surface heat flow (Demetrescu and Andreescu 1994), predicted the maximum shear stress localization to coincide with the hypocentres of the intermediate-depth seismicity (Fig. 1), and stress orientations to be in a good agreement with the stress regime defined from fault-plane solutions for the intermediate-depth earthquakes.

Understanding stress re-distribution after earthquakes have been improved for the last few decades. Using the Coulomb failure criterion King et al. (1994) explored how changes in Coulomb stress conditions associated with an earthquake may trigger subsequent earthquakes (aftershocks). An earthquake alters the shear and normal stress on surrounding faults, and small sudden stress changes cause large changes in seismicity rate. These or relevant studies of tectonic stress and its

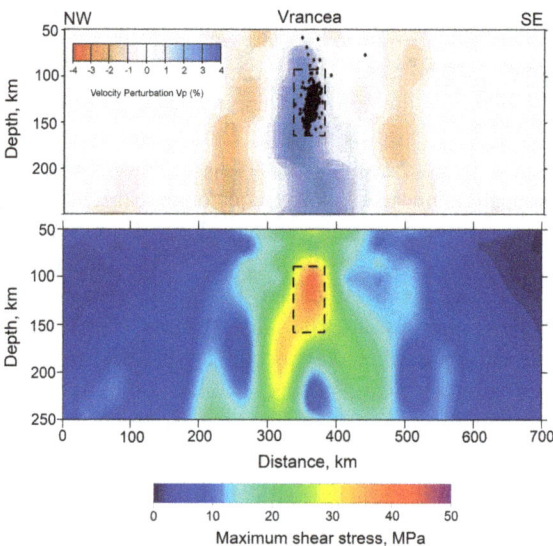

Fig. 1 Seismic velocity anomalies, earthquake hypocentres, and predicted tectonic stresses for the Vrancea region. Upper panel: *P*-wave velocity tomography image across NW-SE section through the south-eastern Carpathians (Martin et al. 2006) and the projection onto this cross section of the hypocentres of the Vrancea intermediate-depth earthquakes from 1995 to 2005. Lower panel: predicted maximum shear stress for the same cross-section. The dashed boxes delineate the area of hypocentres and maximum shear stress. After Ismail-Zadeh et al. (2005)

distribution before and after earthquakes provide important information on the localization of stresses and stress changes, which can be used in hazard assessment. Meanwhile, quantitative earthquake simulations at a fault or a system of faults can provide an insight into tectonic stress release at sites that have not been ruptured (or their ruptures in the past have not been recorded).

3   Quantitative Earthquake Simulations

Studying seismicity using the statistical and phenomenological analysis of earth-quake catalogues has the disadvantage that instrumental observations cover a short time interval compared to the duration of the tectonic processes responsible for earthquakes. The patterns of earthquake occurrence identifiable in a catalogue may be apparent and yet may not be repeated in the future. Meanwhile, historical data on seismicity are usually incomplete. Numerical modelling of seismic processes, including tectonic stress localisation and its release in earthquakes, allows gener-ating synthetic earthquake catalogues covering long time intervals and provides a basis for reliable estimates of the parameters of the earthquake occurrences (e.g. Soloviev and Ismail-Zadeh 2003).

Earth-specific quantitative earthquake simulators help to study seismicity in a system of faults (e.g. Gabrielov et al. 1990; Soloviev and Ismail-Zadeh 2003; Rundle et al. 2006). Particularly, a block-and-fault dynamics (BAFD) model by Gabrielov et al. (1990) can answer the following questions: how upper crustal (or lithospheric) blocks react to the plate motions and to a flow of the lower ductile crust (or highly viscous asthenosphere); how earthquakes cluster in the system of major regional faults; at which part of a fault system large events can occur, and what is the occurrence time of the extreme events; how the properties of the frequency-magnitude relationship change prior extreme events; and how fault zones properties influence the earthquake clustering, its magnitude and fault slip rates. The BAFD model was applied to several earthquake-prone areas. A recent review of the model and its applications can be found in Ismail-Zadeh et al. (2017b). Here we discuss briefly the application of the BAFD model to the Vrancea, Caucasus, and Tibet-Himalayan regions.

Vrancea. Large intermediate-depth earthquakes in Vrancea caused destruction in Bucharest (Romania) and shook central and eastern European cities several hundred km away from the hypocentres of the events. The earthquake-prone Vrancea region is situated at the bend of the south-eastern Carpathians. Epicentres of the intermediate-depth earthquakes are concentrated within a very small volume in the mantle extending to a depth of about 180 km. This seismicity is proposed to be associated with a relic part of the oceanic lithosphere sinking in the mantle (McKenzie 1972), and detached from or weakly linked to the continental crust (Fuchs et al. 1979). Seismic tomography imaged a high-velocity body beneath the Vrancea region (e.g. Bijward and Spakman 2000; Wortel and Spakman 2000; Martin et al. 2006; Raykova and Panza 2006), which can be interpreted as a dense lithospheric slab. A detailed review on geology, geodynamics, seismicity, and related studies in the Vrancea region can be found in Ismail-Zadeh et al. (2012).

The BAFD model was applied to study the dynamics of the lithosphere and intermediate-depth large earthquakes in the Vrancea region (Panza et al. 1997; Soloviev et al. 1999, 2000). Figure 2a shows the pattern of faults on the upper plane of the BAFD structure used to model the region. The catalogue of synthetic seismicity was computed for the period of 7000 years. The maximum value of the magnitude in the catalogue of model events is 7.6, close to the magnitude Mw = 7.7 earthquake occurred in Vrancea in 1940. The observed seismicity is shown in Fig. 2b, and the distribution of epicentres from the catalogue of synthetic events in Fig. 2c. A simple BAFD model, consisting of only three lithospheric blocks, was capable to reproduce the main features of the observed seismicity in space. Also, the modelling showed an irregularity in the time distribution of strong synthetic seismic events. For example, groups of large earthquakes occur periodically in the time interval from about 500 to 3000 model years, with a return period of about 300–350 years (Fig. 2d). The periodic occurrence of a single large earthquake with a return period of about 100 years is typical of the interval from 3000 to 4000 years. There is no periodicity in the occurrence of large earthquakes in the remaining parts of the catalogue of synthetic events. These results

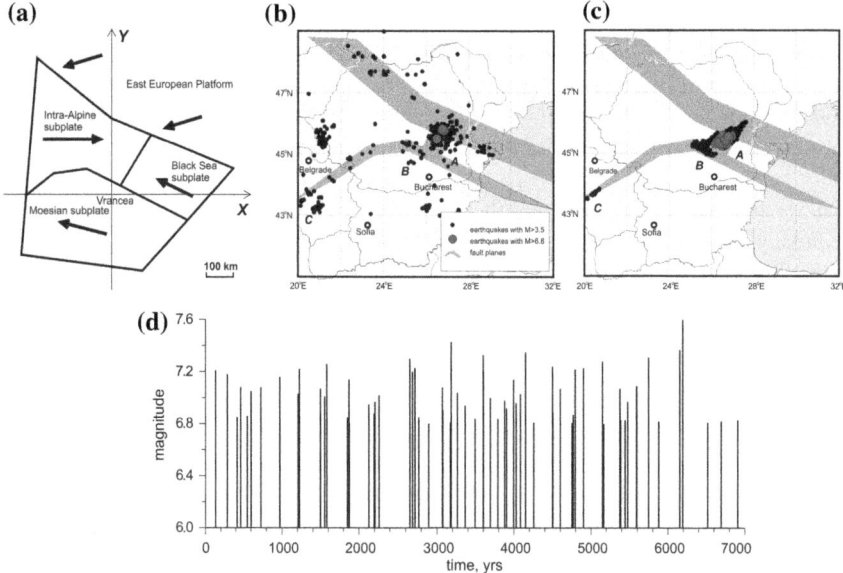

Fig. 2 BAFD model for the Vrancea region. **a** Block structure used for earthquake simulation in Vrancea. Arrows outside and inside the BAFD structure indicate the movements of the model blocks and of the sub-lithospheric mantle, respectively. Maps of observed seismicity in Vrancea in the period 1900–1995 (**b**) and modelled seismicity for 7000 years (**c**). Grey areas are the projections of fault planes on the upper plane. **d** Temporal distribution of large (M > 6.8) synthetic earthquakes for 7000 years. Modified after Soloviev and Ismail-Zadeh (2003)

demonstrate the importance of a careful estimation of the duration of seismic cycles to predict the occurrence of a future large earthquake.

Ismail-Zadeh et al. (1999) introduced a mantle flow into a BAFD model of the Vrancea region. The rate of the motion of the lithospheric blocks was determined from a model of mantle flow induced by a sinking slab beneath the Vrancea region (Ismail-Zadeh et al. 2000). It was shown that changes in modelled seismicity was controlled by small changes in the lithospheric slab's descent, e.g., slab position or dip angle (Ismail-Zadeh et al. 1999).

Caucasus. Earthquakes in Caucasia are associated with the Alpine-Himalayan seismic belt and collision between Eurasia and Arabia. The effect of this collision propagated into the Caucasus region in the early Pliocene (e.g., Philip et al. 1989). Regional deformation is quite complicated and includes lateral transport and rotation of crustal blocks along strike-slip faults (e.g., Reilinger et al. 2006). Most of deformations in Caucasus occurs within the Greater Caucasus Mountains (Jackson et al. 2002). The 1991 M7.0 Racha, Georgia, earthquake occurred in the western Greater Caucasus, and several destructive historical earthquakes occurred near the city of Shamakhi in the eastern Greater Caucasus in 1667, 1859, and 1902 (Kondorskaya et al. 1982).

The BAFD model was used to study earthquake occurrences in the Caucasian region (Ismail-Zadeh et al. 2017b; Soloviev and Gorshkov 2017). The movement of the block structure was constrained by the regional geodynamic models (Philip et al. 1989) and geodetic measurements (Reilinger et al. 2006). The BAFD experiments covered the time interval of 8000 years, which is by factor of about 80 larger than the earthquake catalogue. The synthetic earthquakes mimic the regional seismicity but with some exceptions: synthetic events occurred at some faults segments where no earthquakes have been previously recorded (instrumentally or historically). Meanwhile, the slope of the frequency-magnitude plot for synthetic events shows a good agreement with that for observed regional seismicity in the magnitude range from 4.5 to 7.

Tibet-Himalayas. Following the closure of the Mesozoic Tethys Ocean, the India-Asia collision initiated the development of the Himalayan range and the Tibetan plateau and induced widespread strain in south-eastern Asia and China. Ismail-Zadeh et al. (2007) developed a BAFD model for the region based on a model structure (Fig. 3a) made of six major blocks delineated by Replumaz and Tapponnier (2003). The crustal blocks were separated by thrust and strike-slip faults hundred km long. The movement of the blocks was specified with the rate constrained by the present rate of convergence between India and Asia (Bilham et al. 1997). Using the BAFD model, Ismail-Zadeh et al. (2007) performed a number of numerical experiments generating catalogues of synthetic earthquakes to analyse the earthquake clustering, frequency-to-magnitude relationships, earthquake focal mechanisms, and fault slip rates in the Tibet-Himalayan region. Each BAFD-generated catalogue contains crustal (down to 30 km) events occurring at the same fault planes introduced in the model and covers the time interval of 4000 years. The distribution of maximum magnitudes $M_{max, BAFD}$ of earthquakes from the merged three catalogues along the model faults is presented in Fig. 3b.

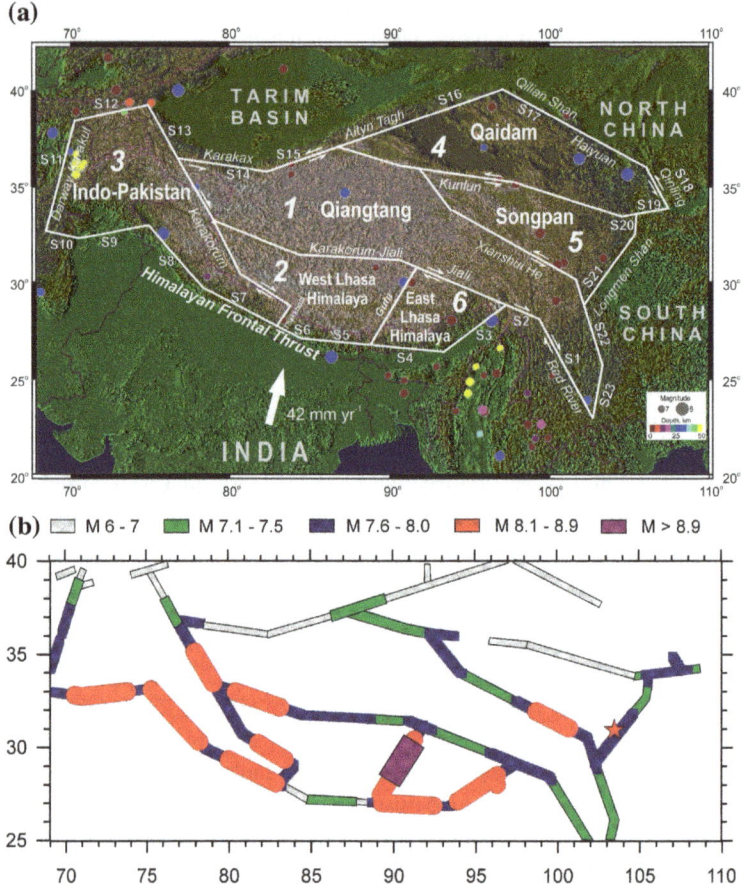

Fig. 3 BAFD model for the Tibet-Himalayan region. **a** Geometry of the block-and-fault structure and spatial distribution of observed seismicity ($M \geq 7.0$) from 1902 to 2000 (after Ismail-Zadeh et al., 2007). White bold lines (model faults) delineate the structural geological elements, and the white arrow indicates the motion of India relative to Eurasia. Earthquake epicentres are marked by coloured (depending on the depth of the hypocentres) circles. **b** Distribution of maximum magnitudes $M_{\mathrm{max},BAFD}$ of the earthquakes predicted by the three chosen BAFD model experiments. Modified after Sokolov and Ismail-Zadeh (2015)

The numerical results demonstrate that large events localize only on some of the faults, and this illustrates the fact that the BAFD model describes the dynamics of a network of crustal blocks and faults rather than the dynamics of individual fault planes. As an example, a cluster of large modelled events (M7.6–8.0) along the Longmen Shan fault was identified by the BAFD model (Fig. 3b). The 2008 M = 7.9 Sichuan (Wenchuan) earthquake (red star in Fig. 3b) occurred along this fault killing about 70,000 in addition to about 400,000 injured and about 20,000 missing people. Ismail-Zadeh et al. (2007) also analysed the focal mechanisms of the synthetic earthquakes computing the angle between the slip direction (in the fault plane) and the fault line and showed a reasonably good agreement between the focal mechanisms of the synthetic and observed earthquakes. Namely, in both cases (in the model and reality) most thrust faulting events occur on the Himalayan seismic belt and normal faulting events on the Gulu rift.

4 Earthquake Hazard Analysis

Seismic hazard can be defined as potentially damaging earthquake, which "may cause loss of life, injury or other health impacts, property damage, social and economic disruption or environmental degradation" (UN 2017). Meanwhile seismologists and earthquake engineers define seismic hazard in terms of engineering parameters of strong ground motion, namely, peak ground velocity/acceleration or seismic intensity. Seismic hazard assessment (SHA) is then based on the information about the features of excitation of seismic waves at the source, seismic wave propagation (attenuation), and site effect in the region under consideration and combines the results of seismological, geomorphological, geological, and tectonic investigations and modelling (e.g. Ismail-Zadeh 2014).

Two principal methods are intensively used in seismic hazard assessment: probabilistic and deterministic SHA. The probabilistic analysis deals with the rates of exceeding various levels of ground motion estimated over a specified time period (Cornell 1968). The probabilistic assessment considers uncertainties in earthquake source, path, and site conditions. The uncertainties are classified as epistemic and aleatory. Epistemic uncertainties reflect the incomplete knowledge about input model parameters to the assessment and variability of interpretations of available data, whereas aleatory uncertainties consider the inevitable unpredictability of the parameters (the uncertainties are mainly quantified using the standard deviation of the scatter around the mean values). The deterministic (or earthquake scenario-based) assessment model analyses the attenuation of seismic energy with distance from a specified earthquake to determine the level of ground motion at a particular site. Ground motion calculations consider the effects of local site conditions and use the available knowledge on earthquake sources and wave propagation processes. Namely, attenuation relationships are used for a given earthquake magnitude to calculate ground shaking demand for rock sites, which is then amplified by factors based on local soil conditions. Although the occurrence frequency of the ground motion is usually not addressed in the deterministic SHA, the method is robust for an assessment of seismic hazard and remains useful in decision-making (e.g. Babayev et al. 2010).

Compared to deterministic seismic hazard maps, probabilistic maps do not present the ground shaking at site, but instead present a level of ground shaking, which can be exceeded with a certain probability within a certain period of time. The probabilistic seismic hazard maps provide a low bound of seismic hazard useful for engineering purposes. Meanwhile non-expert scientists and other stakeholders dealing with hazard assessments try sometimes to associate the colours in probabilistic maps with potential ground shaking, and become surprised, if the real ground shaking is higher than predicted by the probabilistic maps. This reveals a weakness in mapping and interpretations of probabilistic SHA results.

An alternative approach to SHA is based on computations of realistic synthetic seismograms (Panza et al. 2001) and employs the knowledge of the crust and the lithosphere, seismic sources, and regional seismicity. The synthetic seismograms quantify peak values of acceleration, velocity and other ground motion parameters relevant to earthquake engineering. Considering a wide set of scenario events, including maximum credible earthquake, as well as geological and geophysical data, this approach offers the envelope of values of earthquake ground motion parameters (Panza 2017).

In many cases, large earthquakes are not accounted in the SHA due to the lack of information about them and unknown reoccurrence time of the extremes. Our present knowledge about characteristics of seismicity is based on observed (recorded) data and available historical data (obtained from palaeo seismological and archaeological studies, written stories about intensities of large earthquakes and some other sources). The information about large events in a particular region is incomplete as they are rare. Modelling of seismic events using earthquake simulators can overcome the difficulties in SHA by combination of observations, historic data and modelled results.

Sokolov and Ismail-Zadeh (2015) developed a new approach to a Monte-Carlo probabilistic SHA combining the observed regional seismicity with large magnitude synthetic events obtained by BAFD simulations. Three catalogues of synthetic events from Ismail-Zadeh et al. (2007) were chosen. The choice of these catalogues was based on the proximity of observed and simulated values of the following physical or statistical parameters: the slip rate at major faults, the orientation of crustal movements, the earthquake focal mechanism, the rate of seismic moment release, and the slope of the frequency-magnitude relations for the observed and modelled earthquakes. Also, the catalogues were chosen to minimize the difference between the annual rate of earthquake occurrence in the BAFD model and that of observed seismicity. Earthquake scenarios for hazard assessment are generated stochastically to sample the magnitude and spatial distribution of seismicity, as well as the distribution of ground motion for each seismic event.

This approach was employed for seismic hazards analysis in the Tibet-Himalayan region. Figure 4 presents a comparison of the results of the hazard assessment performed by the standard probabilistic SHA (the Global Seismic Hazard Assessment Program (GSHAP) model, Giardini et al. 1999) and the data-enhanced SHA approach accounting for large synthetic events (DESHA model, Sokolov and Ismail-Zadeh 2015). It is evident that the difference in ground shaking predicted by the two models, or, strictly speaking, the difference $\Delta_{PGA} = \log_{10}(\text{PGA}_{DESHA}/\text{PGA}_{GSHAP})$ between the relevant peak ground acceleration (PGA) estimates obtained by the DESHA model (PGADESHA) and by the GSHAP model (PGAGSHAP), is significant $(\text{PGA}_{\text{DESHA}} \geq 1.5\text{PGA}_{GSHAP}$, i.e. $\Delta_{PGA} \geq 0.176)$ for several areas including the area of the 2008 Wenchuan earthquake. Therefore, the DESHA model by Sokolov and

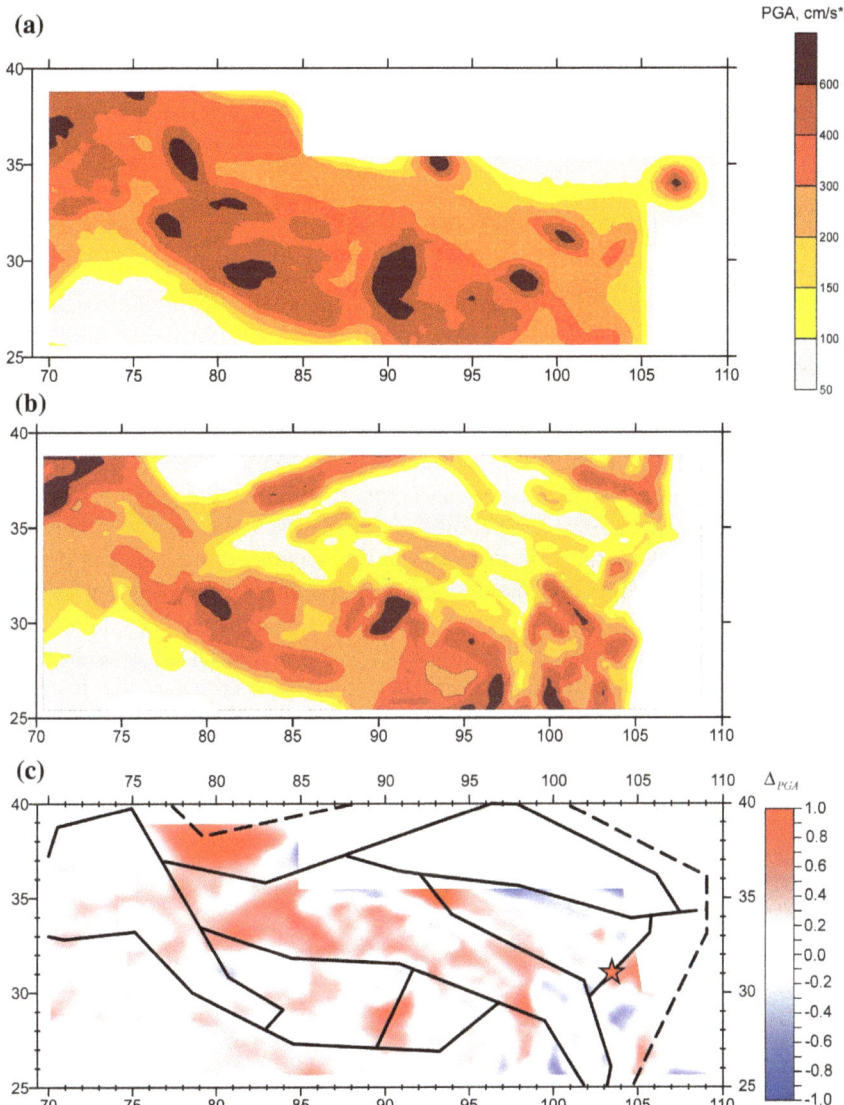

Fig. 4 Probabilistic seismic hazard maps of the Tibet-Himalayan region obtained (**a**) in this study and (**b**) from the GSHAP data (Giardini et al. 1999). The maps present peak ground accelerations, which are expected to be exceeded at least once in 50 years with probability 0.1 (with the average return period of 475 years). **c** The difference Δ_{PGA} between two ground motion assessments defined in the text. Black lines are the fault system used in the BAFD models. Due to the lack of information on earthquake source zones for the area located to the north from 35°N and to the east from 85°E, the PGA values are not calculated for this area After Sokolov and Ismail-Zadeh (2015)

Ismail-Zadeh (2015) allows for better understanding of ground shaking and could be useful for earthquake risk assessment, engineering purposes, and emergency planning.

Current probabilistic SHA methods are based on point-wise (site by site) assessments of ground shaking. Sokolov and Ismail-Zadeh (2016) analysed some features of multiple-site (MS) probabilistic SHA, i.e. the annual rate of ground-motion level exceedance in at least one site of several sites of interest located within an area or along a linear extended object, and showed that the expected ground motion level in selected area (multiple sites) are higher than that at individual sites. To assess the difference between point-wise and the multiple-site estimations of seismic hazards, Sokolov and Ismail-Zadeh (2016) considered an area located near the epicentre of the 2008 Wenchuan earthquake or stretched along the causative fault. It was assumed that the entire area is characterized by the same design PGA value (PGAPNT) and that there are several sites (e.g. strong-motion stations or critical facilities), where the multiple-site hazard (PGAMLT) should be evaluated. Also, a low level of the ground motion correlation (e.g. the correlation distance of 5 km) was assumed and allows for a high difference (variances) between the expected ground-motion parameters even for neighbouring sites. The dependence PGAMLT/PGAPNT on the area size, the number of sites, and the return period is shown in Fig. 5. The greater is the area (and the larger is the number of considered sites), the greater is the difference between the hazard estimates for individual sites and the multiple sites. In the considered case, the level of multiple-site hazard estimated for the 2475-year return period is larger than the highest recorded level of ground motion even for the relatively small area and for a

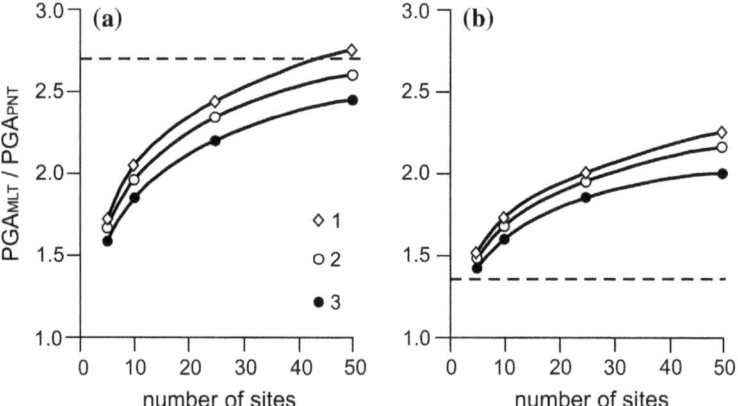

Fig. 5 Relation between the peak ground acceleration obtained from multiple-site (PGA$_{MS}$) and point-wise (PGA$_{PW}$) hazard assessments for the area of the 2008 Wenchuan earthquake for the 475-year (**a**) and 2475-year (**b**) return periods. The dashed line shows ratio between the maximum PGA recorded during the earthquake (about 810 cm s^{-2}) and the estimated design ground motion PGA$_{475}$ $\sim$ 300 cm s^{-2} (**a**) and PGA$_{2475}$ $\sim$ 600 cm s^{-2} (**b**). 1: area 600 km^2; 2: 400 km^2; 3: 100 km^2. Modified after Sokolov and Ismail-Zadeh (2015)

small number of sites. Sokolov and Ismail-Zadeh (2016) proposed a multi-level approach to probabilistic SHA considering fixed reference probability of exceedance (e.g. 10% in 50 years): (i) a standard point-wise hazard assessment to be performed in a seismic-prone region, and (ii) this analysis should be supplemented by a multiple-site hazard assessment for urban and industrial areas, or zones of an economic and social importance. This multi-level approach can provide better assessment of expected ground motion in a region of high vulnerability and/or exposed values, and hence enhance SHA.

5 Forecasting Seismic Hazard Events

The abruptness along with apparent irregularity and infrequency of large earthquake occurrences perpetuate the perception that earthquakes are random unpredictable phenomena (Ismail-Zadeh 2013). Earthquake prediction research has been widely debated, and opinions on the possibilities of prediction vary from the statement that earthquake prediction is intrinsically impossible (Geller et al. 1997) to the statement that prediction is possible, but difficult (Knopoff 1999; Keilis-Borok et al. 2001). Although many observations reveal unusual changes of geophysical fields at the approach of a large earthquake (e.g. animal behaviour, ground elevation, water level in boreholes, radon emission), most of them report a unique case history and lack a systematic description (e.g. Wyss 1991).

To predict an earthquake, one must "specify the expected magnitude range, the geographical area within which it will occur, and the time interval within which it will happen with sufficient precision so that the ultimate success or failure of the prediction can readily be judged. ... Moreover, scientists should also assign a confidence level to each prediction" (Allen et al. 1976). A prediction of an earthquake of certain magnitude range can be identified by (1) duration of time interval (long-term or a decadal time scale, intermediate-term or one to several years, short-term or weeks to months, and immediate or seconds to hours), and/or (2) territorial specificity based the rupture size of the incipient earthquake.

Earthquake forecasting based on monitoring of precursor(s) of earthquakes (that is, physical, chemical or biological signals, which indicate that a large earthquake can be anticipated) issues an alarm at the time of the abnormal behaviour of the precursor (the so-called time of increased probability of large event occurrence). Sometimes such forecasting is referred to as 'alarm-based earthquake prediction' (Ismail-Zadeh 2013). Several alarm-based intermediate-term earthquake prediction methods have been developed for the last decades (e.g., Keilis-Borok and Kossobokov 1990; Shebalin et al. 2006; among others). The intermediate-term earthquake prediction method (M8 algorithm; Keilis-Borok and Kossobokov 1990) aims to forecast large (magnitude 8 and greater) earthquakes by monitoring and analysis of several parameters of the seismic activity in a region. This prediction algorithm has received a fair amount of attention due to on-going real-time experimental testing unprecedented in rigor and global coverage (Ismail-Zadeh and

Kossobokov 2011). The accumulated statistical data of this experiment confirm intermediate-term predictability of large earthquakes with middle- to exact-range of location (Kossobokov 2013). Independent assessments of the M8 algorithm performance confirm that the method is non-trivial to predict large earthquakes (Zechar and Jordan 2008; Molchan and Romashkova 2011).

A short-term prediction of the devastating 1975 Haicheng (China) Ms = 7.0 earthquake by Chinese seismologists was based on monitoring anomalies in land elevation, in ground water level, and in seismicity prior to the large event and on the observations of peculiar behaviour of animals (Zhang-li et al. 1984). The success of this prediction stimulated further design of methods for diagnosis of an approaching large earthquake. Unfortunately, other prediction methods suggested at that time were not confirmed in the following years. The catastrophic 1976 Tangshan (China) Ms = 7.4 earthquake, which caused hundreds of thousands of fatalities, was not predicted.

A method of short-term earthquake prediction (the VAN method) was proposed in the 1980s; it was based on detection of characteristic changes in the geoelectric potential (so called "seismic electric signals", SES) via a telemetric network of conductive metal rods inserted in the ground (e.g. Varotsos et al. 1986). The anomaly pattern is continually refined as to the manner of identifying SES from within the abundant electric noise the VAN sensors are picking up. Despite many years of investigations and some progress for the last three decades (Lazaridou-Varotsos 2013), the short-term prediction by the VAN method is still controversial.

Another type of short-term prediction is based on calculating the probabilities of target events within future space-time domains, e.g., the short-term earthquake probability (STEP) method developed by Gerstenberger et al. (2005). The STEP method uses aftershock statistics to make hourly revisions of the probabilities of strong ground motion. The probability-based forecasts are the mean for transmitting information about probabilities of earthquakes in the region under monitoring. While the probability gains of short-term forecasts can be high, the probabilities of potential destructive earthquakes remain much smaller than 0.1 as the forecasting intervals are much shorter than the recurrence intervals of large earthquakes.

Although earthquake prediction methods are improving along with seismological data analysis, the current quality and accuracy of earthquake forecasting is significantly low compared to those of weather forecasting (Bauer et al. 2015). Our knowledge of earthquake physics and earthquake dynamics is still limited to predict large earthquakes with a relatively high accuracy. There is no strict mathematical description of non-linear dynamics of fault systems, earthquake generation process, and earthquake rupturing. This situation is unlike meteorology, where huge observations, a set of well-known mathematical equations describing atmospheric flow, and data assimilation techniques allow weather forecast with a relatively high accuracy for time scales ranging from a few hours to a few days. Success in earthquake hazard forecasting can be achieved by enhancement in (i) studies of non-linear tectonic stress generation, its localization and release in earthquakes as well as the rupture dynamics and earthquake mechanics; (ii) more geophysical, seismological and geodetic observations and data on fault and fault network geometry and their interaction; (iii) a mathematical description of the processes leading to earthquakes and methods for earthquake analysis (e.g., a set of governing equations and relevant conditions describing a transition to an earthquake; data assimilation and ensemble forecasting in stress and earthquake research); (iv) development of earthquakes models and powerful simulators (incl. numerical methods and supercomputer power to allow interactions in fault networks, at the scale of about 50 m).

Meanwhile, even current level of earthquake prediction capacity can be useful for seismic risk assessment and disaster preparedness. How the available basic scientific knowledge and earthquake-forecasting strategies could be linked to risk reduction strategies to make cost effective mitigations? "Optimized earthquake prediction algorithms will greatly aid disaster managers and decision makers in their preparations once a prediction is made. The loss functions help to develop a greater understanding between earthquake prediction research and disaster preparedness implementation, allowing for future improvements in earthquake disaster prevention" (Davis 2012).

6 Conclusion

Although the origin of seismology could be dated back to the Eastern Han Dynasty (25–220 AD) in China and great progress is achieved in this branch of science, there still exist several challenging problems (Forsyth et al. 2009), and among them an important question: how do faults interact and slip to generate an earthquake? The fundamental difficulty in solving the challenging problems is that no earthquake (process of rupture initiation at depth) has ever been observed directly and just a few of them were subject to an in situ verification of their physical parameters. Many of the extreme seismic events of the beginning of this century are linked in a chain of subsequent events that produce a disaster (the chain of events is called also 'concatenated events'). The 2004 great Sumatra-Andaman earthquake was following the Indian Ocean tsunami, which affected the coastal regions by flooding, contaminated water resources, and influenced the tourism business in the entire region. The 2008 Wenchuan earthquake was followed by the number of big landslides with a significant damage and human/economic losses. The 2011 Great East Japan earthquake followed by extreme tsunami waves, significant inundation, technological accident in the nuclear power plant, an environmental pollution with food and health security problems in the country and abroad.

Seismic hazards have been recognized as a grant challenge long time ago, and seismological and engineering communities concentrated their efforts on solving the challenging problem with a significant progress achieved. Recent advances are associated with neo-deterministic approach (Panza et al. 2010) and data-enhanced probabilistic approach (Sokolov and Ismail-Zadeh 2015) to SHA. Particularly, tectonically-realistic earthquake simulators help to generate seismicity for a significant duration of time and to employ large synthetic seismic events for hazard assessment. New scientific methods and approaches can enhance understanding of natural extreme events and vulnerabilities assessments at all levels including the gathering of a wide range of measures for increasing the resilience of society to seismic hazards.

With time, it was recognized that there are more challenging problems requiring co-produced research on disaster risks associated with earthquakes and other natural hazards that can enable understanding of the roots of potential disasters. Some challenging problems are analysed and discussed in the accompanying chapter "Earthquake risk assessment for seismic safety and sustainability" along with major components of earthquake risk assessment (seismic hazards, vulnerability and exposure), preventive measures to mitigate earthquake disasters, and the progress in disaster risk science in the framework of sustainability.

Acknowledgements The author acknowledges a support from the German Science Foundation (DFG grant IS-203/4-1).

References

Allen CR, Edwards W, Hall WJ, Knopoff L, Raleigh CB, Savit CH, Toksoz MN, Turner RH (1976) Predicting earthquakes: a scientific and technical evaluation—with implications for society. Panel on Earthquake Prediction of the Committee on Seismology, National Research Council. U.S. National Academy of Sciences, Washington DC

Aoudia A, Ismail-Zadeh A, Romanelli F (2007) Buoyancy-driven deformation and contemporary tectonic stress in the lithosphere beneath Central Italy. Terra Nova 19(6):490–495

Babayev G, Ismail-Zadeh A, Le Mouël J-L (2010) Scenario-based earthquake hazard and risk assessment for Baku (Azerbaijan). Nat Hazard Earth Syst Sci 10:2697–2712

Bauer P, Thorpe A, Brunet G (2015) The quiet revolution of numerical weather prediction. Nature 525:47–55

Bijwaard H, Spakman W (2000) Non-linear global P-wave tomography by iterated linearized inversion. Geophys J Int 141:71–82

Bilham R, Larson K, Freymueller J, Project Idylhim members (1997) GPS measurements of present-day convergence across the Nepal Himalayas. Nature 386:61–64

Braile LW, Hinze WJ, Keller GR, Lidiak EG, Sexton JL (1986) Tectonic development of the New Madrid rift complex, Mississippi Embayment, North America. Tectonophysics 131:1–21

Cornell CA (1968) Engineering seismic risk analysis. Bull Seismol Soc Amer 58:1583–1606

Cutter S, Ismail-Zadeh A, Alcántara-Ayala I, Altan O, Baker DN, Briceño S, Gupta H, Holloway A, Johnston D, McBean GA, Ogawa Y, Paton D, Porio E, Silbereisen RK, Takeuchi K, Valsecchi GB, Vogel C, Wu G (2015) Pool knowledge to stem losses from disasters. Nature 522:277–279

Davis CA (2012) Loss functions for temporal and spatial optimizing of earthquake prediction and disaster preparedness. Pure appl Geophys 169(11):1989–2010

Demetrescu C, Andreescu M (1994) On the thermal regime of some tectonic units in a continental collision environment in Romania. Tectonophysics 230:265–276

Dilley M, Chen RS, Deichmann W, Lerner-Lam AL, Arnold M (2005) Natural disaster hotspots: a global risk analysis. The World Bank, Washington DC

Forsyth DW, Lay T, Aster RC, Romanowicz B (2009) Grand challenges for seismology. EOS Trans AGU 90(41):361–362. https://doi.org/10.1029/2009EO410001

Fuchs K, Bonjer K, Bock G et al (1979) The Romanian earthquake of March 4, 1977. II. Aftershocks and migration of seismic activity. Tectonophysics 53:225–247

Gabrielov AM, Levshina TA, Rotwain IM (1990) Block model of earthquake sequence. Phys Earth Planet Inter 61:18–28

Geller RJ, Jackson DD, Kagan YY, Mulargia F (1997) Earthquakes cannot be predicted. Science 275:1616–1617

Gerstenberger MC, Wiemer S, Jones LM, Reasenberg PA (2005) Real-time forecasts of tomorrow's earthquakes in California. Nature 435:328–331

Giardini D, Grünthal G, Shedlock KM, Zhang P (1999) The GSHAP global seismic hazard map. Ann Geofis 42:1225–1228

Gupta HK, Purnachandra Rao N, Rastogi BK, Sarkar D (2001) The deadliest intraplate earthquake. Science 291:2101–2102

Gutenberg B, Richter CF (1944) Frequency of earthquakes in California. Bull Seismol Soc Amer 34:185–188

Ismail-Zadeh AT, Keilis-Borok VI, Soloviev AA (1999) Numerical modelling of earthquake flows in the southeastern Carpathians (Vrancea): effect of a sinking slab. Phys Earth Planet Inter 111:267–274

Ismail-Zadeh AT, Panza GF, Naimark BM (2000) Stress in the descending relic slab beneath Vrancea, Romania. Pure appl Geophys 157:111–130

Ismail-Zadeh A, Mueller B, Schubert G (2005) Three-dimensional modelling of present-day tectonic stress beneath the earthquake-prone southeastern Carpathians based on integrated analysis of seismic, heat flow, and gravity observations. Phys Earth Planet Inter 149:81–98

Ismail-Zadeh AT, Le Mouël JL, Soloviev A, Tapponnier P, Vorobieva I (2007) Numerical modelling of crustal block-and-fault dynamics, earthquakes and slip rates in the Tibet-Himalayan region. Earth Planet Sci Lett 258:465–485

Ismail-Zadeh A, Schubert G, Tsepelev I, Korotkii A (2008) Thermal evolution and geometry of the descending lithosphere beneath the SE-Carpathians: an insight from the past. Earth Planet Sci Lett 273:68–79

Ismail-Zadeh A, Aoudia A, Panza G (2010) Three-dimensional numerical modelling of contemporary mantle flow and tectonic stress beneath the Central Mediterranean. Tectonophysics 482:226–236

Ismail-Zadeh AT, Kossobokov VG (2011) Earthquake prediction M8 algorithm. In: Gupta H (ed) Encyclopaedia of solid earth geophysics. Springer, Heidelberg, pp 178–182

Ismail-Zadeh A, Matenco L, Radulian M, Cloetingh S, Panza G (2012) Geodynamic and intermediate-depth seismicity in Vrancea (the south-eastern Carpathians): current state-of-the-art. Tectonophysics 530–531:50–79

Ismail-Zadeh A (2013) Earthquake prediction and forecasting. In: Bobrowsky PT (ed) Encyclopedia of natural hazards. Springer, Dordrecht, pp 225–231

Ismail-Zadeh A (2014) Extreme seismic events: from basic science to disaster risk mitigation. In: Ismail-Zadeh A, Urrutia Fucugauchi J, Kijko A, Takeuchi K, Zaliapin I (eds) Extreme natural hazards, disaster risks and societal implications. Cambridge University Press, Cambridge, pp 47–60

Ismail-Zadeh A, Cutter SL, Takeuchi K, Paton D (2017a) Forging a paradigm shift in disaster science. Nat Haz 86(2):969–988

Ismail-Zadeh A, Soloviev A, Sokolov V, Vorobieva I, Muller B, Schilling F (2017b) Quantitative modelling of the lithosphere dynamics, earthquakes and seismic hazard. Tectonophysics. https://doi.org/10.1016/j.tecto.2017.04.007

Jackson J, Priestley K, Allen M, Berberian M (2002) Active tectonics of the South Caspian Basin. Geophys J Int 148:214–245

Keilis-Borok VI, Kossobokov VG (1990) Premonitory activation of earthquake flow: algorithm M8. Phys Earth Planet Inter 61:73–83

Keilis-Borok V, Ismail-Zadeh A, Kossobokov V, Shebalin P (2001) Non-linear dynamics of the lithosphere and intermediate-term earthquake prediction. Tectonophysics 338(3–4):247–259

King GCP, Stein RS, Lin J (1994) Static stress changes and the triggering of earthquakes. Bull Seismol Soc Amer 84:935–953

Knopoff L (1999) Earthquake prediction is difficult but not impossible. Nature debates. http://www.nature.com/nature/debates/earthquake. Accessed on 28 Aug 2017

Kondorskaya NV, Shebalin NV, Khrometskaya YA, Gvishiani AD (1982) New catalog of strong earthquakes in the USSR from ancient times through 1977, Report SE-31, World Data Center A for Solid Earth Geophysics, NOAA, National Geophysical Data Center, Boulder, Colorado, USA, 608 p

Kossobokov VG (2013) Earthquake prediction: 20 years of global experiment. Nat Haz 69:1155–1177. https://doi.org/10.1007/s11069-012-0198-1

Lay T, Kanamori H (2011) Insights from the great 2011 Japan earthquake. Phys Today 64(12):33–39

Lazaridou-Varotsos MS (2013) Earthquake prediction by seismic electric signals: the success of the VAN method over thirty years. Springer, Heidelberg

Martin M, Wenzel F, the CALIXTO working group (2006) High-resolution teleseismic body wave tomography beneath SE-Romania—II. Imaging of a slab detachment scenario. Geophys J Int 164:579–595

McKenzie DP (1972) Active tectonics of the Mediterranean region. Geophys J Royal Astron Soc 30:109–185

Molchan G, Romashkova L (2011) Gambling score in earthquake prediction analysis. Geophys J Int 184:1445–1454

Panza GF (2017) NDSHA: Robust and reliable seismic hazard assessment. In: Proceedings, international conference on disaster risk mitigation, Dhaka, Bangladesh, September 23–24. arXiv:1709.02945

Panza GF, Soloviev AA, Vorobieva IA (1997) Numerical modelling of block-structure dynamics: application to the Vrancea region. Pure appl Geophys 149:313–336

Panza GF, Romanelli F, Vaccari F (2001) Seismic wave propagation in laterally heterogeneous anelastic media: theory and applications to seismic zonation. Adv Geophys 43:1–95

Panza GF, Irikura K, Kouteva M, Peresan A, Wang Z, Saragoni R (2010) Advanced seismic hazard assessment. Pure Appl Geophys 168. https://doi.org/10.1007/s00024-010-0179-9

Philip H, Cisternas A, Gvishiani A, Gorshkov A (1989) The Caucasus: an actual example of the initial stages of continental collision. Tectonophysics 161:1–21

Raykova RB, Panza GF (2006) Surface waves tomography and non/linear inversion in the southeast Carpathians. Phys Earth Planet Inter 157:164–180

Reilinger R, McClusky S, Vernant P, Lawrence S, Ergintav S, Cakmak R, Ozener H, Kadirov F, Guliev I, Stepanyan R, Nadariya M (2006) GPS constraints on continental deformation in the Africa-Arabia-Eurasia continental collision zone and implications for the dynamics of plate interactions. J Geophys Res 111:BO5411. https://doi.org/10.1029/2005jb004051

Replumaz A, Tapponnier P (2003) Reconstruction of the deformed collision zone between Indian and Asia by backward motion of lithospheric blocks. J Geophys Res 108:2285. https://doi.org/10.1029/2001JB000661

Rundle PB, Rundle JB, Tiampo KF, Donnellan A, Turcotte DL (2006) Virtual California: fault model, frictional parameters, application. Pure appl Geophys 163:1819–1846

Shebalin P, Keilis-Borok V, Gabrielov A, Zaliapin I, Turcotte D (2006) Short-term earthquake prediction by reverse analysis of lithosphere dynamics. Tectonophysics 413:63–75

Sokolov V, Ismail-Zadeh A (2015) Seismic hazard from instrumentally recorded, historical and simulated earthquakes: application to the Tibet-Himalayan region. Tectonophysics 657:187–204

Sokolov V, Ismail-Zadeh A (2016) On the use of multiple-site estimations in probabilistic seismic hazard assessment. Bull Seismol Soc Amer 106(5):2233–2243. https://doi.org/10.1785/01201503062016

Soloviev AA, Gorshkov AA (2017) Modeling the dynamics of the block structure and seismicity of the Caucasus. Izv Phys Solid Earth 53(3):321–331

Soloviev AA, Vorobieva IA, Panza GF (1999) Modeling of block-structure dynamics: parametric study for Vrancea. Pure appl Geophys 156:395–420

Soloviev AA, Vorobieva IA, Panza GF (2000) Modelling of block structure dynamics for the Vrancea region: source mechanisms of the synthetic earthquakes. Pure appl Geophys 157: 97–110

Soloviev AA, Ismail-Zadeh AT (2003) Models of dynamics of block-and-fault systems. In: Keilis-Borok VI, Soloviev AA (eds) Nonlinear dynamics of the lithosphere and earthquake prediction. Springer, Heidelberg, pp 69–138

Turcotte DL, Schubert G (2014) Geodynamics, 3rd edn. Cambridge University Press, Cambridge

UN (2017) Report of the open-ended intergovernmental expert working group on indicators and terminology relating to disaster risk reduction. http://www.preventionweb.net/files/50683_oiewgreportenglish.pdf. Accessed on 28 Aug 2017

Varotsos P, Alexopoulos K, Nomicos K, Lazaridou M (1986) Earthquake predictions and electric signals. Nature 322:120

Wortel MJR, Spakman W (2000) Subduction and slab detachment in the Mediterranean-Carpathian region. Science 290:1910–1917

Wyss M (ed) (1991) Evaluation of proposed earthquake precursors. Special Publication 32. American Geophysical Union, Washington DC

Zechar JD, Jordan TH (2008) Testing alarm-based earthquake predictions. Geophys J Int 172: 715–724

Zhang-li C, Pu-xiong L, De-yu H, Da-lin Z, Feng X, Zhi-dong W (1984) Characteristics of regional seismicity before major earthquakes. Earthquake Prediction. UNESCO, Paris, pp 505–521

Catalogue of Earthquake Mechanism and Correlation with the Most Active Seismic Zones in South-Eastern Part of Romania

Andrei Bălă, Mircea Radulian, Emilia Popescu and Dragoş Toma-Danilă

Abstract Earthquake mechanism and fault plane solution information is fundamental to determine the stress field and to define seismogenic and active tectonic zones. At the same time, it is a basic input to compute seismic hazard by deterministic approach. The purpose of this paper is to update the catalogue of the fault plane solutions for Romanian earthquakes for the time interval 1998–2012. The catalogue is limited geographically to the Carpathians Orogeny and extra-Carpathians area located in the south—eastern part of Romania. The catalogue comprises 259 intermediate-depth seismic events and 90 crustal seismic events, covering the study time interval. All the existing information is considered and revised. The fault plane solutions of the Vrancea earthquakes generated in a confined sinking plate in the mantle reflect the dominant geodynamic process in the study region. The typical features revealed by all the previous studies on the subcrustal seismic activity (predominant dip-slip, reverse faulting, characterizing both the weak and strong earthquakes) are reproduced as well by our investigation. As concerns the earthquake activity in the crust, a few new refined aspects are highlighted in the present work: (1) a deficit of the strike-slip component over the entire Carpathians foredeep area, (2) different stress field pattern in the Făgăraş—Câmpulung zone as compared with the Moesian Platform and Pre-Dobrogean and Bârlad Depressions, (3) a larger range for the dip angle of the nodal planes in the Vrancea subcrustal source, $\sim 40^0$–70^0 against $\sim 70^0$, as commonly considered.

Keywords Earthquake mechanism · Active tectonic zones · Geodynamic processes

A. Bălă (✉) · M. Radulian · E. Popescu · D. Toma-Danilă
National Institute CD for Earth Physics, Magurele, Romania
e-mail: andrei_bala@infp.ro

M. Radulian
e-mail: mircea@infp.ro

E. Popescu
e-mail: epopescu@infp.ro

D. Toma-Danilă
e-mail: toma@infp.ro

© Springer International Publishing AG, part of Springer Nature 2018
R. Vacareanu and C. Ionescu (eds.), *Seismic Hazard and Risk Assessment*,
Springer Natural Hazards, https://doi.org/10.1007/978-3-319-74724-8_2

1   Introduction

The present paper is an extension of previous studies dealing with the focal mechanism and related stress characteristics for the earthquakes recorded in Romania (e.g., Radulian et al. 1999, 2002; Bălă et al. 2003). Thus, the catalogue of the fault plane solutions built up until 1997 is updated and expanded for the time interval 1998–2012 and subsequently analyzed in correlation with specific cluster of events and active faults.

The earthquake mechanisms are computed in all cases using the SEISAN algorithm (Havskov 2001) and the polarities of the first P-wave arrivals. The catalogue of fault plane solutions, presented in the Appendix A and accessible on line at www.infp.ro, comprises 259 intermediate-depth seismic events and 90 crustal seismic events, recorded in the time interval 1998–2012. The location of the events is presented in Fig. 1 together with the seismogenic zones, as they were defined by Radulian et al. (2000) and slightly changed here. We selected only the solutions with minimum 10 reliable polarities and acceptable stations coverage.

Most of the study crustal earthquakes belong to the background seismicity, with magnitudes M_w below 4, except the event of 3.10.2004 occurred in Northern Dobrogea with $M_w = 4.9$. Almost three thirds of earthquakes occurred in the Vrancea region (VNI) in the mantle range (64–167 km depth, Table 1). Most of

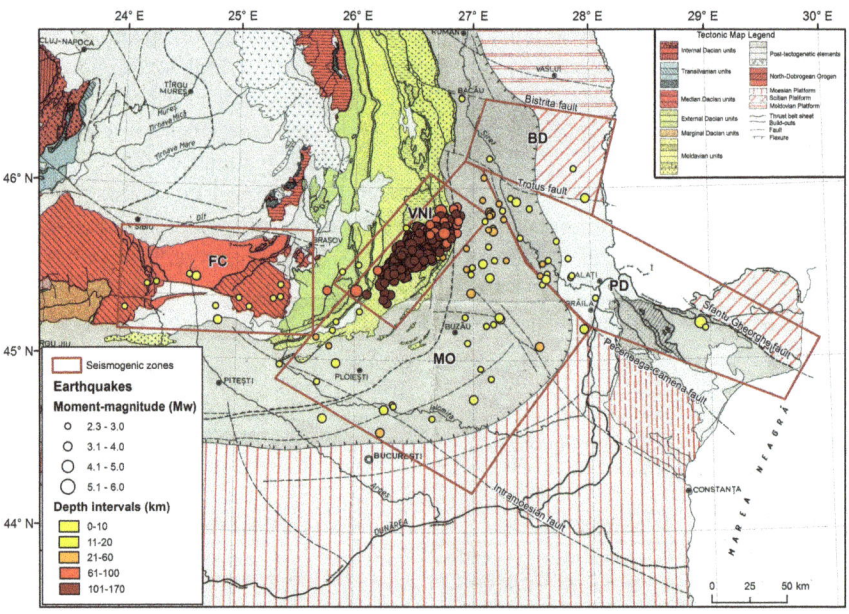

Fig. 1 Tectonic map of the south-eastern part of Romania (after Sandulescu 1984) and the epicentre location for the earthquakes considered in this paper (in agreement with the ROMPLUS Catalogue)

Table 1 Number of earthquakes with mechanism and their location analyzed for the period 1998–2012

Seismogenic zone	No. events	Mw	Depth [km]	No. stations
Moesian platform—MO	42	2.3–3.3	2–38	10–51
Barlad depression & Pre-Dobrogean depression—PD-BD	28	2.3–4.9	0–46	10–25
Fagaras—Campulung—FC	20	2.3–3.3	0–25	10–34
Vrancea intermediate depth zone—VNI	259	2.7–6.0	64–167	10–60

them have magnitudes M_w below 5, with only 5 earthquakes with magnitudes $M_w \geq 5$, the most important being the earthquake of 27.10.2006 with $M_w = 6$ (Fig. 1). The crustal earthquakes follow closely the areal distribution of the seismogenic zones, which were set on the basis of seismicity trends as covered up by the entire catalogue of earthquakes in Romania and on the basis of geotectonic grounds.

We limit geographically our data set to the Carpathians Orogeny and extra-Carpathians area located in the south-eastern part of Romania. Similar investigations, carried out by Oros et al. (2008), were focused on the seismogenic zones located in the western part of Romania: Danubian zone (DA) and Banat area (BA). The authors find some 140 earthquakes mechanisms, using different sources in DA and BA, in the time interval 1959–2006.

2   Seismicity Distribution

The seismic activity in the crust is dispersed over the Carpathians orogeny and foreland with significant enhancements in several seismogenic zones, as defined first by Radulian et al. (2000). First it is defined the Vrancea intermediate depth seismogenic zone (VNI) with earthquakes that occur under 60 km depth and with moment magnitudes that might be over 7.

Then we have the seismogenic zones that cover some tectonic provinces of Romania at crustal level. In Fig. 1 we focus our attention on the seismogenic zones located in the south-eastern part of Romania. We make some adjustments relative to the previously defined seismogenic zones in order to include all the earthquakes in the catalogue and to keep at the same time their adherence to a specific tectonic province. Given that the seismogenic areas situated in front of the Carpathians Arc Bend, south of the Peceneaga-Camena fault, in the Moesian Platform, do not differ notably, we prefer to consider in our subsequent analysis a single area (MO) in this region (Fig. 1). The seismogenic area MO covers a great part of the eastern Moesian Platform, almost until the Danube to the east, and to the west occupies a zone that include Intramoesian fault, almost to the Argeş river (Fig. 1).

We slightly adjusted the Pre-Dobrogean zone, which was extended to the south-east in order to cover also the North Dobrogean Orogen, to have a common margin with Barlad Depression zone (along Trotuş fault) and to be extended over the northern flank of Sf. Gheorghe fault (Fig. 1). From the tectonic point of view this is proposed by Hippolyte (2000) which suggested that all the area between Sf. Gheorghe fault and Peceneaga-Camena fault should be considered a single tectonic unit belonging to the North Dobrogean Orogen.

In the same way, we combined the Pre-Dobrogean depression and Bârlad depression zones into a single area (PD-BD).

The Vrancea intermediate zone (VNI) and Făgăraş-Câmpulung zone (FC) are the same as defined in Radulian et al. (2002). The crustal earthquakes are commonly small to moderate ($M_w < 6$). Only in the FC zone a few shocks of magnitude above 6 ($M_{max} = 6.5$) were reported in the Romanian catalogue in about one millennium time interval (Oncescu et al. 1999). The crustal seismicity is generally associated with the basement fracture systems (Bala et al. 2015) that might be in some cases as deep as the crust thickness.

2.1 Fault Plane Solutions

We follow the convention of Aki and Richards (1980) to define the nodal plane parameters (strike, dip and rake). Fault strike is the direction of a line created by the intersection of a fault plane and a horizontal surface, measured relative to North ($0°$–$360°$). Strike is defined such as the fault dips always to the right side of the trace when moving along the trace in the strike direction.

Fault dip is the angle between the fault and the horizontal plane (Earth's surface) measured from the horizontal plane ($0°$–$90°$). Rake is the direction in which the hanging wall block moves relative to the foot wall during rupture, as measured on the plane of the fault. It is measured relative to fault strike, $\pm 180°$.

The fault plane solutions are computed on the basis of the polarities of the first P-wave arrivals using the SEISAN algorithm (Havskov and Ottemöller 2001). The regional velocity model used in the calculations is common for crustal and sub-crustal earthquakes and it is used today to compute the earthquake location in the Romplus catalog (www.infp.ro).

2.2 Statistical Analysis

One of our goals is to determine the main features of the parameters of the fault plane solutions computed in this paper and to compare the results with previous investigations. For this purpose, we used graphical tools able to emphasize statistically representative features in our data set for each seismogenic area considered

Fig. 2 Classification
diagram. N: Normal; N-SS:
Normal—Strike-slip; SS-N:
Strike-slip—Normal; SS:
Strike-slip; SS-R: Strike-slip
—Reverse; R-SS: Reverse—
strike-slip; R: Reverse (after
Álvarez-Gómez 2015)

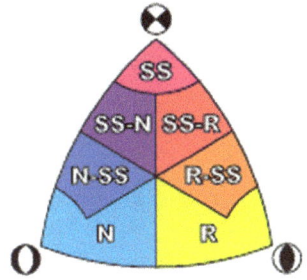

in the study. In this way, simple polar diagrams were employed to describe strike and dip behavior of the events in each active area.

In order to classify the type of faulting, we represent the distribution of the principal axes P, T and B using Kaverina et al. (1996) projection, which improves the Frohlich and Apperson (1992) ternary diagram. A dedicated program is employed for this purpose, which is described by Álvarez-Gómez (2014, 2015). According to this procedure, we can classify earthquakes in seven classes represented by specific rupture types: (1) Normal; (2) Normal—Strike-slip; (3) Strike-slip—Normal; (4) Strike-slip; (5) Strike-slip—Reverse; (6) Reverse—strike-slip and (7) Reverse (Fig. 2).

3   Discussion of the Results in the Main Seismogenic Areas

3.1   *Vrancea Subcrustal Source—VNI*

Seismicity in Romania is concentrated at the Carpathians Arc bend in the Vrancea region. Here, an isolated lithospheric slab down going in the mantle is permanently releasing seismic energy in an extreme narrow volume. In average, 3 earthquakes with magnitude above 7 were reported each century for a time span of six centuries. The origin of intermediate-depth seismicity in the Vrancea area is still an ongoing debate. The lithospheric volume which is seismically active can be approximated by a prism vertically oriented between 60 and 170 km depth, with a horizontal cross section of 30×70 km^2. Above 60 km and below 170 km the seismicity is suddenly cut off, although the high-velocity body, as determined by seismic tomography, extends notably beyond these limits (Ren et al. 2012).

The polar diagrams for the azimuthal distribution of one of the nodal planes (A) and for the plunge angle are plotted in Fig. 3. Note the predominance of nodal planes oriented parallel to the Carpathians Arc bend (40°–50° or the opposite azimuths) and at 50°–60° plunge angle. It is worth mentioning that this geometry of the faulting system corresponds with the orientation of the rupture faults for the largest Vrancea events. Taking into account the difference in scale between a moderate earthquake ($M_w = 4$–5) and a major earthquake ($M_w > 7$), it is not trivial

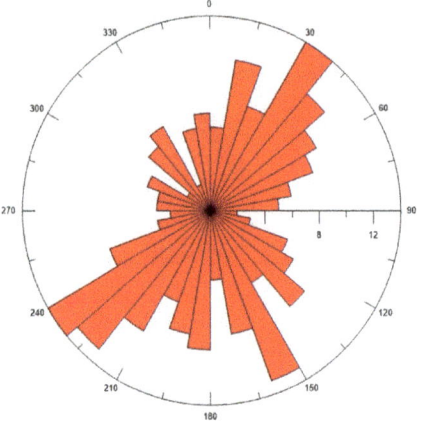
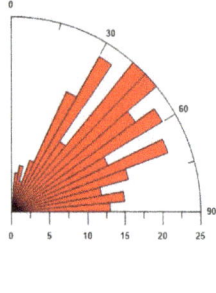

Fig. 3 Angular diagrams for azimuth and dip angles for Plane A—Vrancea subcrustal source (VNI)

such a feature. We may assume that the process of generating moderate earthquakes is controlled largely by the same tectonic forces as for generating the largest shocks.

The analysis of the Fig. 3 reveals also a secondary tendency for the azimuthal orientation of the nodal planes, roughly perpendicular to the Carpathians Arc bend, respectively. In this case, similar focal mechanisms are noticed for some of the Vrancea events of moderate-to-large magnitudes ($M_w = 5$–7).

It looks like the rupture orientation and length scale with the seismicity geometry (elongated along NE-SW, see Fig. 1): the rupture process propagates always along NE-SW plane for the large events, while for the smaller events propagate occasionally along a perpendicular direction (NW-SE). However, such a hypothesis remains to be checked for future events or through advanced modelling.

The distribution of the principal axes B, T and P, is represented in Fig. 4 for the entire data set (259 Vrancea intermediate-depth earthquakes). In order to classify the type of faulting, we represent the distribution of the principal axes using Kaverina's ternary diagram (Kaverina et al. 1996). A dedicated program is employed for this purpose, which was assembled in Python and which is described in Álvarez-Gómez (2014).

The distribution of the principal axes B, T and P, is represented in the Fig. 5a only for a selection of earthquakes mechanisms that are computed with at least 30 polarities. That is possible as in the last years more seismic stations with online recording were installed in Romania. In Fig. 5b only 15 events with Mw > 5 from Vrancea subcrustal source (VNI) are represented in the same kind of diagram (from Radulian 2014).

Considering Fig. 4 it is clear that the reverse faulting is dominating in the Vrancea subcrustal source, independently of depth or magnitude ranges. In Fig. 5a practically none of the events has well-defined normal or strike-slip faulting; two events, located at the bottom edge of the descending seismically active body, are

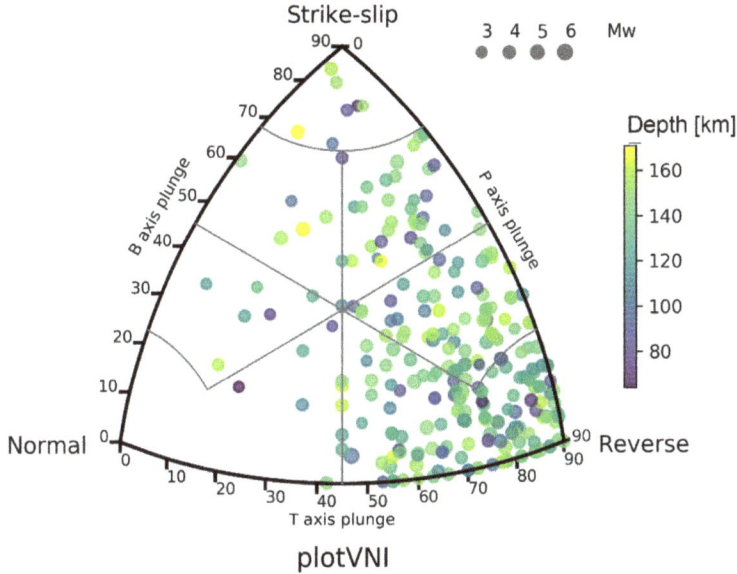

Fig. 4 Diagram for B, T and P principal axes distribution for all the events considered in Vrancea intermediate zone (VNI). The dots are proportional to M_w; to the right, the color scale represents the depths of the events in km

characterized by strike-slip with normal components, and four events by strike-slip with reverse components. Whereas, most of the events have pure reverse faulting. Figure 5b shows that for events with Mw > 5, the characteristics are of pure reverse mechanism independently of the depths.

Several papers describe the mechanism of some 15 stronger earthquakes in the area, with Mw > 5. The nodal plane typically striking NE-SW (220°) and dipping 60°–70° to the NW is ascribed to be the rupture plane (Oncescu and Bonjer 1997; Wenzel et al. 2002). In all cases, the rake angle is close to 90° (i.e., up dip). Reverse faulting with down-dip extension is predominant along the entire depth range (Radulian 2014) for these stronger earthquakes in VNI (Figs. 5b and 6). Although this is true for stronger earthquakes (Mw > 5), it appears that the earthquakes in the catalogue show a much broader interval for the fault directions, as well as for the dip of the fault plane.

3.2 Eastern Moesian Platform Seismogenic Zone—MO

The eastern sector of the Moesian Platform, located between the Intramoesian fault and Peceneaga-Camena fault is more seismically active as compared with the sector located west of the Intramoesian fault, which is almost aseismic. The seismicity

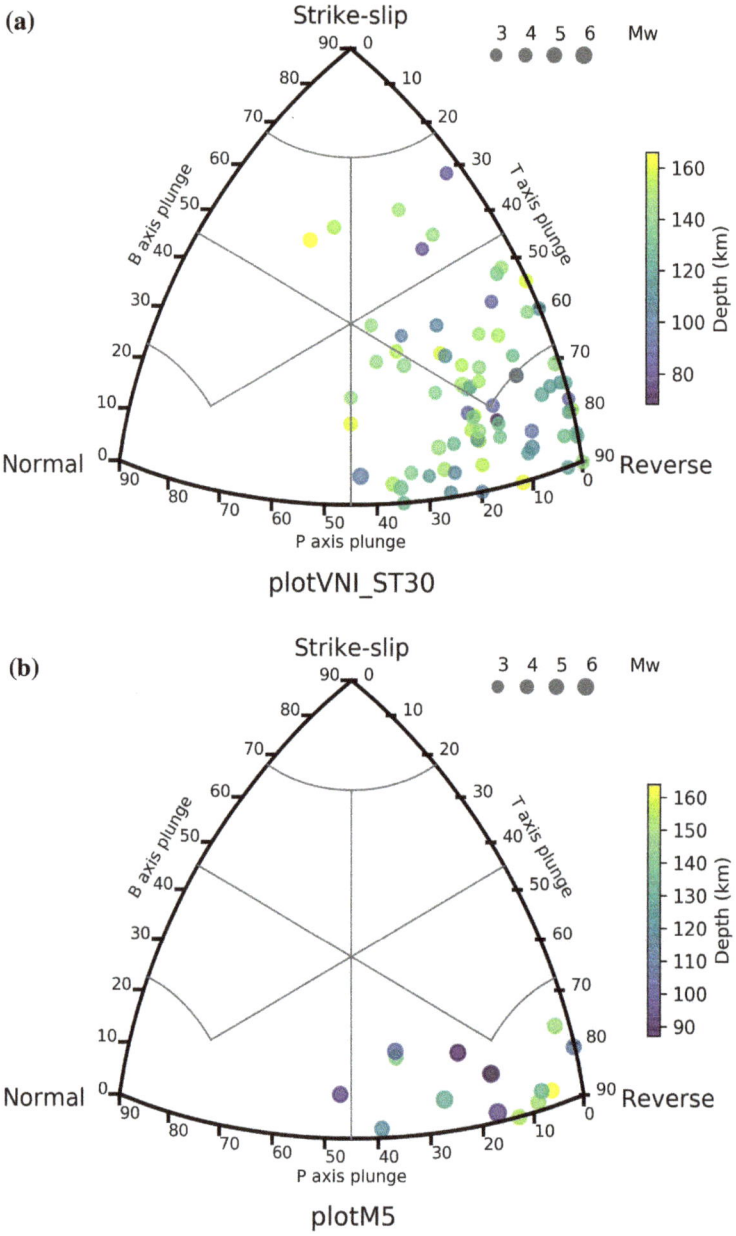

Fig. 5 a Diagram for B, T and P principal axes only for the events in Vrancea zone with mechanisms computed from at least 30 polarities; **b** diagram for B, T and P only for 15 events with Mw > 5. The dots are proportional to M$_w$ magnitude; to the right, the color scale represents the depths of the events

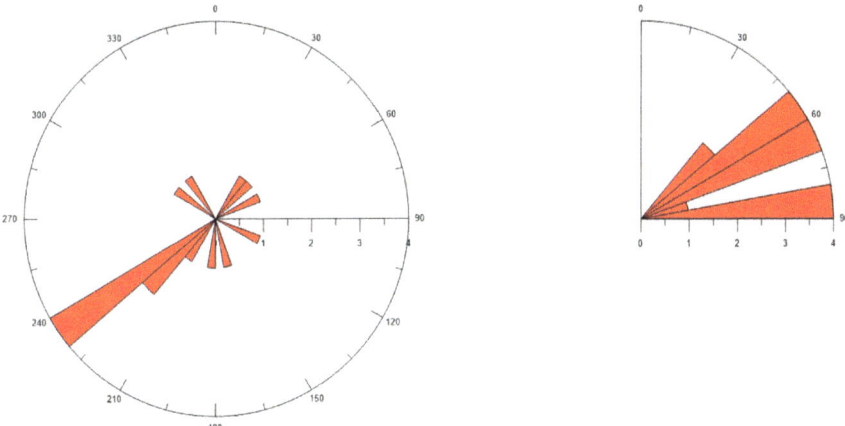

Fig. 6 Angular diagrams for azimuth and dip angles for Plane A; 15 events with Mw > 5 from Vrancea subcrustal source (VNI)

concentrates close to the Carpathians Arc Bend, overlapping to some extend the epicentral area of the Vrancea subcrustal earthquakes.

The earthquakes with computed focal mechanism that we consider to fit in the eastern Moesian seismogenic zone are spread over a wide area, both to the south-east and south of the Vrancea region (VNI) and covering almost all the eastern Moesian Platform (Wallachian sector), from Peceneaga Camena fault almost to the Danube line and to the west to Intramoesian fault, covering both flanks of this fault (Fig. 1). The events are of small-to-moderate magnitude ($2 \leq M_w \leq 3.6$) belonging to the background seismicity. Most of them (36) are in connection with the Vrancea seismogenic area, while a few of them (6) can be rather associated to the Intramoesian fault area.

The diagram for B, T and P axes (Fig. 7) shows an equal distribution among normal and reverse faulting and absence of pure strike-slip faulting. According to our results, mechanisms of subsidence and folding are prevailing in the region against transcurrent mechanisms.

3.3 Pre-dobrogean Depression and Bârlad Depressions—PD-BD

The Predobrogean depression and the North Dobrogean Orogen were considered only one seismogenic zone in our map (Fig. 1) as it was suggested by Hippolyte (2002). The catalogue of earthquakes with computed focal mechanism contains 24 events located in the Predobrogean Depression and North Dobrogean orogen. More than that, with only 4 events with compute mechanism located in the Bârlad

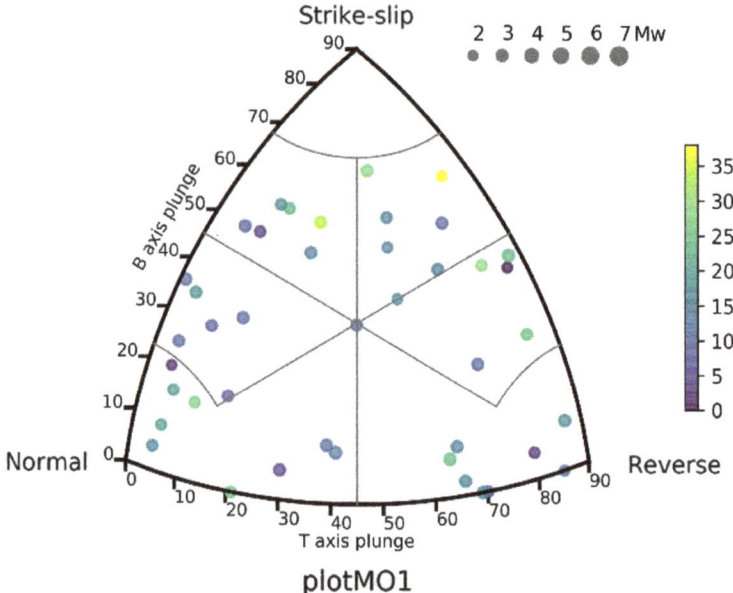

Fig. 7 Ternary diagram for B, T and P principal axes distribution for Moesian Platform seismogenic area

Depression, we consider a single seismogenic zone by merging the Predobrogean depression and Bârlad depression—PD-BD.

The diagram for B, T and P axes (Fig. 8) for PD-BD zone shows broadly the same features as for the Moesian Platform (Fig. 7): an equal probability for normal and reverse faulting and the almost total lack of pure strike-slip faulting. This result suggests that despite the lateral differences among the tectonic units acting in the foredeep area of the Carpathians and in front of it, the faulting processes in the crust are quite similar, with prevalent subsidence and folding mechanisms.

The polar diagrams for the azimuthal distribution of the plane A and for the plunge angle are plotted in the Fig. 9 for both crustal seismogenic regions MO and PD-BD. The azimuthal distribution of the nodal planes does not suggest a preferential orientation, but only a slight tendency to align perpendicularly to the Carpathians Arc bend. The distribution of dip angles in Fig. 9 is close to the distribution find for 104 events in the same crustal region after the previous catalog of earthquake mechanisms until 1997 analyzed by Bala et al. (2003), while the strike angles show the same almost equal distribution to all directions.

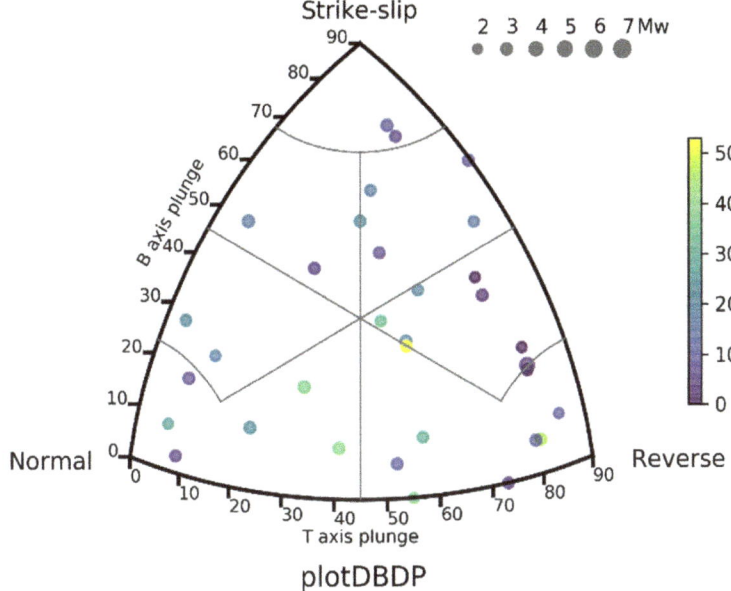

Fig. 8 Diagram for B, T and P principal axes distribution for Pre-Dobrogean and Bârlad Depressions (PD-BD)

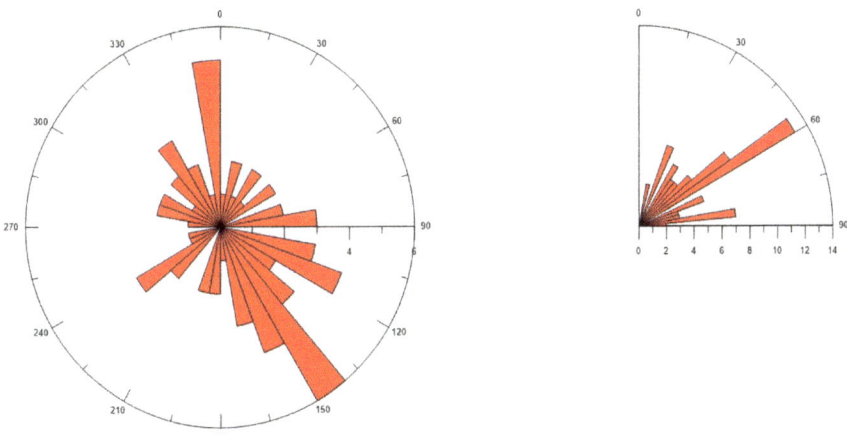

Fig. 9 Angular diagrams for azimuth and dip angles of the nodal plane A—Moesian Platform area (MO) and PD-BD area

3.4 Făgăraş—Câmpulung Seismogenic Zone—FC

The catalogue of earthquakes with computed focal mechanism belonging to the Făgăraş-Câmpulung seismogenic area includes 20 events, most of them located in the southern side, towards the contact between Southern Carpathians and Getic domain. All the events occurred in the 1998–2012 time interval belong to the background seismicity (magnitudes are no greater than 3.3). The statistics shown by polar diagrams is too low to provide reliable trends in nodal plane orientation. Apparently, the fault planes are equally distributed on azimuth. The picture shown by the diagram for B, T and P axes (Fig. 10) for the Făgăraş—Câmpulung area is somewhat different from that pointed out for the other seismogenic areas in the crust. Although the events are shallow (h < 35 km) the distribution is closer to the typical distribution of the Vrancea subcrustal earthquakes, with a particular emphasis on reverse faulting component.

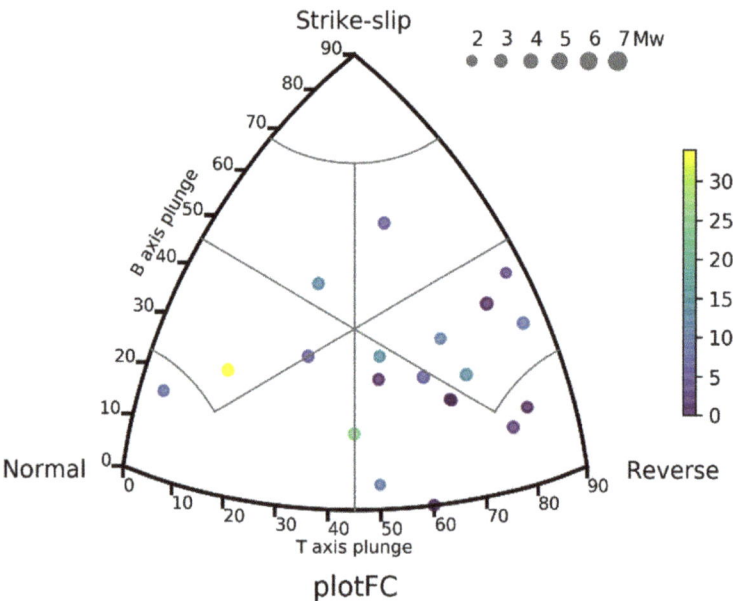

Fig. 10 Diagram for B, T and P principal axes distribution for the Făgăraş—Câmpulung seismogenic zone

4 Conclusions

The catalogue of the focal mechanism for the earthquakes recorded in Romania, available until 1997 (Radulian et al. 2002; Bala et al. 2003) is updated for the period 1998–2012. Thus, a new catalogue of the fault plane solutions for 259 intermediate-depth earthquakes (Vrancea source) and 90 crustal earthquakes (Moesian Platform, Predobrogean Depression, Bârlad Depression and Făgăraş—Câmpulung) is added to the existing catalogue (appendix A on www.infp.ro).

The seismic activity that took place during 1998–2012 time frame was limited to moderate-size events. The largest event was generated on 27 October 2006 (M_w = 6.0) in the Vrancea subcrustal source. The largest event generated in the crust occurred on 3 Oct. 2004 (M_w = 4.9) in the Pre-Dobrogean Depression. For this magnitude range, the fault plane solutions of the shallow earthquakes is more likely related to local secondary system of faults than to the major faults located in the Carpathians foredeep region. We can simply explain on this line the variety of solutions in the Moesian Platform, Pre-Dobrogean Depression and Bârlad Depression zones. As a general trend, which is statistically consistent, note the deficit of strike-slip faulting. This result suggests the prevalence of subsidence and folding processes as stress release mechanisms in the entire Carpathians and foredeep regions. The hypothesis raised by a few authors in the last century on a trans-current movement along the major faults crossing the aria situated between the Black Sea and Vrancea (Airinei 1977; Constantinescu and Enescu 1984) is not supported by our results.

Reverse faulting with down-dip extension is predominant along the entire depth range for the stronger earthquakes in VNI (Radulian 2014). Although this is true for stronger earthquakes (Mw > 5), it appears that the earthquakes in the catalogue show a much broader interval for the fault directions, as well as for the dip of the fault planes.

Vrancea intermediate depth show a larger range for the dip angle of the nodal planes in the Vrancea subcrustal source, $\sim 40°-70°$ against $\sim 70°-80°$ as commonly considered. In the same time the azimuth angles of the fault planes are presenting 3 rather scattered domains, not only one as it was pointed out for the strong intermediate earthquakes.

The earthquake mechanism in the eastern Moesian Platform (MO) looks the same as in the neighbor tectonic provinces, Pre-Dobrogean Depression and North Dobrogean orogen (PD-BD). The ternary diagrams illustrate that there is no preferred stress direction for the entire area. The earthquakes are occurring on all faults present in the area, which separate different blocks, in the area in front of Eastern Carpathians Bend. However a deficit of pure strike-slip component appears over the entire Eastern Carpathians foredeep area. There are no evidence for strike-slip earthquakes along the major faults in the area, like Peceneaga—Camena fault and Capidava—Ovidiu fault, and they show no signs of mobility for the interval of time considered. More than that in his study of paleostress resulted from a great number of fault observations in Dobrogea, Hippolite (2002) considered that after Paleogene the Peceneaga-Camena fault, as other faults in Dobrogea, were reactivated at certain moments only as reverse faults without any strike-slip component until present.

In the Făgăraş—Câmpulung zone (FC) different stress field pattern occur as compared with the Moesian Platform and Pre-Dobrogean and Bârlad depression. Here the earthquakes took place on some shallow faults, with almost the same direction SSW-NNE and the preferred mechanism is reverse. The same mechanism for tectonic faults in the area is established by Werner and Seghedi (2005). The sediments unconformably overlie a broadly folded anticline that involves late Middle Miocene rocks of the Transylvanian Basin (Garbacea and Frisch 1998). With the Olt river running parallel to the northern margin of the basin, the overall geometry suggests that the southern border is an active N-dipping monoclinal structure (Werner and Seghedi 2005).

The crustal seismic activity in this region can be rather seen as a response of the intense processes taking place beneath the Carpathians Arc bend (Vrancea intermediate depth zone) and materialized by a triple number of earthquakes occurred at intermediate depth in the same time interval. Crustal earthquakes appear not to reflect the significant transcurrent motions between the tectonic units in the region along the main faults, but they rather express the moving and re-positioning of different crustal blocks in the area, delimited by either the main fault system, which runs parallel to the Carpathians or by secondary faults, distributed to a general direction perpendicular to the direction of the main fault system.

The focal mechanism patterns obtained in the present study reproduce to a large extent the stress field characterization carried out by Radulian et al. (2002) and Bala et al. (2003) using essentially an independent catalogue of fault plane solutions. However, the considerable increase of fault plane solutions accuracy for the earthquakes recently recorded (due to the significant improvement of the Romanian seismic network and of their statistical representativeness, allows us to individualize a few refined aspects not obvious in the previous investigations, such as: (1) deficit of the strike-slip component over the entire Carpathians foredeep area, (2) different stress field pattern in the Făgăraş—Câmpulung zone as compared with the Moesian Platform and Pre-Dobrogean and Bârlad Depressions, (3) a larger range for the dip angle of the nodal planes in the Vrancea subcrustal source, $\sim 40°-70°$ against $\sim 70°$, as commonly considered.

Acknowledgements The ternary diagrams in this paper are realized with a dedicated program in *Python* which was kindly made available by the author and which is described in Álvarez-Gómez (2014, 2015). The work in this paper was realized in the frame of NUCLEU Program, project no. PN 16.35.01.08 and PN 16.35.01.03 in 2017.

References

Airinei S (1977) Lithospheric Microplates on the Romanian Territory Reflected by Regional Gravity Anomalies (in Romanian). St Cerc Geol Geogr Geofiz 15:19–30
Álvarez-Gómez JA (2014) FMC: a one-liner Python program to manage, classify and plot focal mechanisms. Geophys Res Abstr 16, EGU2014-10887

Álvarez-Gómez JA (2015) FMC A program to manage, classify and plot focal mechanism data, version 1.01, March 2015, Faculty of Geology. Universidad Complutense de Madrid, Spain

Aki K, Richards PG (1980) Quantitative seismology. Theory and methods. W.H Freeman, San Francisco

Bala A, Radulian M, Popescu E (2003) Earthquakes distribution and their focal mechanism in correlation with the active tectonic zones of Romania. J Geodyn 36:129–145

Bala A, Raileanu V, Dinu C, Diaconescu M (2015) Crustal seismicity and active fault systems in Romania. Rom Rep Phys 67(3):1176–1191

Constantinescu L, Enescu D (1984) A tentative approach to possibly explaining the occurrence of the Vrancea earthquakes. Rev Roum Geol Geophys Geogr Ser Geophys 28:19–32

Frohlich C, Apperson KD (1992) Earthquake focal mechanisms, moment tensors, and the consistency of seismic activity near plate boundaries. Tectonics 11:279–296

Garbacea R, Frisch W (1998) Slab in the wrong place: lower lithospheric mantle delamination in the last stage of the eastern Carpathians subduction retreat. Geology 26:611–614

Havskov J, Ottemöller L (2001) SEISAN: The earthquake analysis software, Version 7.2, University of Bergen, Norway, 256 p

Hippolyte JC (2002) Geodynamics of Dobrogea (Romania): new constraints on the evolution of the Tornquist-Teisseyre line, the Black Sea and the Carpathians. Tectonophysics 357:33–53

Kaverina AN, Lander AV, Prozorov AG (1996) Global creepex distribution and its relation with earthquake—source geometry and tectonic origin. Geophys J Int 125:249–265

Oncescu MC, Bonjer KP (1997) A note on the depth recurrence and strain release of large Vrancea earthquakes. Tectonophysics 272:291–302

Oncescu MC, Marza V, Rizescu M, Popa M (1999) The Romanian earthquake catalogue between 983-1996. In: Wenzel F, Lungu D, Novak O (eds) Vrancea earthquakes: tectonics, hazard and risk mitigation. Kluwer Academic Publishers, Dordrecht, pp 43–49

Oros E, Popa M, Popescu E, Moldovan IA (2008) Seismological database for BANAT seismic region (ROMANIA)—Part 2: the catalogue of the focal mechanism solutions. Rom J Phys 53(7–8):965–977

Radulian M, Mandrescu N, Popescu E, Utale A, Panza GF (1999) Seismic activity and stress field in Romania. Rom J Phys 44(9–10):1051–1069

Radulian M, Mandrescu N, Panza GF, Popescu E, Utale A (2000) Characterization of seismogenic zones of Romania. Pure Appl Geophys 157:57–77

Radulian M, Popescu E, Bala A, Utale A (2002) Catalog of the fault plane solutions for the earthquakes occured on the Romanian territory. Rom J Phys 47(5–6):663–670

Radulian M (2014) Mechanisms of earthquakes in Vrancea. In: Beer M, Patelli E, Kougioumtzoglou I, Siu-Kui Au I (eds) Encyclopedia of earthquake engineering. Springer, Berlin, Heidelberg. https://doi.org/10.1007/978-3-642-36197-5_302_1

Ren Y, Stuart GW, Houseman GA, Dando B, Ionescu C, Hegedüs E, Radovanović S, Shen Y (2012) Upper mantle structures beneath the Carpathian-Pannonian region: Implications for the geodynamics of continental collision. Earth Planet Sci Lett 349–350:139–152

Werner F, Seghedi I (2005) Late Miocene-Quaternary volcanism, tectonics and drainage system evolution in the East Carpathians, Romania. Tectonophysics 410:111–136

Wenzel F, Sperner B, Lorenz F, Mocanu V (2002) Geodynamics, tomographic images and seismicity of the Vrancea region (SE-Carpathians, Romania). EGU Stephan Mueller special publication series, 95–104, ISSN: 1868-4556

The Space-Time Distribution of Moderate- and Large-Magnitude Vrancea Earthquakes Fits Numerically-Predicted Stress Patterns

Mirela-Adriana Anghelache, Horia Mitrofan, Florina Chitea, Alexandru Damian, Mădălina Vişan and Nicoleta Cadicheanu

Abstract Each of the three major earthquakes ($M_w \geq 6.9$) recorded within the Vrancea seismogenic body in the years 1977, 1986 and 1990 may have been the result of long-range interactions. The latter seemed to be initiated in coincidence with a moderate ($4.7 \leq m_b \leq 4.9$) shock that systematically occurred in the 160–175 km depth-range, 3–4 years in advance of the major earthquake. In addition, each corresponding pair of moderate/major events systematically exhibited, in terms of focal mechanisms, a particular pattern: the latter complied with predictions of a numerical model that had addressed along-strike break-off experienced by a near-vertical slab, which—at the same time—was strongly coupled with its overriding plate along a steeply dipping contact extending on a significant vertical length. It was thus suggested that the Vrancea moderate events of thrust-fault type occurred in the 160–175 km depth-range could correspond to the along-strike propagation of a detachment horizon tip; while the major earthquakes subsequently occurred at shallower depths (85–135 km), were possibly caused by the shearing

M.-A. Anghelache (✉) · H. Mitrofan · F. Chitea · M. Vişan · N. Cadicheanu
"Sabba Ştefănescu" Institute of Geodynamics of the Romanian Academy,
19-21 Jean-Louis Calderon St., 020032 Bucharest, Romania
e-mail: mirelaadrianaa@yahoo.com

H. Mitrofan
e-mail: horia.mitrofan@geodin.ro

F. Chitea
e-mail: florina.tuluca@gg.unibuc.ro

M. Vişan
e-mail: danamadalina@yahoo.com

N. Cadicheanu
e-mail: nicoleta_cadichian@yahoo.com

F. Chitea · A. Damian
Department of Geophysics, Faculty of Geology and Geophysics,
University of Bucharest, 020956 Bucharest, Romania
e-mail: alex879ro2007@yahoo.com

© Springer International Publishing AG, part of Springer Nature 2018   39
R. Vacareanu and C. Ionescu (eds.), *Seismic Hazard and Risk Assessment*,
Springer Natural Hazards, https://doi.org/10.1007/978-3-319-74724-8_3

forces that acted, at the upper-plate/underlying-plate interface, in response to the increased downward pull experienced by the Vrancea slab as the break-off was propagating laterally.

Keywords Focal mechanism · Lithospheric slab · Lateral break-off
Long-range interaction

1 Introduction

Within the Vrancea intermediate-depth seismogenic body, long-range interactions might operate between earthquakes that both in time, and in space, are widely separated from one another.

This issue has been previously investigated by computing (Heidbach et al. 2007; Ledermann and Heidbach 2007; Ganas et al. 2010) how a particular Vrancea shock, having magnitude larger than 6, could alter the Coulomb Failure Stress distribution within the seismogenic body: it was next checked if the subsequent strong shocks nucleated in a region where—according to the computed static stress changes—seismogenic processes were favoured. Yet in spite of adopting similar computational approaches, based on linear elastic rheologies, the ensemble of the above-indicated authors failed to reach convergent conclusions on the possible existence of long-range interactions between various strong earthquakes occurred within the Vrancea intermediate-depth seismogenic body.

An alternative approach toward investigating such possible long-range interactions may start from the observation that the sub-crustal Vrancea earthquakes occurred exclusively (e.g., Heidbach et al. 2007; Ganas et al. 2010) within a high seismic-velocity body. The latter was interpreted to represent a relic of a subducted lithospheric slab which was, moreover, assumed to undergo (Wortel and Spakman 2000; Heidbach et al. 2007; Bonjer et al. 2008) a process of lateral (along-strike) progressive break-off. If, indeed, such a setting occurred, then earthquakes generated at the tip of the laterally propagating fracture could mirror episodes of detachment progress. The consequent increase in the vertical pull experienced by the slab section that was still "intact" (i.e., not yet detached) could correspond to a process of viscoelastic stress transfer, liable to subsequently trigger—above the detachment horizon—other earthquakes.

Such possible long-range interactions—between earthquakes occurred at the tip of the laterally advancing detachment fracture and earthquakes occurred at shallower levels within the slab—have not been considered by previous investigators of the Vrancea intermediate-depth seismicity. Criteria for discriminating between the distinct stress regimes that corresponded to those two categories of earthquake are provided by a series of numerical simulations (Yoshioka and Wortel 1995), which had taken into account a near-vertical slab and had assumed that:

- the slab underwent, in its lower section, along-strike break-off; the failure regime corresponding to the horizontal propagation of the detachment fracture was thrust faulting (hereafter TF), with the compression (hereafter P) axes oriented roughly *parallel to the slab strike*;
- the slab was strongly coupled, in its upper part, with a rather long section of the overriding plate (for instance, if a wedge of that upper tectonic plate had been dragged downwards, in a so called "two-sided collision" setting—e.g., Faccenda et al 2008; Li et al. 2011); as the slab lateral break-off advanced, the resultant progressive increase in the vertical pull experienced by the slab section which was still "intact" (i.e., not yet detached), was liable to trigger—in the slab region experiencing increased shearing forces at the contact with the overriding plate—thrust faulting with the P axes oriented *perpendicular to the slab strike*.

This particular model of Yoshioka and Wortel (1995) should be applicable, to a large extent, to the Vrancea intermediate-depth seismogenic slab: especially, when one considers that the two tectonic plates—Moesian and Intra-Alpine, which had been involved in the Vrancea pre-Miocene subduction process, while now they are subject only to continent-continent collision—could be strongly coupled to each other over a vertical length of more than 100 km; such a setting was invoked by Cloetingh et al. (2004) in order to explain why all three most recent strong ($M_w \geq 6.9$) Vrancea earthquakes had displayed sub-horizontal P axes that were approximately *perpendicular to the inferred direction of the Vrancea slab*.

For testing possible compliance with the Yoshioka and Wortel (1995) model, we have searched for evidence of long-range interactions that might have been operating between:

- seismic events displaying TF focal mechanisms with P axes *parallel to the Vrancea slab strike*, and located below 150 km depth (in a region where the recent seismic tomographic images of Zhu et al. (2012), had suggested that a horizontal detachment process might concern the slab), and
- shallower earthquakes—but still located in the intermediate-depth range—which had TF focal mechanisms with the P axes oriented *orthogonal to the Vrancea slab strike* (in particular, this focal mechanism type was common, as already specified above, to all three most recent major earthquakes occurred—in the years 1977, 1986 and 1990—in Vrancea region).

2 Data Processing

In the following, the data-processing procedures will be only briefly addressed, since they had been discussed in detail in Mitrofan et al. (2016). The considered earthquakes' selection was, nevertheless, updated, by additionally including the seismic events occurred in the years 2012, 2013 and 2014, for which, in the meantime, the necessary information became available.

Specifically, there have been considered the Vrancea earthquakes:

a. which had occurred at intermediate-depths (>60 km);
b. whose magnitudes fell in a range for which complete event-catalogues could be extracted, both from the ISC, and from the PDE (USCGS-NEIS-NEIC) data-bases.

We have consequently adopted a composite magnitude criterion, according to which the body waves' magnitude which we ascribed to a particular Vrancea event, m_{b-asc}, was the lowest of the two distinct magnitude values indicated for that event by the ISC and the PDE data-bases (m_{b-ISC} and m_{b-PDE} respectively):

$$m_{b-asc} = \mathrm{Min}(m_{b-ISC}, m_{b-PDE}) \qquad (1)$$

It thus resulted that 84 Vrancea intermediate-depth events with $m_{b-asc} \geq 4.5$ have been recorded within the time-interval 1965–2014 (1965 being the first year when there was reported an intermediate-depth Vrancea earthquake for which both the ISC, and the PDE catalogues indicated $m_b \geq 4.5$; while 2014 was the most recent year for which seismological data reviewed by the ISC were available).

The frequency-magnitude diagram corresponding to the accordingly selected data-set is illustrated in Fig. 1: it suggests that the adopted earthquakes-selection was complete only down to $m_{b-asc} = 4.7$ (in the following, the m_{b-asc} notation will be replaced simply by m_b).

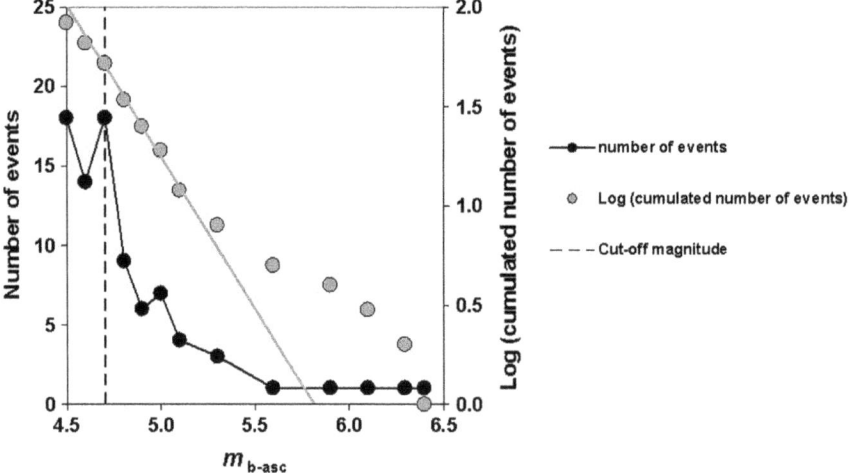

Fig. 1 Frequency-magnitude diagram of intermediate-depth (>60 km) Vrancea earthquakes occurred during the 1965–2014 interval. There are included all the events for which both the ISC, and the PDE catalogues indicated $m_b \geq 4.5$, to each earthquake being ascribed the lowest (m_{b-asc}) of the two distinct m_b values separately provided by each of the two indicated catalogues

The accordingly-retained events with $m_b \geq 4.7$ have next been plotted (by taking into account the very confined epicentral area—of only about 25×70 km—associated to the intermediate-depth Vrancea earthquakes) on a vertical plane traced along the main direction (N50°E) of the seismogenic body. A visual inspection of the resulting cross section (Fig. 2) suggests that the selected earthquakes hypocenters are distributed into 4 distinct groups:

- an "Uppermost group"; it includes only 2 events having occurred around 70 km depth;

next, within the cross-section central part, the hypocenters occupy two distinct lineaments, both of which display an obvious dip from NE to the SW:

- an "Upper diagonal lineament"; it extends between 80 and 160 km depth and it includes, besides moderate magnitude earthquakes, also all the analyzed time-period major events ($m_b > 6$, occurred in the years 1977, 1986 and 1990);

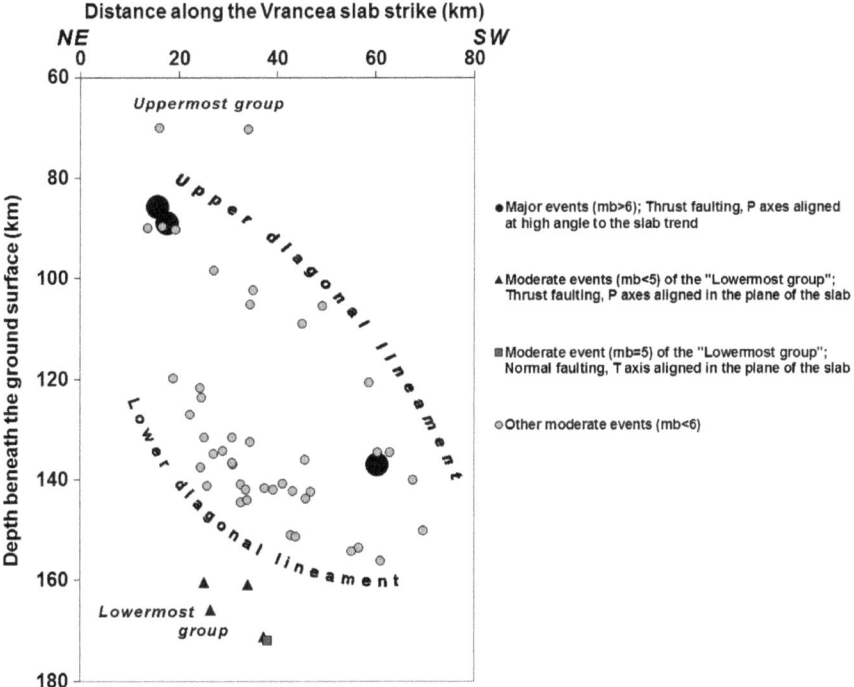

Fig. 2 Cross-section along the inferred strike (N50°E) of the Vrancea slab. The hypocenters of the intermediate-depth (>60 km) Vrancea earthquakes occurred during the 1965–2014 time-interval and for which both the ISC and the PDE catalogues indicated $m_b \geq 4.7$, are projected perpendicular to the vertical plane of the transect. The original coordinates of the projected hypocenters were retrieved from the ISC catalogue. The maximum distance between a hypocenter and its projection on the transect plane is 11 km

- a "Lower diagonal lineament"; it extends between 120 and 160 km depth and it exhibits the largest abundance of events, yet all of them of moderate magnitude;

and finally:

- in the bottom section of the seismogenic body, in the 160–175 km depth range, there occurs the "Lowermost group", which is separated from its closest neighbouring hypocenters of the "Lower diagonal lineament" by a linear—essentially vertical—gap of about 20 km.

The appropriateness of the adopted categorization has been tested as follows: for each hypocenter of a particular group, the distance to the nearest hypocenter within its own group has been compared with the distance to the nearest hypocenter of the group situated immediately above; then, a similar comparison was conducted by considering the distance to the nearest hypocenter in the group situated immediately below.

Reciprocal plots of the indicated pairs of distances are illustrated in Fig. 3 (which was prepared by considering the hypocenters' locations provided by the ISC catalogue). A line (dashed) of slope 1:1, and another one (solid) of slope 1:2, are also plotted for reference in each diagram. It can be accordingly noticed that all the data points plot below the 1:1 line, and most of them even below the 1:2 line. This fact confirms that the distances between two adjoining hypocenters of a particular group are always smaller than the distances to the nearest hypocenter of any adjacent group; and in fact, most such distances between adjoining hypocenters of one and the same group, are even less than half the distances to the nearest hypocenters of any adjacent group. Similar plots (not shown) were obtained by using hypocenters locations provided by the PDE catalogue.

The distinction between the groups which we have thus defined is also illustrated by the fact that the average distance between adjoining hypocenters of a particular group is, systematically, much smaller (last two columns of Table 1) than the average distance to the nearest hypocenter of any adjacent group (previous four columns of Table 1). Table 1 also outlines that separate computations, based on hypocenters locations provided by the ISC and the PDE catalogues respectively, led to comparable results.

3 Particular Patterns Associated to the "Lowermost Group" Seismic Activity

We focused primarily on the "Lowermost group", which by comparison to both the "Upper" and the "Lower" diagonal lineaments, appeared to be far less seismically active: only five earthquakes (all of them in the range $4.7 \leq m_b \leq 5.0$) have been recorded in the "Lowermost group" over the time-interval 1965–2014. A remarkable fact is that out of those five events, four have occurred between the years 1973–1987, and have systematically displayed a TF mechanism with the

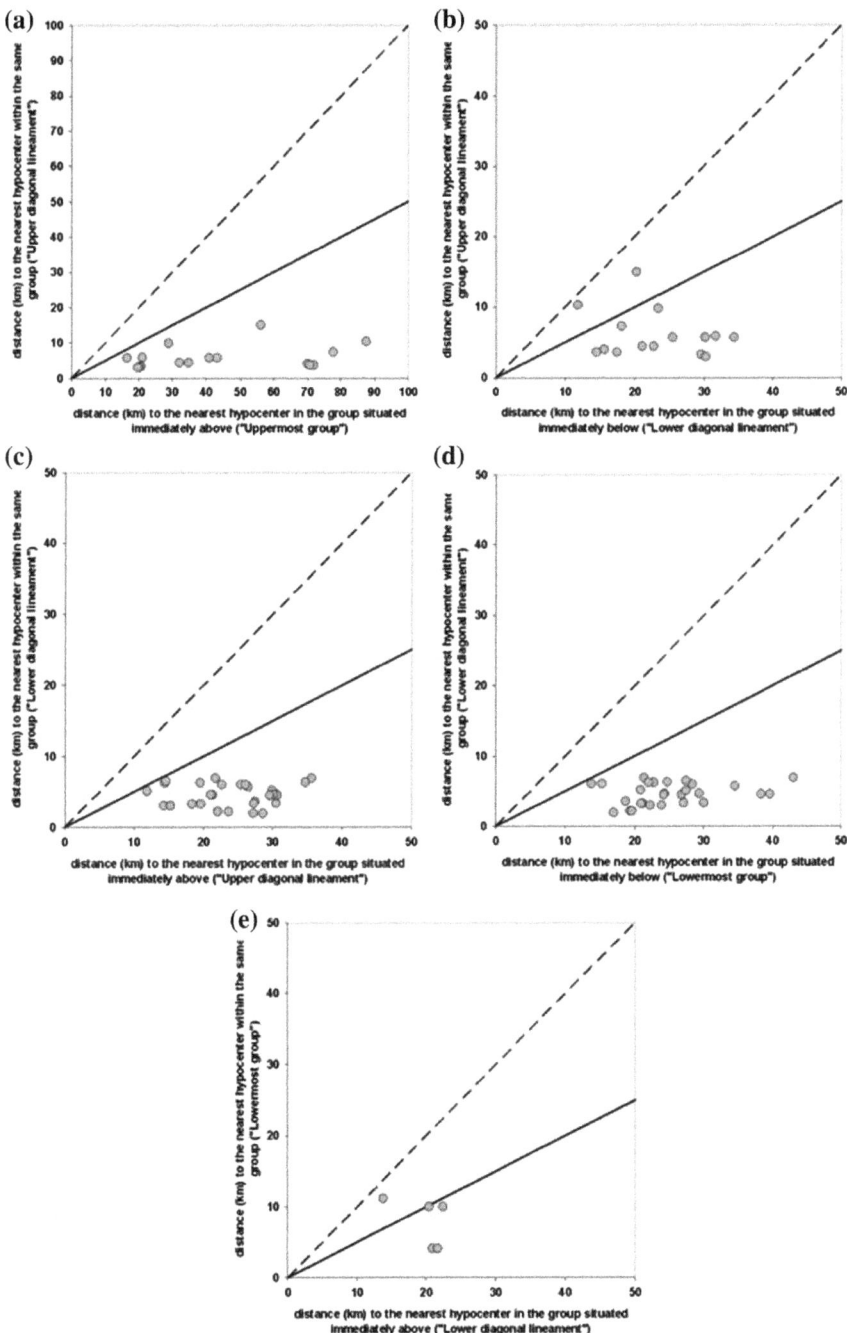

Fig. 3 The distances from each particular hypocenter of a specific group ("Upper diagonal lineament"—**a**, **b**; "Lower diagonal lineament"—**c**, **d**; "Lowermost group"—**e**), to its nearest hypocenter within the same group, are plotted against the distances between the same particular hypocenter and the nearest hypocenter of the group situated immediately above (**a**, **c**) and immediately below (**b**, **d**, **e**). Dashed-line of slope 1:1; solid-line of slope 1:2

Table 1 Average distance from a hypocenter of a particular group to the nearest hypocenter

| | Average distance (km) to the nearest hypocenter | | | | | |
| | Of the group above | | Of the group below | | Within the same group | |
	ISC	PDE	ISC	PDE	ISC	PDE
Upper diagonal lineament	–	–	23.5 ± 6.9	22.4 ± 8.7	5.9 ± 3.3	8.3 ± 2.9
Lower diagonal lineament	24.1 ± 6.4	24.0 ± 6.1	24.9 ± 7.1	25.8 ± 7.7	4.5 ± 1.6	5.2 ± 2.5
Lowermost group	19.9 ± 3.5	18.5 ± 4.5	–	–	7.8 ± 3.5	8.5 ± 1.9

P axes striking *parallel to the Vrancea slab direction*; while only one earthquake—occurred much later, in 2002—displayed a normal-fault (NF) focal mechanism (for each of the five considered events, the indicated fault plane solution type was consistently similar, irrespective of which independent catalogue—e.g., Oncescu 1987; Mostrioukov and Petrov 1994; Radulian et al. 2002; Sandu and Zaicenco 2008, etc.—had provided it).

It is at the same time worth noticing that the time interval 1973–1987—when all the "Lowermost group" moderate-magnitude events due to thrust faulting induced by slab-parallel compression have occurred—is, to a large extent, overlapping the period 1977–1990, when the most recent three major Vrancea earthquakes ($M_w \geq 6.9$) have been recorded (Fig. 4). Moreover, each of the latter three strong

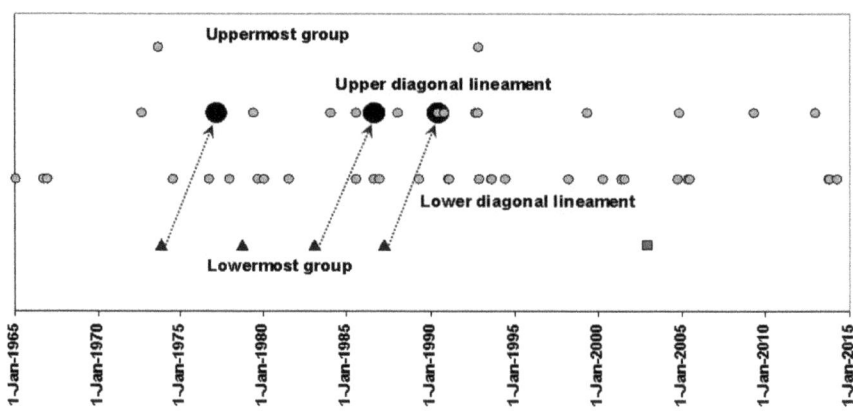

Fig. 4 Time-distribution of the considered Vrancea intermediate-depth earthquakes, as a function of their location in one of the four sub-regions of the slab. Symbols as specified in Fig. 2. Dotted arrows indicate possible long-range interactions operating between a specific moderate event of the "Lowermost group", and the major earthquake which systematically followed it after a time interval of 3–4 years

shocks (whose depth had been much shallower—ca. 85–135 km, Fig. 2) has been preceded—within a time-interval that ranged strictly between 3 and 4 years—by one of the aforementioned TF moderate events occurred in the "Lowermost group"; and since this pattern failed to be obeyed by only one of the four TF events having occurred within the "Lowermost group", it seems that some kind of long-range interaction could be operating within the seismogenic body.

It is worth mentioning, in this respect, that there has been estimated a $\sim 1:4600$ probability that the correspondence noticed between the indicated moderate and major Vrancea earthquakes was due to pure chance (the related statistic computations were conducted according to the approach described in Mitrofan et al. (2016), while yet considering the presently updated selection of Vrancea earthquakes— namely a total of 52 events with $m_b \geq 4.7$, occurred over a period of 50 years).

The cross-section in Fig. 5 illustrates the collisional setting inferred to presently exist in Vrancea area: the edge of the overriding, Intra-Alpine plate has been dragged downwards by the underlying, Moesian plate; so that now, the two plates are coupled to each other along a near-vertical contact that extends over more than

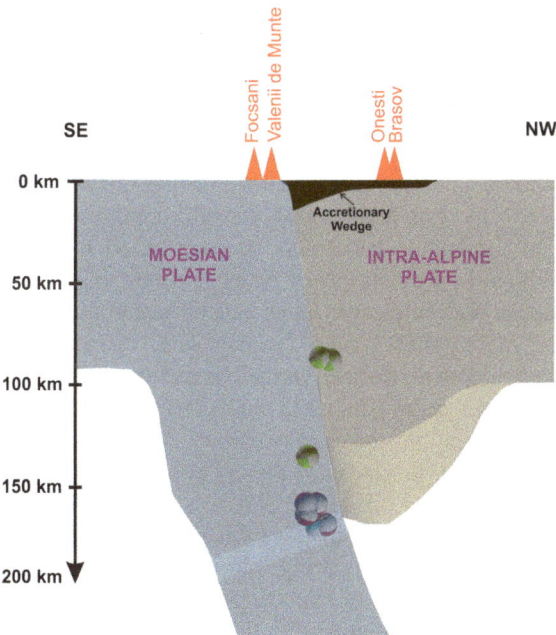

Fig. 5 View (facing SW) of the Vrancea intermediate-depth seismicity domain, at the collisional contact between the Intra-Alpine plate and the underlying Moesian plate; the latter experiences, below 160 km depth, along-strike (away from the viewer's plane) lateral break-off. Focal mechanisms are indicated for: the concerned period Vrancea strongest ($M_w \geq 6.9$) earthquakes (TF, compression approximately perpendicular to the plates contact), the "Lowermost group" TF earthquakes (compression approximately parallel to the plates contact), and the "Lowermost group" NF earthquake (tension approximately parallel to the plates contact)

100 km depth (as already proposed by the model of Cloetingh et al. 2004). The Moesian plate is, on the other hand, assumed to experience along-strike detachment in the depth-region below 160 km: there, seismic tomography (Zhu et al. 2012) has outlined a low velocity zone, which partly interrupted the continuity of the Vrancea "fast" slab; and the "Lowermost group" earthquakes are, as well, concentrated in the same depth-domain.

A largely similar setting has been taken into account by the numerical model of Yoshioka and Wortel (1995), which provided simulated stress distributions (their Fig. 5) that remarkably fitted those suggested by the considered TF Vrancea earthquakes (our Fig. 5) focal mechanisms: the latter indicated, in the inferred lateral detachment region, compression *parallel to the slab strike*; while next to the two plates coupling region, they indicated *normal to the slab strike* compression— possibly due to the shearing forces acting at the upper-plate/underlying-plate interface, subject to the increased downward pull experienced by the slab as break-off was propagating laterally.

Hence the "Lowermost group" events relative space-time distribution, correlated to that of the major Vrancea earthquakes, seems to comply with a process of lateral (along-strike) detachment presently undergone by the corresponding lithospheric slab.

Still the ensemble of the so far discussed considerations provides no clue about the direction in which the slab detachment advances. Convergent diagnostics on this matter could, however, be derived from two independent sets of observations.

1. There was postulated (e.g., Wortel and Spakman 2000) that the present-day topography should undergo differential vertical displacements, consisting of subsidence above the slab section that was still continuous (and which hence experienced, as the lateral detachment propagated, an increased gravitational pull), and uplift above the slab section which, ensuing to the break-off progress, became detached. With reference to that general pattern, it was noticed that the cluster of the "Lowermost group" earthquakes' epicentres was positioned (Fig. 6) between a domain (located in the NE section of the Vrancea epicentral area) where cf. GPS records (Heidbach et al. 2007), present-day uplift occurred, and another domain (in the Vrancea epicentral area SW part) which experienced present-day subsidence. It was consequently possible that break-off advancing along the slab strike, from NE to the SW, was responsible for inducing those differential vertical movements of the local topography, the tip of the corresponding detachment horizon being positioned in the region where the "Lowermost group" earthquakes hypocenters were located.

2. The maximum shear stress outlined by the numerical model of Yoshioka and Wortel (1995) occurs (their Plate 5a) in a slab region situated above the detachment horizon; that maximum shear stress region extends obliquely upwards from the break-off fracture tip, toward the slab margin where the lateral detachment had been initiated. One cannot exclude the possibility that the Vrancea earthquakes hypocentres "diagonal lineaments" (Fig. 2) actually correspond to such a maximum shear stress region. There can be inferred in this

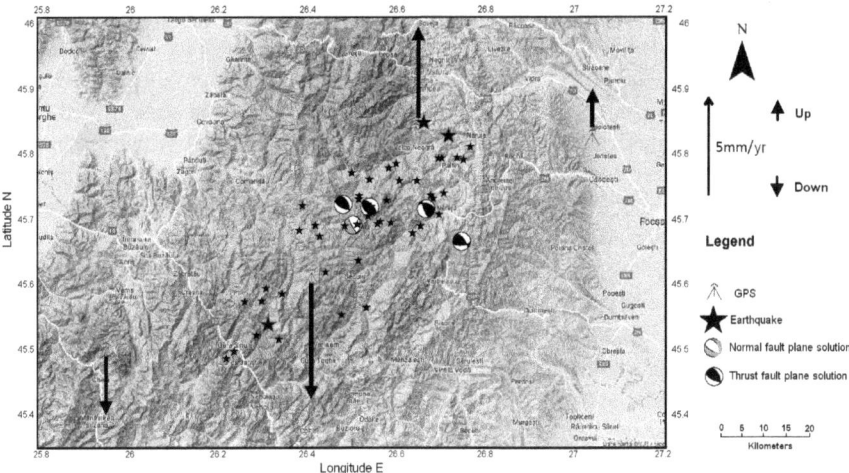

Fig. 6 Location of the "Lowermost group" earthquakes epicentres (beach-balls), with respect to the present-day topography vertical displacements recorded by the GPS measurements reported in Heidbach et al. (2007). Stars designate the epicentres of all other earthquakes indicated in Figs. 2 and 4, larger symbols designating the three major events (Mw ≥ 6.9) occurred during the considered time-interval

case (based on the "diagonal lineaments" position in Fig. 2) that the lateral detachment had been initiated at the NE margin of the Vrancea slab, then it propagated toward the SW—a setting which also agreed with the break-off lateral progression direction indicated by the present-day topography differential vertical movements.

4 Conclusions

During the 50 years time-interval spanning from 1965 to 2014, in the Vrancea intermediate-depth seismicity domain there have occurred 52 earthquakes for which both the ISC and the PDE catalogues have indicated $m_b \geq 4.7$. The corresponding hypocenters are not uniformly distributed: they are included into four distinct clusters, which within the Vrancea near-vertical seismogenic body (that in the 60–180 km depth-range, extends roughly 70×25 km on the horizontal) are clearly separated from one another.

Between two of those hypocenter-clusters, long-range interactions seem to operate. Specifically, each of the three major earthquakes ($M_w \geq 6.9$) of the concerned period—all of which were located in the 85–135 km depth range, being thus included in the so-called "Upper Diagonal Lineament"—has been preceded by a moderate ($4.7 \leq m_b \leq 4.9$) event that systematically occurred 3 to 4 years in advance, and was located in the so-called "Lowermost group". Moreover, while the major event focal mechanism was always of TF type with the *P* axes oriented approximately *perpendicular to the Vrancea slab strike*, the preceding moderate event was systematically of TF type with the *P* axes striking approximately *parallel to the Vrancea slab direction*. This pattern is compliant with the stress distribution predicted by a numerical model (Yoshioka and Wortel 1995) which addressed along-strike break-off occurring within a near-vertical slab that was also strongly coupled—along a significant vertical distance—with its overriding plate: it was thus suggested that the Vrancea "Lowermost group" moderate events of TF type could mirror the along-strike progression of a detachment horizon tip; while the subsequently occurred major earthquakes were caused by the shearing forces acting at the upper-plate/underlying-plate interface, subject to the increased downward pull experienced by the Vrancea slab as the break-off was propagating laterally. A break-off process advancing from NE to the SW along the Vrancea slab strike would also be consistent with the present-day topography differential vertical displacements which had been outlined (Heidbach et al. 2007) by GPS records.

Acknowledgements We are grateful to Luminiţa Ardeleanu, Mihaela Popa, Maria Tumanian, Mircea Radulian and Alin Tudorache for the kind assistance they provided us for enlarging our necessary database of scientific publications.

References

Bonjer KP, Ionescu C, Sokolov V, Radulian M, Grecu B, Popa M, Popescu E (2008) Ground motion patterns of intermediate-depth Vrancea earthquakes: the October 27, 2004 event. In: Zaicenco A, Craifaleanu I, Paskaleva I (eds) Harmonization of seismic hazard in Vrancea zone. Springer, Dordrecht, pp 47–62

Cloetingh SAPL, Burov E, Matenco L, Toussaint G, Bertotti G, Andriessen PAM, Wortel MJR, Spakman W (2004) Thermo-mechanical controls on the mode of continental collision in the SE Carpathians (Romania). Earth Planet Sci Lett 218:57–76

Faccenda M, Gerya TV, Chakraborty S (2008) Styles of post-subduction collisional orogeny: influence of convergence velocity, crustal rheology and radiogenic heat production. Lithos 103:257–287

Ganas A, Grecu B, Batsi E, Radulian M (2010) Vrancea slab earthquakes triggered by static stress transfer. Nat Hazards Earth Syst Sci 10:2565–2577

Heidbach O, Ledermann P, Kurfeß D, Peters G, Buchmann T, Maţenco L, Neguţ M, Sperner B, Müller B, Nuckelt A, Schmitt G (2007) Attached or not attached: slab dynamics beneath Vrancea, Romania. In: International symposium on strong Vrancea earthquakes and risk mitigation, Bucharest, Romania, 4–6 Oct 2007

Ledermann P, Heidbach O (2007) Stress transfer modeling of the strong earthquake sequence at intermediate depths in the Vrancea area Romania. Geophys Res Abstr 9:02161

Li ZH, Xu ZQ, Gerya TV (2011) Flat versus steep subduction: contrasting modes for the formation and exhumation of high- to ultrahigh-pressure rocks in continental collision zones. Earth Planet Sci Lett 301:65–77

Mitrofan H, Anghelache MA, Chitea F, Damian A, Cadicheanu N, Vişan M (2016) Lateral detachment in progress within the Vrancea slab (Romania): inferences from intermediate-depth seismicity patterns. Geophys J Int 205:864–875

Mostrioukov AO, Petrov VA (1994) Catalogue of focal mechanisms of earthquakes 1964–1990. In: Materials of the world data center B, Moscow. http://www.ifz.ru/open_data/focal-mechanisms. Accessed 2 Aug 2017

Oncescu MC (1987) On the stress tensor in Vrancea region. J Geophys 62:62–65

Radulian M, Popescu E, Bala A, Utale A (2002) Catalog of fault plane solutions for the earthquakes occurred on the Romanian territory. Rom J Phys 47:663–685

Sandu I, Zaicenco A (2008) Focal mechanism solutions for Vrancea seismic area. In: Zaicenco A, Craifaleanu I, Paskaleva I (eds) Harmonization of seismic hazard in Vrancea Zone. Springer, Dordrecht, pp 17–46

Wortel MJR, Spakman W (2000) Subduction and slab detachment in the Mediterranean-Carpathian region. Science 290:1910–1917

Yoshioka S, Wortel MJR (1995) Three-dimensional numerical modeling of detachment of subducted lithosphere. J Geophys Res 100:20223–20244

Zhu H, Bozdağ E, Peter D, Tromp J (2012) Structure of the European upper mantle revealed by adjoint tomography. Nat Geosci 5:493–498

The Seismogenic Sources from the West and South-West of Romania

E. Oros, M. Popa and M. Diaconescu

Abstract The study region is the most important seismic region of Romania when we refer to the crustal seismicity as a source of seismic hazard. So far there have been recorded 91 seismic events that produced significant effect in buildings (Io $\geq$ 6 EMS), some of them resulting in severe damage and even casualties (Io $\geq$ 7 EMS). In this paper we modelled the seismogenic sources in the region using a new seismotectonic model constructed on new earthquakes and focal mechanisms catalogues basis. This model was elaborated starting from the relationship between geology and historical and instrumental seismicity and then it was better constrained by geophysical, neotectonic, geodetic data and particularly by active stress field features. The stress tensor parameters and the stress regime have been determined by formal inversion of the focal mechanisms solutions. Our study provides evidence of at least seven different deformation domains with different tectonic regimes as a realistic support for assessing the seismogenic potential of the geological structures. Each seismogenic source is characterized by completeness magnitude (Mcomp), maximum probable magnitude (Mmax) and magnitude—recurrence parameters. The probabilistic hazard maps produced in terms of PGA using the new seismic sources highlights the importance of their configuration on the hazard parameter values and their spatial distribution.

Keywords Seismogenic sources · Seismotectonics · Seismicity
Mmax · Stress field

E. Oros (✉) · M. Popa · M. Diaconescu
National Institute for Earth Physics, Magurele, Romania
e-mail: eugeno@infp.ro

M. Popa
e-mail: mihaela@infp.ro

M. Diaconescu
e-mail: diac@infp.ro

© Springer International Publishing AG, part of Springer Nature 2018
R. Vacareanu and C. Ionescu (eds.), *Seismic Hazard and Risk Assessment*,
Springer Natural Hazards, https://doi.org/10.1007/978-3-319-74724-8_4

53

1 Introduction

The seismic hazard analyses, probabilistic or deterministic, use as the key input data the earthquake source models which describe the seismicity of the location of interest and reflect the architecture of active fault systems. These models, so-called *seismic source*, are associated with a geological structure and are defined in different configurations depending on the quality and resolution of geological, seismological, tectonic, geophysical and geodetic available data. The seismic sources are usually characterized by their potential to produce strong earthquakes (Mmax) and their ability to produce such earthquakes repeatedly (recurrence parameters). These characteristics are estimated using earthquakes catalogues that cover most of the time only a very short period of a seismic cycle. Thus, if there are no data about the seismic activity some geological structures can be neglected in the process of modelling of seismic sources even if their geometry is favourable to be reactivated in a particular stress field. Therefore the seismic sources models require detailed geological and geophysical constraints to ensure realism and reliability to the final hazard models. The seismic hazard analysts use different types of seismic sources the most common being faults, localizing structures and seismotectonic units (Reiter 1991). Recently have been defined Individual Seismogenic Sources (ISS), Composite Seismogenic Sources (CSS), Seismogenic Areas (SA) and Debated Seismogenic Sources (DSS) (e.g. Basili et al. 2008).

Radulian et al. (2000) defined the Seismogenic Zones in Romania using seismicity and tectonics data and their correlation with morpho-structural units. They described two seismogenic zones in the study region, *Banat Seismogenic Zone* (*BSZ*) in the North and *Danube Seismogenic Zone* (*DSZ*) in the South, respectively. This seismic zoning was used in many seismic hazard studies, (e.g. Ardeleanu et al. 2005; Moldovan et al. 2008; Oros 2011; Simeonova et al. 2006).

The study region can be considered the most important region of Romania if we refer to the seismic hazard associated with crustal seismicity. The last earthquakes catalogues (e.g. Oncescu el al. 1999; Oros 2011; Stucchi et al. 2013) contain a lot of earthquakes which produced significant engineering effects (Io $\geq$ VI EMS, *EMS is European Macroseismic Scale*) and even heavy damages and casualties (Io $\geq$ VII EMS). Generally the seismicity is diffuse but there is a clear clustering trend for both historical and instrumental periods (Oros and Diaconescu 2015). The region is located at the contact between the Carpathians and Pannonian Depression where three geodynamic units develop and control the seismic activity (Zugravescu and Plolonic 1997). Several crustal blocks characterize the basement tectonics. These are covered by sedimentary formations and fragmented by neo-structures (Polonic 1985; Sandulescu 1984). All structural units are bounded by faults systems of different ages that were reactivated or blocked in a variable stress field controlled by NE Adria Microplate pushing and the basin inversion processes (e.g. Bada et al. 2007; Bala et al. 2015).

In this paper we present a new seismic zoning of the study region. The definition of seismogenic source as a geological structure reactivated or with reactivation potential in a particular stress field, with or without associated seismicity is the basic concept of the applied methodology. The seismogenic sources were identified and defined using a new seismotectonic model. This model has been constructed at different scales on the relationship between geology, seismicity and active stress field basis. Geophysical, geodetic and neotectonic data constrained the final version of the model. Thus, identifying, defining and characterization the earthquake sources rely (1) on knowing the geology and tectonics of the region and (2) then the understanding of their relationship to active stress field and seismicity. Each seismogenic source is characterized by completeness magnitude of the catalogue, magnitude—recurrence parameters and *Mmax*.

2 Geology and Geotectonic Setting

The study region is located on the south-eastern border of the Pannonian Basin at the contact between Carpathians and Pannonian Depression (Fig. 1, top). The Pannonian Basin sensu stricto, is a Miocene-Quaternary back-arc extensional basin formed after pre-Neogene orogeny. The present structure of the region consists of (1) *basement pre-Neogene units* (nappes, suture zones, close rifts, magmatic bodies, sedimentary formations, etc.) bordered by faults systems (thrusts) that were successively reactivated during Alpine history under different stress regimes and (2) *neo-structures* (deep basins, grabens, horsts, depressions) having different structural positions controlled by normal/listric faults (Fig. 1, bottom) (Polonic 1985). Several geotectonic units compose the basement and are described by Sandulescu (1984) as *Inner Dacides* (*ID*), *Median Dacides* (*MD*), *Marginal Dacides* (*mD*) *and Transylvanides* (*T*). These units are structural components of the Pannonian, Geto-Danubian and Moesian Geodynamic Blocks (Zugravescu and Polonic 1997). Dacides and Transilvanides structures are segmented by faults and neo-structures that intersect them more or less orthogonal. Two main faults systems can be defined in the region. One, called here as *Carpathian system*, is NE–SW to EW oriented and characterizes the basement and orogenic structures bordering the major geotectonic and geodynamic units (Fig. 1, bottom). These are either thrusts that delineate SSE—verging nappes within Inner Dacides (e.g. Sinnicolau Mare-Arad and Jimbolia-Lipova Thrusts) and WNW—verging nappes within Median and Marginal Dacides (e.g. Sichevita-Retezat Thrust) either vertical trans-crustal faults (e.g. Oravita-Moldova Noua Fault system). The other one (*Pannonian system*) controlled the neotectonic activity in the region and have predominant NW–SE to NS and NNE–SSW directions being generally normal and low angle faults (Sandulescu 1984; Polonic 1985).

Magmatic plutons (Cretaceous *banatites*) and volcanic intrusions (e.g. Lucaret and Gataia Neogene basalts) are distributed through the basement of the study region being associated with continental subduction zones and deep faults systems.

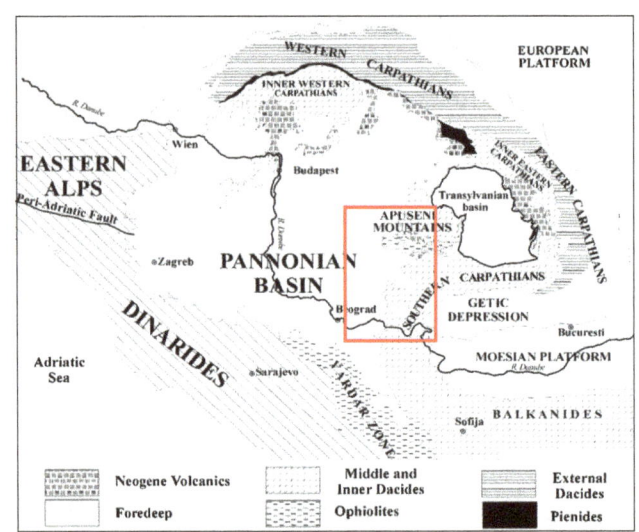

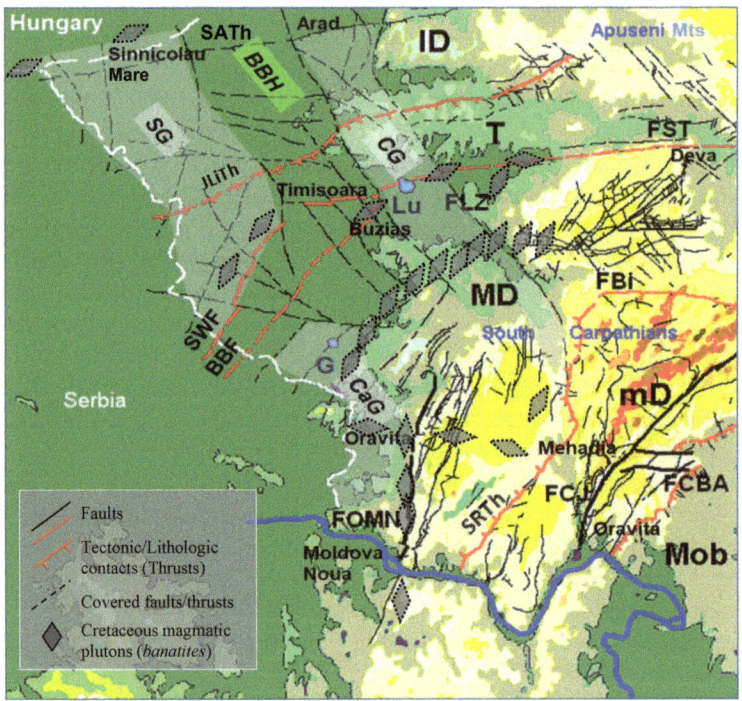

◄**Fig. 1** Upper Carpatho-Pannonian system and the location of the study area (red polygon). Bottom map Syntetic map of geotectonic units and major faults sytems (topografic background). ID Inner Dacides, T Transilvanides, MD Median Dacides, mD Marginal Dacides, Mob Moesian Bazin, SG Sinnicolau Mare-Szeged Grabens system, CG Caransebes Graben, CaG Caras Graben, BBH Battonya-Buzias High, Lu/G Lucaret and Gataia Quaternary basalts, SATh Sinnicolau Mare-Arad Thrust, JliTh Jimbolia–Lipova Thrust, SRTh Sichevita-Retezat Thrust, FST South Transilvanian Fault system, BBF Banloc-Buzias Fault system, SWF SW Timisoara Fault system, FLZ Lugoj-Zarand Fault system, FBi Bistra Fault (segment of South Carpathian Fault system), FOMN Oravita-Moldova Nouă Fault system, FCJ Cerna-Jiu Fault system, FCBA Closani-Baia de Arama Fault system. Some of symbols for magmatic plutons only localize the structures without defining their limits (top map reproduced after Ciulavu et al. 2000, bottom map compilation after Ciobanu et al. 2002; Polonic 1985; Sandulescu 1984; Institutul Geologic Roman 1968)

3 Seismicity

The seismicity map of the region is presented in Fig. 2a, b. These maps are constructed using a compilation of the revised catalogues elaborated by Oros et al. (2008a, b) and Oros (2011) that was completed with data from the catalogues of Oncescu et al. (1999) (updated version on www.infp.ro), Stucchi et al. (2013) and Grunthal et al. (2013). This compilation contains over 8500 earthquakes with Mw = 0.4 − 5.6 that occurred between 1443 and 2016. A number of *91* earthquakes with Io ≥ VI EMS (Mw ≥ 4.0) have been catalogued. 32 of them had Io ≥ VII EMS (Mw ≥ 4.7) producing heavy damages and even casualties. In the known seismic history of the study region there are *2 major seismic crisis* well documented by Oros (2011). The first was recorded between 1879 and 1880 and is characterized by two sequences located in two areas: (1) one in DSZ at Moldova Noua (10.10.1879, Io = VIII EMS/ Mw = 5.8 and 11.10.1879, Io = VII EMS/Mw = 5.3) and (2) the second in BSZ at

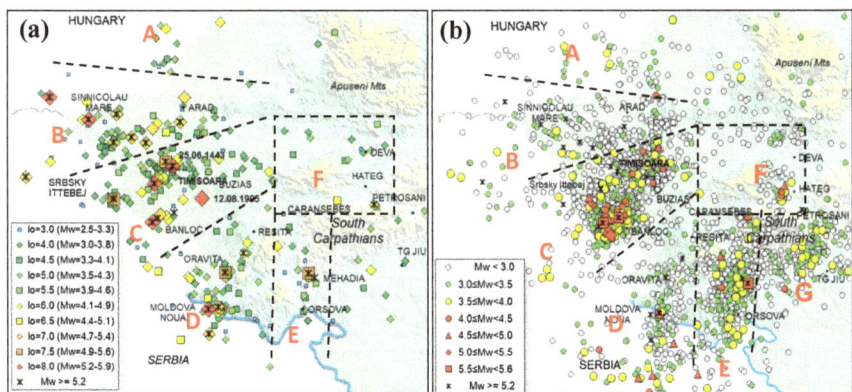

Fig. 2 Seismicity maps of the study region. **a** Macroseismic epicentres (1443–1979), conversion relationship Mw = 0.53Io + 1.2 log *h* + 0.11 for h = 5–20 km from Oros (2011); **b** instrumental epicentres (1980–2015). The events with Mw ≥ 5.2 (I = VIII EMS) recoded from the whole period are displayed on both maps. Black dashed lines delineate the zones with particular grouped seismicity (A–G)

Tomnatic (16.04.1879, Io = VI − VII/Mw = 5.2), Sinnicolau Mare (31.10. 1879, Io = VII EMS/Mw = 5.3) and Sacalaz-Timisoara (19.11.1879, Io = VII EMS/Mw = 5.2). The second crisis occurred in 1991 and lasted several years. It also had two different sequences and started at Banloc in BSZ (12.07.1991, Io = VIII EMS/Mw = 5.6). It was followed on 18.07.1991 by a strong earthquake (Io = VII EMS/Mw = 5.6) located at Mehadia in DSZ. On 2nd December 1991 other strong event occurred at Voiteg (Io = VIII EMS/Mw = 5.5).

Two main groups of epicentres can be defined, one in North (BSZ) and the other in South (DSZ), respectively (Fig. 2). Inside the two zones there is a clear tendency of grouping the historical epicentres in several groups, called from North to South (Fig. 1a): *A*, Bekes (at the Romanian border with Hungary), *B*, Sinnicolau Mare-Arad with two almost EW—orientated alignments, parallel with Inner Dacides structures; *C*, Timisoara-Banloc, with clusters at Timisoara, Banloc, Sirbsky Ittebej at the Romanian border with Serbia and at North of Buzias; *D*, Moldova Noua-Resita with groups at Moldova Noua, Oravita, Oravita-Resita; *E*, Orsova-Caransebes, *F*, Hateg-Deva-Mures Valley. The zone *G*, South Petrosani is outlined only on the instrumental seismicity map (Fig. 2b). These areas have been defined so that they can be characterized by at least one major earthquake with Mw ≥ 5.2, except G zone located outside the study region. The instrumental seismicity displays a roughly identical model. However some differences can be noted: (1) the boundaries between A, B and C zones are more poorly defined by Mw ≤ 3.0 earthquakes the distribution; (2) the distribution of smaller events (Mw ≤ 3.5) defines more clearly small clusters and alignments (e.g. between Arad and Timisoara, Oravita, Moldova Noua, Hateg, Deva and South Petrosani). The depth distribution shows a layering model with three levels (Oros 2011): h_1 = 2–10 km ($h_{average}$ = 8.0 ± 1.6 km), h_2 = 10–15 km ($h_{average}$ = 12.5 ± 1.6 km) and h_3 = 15–20 km ($h_{average}$ = 17.2 ± 1.3 km). There are also a few deeper hypocentres (Oros and Oros 2009).

Seismicity is usually described by the Gutenberg–Richter relationship which coefficients, *a* and *b*, quantify the seismic activity rate and the ratio between small and large magnitudes, respectively. The space distribution of b-value is useful to mapping the state of stress and its heterogeneities within seismic active areas (Scholz 2015). The values of b vary significantly if the stress changes are large and there are some structural conditions (e.g. geometric heterogeneities) that may affect the seismogenic potential of fault zones and fractured structures. Thus *b-value variations* can constrain the limits of areas where particular correlations between seismicity, tectonics and stress field are defined. We present in Fig. 3 the 2D the distribution of *b*-values along with several parameters that shows their statistical quality and support the reliability of the analyses that rely on them, e.g. goodness of fit to power law, R as the percentage of the distribution frequency—magnitude which can be modelled by power law type; the data set can be considered reliable if R = 80–90%, (Wiemer 2001), standard deviation of b-value and completeness magnitude, Mcomp. The maps were constructed using the Zmap code by Wiemer (2001), samples with N = 250 events (Nmin = 50) and a grid with cells of 0.1 × 0.1 degree (Oros 2011).

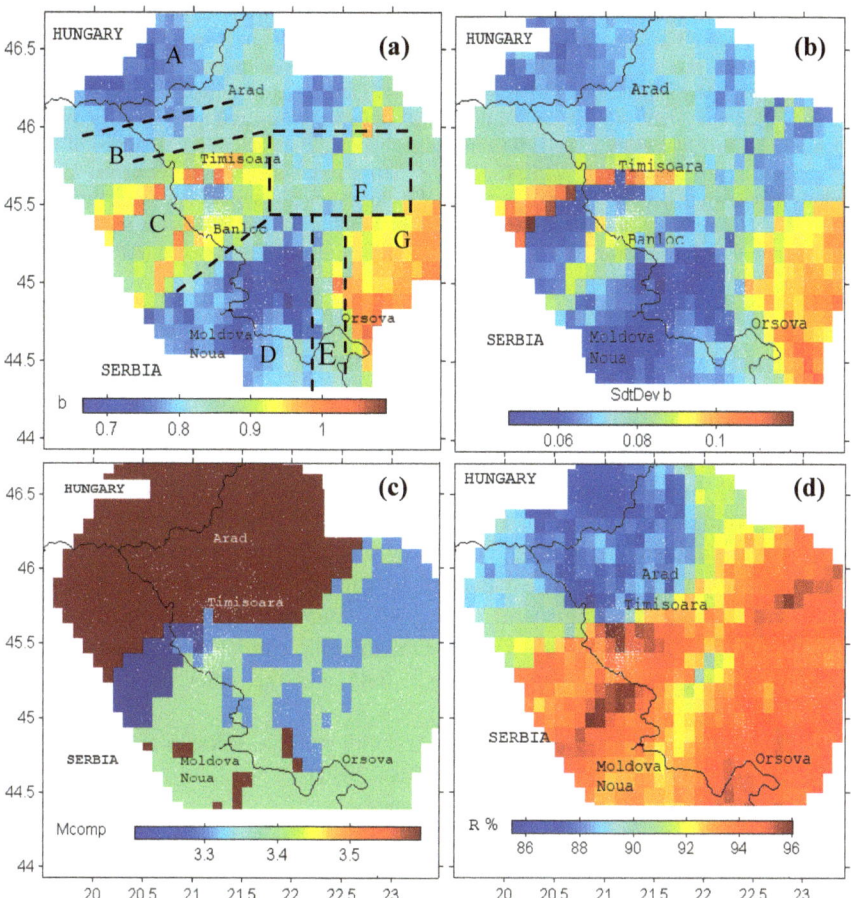

Fig. 3 **a** 2D distribution of b-value, **b** Standard deviations of b-value, **c** Completeness magnitude Mcomp, **d** Goodness of fit to power law (R), (reproduced from Oros 2011). *Black dashed lines* delineate the zones with particular grouped seismicity and nearly uniform b-values (A–G)

The b-values indicate wide variation (0.65 < b < 1.05) well correlated with geotectonic structures in the study region. Their distribution suggests a high heterogeneous state of stress highlighted through an alternation between: (1) *low b values* (b ≤ 0.8) within A, B, and D zones and partly C, E and F zones from Fig. 2 what does it means that high stress concentration associated with strain accumulation exists in the region and (2) *higher b-value* (b > 1.0) within small areas where very active faults are descried (e.g. Bala et al 2015; Oros 2011). These variations suggest a high level of stress heterogeneity or high stress drop associated with high seismic activity. Strong earthquakes (Mw ≥ 5.2) occurred recently in these zones, e.g. Timisoara (27.05.1959, Mw = 5.3), Banloc (07.12.1991, Mw = 5.6) and Mehadia (18.07.1991, Mw = 5.7). All areas from A to G, defined in Fig. 2, can be

Fig. 4 a Modelled principal stress direction (black bars) with inversion variance constructed ▶ using Zmap code (Wiemmer 2001); dashed white lines delineate the zones A–G from Fig. 1, b Seismotectonic sketch. *Inset* (right upper corner): regional stress tensor; *1* Inner Dacides, *2* Transilvanides *3* Median Dacides *4* Marginal Dacides *5* Outer Dacides (rift structures) *6* Moesian Paleozoic Platform *7* faults (black/dashed thin lines are faults/thrusts; red thick lines, neotectonic structures) *8* Late Quaternary basalts, 9 Cretaceous plutons ("Banatite") *10* trajectories of recent stress *11* earthquakes with Io ≥ VI EMS (Mw ≥ 5.0); *12* vertical recent movements; *13* stress symbols (black/white bars are Shmax/Shmin, small coloured circles symbolizes stress regime and strain style: N, SS, T are normal, strike slip and thrust faulting) (Inset and stress tensors—large symbols reproduced from Oros et al. 2016; tectonics and stress model reproduced from Sandulescu 1984 and Bada et al. 2007; vertical movement from Horvath et al. 2006)

characterized by nearly uniform b-values. However there are some differences regarding their limits, dimensions and shapes that better correlate with geology, e.g. (1) *A* zone can be extended towards South including the Sinnicolau Mare epicentral area from B zone that is well correlated with SATh in the SG graben; (2) *F* zone with generally low b-value is divided by a N–S oriented narrow area of higher b-value resulting two sub-zones, Hateg in East and Caransebes in West, respectively. Analysed at smaller scale almost every seismic zone may be divided on the basis of sharply variations in the values of b. Thus, within C zone there are three areas with b = 1.05 (Timisoara), b = 0.92 (Banloc) and b = 0.75 (South Timisoara), respectively. Moldova Noua-Oravita zone can be also divided in two sub-zones, one in South having b = 0.8 − 0.85 and the other in North where b = 0.65 − 0.75.

The low b-values areas overlap with zones with positive velocity anomalies of P waves (Vp) computed by Zaharia et al. (2010, 2017).

4 Active Stress Field and Seismotectonic Model

Stress field pattern helps to identify deformation zones and tectonically active structures at different scales both in space and time. Usually, the stress inversion using earthquake focal mechanisms allows estimating the orientation of the principal stress axes (S1 > S2 > S3), the stress ratio [R = (S2 − S3)/(S1 − S3)] and the derived stress regime index R' (R' = 0–1 for normal faulting, R' = 1–2 for strike-slip faulting and R' = 2–3 for reverse faulting) (Delvaux and Sperner 2003). We used 438 focal mechanisms solutions determined by Oros et al. (2008b, 2016) to compute the reduced stress tensor. Two analyses were conducted to obtain reliable data to a realistic constraint of seismogenic sources defined on seismicity basis. First, to identify characteristics of stress field in relation to local tectonics and seismicity we computed and mapped the 2D distribution of the principal stress orientation (S1) and the variance of inversion using the Zmap code (Wiemer 2001) (Fig. 4a). Second, we investigate the horizontal stresses (SHmax and Shmin) and

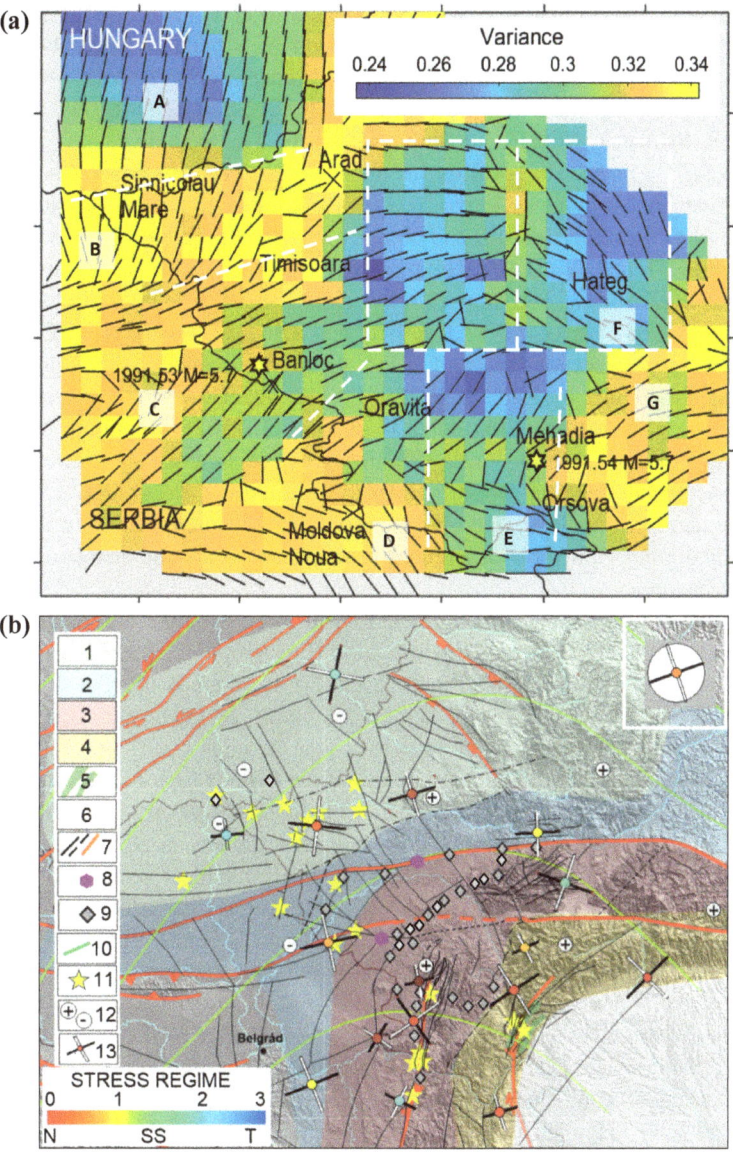

tectonic regime for every seismic zone defined previously in order to identify the structures with seismogenic potential (Fig. 4b). For this purpose we analysed the preliminary stress field modelled by Oros et al. (2016) and we used Win-Tensor program of Delvaux and Sperner (2003) to compute reduced stress tensors for different smaller data sets selected in areas with large variations of seismicity, S1 directions and homogeneity and with particular tectonics (small symbols in Fig. 4b).

The regional stress field was investigated by Oros et al. (2016, 2017) using 1132 focal mechanisms from Intra-Carpathian region of Romania. They obtained the following data that are comparable with the first order stress field described by Bada et al. (2007): S1 (azimuth/plunge) = 234°/45°, S2 = 72°/44°, S3 = 333°/9°, R = 0.48, SHmax = N60°E and Shmin = N15°E. The tectonic regime is oblique extensive (R' = 0.48). The misfit is 56° meaning that the stress field in the region is spatially very heterogeneous. Oros (2011) used for the stress tensor inversion in West Romania 140 focal mechanisms and obtained similar parameters: SHmax = N71°E, Shmin = 161°E and R = 0.40 (SHmax = N67°E, R = 71 and extensional strike slip regime in BSZ and N52°E, R = 77, extensional regime in DSZ). Radulian et al. (2000) computed SHmax = N87°E in BSZ and SHmax = N70°E in DSZ.

Generally, S1 directions mapped in Fig. 4a are comparable to the regional stress trajectories modelled by Bada et al. (2007) (Fig. 4b). However, the stress field is heterogen ($\sigma \geq 0.2$—threshold after Wiemer et al. 2002) throughout the region. The pattern displays a short—scale variations of these parameters suggesting strong disturbance of the regional stress field by local sources of stress, like active faults, contrast density, rheology, etc. It resembles the seismicity zones presented in Figs. 2 and 3 but with some differences in shape and limits and also correlate with tectonic features.

The stress regime and SHmax orientations disclose two different major trends in the active stress field and crustal deformation (Fig. 4b): (1) SHmax is parallel with SHmax regional modelled by Bada et al. (2007) and there is a strike slip faulting, with a large extensional component at the limit between the Transilvanides (ophiolites) and the Median Dacides (South Transilvanian Fault) and (2) SHmax is oblique up to perpendicular to the direction of the SHmax regional and transtensive to pure extensional faulting are defined (Sinnicolau Mare-Arad, Moldova Noua Oravita si South Petrosani). The stress field heterogeneity is also noted by stress regime and strain style changes both at regional and local scale, e.g. there are significant differences between BSZ (uppermost extensional strike slip) and DSZ (extensional) and between small areas within B and D zones. The most heterogeneous area is located between Resita and Moldova Noua prolonged in Serbia where the average SHmax is almost NS-oriented, in opposite sense towards the DSZ general direction. This is most likely due to the influence of the focal mechanism form the South of the Danube located in a different tectonic setting. In this area as well as throughout the study region the stress field features and seismicity can be associated with magmatic intrusions of different sizes and shapes distributed on alignments with different orientations (e.g. Sinnicolau Mare, Caras and Caransebes grabens Moldova Noua-Oravita seismic area). Interesting to the N–NW, close to these magmatic and volcanic structures located in DSZ, an area with low seismic activity develops between BSZ and DSZ as transition zone.

The geodetic model of crustal deformations described in the NATO report (2011) is characterized by a large variation of horizontal velocity vectors, both in direction and size suggesting a very complex geodynamic ongoing context and actives tectonic setting. Thus in the North of region they have opposite directions, towards North in the Pannonian Basin area and towards South on the western border of Apuseni Mountains, respectively. This model could displays a share zone between Carpathian and Pannonian Basin structures that overlap an area with high horizontal gradient of recent vertical movements. In the southern region horizontal velocity vectors are also oriented towards South between roughly 21°E and 22°E but they converged on directions around the East and West borders. Thus, they show crustal movements towards East in the Pannonian Basin area and towards West in the Moesian Basin.

A seismotectonic sketch is presented in Fig. 4b. We highlight several features of the stress-tectonic-seismic activity relationship that support the seismogenic potential of the known geological structures: (1) there is a strong correlation between major earthquakes (Mw $\geq$ 5.0) and the faults intersections; (2) the largest number of these significant events appear in the North of the study region (Sinnicolau Mare-Arad area, BSZ) where a high contrast between negative and positive vertical movements can be observed and opposite horizontal GPS velocity was recorded (NATO report, 2011); (3) the stress field parameters seem to be controlled by major structural peculiarities (e.g. South Transylvania, Oravita-Moldova Noua, Cerna-Jiu faults systems and thrusts) with strong local influences (e.g. magmatic plutons, faults geometry and fault populations, thermal and rheological anomalies); (4) faults systems in the seismic active zones have favourable geometries to be reactivated within particular stress field through normal, strike slip and thrust faulting, e.g. Shmin, as the characteristic axis for the extensional stress regime, is perpendicular or oblique to the thrusts and major faults in the region, suggesting a high probability of their reactivation as low angle normal faults; SHmax is parallel to almost EW-oriented faults systems from the western borders of the region meaning favourable relationship to their reactivation as strike slip faults, (5) geodetic model of recent crustal deformations is generally non-conforming to that of stress suggesting possible decupling between different layers of the crust what may substantially alter the seismogenic model of the region.

5 Seismogenic Sources Models. Results and Discussions

To identify, locate and define the seismogenic sources within the study region we started from their definition that we adopted here according to which any geological structure can be reactivated if there is a favourable relation between its geometry and the active stress field. We defined only seismogenic source of type area since missing the essential data necessary for a well-documented investigation of the active faults systems. Thus, there is not available a geological database of faults and faults systems and the seismicity data cannot yet assure an acceptable quality level of correlation seismicity—tectonics throughout the region.

The process of defining the seismogenic sources consists of 3 stages. Every stage is characterized by different criteria and constraints used for the identification and delimitation of seismogenic sources. The process started with a simple model, defined on the basis of the regional structure (blocks geodynamic process) correlated with seismicity (Mw ≥ 5.0). This model was further improved by applying successive some additional constraints. We obtained at each stage different sources models characterized by particular recurrence parameters. Any of these models can be used to seismic hazard assessment depending of the final products of the end users. The recurrence parameters were computed on a declustered catalogue basis (Table 1).

Stage1/Model 1. Starting with the geodynamic and structural models constructed at basement level (geotectonic units, trans-crustal faults) and the seismicity model (Mw ≥ 5.0) we delineated three main areas with reactivated structures under different stress conditions: extensive strike slip in North (BSZ), pure strike slip in ENE (Hateg—Caransebes Seismogenic Zone, HCSZ) and pure extensive in South (DSZ) (Fig. 5a). The earthquakes are located along Carpathian faults system (tectonic contacts between the geotectonic units), especially at their intersections with Pannonian faults systems. BSZ and DSZ are similar with the models defined by Radulian et al. (2000). HCSZ is located in Hateg—Caransebes area where a significant earthquake (M = 5.2) occurred in 1912 and it is characterized by relative homogeneous stress field and strike slip stress regime.

Stage 2/Model 2. We divided BSZ and DSZ in several sub-zones using some geological and geophysical constraints described on smaller scale such as the tectonic limits between the geotectonic units (Dacides and Transilvanides) as well as the 2D distribution of b-value, stress regime and SHmax/Shmin orientation and seismicity pattern (Mw ≥ 4.0). Finally we obtained a more detailed earthquake model with 5 Seismogenic Sources having dimensions of order of dacidic structures segmented by neo-structures: Sinnicolau Mare-Arad (SASZ) and Timisoara-Banloc (TBSZ) within BSZ (extensive strike slip stress regime), Moldova Noua-Oravita

Table 1 Recurrence parameters for the seismogenic sources

Source	Mwcomp	a (yearly)	b
BSZ	4.0	2.64 ± 0.277	0.74 ± 0.057
DSZ	3.0	2.12 ± 0.077	0.68 ± 0.018
HCSZ	3.9	1.99 ± 0.171	0.75 ± 0.038
SASZ	4.0	2.67 ± 0.443	0.82 ± 0.094
TBSZ	3.2	2.66 ± 0.064	0.77 ± 0.015
MNOSZ	4.0	2.35 ± 0.302	0.80 ± 0.064
OCSZ	3.5	1.88 ± 0.141	0.72 ± 0.032
TSZ	3.5	2.49 ± 0.122	0.79 ± 0.028
BVSZ	3.6	1.82 ± 0.142	0.66 ± 0.031
OSZ	4.3	2.37 ± 0.562	0.83 ± 0.122
MNSZ	4.0	1.56 ± 0.286	0.60 ± 0.060
OCSZss	3.9	1.42 ± 0.218	0.63 ± 0.046
BASZ	3.8	2.33 ± 0.304	0.86 ± 0.071
HCSZss	3.0	1.26 ± 0.073	0.59 ± 0.018
DMVSZ	3.1	2.32 ± 0.540	1.04 ± 0.147
SPSZ	3.8	2.60 ± 0.398	0.97 ± 0.117

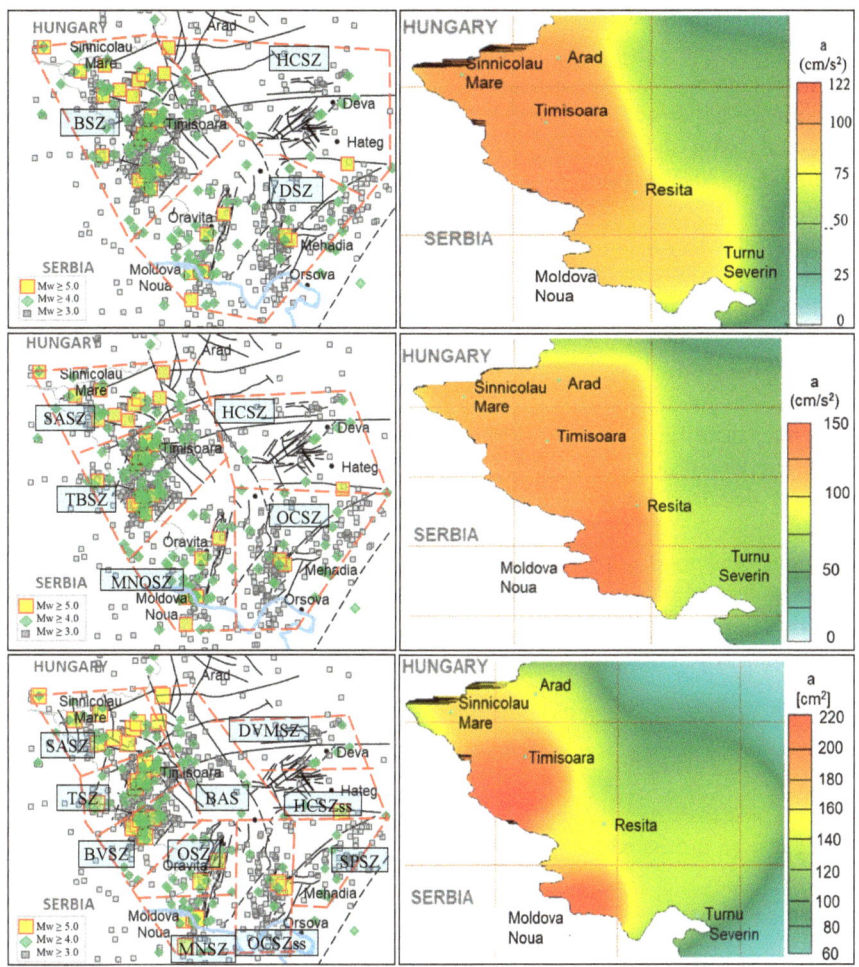

Fig. 5 *Left* Maps of the seismogenic sources defined within the study region (details in text). *Right* Probabilistic Seismic Hazard (PGA) maps corresponding to each seismogenic source model obtained using Crisis 2007 software (Ordaz et al 2007)

(MNOSZ) and Orsova-Caransebes (OCSZ) in DSZ (pure extensive stress regime) and HCSZ as it was before defined. SASZ is characterized by two different stress regimes and a geodetic pattern of crustal deformations that could describe as local sharing area. Thus, there is a strike slip stress regime in West where the velocity vectors are oriented towards the North and an extensive strike slip stress regime in the East where the crustal movements are oriented towards South (NATO report 2011).

Stage 3/Model 3. We improved the previous seismogenic sources model by applying additional constraints. Thus we introduced data about neo-structures and the relationship between their geometry (strike and dip) and SHmax/Shmin direction, 2D principal stress axis orientation and stress regime as well as M ≥ 3.0 seismicity patterns. The new earthquake model contains 9 small seismogenic zones that correlate better with the details of local scale tectonics and having dimensions comparable with those of the active structures and faults systems: Sinnicolau Mare-Arad (SASZ), Timisoara (TSZ), Banloc-Voiteg (BVSZ), Moldova Noua (MNSZ), Oravita (OSZ), Orsova-Caransebes sensu stricto (OCSZss), Buzias-Arad (BASZ), Hateg-Caransebes sensu stricto (HCSZss), Deva-Mures Valley (DMVSZ) and South Petrosani (SPSZ). HCSZ was divided in two sub-zone on SHmax direction and stress regime basis. Thus HCSZss has a pure strike slip stress regime with SHmax oriented towards NW–SE and DMVSZ is characterized by strike slip stress regime with a large extensive component and SHmax oriented on EW direction. This model appears to be more realistic and should be used for estimation of seismic hazard in the study region.

Each model described above present a lot of particular characteristics, e.g. structural details, stress parameters and tectonic regime, possible local stress sources, tectonics, segmentation with locked structures under particular stress conditions, seismogenic potential of each dominant structure.

Mmax, defined simply as the maximum magnitude that may occur under specific geological conditions (the greatest possible magnitude), was computed for the study region using Kijko-Sellevol parametric estimator (Kijko 2004):

$$\Delta = \frac{E1(n2) - E1(n1)}{\beta \exp(-n2)} + m\min \exp(-n) \tag{1}$$

where n1 = n/{1 − exp [−β(mmax − mmin)]}, n2 = n1 exp[−β(mmax − mmin)], and E1 can be estimated as follows:

$$E1(z) = \frac{z^2 + a1z + a2}{z(z^2 + b1z + b2)} \exp(-z) \tag{2}$$

with a1 = 2.334733, a2 = 0.250621, b1 = 3.330657, b2 = 1.681534. Then solve the equation: Mmax = mmaxObs + Δ.

Appling these relationships we calculated Mmax = 6.1 ± 0.3. This value is similar to that computed from Mmaxobs (5.7 + 0.5 = 6.2). Oros (2011) computed the thickness of the seismogenic layer obtaining h = 11.6 ± 4.4 km. Using the relationship between magnitude and the rupture width (Wells and Coppersmith 1994) he obtained Mw = 6.5 for h = 11.6 km.

We computed Probabilistic Seismic Hazard in terms of PGA (cm/s^2) to investigate the effects of the three Seismogenic Sources Models on the space hazard distribution and its values (Fig. 5, right). All hazard models display a distribution of PGS values concentrated in the proximity of seismogenic sources. However the first two models have a high level of smoothing of hazard distribution with overestimating of the PGA values in areas with low seismic activity or with structures without seismogenic potential. It appears in the context that the seismic hazard computed using the "model 3" of seismogenic sources is the most reliable, both as PGA values and their space distribution.

6 Conclusions

We elaborated three models of seismogenic sources as input for the seismic hazard assessment of the western and southwestern region of Romania. They are constructed on new seismotectonic features basis resulted from detailed analysis of the relationship between geology, active/recent stress field and seismicity. We used new earthquakes and focal mechanisms catalogues, stress field computed through formal inversion of the focal mechanisms. The seismogenic sources have been defined by a successive division of regional geological structures at different scales (seismogenic structures) and applying realistic constraints (e.g. stress field properties, 2D distribution of a − b coefficients of Gutemberg Richter relationship, geophysical anomalies, geodetic model of crustal deformations). The detailed seismic zoning of the study region with many smaller seismogenic sources avoid excessive smoothing of seismic hazard that happen in the case of greater seismogenic zones. Each source was characterized by magnitude—recurrence parameters and a unic, regional, maximum magnitude computed by an analytical method and validated using observational (Mmax observed) and seismo-geological (thickness of seismogenic layer) data.

Some seismotectonic peculiarities pointed out by our study may be useful in tectonic, geodynamic, neotectonic studies or deterministic assessment of seismic hazard and risk. Thus, we noted (1) the strongest earthquakes recorded in the study region occurred preferentially at the intersections of almost orthogonal faults systems; (2) there are several local sources of stress that can significantly influence the seismotectonics of the region with direct impact on hazards and seismic risk.

New high quality and resolution data about geological faults, 3D models of seismicity, local stress features are required to realistically model the seismogenic sources as faults.

The methodology used in this study will be applied to all crustal seismic zones from Intra- Carpathian region of Romania.

Acknowledgements This paper was carried out within Nucleu Program supported by ANCSI, partly the Project 30°N/27.02.2009/PN 09 30–01 06 and partly Project PN 16 35 01 05 and PN 16 35 01 12.

References

Ardeleanu L, Leydecker G, Bonjer K, Busche H, Kaiser D, Schmitt R (2005) Probabilistic seismic hazard map for Romania basis for new building code. Nat Hazards Earth Syst Sci 5:679–684

Bada G, Horvath F, Doveny P, Szafian P, Windhoffer G, Cloetingh S (2007) Present-day stress field and tectonic inversion in the Pannonian basin. Glob Planet Change 58:165–180

Bala A, Raileanu V, Dinu C, Diaconescu M (2015) Crustal seismicity and active fault systems in Romania. Romanian Rep Phys 67(3):1176–1191

Basili R, Valensise G, Vannoli P et al (2008) The Database of Individual Seismogenic Sources (DISS) version 3: summarizing 20 years of research on Italy's earthquake geology. Tectonophysics 453:20–43

Ciobanu CL, Cook NJ, Stein H (2002) Regional setting and geochronology of the late cretaceous banatitic magmatic and metallogenetic belt. Mineral Deposita 37:541–567

Ciulavu D, Dinu C, Szakacs A, Dordea D (2000) Neogene kinematics of the Transylvanian basin (Romania). AAPG Bull 84(10):1589–1615

Delvaux D, Sperner B (2003) Stress tensor inversion from fault kinematic indicators and focal mechanism data: the TENSOR program. In: Nieuwland D (ed) New insights into structural interpretation and modelling, vol 212. Geological Society London Special Publication, pp 75–100

Grünthal G, Wahlström R, Stromeyer D (2013) The SHARE European Earthquake Catalogue (SHEEC) for the time period 1900–2006 and its comparison to the European-Mediterranean Earthquake Catalogue (EMEC). J Seismolog 17(4):1339–1344

Horvath F, Csontos L, Dovenyi P, Fodor L, Greneczy G, Sikhegyi F, Szafian P, Szekely B, Timar G, Toth L, Toth T (2006) A Pannon-medence jelenkori geodinamikajanak atlasza: Euro-konform terkepsorozat es magyareze. Magyar Geofizica 47:133–137

Institutul Geologic Roman (1968) Harta Geologica a Romaniei. scale 1:200 000

Kijko A (2004) Estimation of the maximum magnitude. Pure appl Geophys 161:1655–1681

Moldovan IA, Popescu E, Constantin A (2008) Probabilistic seismic hazard assessment in Romania: application for crustal seismic active zones. Rom J Phys 53(3–4):575–591

NATO SfP Project 983054 (2011) Harmonization of seismic hazard maps for the Western Balkan Countries (BSHAP). Final report; http://www.msb.gov.ba/dokumenti/AB38745. Last Accessed 17.07.2017

Oncescu MC, Marza V, Rizescu M, Popa M (1999) The Romanian earthquakes catalogue, 984-1997. In: Wenzel Lungu (ed) Vrancea earthquakes: tectonics, hazard mitigation. Kluwer Publication, The Netherlands, pp 43–47

Ordaz M, Aguilar A, Arboleda J (2007) Crisis program for computing seismic hazard. Instituto de Ingeneria, UNAM, Mexico

Oros E (2011) Researches about seismic hazard for Banat Region. PhD dissertation, University of Bucharest

Oros E, Diaconescu M (2015) Recent vs. historical seismicity analysis for banat seismic region. Math Model Civ Eng 11(1):24–32. https://doi.org/10.1515/mmce-2015-0001

Oros E, Oros V (2009) New and updated information about the local hazard seismic sources in the Banat Seismic Region. In UTCB (ed) Hazard, vulnerability and risk. 4th national conference of earthquake engineering, vol 1, pp 133–139. CONPRESS, Bucharest, October 2009

Oros E, Popa M, Moldovan IA (2008a) Seismological database for banat seismic region (Romania)—part 1: the parametric earthquake catalogue. Rom J Phys 53(7–8):955–964

Oros E, Popa M, Moldovan IA, Popescu M (2008b) Seismological database for banat seismic region (Romania) - Part 2: the catalogue of the focal mechanism solutions. Rom J Phys 53(7–8):965–977

Oros E, Popa M, Ghita C, Rogozea M, Rau-Vanciu A, Neagoe C (2016) Catalogue of focal mechanism solutions for crustal earthquakes in intra-carpathian region of Romania. Paper presented at the 35th General Assembly of the European Seismological Commission, Italy, 24–11 Sept 2016

Oros E, Constantinescu EG, Diaconescu M, Popa M (2017) Stress field, seismicity and seismotectonic features in the Apuseni Mts area. In: Proceedings of 17th International Scientific GeoConference SGEM 2017, Science and technologies in geology, exploration and mining, Issue 14. Appl Environ Geophys Oil Gas Explor 17:421–428

Polonic G (1985) Neotectonic activity at the eastern border of the Pannonian depression and its seismic implications. Tectonophysics 47:109–115

Radulian M, Mandrescu N, Panza GF, Popescu E, Utale A (2000) Characterization of seismogenic zones of Romania. Pure Appl Geophys 157:57–77

Reiter L (1991) Earthquakes hazard analysis: issues and insights. Colombia University Press, New York

Sandulescu M (1984) Geotectonica Romaniei. Tehnical Publishing House, Bucharest

Scholz CH (2015) On the stress dependence of the earthquake b value. Geophys Res Lett 42:1399–1402. https://doi.org/10.1002/2014GL062863

Simeonova SD, Solakov DE, Leydecker G, Busche H, Schmitt T, Kaiser D (2006) Probabilistic seismic hazard map for Bulgaria as a basis for a new building code. Nat Haz Earth Syst Sci 6:881–887

Stucchi M, Rovida A, Gomez Capera AA et al (2013) The SHARE European Earthquake Catalogue (SHEEC) 1000–1899. J Seismol 17:523–544

Wells DL, Coppershmith KJ (1994) New empirical relationships among magnitude, rupture length, width, area and displacement. Bull Seismol Soc Am 84(4):974–1002

Wiemer S (2001) Software package to analyse seismicity: ZMAP. Seismol Res Lett 72:374–383

Wiemer S, Gerstenberger MC, Hauksson E (2002) Properties of the 1999, Mw7.1, Hector Mine earthquake: implications for aftershock hazard. Bull Seismol Soc Am 92:1227–1240

Zaharia B, Oros E, Popa M, Radulian M (2010) Tomographic research in Banat area using local earthquake data. Paper presented at the 32nd General Assembly of ESC, Montpellier, France, 6–10 Sept 2010

Zaharia B, Grecu B, Popa M, Oros E, Radulian M (2017) Crustal structure in the western part of Romania from local seismic tomography. Paper presented at the world multidisciplinary earth sciences symposium, WMESS 2017, Praga, Czech Republic, 11–15 Sept 2017

Zugravescu D, Polonic G (1997) Geodynamic compartments and present-day stress state on the Romanian territory. Revue Roumaine de Geophysique 41:3–24

NATO SfP Project 983054 (2011) Harmonization of seismic hazard maps for the Western Balkan Countries (BSHAP). Final report. http://www.msb.gov.ba/dokumenti/AB38745. Last Accessed 17.07.2017

Seismic Intensity Estimation Using Macroseismic Questionnaires and Instrumental Data—Case Study Barlad, Vaslui County

Iren-Adelina Moldovan, Bogdan Grecu, Angela Petruta Constantin,
Andreita Anghel, Elena Manea, Liviu Manea,
Victorin Emilian Toader and Raluca Partheniu

Abstract In the last decade, many efforts were done to predict the macroseismic intensity in case of felt Vrancea earthquakes and additionally an online environment was developed for the automatic approximation of the intensity from peoples' feedback. Besides the extended scientific studies, the near real-time estimation of the macroseismic intensity recently became mandatory for the insurance companies to cover some of the losses and damages that earthquakes might cause to houses, belongings, and other structures. Due to the insurance companies' requests, the macroseismic questionnaires method was doubled by the seismic intensity determination using instrumental data, as recommended in the Romanian Seismic

I.-A. Moldovan (✉) · B. Grecu · A. P. Constantin · A. Anghel · E. Manea
L. Manea · V. E. Toader · R. Partheniu
National Institute of Research and Development for Earth Physics,
Magurele, Romania
e-mail: irenutza_67@yahoo.com

B. Grecu
e-mail: bgrecu@infp.ro

A. P. Constantin
e-mail: angela@infp.ro

A. Anghel
e-mail: andreita.anghel@infp.ro

E. Manea
e-mail: flory.manea88@gmail.com

L. Manea
e-mail: mlmarius@yahoo.com

V. E. Toader
e-mail: victorin@infp.ro

R. Partheniu
e-mail: raluca@infp.ro

© Springer International Publishing AG, part of Springer Nature 2018   71
R. Vacareanu and C. Ionescu (eds.), *Seismic Hazard and Risk Assessment*,
Springer Natural Hazards, https://doi.org/10.1007/978-3-319-74724-8_5

Intensity Scale Standard (STAS 3684-71). In the present study, the procedure is shown, for the last earthquakes with M_L larger than 5.0, occurred in Vrancea zone, and felt on the extra-Carpathian area. We have selected the case study in Barlad, Vaslui county, because there have been recorded the largest accelerations (122 cm/s^2) and have been reported the largest MSK intensities (VI) from Romania during the Mw 5.5 September 24, 2016 earthquake. The results obtained using the two approaches (macroseismic and instrumental data) have been compared and some differences have been found.

Keywords Intensity data points (IDPs) · Vrancea earthquakes · Peak ground acceleration (PGA)

1 Introduction

Romania is one of the most seismic-prone countries in Europe due to the periodically occurrence of strong intermediate-depth earthquakes (Marmureanu et al. 2011). These types of earthquakes, unique in Europe, appear in Vrancea seismogenic zone. This area is located beneath the South–Eastern Carpathian Arc bend, at the contact between the East-European plate and the Intra-Alpine and Moesian sub-plates (continental collision). An intense seismic activity is recorded in the mantle, within a narrow, almost vertical descending volume between 60 and 180 km depth (Radulian et al. 2000). In the last decade, many efforts were done to predict the macroseismic intensity (I) in case of felt Vrancea earthquakes with Mw > 5.0 (Io > IV) and additionally an online environment was developed for the automatic approximation of the intensity from people feedback.

Earthquakes with macroseismic effects exceeding IV MSK degrees, on extended populated area (with magnitudes larger than 5.0 M_L) occur in Vrancea intermediate depth seismic zone, with a expected return period of 2.5 years (Jianu and Pantea 1994), or 2 years (Moldovan 2007).

Despite the estimated return period, during the last 12 months, three largely felt earthquakes (two of them with Mw > 5.0) occurred in September and December 2016 and February 2017. This is not a single case, because similar behaviour was observed in 2005, as well (Moldovan et al. 2016, 2017).

On September 23, 2016, 23:11:20 UTC time, a crustal earthquake with M_L = 5.8 (Eq2-Table 1; Fig. 1) occurred at a depth of 92 km. The seismic moment calculated by the direct method has the value M_0 = 3.37E + 17 Nm, and the magnitude (Mw) determined from seismic moment is 5.5. The earthquake epicentre was located near the following cities: Covasna (38 km), Focsani (44 km), Targu Secuiesc (49 km), Ramnicu Sarat (50 km), Marasesti (51 km) and felt on the extra-Carpathian area. The quake intensity in the epicenter area was V degree on the MSK scale, but there were reported higher intensities outside this area due to local

Table 1 Recent seismicity ($M_L > 5.0$) from Vrancea zone (after Romplus and Constantin et al. 2016)

Earthquake No	Date	Time (UTC)	Latitude	Longitude	Depth	M_L	Mw	Io	I_{Bd}
Eq1	2014/ 11/22	19:14:17.11	45.8683	27.1517	40.9	5.7	5.4	VI	IV–V
Eq2	2016/ 09/23	23:11:20.06	45.7148	26.6181	92.0	5.8	5.5	VI	VI
Eq3	2016/ 12/27	23:20:55.94	45.7139	26.5987	96.9	5.8	5.6	VI	V
Eq4	2017/ 02/08	15:08:20.89	45.4874	26.2849	123.2	5.0	4.8	IV	IV

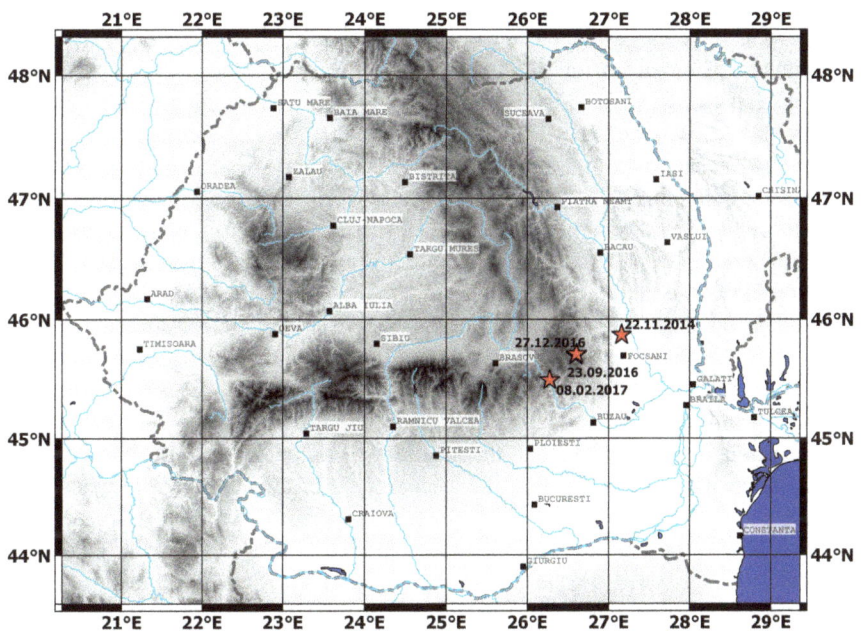

Fig. 1 Location of earthquakes from Table 1

effects. The earthquake was felt in many cities in the country (Barlad, Brasov, Braila, Bucuresti, Buzau, Constanta, Craiova, Galati, Iasi, Sibiu, Suceava, etc.), but also in the Republic of Moldova and Bulgaria.

On December 27, 2016, at 23:20:55 UTC, an intermediate earthquake with $M_L = 5.8$ (Eq3-Table 1; Fig. 1) occurred at a depth of 99 km, in Vrancea area. The seismic moment calculated by the direct method has the value $Mo = 4.2E + 17$ Nm, and the magnitude (Mw) determined from seismic moment is 5.6. The earthquake epicentre was located near the following cities: Covasna

(37 km), Focsani (45 km), Targu Secuiesc (48 km), Ramnicu Sarat (51 km), Marasesti (51 km). The epicentral intensity was VI degrees on the MSK scale. The earthquake was felt in many cities in the country (Bârlad, Braşov, Brăila, Bucureşti, Buzău, Constanţa, Craiova, Galaţi, Iaşi, Sibiu, Suceava, etc.), but also in the Republic of Moldova, Ukraine, Bulgaria and Western Turkey.

On February 8, 2017, at 15:08:20 UTC, an intermediate earthquake with $M_L = 5.0$ (Eq4-Table 1; Fig. 1) occurred at a depth of 127 km, in Vrancea area. The seismic moment calculated by the direct method has the value $Mo = 2.51E + 16$ Nm, and the magnitude (Mw) determined from seismic moment is 4.8. The earthquake epicentre was located near the following cities: Vălenii de Munte (40 km), Covasna (41 km), Săcele (48 km), Braşov (55 km), Plopeni (56 km). The epicentral intensity was IV degrees on the MSK scale. The earthquake was slightly felt in some cities in the country (Bârlad, Bucureşti, Buzău, Constanţa, Galaţi, Iaşi, etc.).

In the last years, in Vrancea seismogenic zone, also occurred the largest instrumental recorded crustal earthquake, on November 22nd, 2014. The crustal earthquake with $M_L = 5.7$ (Eq1-Table 1; Fig. 1) occurred at a depth of 39 km, on November 22, 2014, at 19:14:17 UTC, in Vrancea area. The earthquake epicentre was located at 5 km from Marasesti and 19 km from Focsani, both in Vrancea county. The quake intensity in the epicenter area was VI degrees on the MSK scale, and felt on the extra-Carpathian area. The earthquake was felt in many cities in the country (Barlad, Braila, Bucuresti, Buzau, Constanta, Craiova, Galati, Iasi, Sibiu, Suceava, etc.), but also in the Republic of Moldova.

Besides the extended scientific studies, the near real time estimation of the macroseismic intensity became mandatory for the insurance companies to cover some of the losses and damage that earthquakes might cause to houses, belongings, and other buildings. The estimation of seismic intensities should be done using the macroseismic questionnaires but the insurance companies requires also instrumental intensities derived from the recorded instrumental data (acceleration, velocity and displacement).

In this study, the procedure applied to compute the macroseismic intensity from instrumental data based on the Romanian Standard for Seismic Intensities Scale (STAS 3684-71) is used for the crustal and subcrustal Vrancea earthquakes from Table 1, and showed in detail for September 23, 2016. The values obtained with the above-mentioned procedure are compared with the intensity values from the online feedback, and also from the available equations for conversion of peak ground acceleration in macroseismic intensity and the ground motion predicted equations. The automatic intensity estimation code from the online feedback proposed by Wald et al. (1999a, b) and adopted by Ionescu and Dragoicea (2010) for the Romanian earthquakes was improved and also used for this study.

2   Macroseismic Intensities Obtained from Questionnaires

The macroseismic intensity scale is a scale that can be measured only by the response of the population to some specific questions that gives us opportunity to calculate the size of the seismic event in a given area. The macoseismic questionnaire has been made in conformity with Romanian STAS and follows strictly the descriptions of macroseismic intensity scale. Because *"seismic intensity scale is established based on the effects of seismic actions on humans and environment, on buildings and land crust"*, the macroseismic intensity can be obtained from the response of the population to some specific questions that gives us opportunity to calculate the size of the seismic event in a given area. Population must be aware that this is no constrain, there are just a few minutes of volunteering for a more accurate map of intensity. The macroseismic questionnaire has been made in conformity with Romanian STAS and follows strictly the descriptions of macroseismic intensity scale.

Starting with 23rd September when the intermediate earthquake occurred until November 2016, the National Institute for Earth Physics (INCDFP), collected information from 954 questionnaires (from 205 localities whith almost 300 answers from Bucharest) using two computer and one mobile phone applications, and obtained an extended report on macroseismic effects due to the earthquake To these responses were also added the answers from EMSC site (Fig. 2).

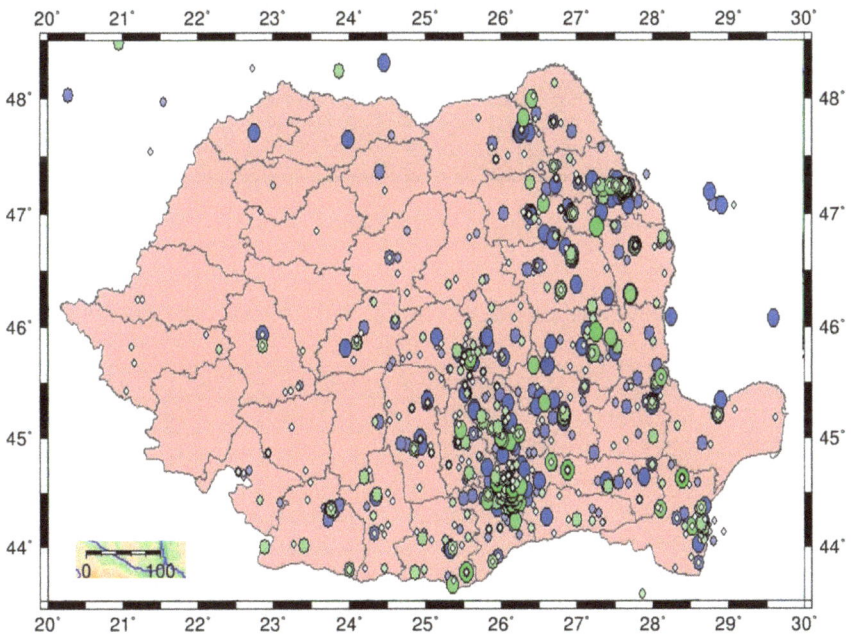

Fig. 2 Distribution of the sites with macroseismic information for September 23rd, 2016 (blue) and December 27th, 2016 (green) earthquakes

From Barlad, Vaslui County, there are 14 questionnaires and 3 quick feedback responses, with intensity values between IV (from 2 replies) and VII (from 1 reply). Most responses (11) indicates a macroseismic intensity equal with V–VI and VI (Fig. 3; Table 2).

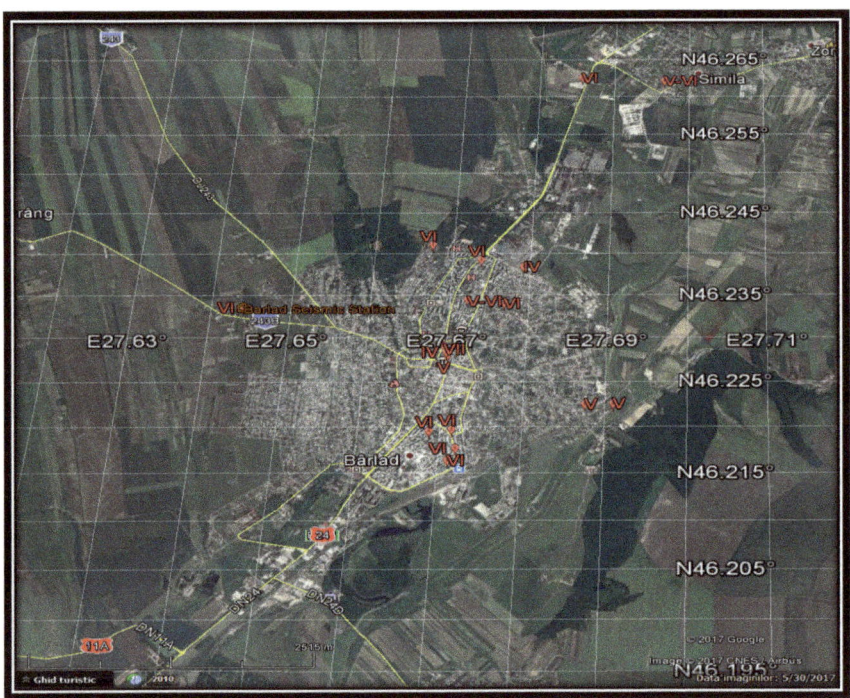

Fig. 3 Reported macroseismic intensities in Barlad, Vaslui County

Table 2 Reported macroseismic intensities in Barlad, Vaslui County, due to 23rd September 2016 earthquake

No	Lat	Long	I	d	No	Lat	Long	I	d
1	46.227459	27.670111	IV	100	10	46.233082	27.676945	VI	99
2	46.237434	27.679137	IV	100	11	46.2155	27.6713	VI	100
3	46.221599	27.687626	V	100	12	46.218644	27.668766	VI	99
4	46.221652	27.69104	V	100	13	46.21884	27.671557	VI	99
5	46.227056	27.670333	V	99	14	46.232655	27.644144	VI	98
6	46.233556	27.672281	V–VI	100	15	46.238277	27.673702	VI	100
7	46.261063	27.696669	V–VI	100	16	46.24012	27.667302	VI	100
8	46.261511	27.685634	VI	97	17	46.228138	27.669732	VII	98
9	46.216809	27.672144	VI	99					

Because the Intensity Scale is a measure that has been used from historical times, before first seismograph, the interrogation method of population is used on a global level and is the only way to find clear information from all areas. In our opinion, this method is more useful, because acceleration/velocity/displacements values can be obtained only in seismic station sites and not from around the country.

3 Macroseismic Intensities Obtained from Instrumental Data

Despite the recommendations taken from STAS, and our previous experience, the insurance companies do not trust the method, due to its subjective character and are requesting instrumental intensities.

The accelerations recorded by the seismic stations of the National Institute for Earth Physics network are shown in Fig. 4a, superimposed on the macroseismic intensity map obtained with the values from the questionnaires, for Romania. Intensities obtained by PGA-I conversion formulas (Bonjer et al. 1999; Enescu and Enescu 1999; Vacareanu et al. 2016; Wald et al. 1999a, b) are also superimposed on the macroseismic intensity map obtained from online reports and presented in Fig. 4b–e. The PGA and I values for Barlad site are presented in Table 3.

Further, the paper will present in detail the method used to obtain the macroseismic intensities from the recorded ground motion parameters (acceleration/velocity/displacements) as it is presented in STAS 3684-71, see Table 4.

In Barlad, the peak ground acceleration (PGA) was 122 cm/s^2. As we can see in Fig. 4, the PGA in Barlad was the highest from Romania. From Table 4, one can see that the correspondence between recorded ground parameters and intensities is given for different frequency domains. For PGA for frequencies from 2 to 10 Hz (0.1–0.5 s), and PGV for frequencies from 0.5 to 2 Hz (0.5–2.0 s).

In the following, we present seismic recordings from Barlad seismic station and the connection between recorded unfiltered acceleration/velocity and computed displacement and filtered recordings in specified domains from Table 4 and macroseismic intensity. This is a must, due to STAS 3684-71 (Annex 1): *"For a quantitative earthquake survey on the territory of Romania it is recommended intensity determination also based on instrumental recordings"*. Fortunately, this is possible in the town of Barlad, because there is a seismic station. In other Romanian cities, this computation is not possible, due to the lack of seismic stations.

Figure 5 represents time diagrams for unfiltered recorded acceleration and computed velocity and displacement, at Barlad seismic station. One can see that PGA at this station was 122 cm/s^2, PGV 7.2 cm/s and peak displacement 1 cm.

To be able to make the conversion from acceleration—to macroseismic intensity, the recorded signal was filtered in the frequency domain specified in Table 3, i.e. 0.1 s < T < 0.5 s (2–10 Hz).

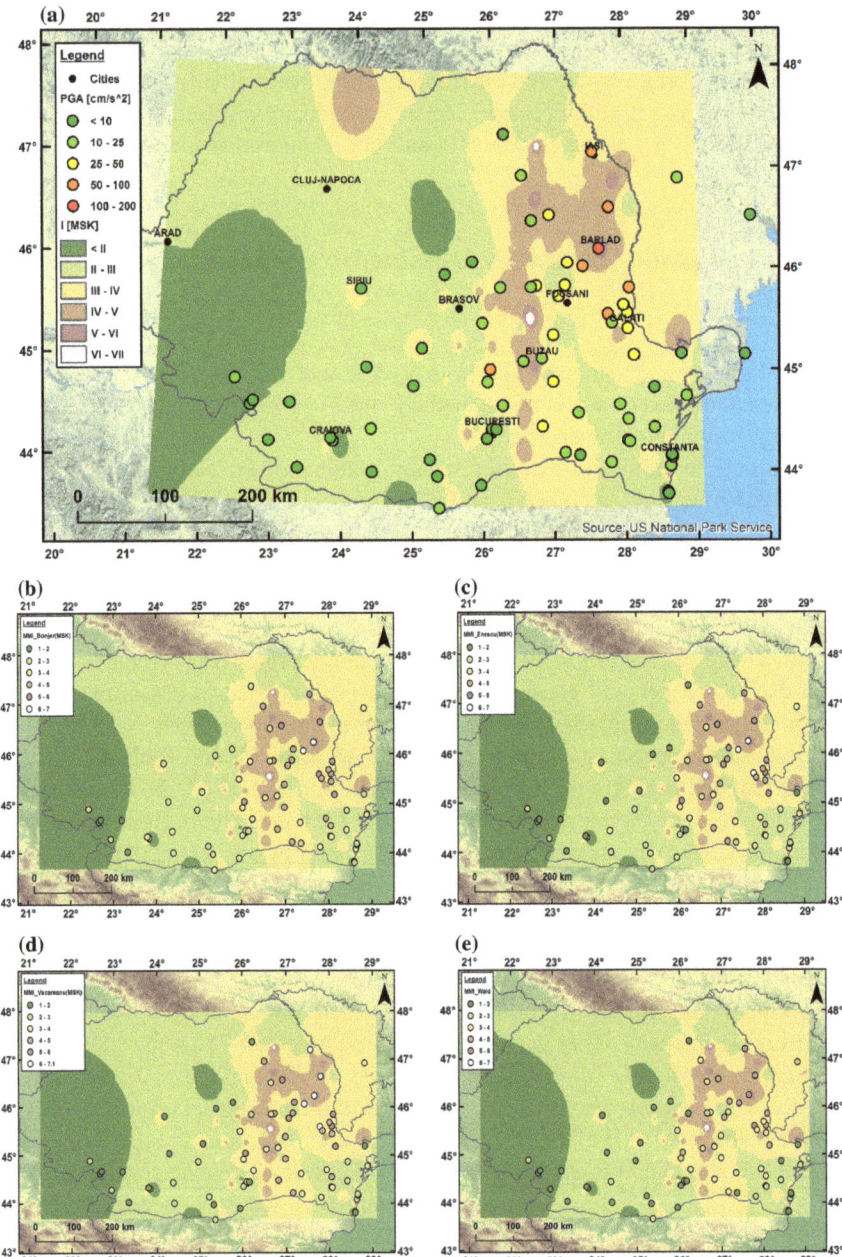

Fig. 4 Macroseismic intensity map for 23rd September, 2016 earthquake and **a** PGA (horizontal components EW) recorded by the national seismic network. **b–e** Intensities obtained using the conversion formulas of Bonjer et al. (1999), Enescu and Enescu (1999), Vacareanu et al. (2016), Wald et al. (1999a)

Table 3 PGA and converted macroseismic intensities for Barlad site due to September 23rd 2016 earthquake

STA	Lat	Long	Depi (km)	PGA (cm/s^2)	I (Bonjer)	I (Enescu)	I (Vacareanu)	I (Wald)
BIR	27.6436	46.2334	134.5	122.2	VI	VII	VII	VI ½

Table 4 Seismic intensity determination based on instrumental data (STAS 3684-71)

Seismic intensity in degrees	α (cm s^{-2})	v (cm s^{-1})	x_0 mm
V	12......25	1.0.......2.0	0.5.......1.0
VI	26......50	2.1.......4.0	1.1.......2.0
VII	51.......100	4.1.......8.0	3.1.......4.0
VIII	101.......200	8.1.......16.0	4.1.......8.0
IX	201.......400	16.1.......32.0	8.1......16.0
X	401.......800	32.1.......64.0	16.1.......32.0

where

α = peak ground acceleration for periods of 0.1 .... 0.5

v = peak ground velocity for periods of 0.5 .... 2.0

x_0 = relative displacement amplitude of the mass center of a pendulum with its own oscillation period of 0.25 s and the logarithmic decay of the damping of 0.5

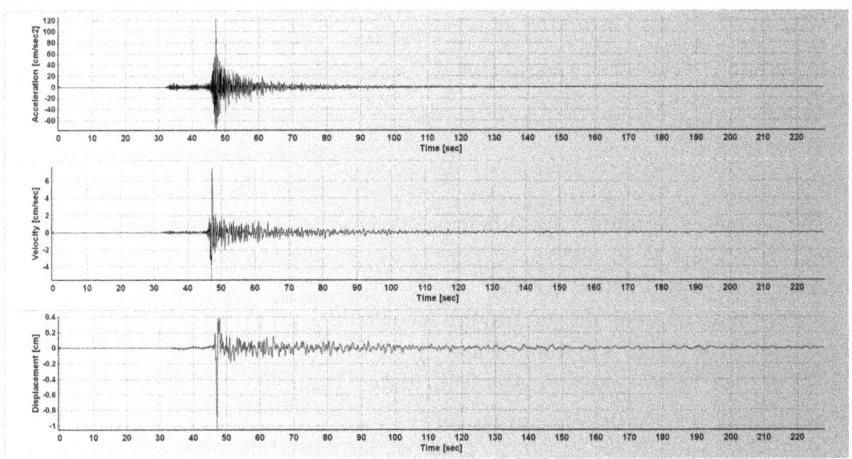

Fig. 5 Unfiltered recorded acceleration, and corresponding computed velocity and displacement

In Fig. 6 are presented the filtered acceleration recordings, and the computed velocity and displacement. In this frequency domain, acceleration drop to 89 cm/s^2 (Table 5).

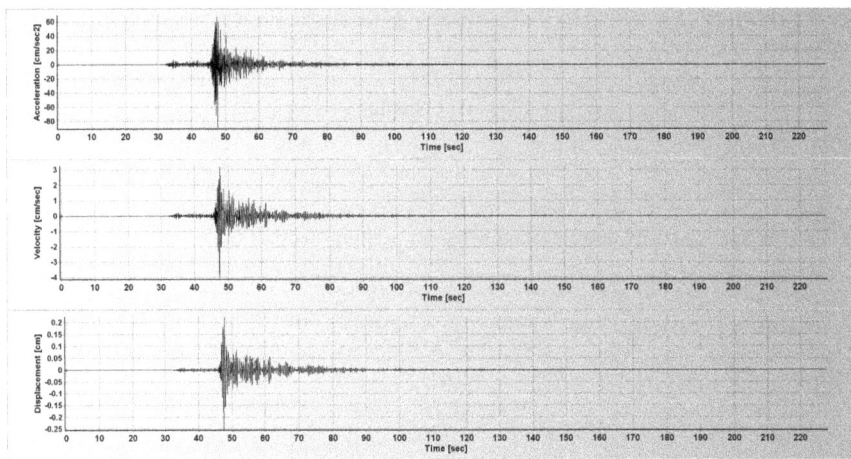

Fig. 6 Recorded acceleration filtered in 0.1–0.5 s domain and corresponding computed velocity, and displacement

Table 5 Peak values for observed acceleration on the East component and computed velocity and displacement

BIR HNE					
Unfiltered values			Filtered values in the frequency domain 2–10 Hz		
Acceleration (cm/s^2)	Velocity (cm/s)	Displacement (cm)	Acceleration (cm/s^2)	Velocity (cm/s)	Displacement (cm)
123.2	7.5	1.0	89.2	4.0	0.2

The response spectrum for filtered accelerogram in the frequency domain 2–10 Hz (0.1–0.5 s) is shown in Fig. 7, to see clearly the amplitudes of the frequency range required in STAS.

Taking into account the maximum acceleration value from Table 5 and the conversion from Table 4 (column 1 and 2), we find that the seismic intensity at the site of Barlad seismic station was VII, higher than the macroseismic intensity obtained from the "subjective" questionnaires.

In order to validate the obtained result, we have used the recorded velocity and filtered it between 0.5 and 2 Hz (Table 4) and represented the results in Fig. 8 and Table 6.

Maximum velocity is 4.2 and 2.7 cm/s on the filtered component, which corresponds to a macroseismic intensity I = VI in the Barlad seismic station site.

The last correspondence between the recorded values and macroseismic intensity was made on the computed displacement, as specified in Table 4, column 4. For this purpose, the displacement response has been made with a dumping of 0.5 and the displacement value was considered of T = 0.25 s (Fig. 9). The obtained value is

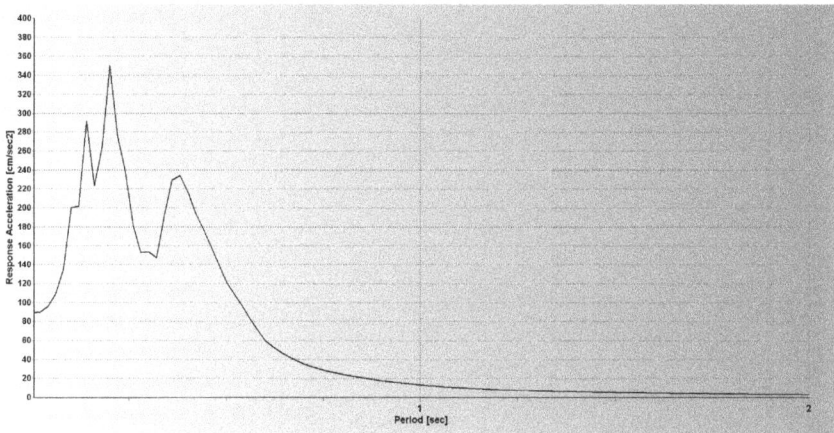

Fig. 7 Acceleration spectrum

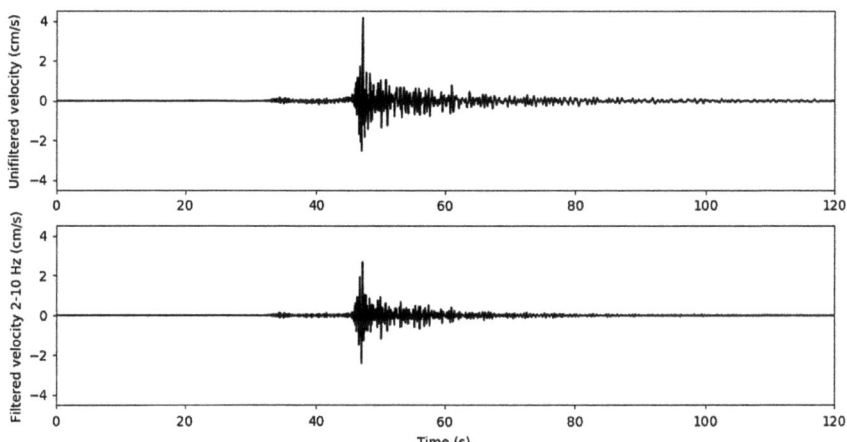

Fig. 8 Recorded velocity and filtered velocity in the frequency domain 0.5–2 Hz, (0.5–2 s)

Table 6 Peak velocity values for the East component

BIR EHE	
Unfiltered values (cm/s)	Filtered values in the frequency domain 2–10 Hz (cm/s)
4.2	2.7

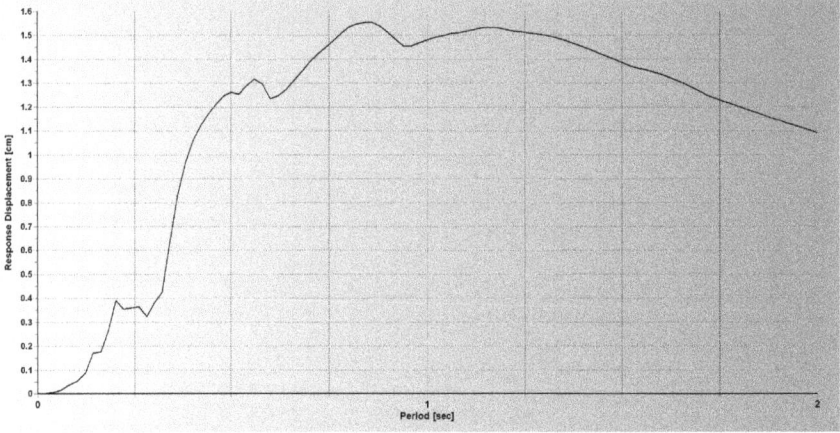

Fig. 9 Displacement response spectrum (T = 0.25 s, damping 5%)

0.37 cm or 3.7 mm, which correspond to a macroseismic intensity in the Barlad seismic station site I = VII.

In conclusion, using the recorded, real and "objective" values, and the conversion Table 4 (STAS), the macroseismic intensity in Barlad Seismic station site is I = VI–VII.

A possible explanation of the difference between I = VI–VII obtained from instrumental recordings and the intensities from the reported observation of the people (Table 2), is that the seismometer is installed on a hill from the Western part of the town, where can be some signal amplification due to topographical characteristics.

4   Conclusions

Using the recorded, real and "objective" values, and the conversion Table from STAS 3684-71, the macroseismic intensity of the 23rd September earthquake (Eq. 2), in Barlad Seismic station site is I = VII (Table 7). Using the conversion formulas of Bonjer et al. (1999), Enescu and Enescu (1999), Vacareanu et al. (2016), Wald et al. (1999a, b) PGA to I, (Table 3) we have obtained intensities between VI and VII. The reported value from the macroseismic questionnaires was VI (Table 1), lower with one degree than the one obtained from conversion of recorded accelerograms. For the case of Eq. 4, the conversion method is returning only the information that the intensity in Barlad city is smaller than V.

From Table 7 one can see the same difference (of one degree) between the reported and the computed macroseismic intensity values for all studied earthquakes: Eq1, Eq2, Eq3 and Eq4 (see Table 1).

Table 7 Peak values at Barlad seismic station for acceleration, velocity and displacement for unfiltered and filtered recordings for earthquakes from Table 1 and the corresponding intensity

Earthquake No	Component	Unfiltered			Filered (2–10 Hz)			Filtered (0.5–2 Hz)			Displacement	Intensity from		
		PGA (cm/s*s)	PGV (cm/s)	PGD (cm)	PGA (cm/s*s)	PGV (cm/s)	PGD (cm)	PGA (cm/s*s)	PGV (cm/s)	PGD (cm)	response spectrum	PGA (2–10 Hz)	PGV (0.5–2 Hz)	D
Eq1	HNE	37.15	1.48	0.24	40.05	1.24	0.04	6.1	0.61	0.13	0.1	VI	<V	V
	HNN	27.08	0.88	0.16	23.55	0.69	0.03	4.77	0.64	0.1	0.06	V	<V	V
	HNZ	16.96	0.51	0.07	14.54	0.35	0.01	2.08	0.25	0.04	0.05	V	<V	V
Eq2	HNE	122.2	7.48	1.03	89.25	4.04	0.25	45.66	5.1	0.67	0.37	VII	VII	VII
	HNN	71.56	3.13	0.36	67.45	3.08	0.16	10.85	1.03	0.17	0.18	VII	V	VI
	HNZ	64.92	2.52	0.14	56.49	2.14	0.09	6.27	0.73	0.11	0.32	VII	<V	VII
Eq3	HNE	43.36	2.09	0.23	37.72	1.24	0.08	9.03	1.03	0.13	0.07	VI	V	V
	HNN	41.98	1.13	0.12	28.94	1.15	0.05	2.75	0.36	0.06	0.08	VI	<V	V
	HNZ	22.98	0.6	0.05	20.21	0.6	0.02	2.02	0.25	0.03	0.05	V	<V	<V
Eq4	HNE	9.45	0.4	0.02	8.48	0.31	0.01	1.33	0.12	0.01	0.016	<V	<V	<V
	HNN	5.64	0.15	0.01	5.71	0.16	0.01	0.45	0.05	0.01	0.014	<V	<V	<V
	HNZ	5.23	0.16	0.01	5.11	0.15	0.01	0.21	0.02	0.01	0.013	<V	<V	<V

A possible explanation of the difference between the intensity obtained from instrumental recordings, and the intensities from the reported observation of the people, is that the seismometer is installed on a hill from the Western part of the town, where can be some signal amplification due to topographical characteristics.

In conclusion, the main method of obtaining the values of macroseismic intensities remains the interpretation of people online questionnaires. This is the only way to have real data about the effects of the seismic ground shaking, in all inhabited areas. However, the number of reports are very small compared with the big number of people that lives in big cities, as there is no obligation to fill this form. For example, from Bacau, Bacau County, with almost 750.000 inhabitants (of which 78% over 15 years old) we had only 18 completed macroseismic form, or even the case like Barlad, where out of 50.000 inhabitants we had 17 completed forms.

In this way, we appeal, that for more reliable results that reflect the reality in the field, useful for insurance companies, we need more involvement of the population to fill the macroseismic forms. Filling the online macroseismic forms or mobile application has already been impose at the institutional level and in the staff of INCDFP, but also to every person who ask for addendum to obtain compensation from insurance companies. However, the number of returned completed forms continues to remain very small.

The quantitative study proposed by STAS has limitations. The most important is due to the necessity of more monitoring equipment. Nowadays, the National Seismic Network of INCDFP, has 142 accelerometers and 125 short period (50) and broad band (75) seismometers installed in Romania in 151 different locations, this number being much smaller than the number of existing localities on the same territory, 320 cities, 2900 towns and more than 13,000 villages. For this reason, the quantitative method can give us information only from accelerometer/seismometer locations.

Acknowledgements This paper was partially carried out within Nucleu Program, supported by ANCSI, projects no. PN 16 35 01 06, PN 16 35 03 01, PN 16 35 01 01 and the Partnership in Priority Areas Program—PNII, under MEN-UEFISCDI, DARING Project no. 69/2014 and a grant of the Romanian National Authority for Scientific Research and Innovation (ANCSI)—UEFISCDI, project number PN-II-RU-TE-2014-4-0701.

References

Bonjer KP, Oncescu MC, Driad L, Rizescu M (1999) A note on empirical site responses in Bucharest, Romania. Vrancea earthquakes: tectonics. Hazard and Risk Mitigation. Springer, Netherlands, pp 149–162

Constantin AP, Moldovan IA, Craiu A, Radulian M, Ionescu C (2016) Macroseismic intensity investigation of the November 2014, M = 5.7, Vrancea (Romania) crustal earthquake. Ann Geophys 59(5):S0542

Enescu BD, Enescu D (1999) The Vrancea earthquake of May 31, 1990 (Mgr = 6.1). Isoacceleration maps of ground movement and macroseismic maps. Rom Journ Phys 44(5–6):645–653

Ionescu C, Dragoicea M (2010) Macroseis: a tool for real-time collecting and querying macroseismic data in Romania. Rom Journ Phys 55(7–8):852–861

Jianu D, Pantea A (1994) Application of a probabilistic model to vrancea intermediate earthquakes. Nat Hazards 10(1–2):73–77

Mărmureanu G, Cioflan CO, Mărmureanu A (2011) Intensity seismic hazard map of Romania by probabilistic and (neo) deterministic approaches, linear and nonlinear analyses. Rom Rep Phys 63(1):226–239

Moldovan IA (2007) Metode si modele statistice in seismologie. Bucuresti, Editura Morosan, p 236

Moldovan IA, Diaconescu M, Popescu E, Radulian M, Toma-Danila D, Constantin A, Placinta A (2016) Input parameters for the probabilistic seismic hazard assessment in the Eastern part of Romania and Black Sea Area. Rom Journ Phys 61(7–8):1412–1425

Moldovan IA, Diaconescu M, Partheniu R, Constantin AP, Popescu E, Toma-Danila D (2017) Probabilistic seismic hazard assessment in the Black Sea Area, Rom Journ Phys 62(5–6):809

Radulian M, Mandrescu N, Popescu E, Utale A, Panza G (2000) Characterization of Romanian seismic zones. Pure appl Geophys 157:57–77

STAS 3684-71 (1971) Seismic intensity scale. Romanian Institute for Standardization, IRS, Bucharest (in Romanian)

Vacareanu R, Aldea A, Lungu D, Pavel F, Neagu C, Arion C, Demetriu S, Iancovici M (2016) Probabilistic seismic hazard assessment for Romania. In: Earthquakes and their impact on society, Springer International Publishing, pp 137–169

Wald DJ, Quitoriano V, Dengler LA, Dewey JW (1999a) Utilization of the internet for rapid community intensity maps. Research Letters, Seis. https://doi.org/10.1785/gssrl.70.6.680

Wald DJ, Quitoriano V, Heaton TH, Kanamori H (1999b) Relationships between peak ground acceleration, peak ground velocity, and modified Mercalli intensity in California. Earthq Spectra 15(3):557–564

Testing the Macroseismic Intensity Attenuation Laws for Vrancea Intermediate Depth Earthquakes

Maria-Marilena Rogozea, Iren Adelina Moldovan,
Angela Petruta Constantin, Elena Florinela Manea,
Liviu Marius Manea and Cristian Neagoe

Abstract The Vrancea seismic region, located at the bending area of the South–Eastern Carpathians in Romania, is the most active zone of seismicity in Europe, producing earthquakes at intermediate depths (60–200 km). The major events (magnitude above 7) are generated at intermediate depth and produce specific patterns of damage over extended areas. In this study we test the macroseismic intensity attenuation laws, using the intensity data point (IDPs) for 8 intermediate depth earthquakes that occurred in Vrancea (between 1738 and 2000). The macroseismic attenuation laws used for testing were Moldovan (Metode si modele statistice in seismologie. Editura Morosan, Bucuresti, pag 236, 2007), Sorensen et al. (Soil Dyn Earthq Eng, 2010), Vacareanu et al. (Macroseismic intensity prediction earthquake for Vrancea intermediate-depth seismic source. Hazards, Nat, 2015). The main purpose of the testing is to determine the best attenuation law that will be used to estimate the expected macroseismic intensity at different sites, and to further use them in the assessment of the seismic hazard and risk of the country and to design the real time shake maps. In conclusion, we have decided that Moldovan (Metode si modele statistice in seismologie. Editura Morosan, Bucuresti, pag 236, 2007) is the best intensity attenuation law for earthquakes located around 90 km depth (the events from March 4, 1977 and May 30, 1990), Sørensen et al. (Soil Dyn Earthq Eng, 2010) law is the best for modelling the macroseismic field due to earthquakes from the lower segment, located around 130 km depth (events from November 10, 1940 and August 30, 1986). For epicentral distances larger than 300 km, Vacareanu et al. (Macroseismic intensity prediction earthquake for Vrancea intermediate-depth seismic source. Hazards, Nat, 2015) law fits best the intensity distribution.

M.-M. Rogozea (✉) · I. A. Moldovan · A. P. Constantin · E. F. Manea
L. M. Manea · C. Neagoe
National Institute for Earth Physics, Magurele, Ilfov, Romania
e-mail: mrogozea@infp.ro

I. A. Moldovan
e-mail: irenutza_67@yahoo.com

A. P. Constantin
e-mail: angela@infp.ro

© Springer International Publishing AG, part of Springer Nature 2018
R. Vacareanu and C. Ionescu (eds.), *Seismic Hazard and Risk Assessment*,
Springer Natural Hazards, https://doi.org/10.1007/978-3-319-74724-8_6

Keywords Vrancea region · Intensity data point · Macroseismic intensity
Macroseismic attenuation laws

1 Introduction

The Vrancea zone is the most active zone in the South–Eastern part of the Carpatian
Arc, situated at the contact between the principal tectonic units: Est-European Plate,
Intra-Alpine subplate and Moesian subplate. Vrancea zone generates at intermediate
depths (60 km < h < 200 km), in a very small area of approximately 30×70 km^2,
2–3 catastrophic earthquakes per century, with magnitudes up to 7.5, which are felt
over an extended area, passing over the national borders. This zone is responsible
for the highest values of seismic hazard in Romania. In the 20th century, three
major earthquakes (1940, 1977, 1986), occurred in this zone.

The purpose of this study is to test the existing macroseismic intensity attenuation laws, using intensity data points (IDPs) from eight major subcrustal earthquakes (with depths between 80 and 150 km), occurred in Vrancea, both in
historical and instrumental time, starting with year 1700. The events studied in the
present work are listed in Table 1.

Several attenuation models and macroseismic intensity predictive equation
(IPE) have been derived along time, for the specific case of the Vrancea intermediate and normal depth events (Pantea 1994; Marza and Pantea 1995; Zsiros 1996;
Lungu et al. 1997; Pantea and Moldovan 2000; Moldovan et al. 2000; Ardeleanu
et al. 2005; Moldovan 2007; Sokolov et al. 2008; Leydecker et al. 2008; Sørensen
et al. 2008, 2010; Vacareanu et al. 2015).

In various papers, the macroseismic intensity attenuation was analyzed as
function of distance and azimuth, sometimes taking into account the rupture
directivity or the local site effects. Some studies have divided the Romanian territory in cells with similar geological and geomorphological conditions, characterized by empirical region coefficients, and others' have considered only the

Table 1 Parameters of studied earthquakes (after Romplus–Oncescu et al. 1999)

Crt No.	Data (dd. mm. yyyy)	Latitude (°N)	Longitude (°E)	Depth (km)	Magnitude (M$_L$)	I$_0$ (MSK)
1.	11.06.1738	45.7	26.6	130	7.7	IX–X
2.	26.10.1802	45.7	26.6	150	7.9	X
3.	23.01.1838	45.7	26.6	150	7.5	IX
4.	10.11.1940	45.8	26.7	150	7.7	IX–X
5.	04.03.1977	45.78	26.8	94	7.4	IX
6.	31.08.1986	45.55	26.62	131.4	7.1	VIII–IX
7.	30.05.1990	45.83	26.89	84	6.9	VIII
8.	31.05.1990	45.88	26.98	90.9	6.4	VII

curvature of the Carpathian Mountains. The intensity data points (IDPs') sets used for obtaining these relations were different from author to author. Some have used the most recent data provided by Kronrod et al. 2013, and other from digitized maps from University of Karlsruhe obtained from Demetrescu (1941), Radu et al. (1990), Radu and Utale (1989), Grigerova et al. (1978), Glavcheva (1983)—from Bulgaria territory.

2 Data and the Attenuation Relations

In this study were used 8 earthquakes (Table 1; Fig. 1) that occurred in Vrancea zone, between 1738 and 1990. The IDPs' for the historical events (1738, 1802 and 1838) were taken from the papers of Rogozea et al. (2013, 2014), Rogozea (2016), and are represented in Figs. 2 and 3. The IDPs' for the instrumental events from the last century: 1940, 1977, 1986, 1990 (30 (1) and 31(2) of May), were taken from SHARE project, macroseismic data integrated in the paper of Kronrod et al. (2013)—and are represented in Figs. 4, 5 and 6.

We used the intensity scale of Medvedev-Sponheuer-Karnik (MSK-64) and Medvedev (1964), because the intensity data points (IDPs) for all events are determined with MSK seismic intensity scale, corresponding to various sites from Romania and neighborhood.

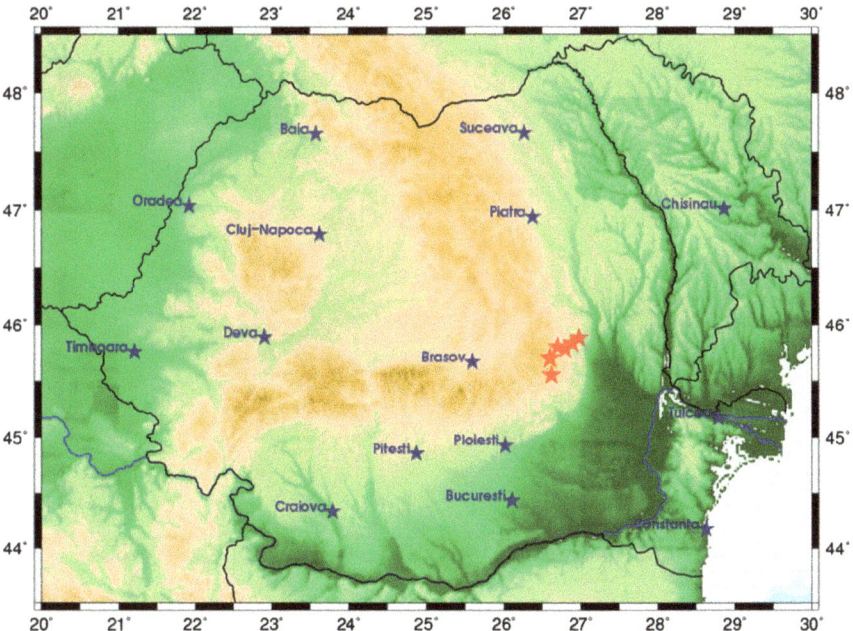

Fig. 1 Location of earthquakes from Table 1 (red stars)

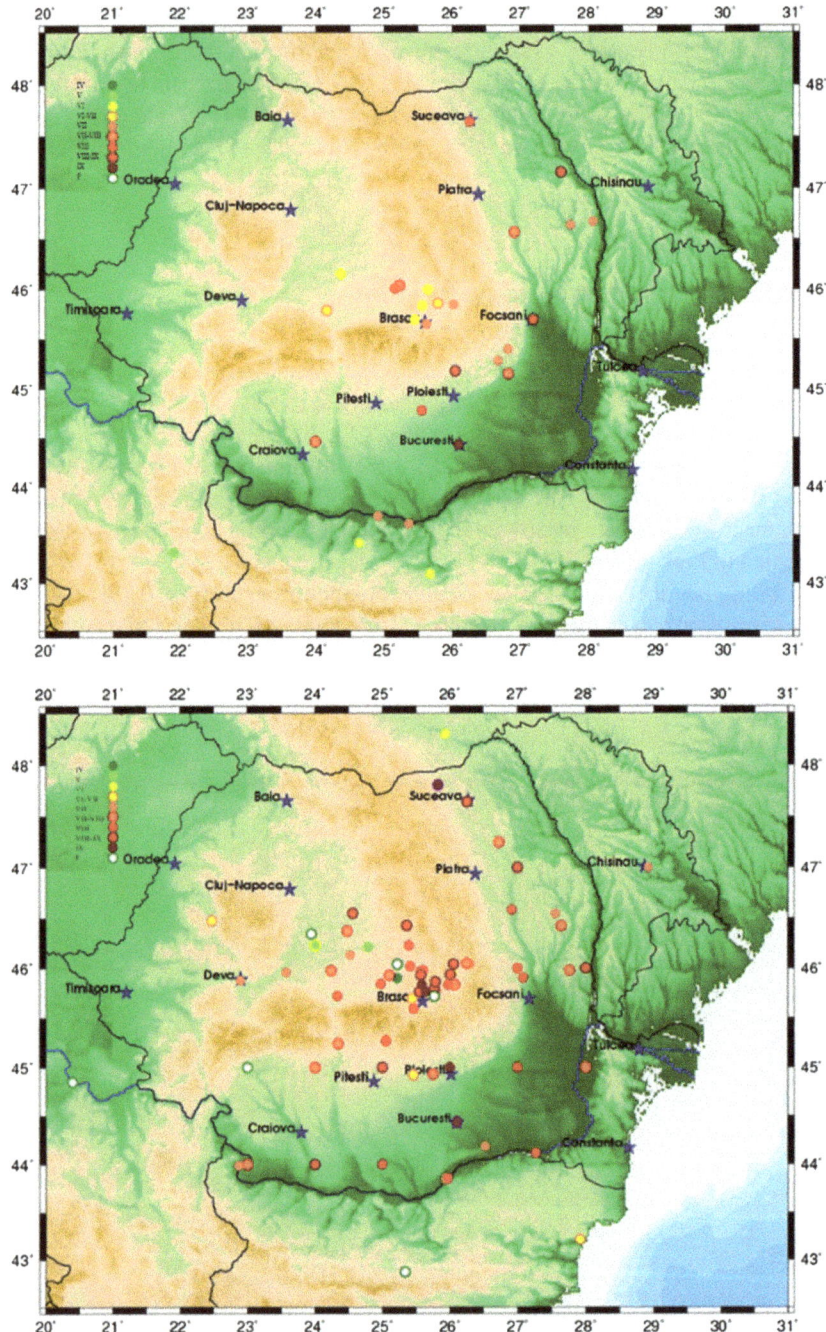

Fig. 2 Intensity map of the events: June 11, 1738 (top) and October 26, 1802 (bottom)

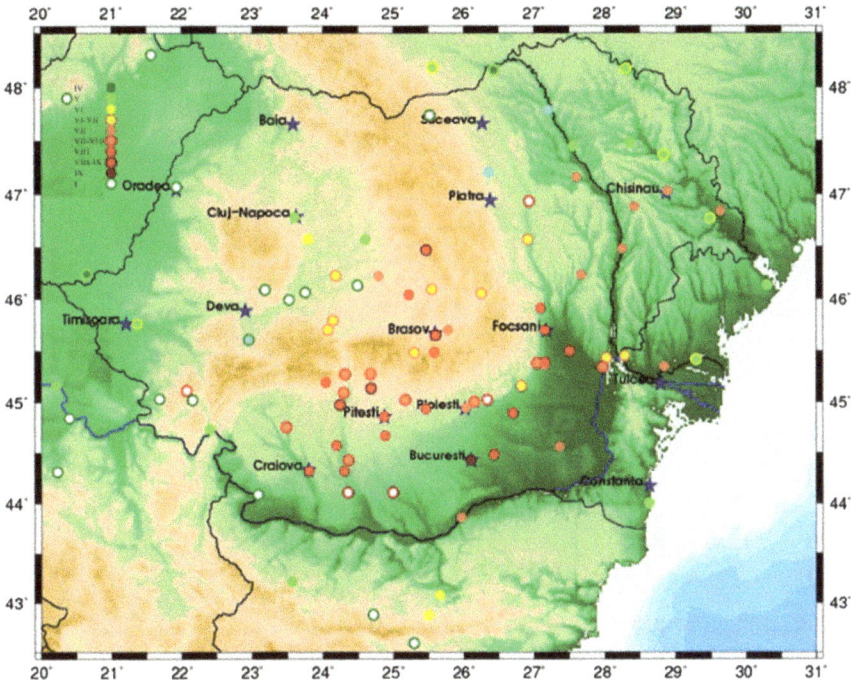

Fig. 3 Intensity map of the event from January 31, 1838

In Figs. 2 and 3 are represented the historical macroseismic information (IDP's) from Rogozea et al. (2014), Rogozea (2016), and in Figs. 4, 5 and 6 are represented the macroseismic information (IDP's) from the electronic supplement of Kronrod et al. (2013). The database contains information from Romania and several neighboring countries like: Republic of Moldavia, Russia, Bulgaria, Ukraine, Serbia, and Turkey.

From the maps one can see that there have been reported large intensity values in Republic of Moldova and Bulgaria, at distances of more than 200 km from the epicenters. The macroseismic information for the historical earthquakes is considerable lower than for instrumental earthquakes occurred after 1900, because the specific problems that appear when collecting information coming from older times when not everyone was able to write and report an event, and the associated effects. In the maps from Figs. 2 and 3 are represented the intensities larger than IV, and in the maps from Figs. 4, 5 and 6 are represented the intensities larger than III.

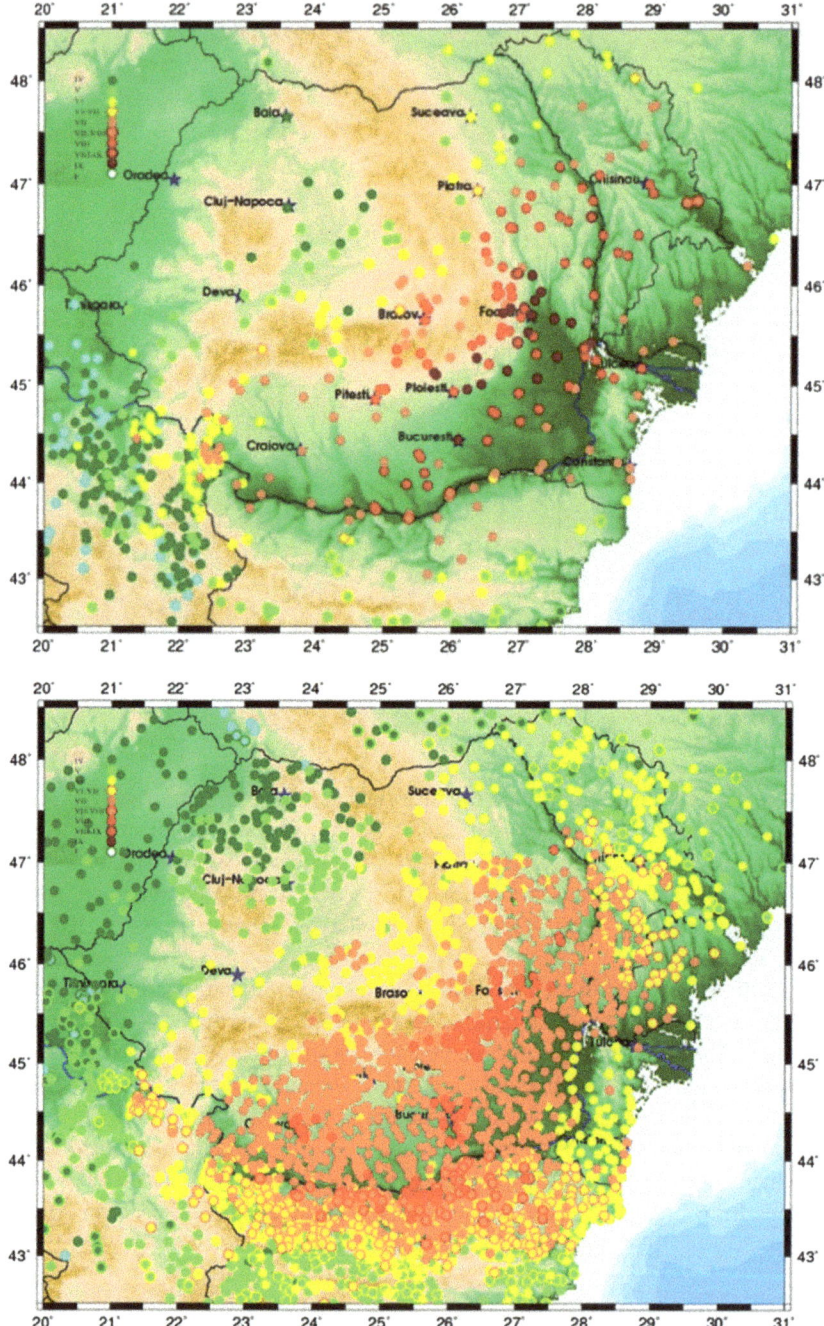

Fig. 4 Intensity map of the events: November 11, 1940 (top) and March 04 1977 (bottom)

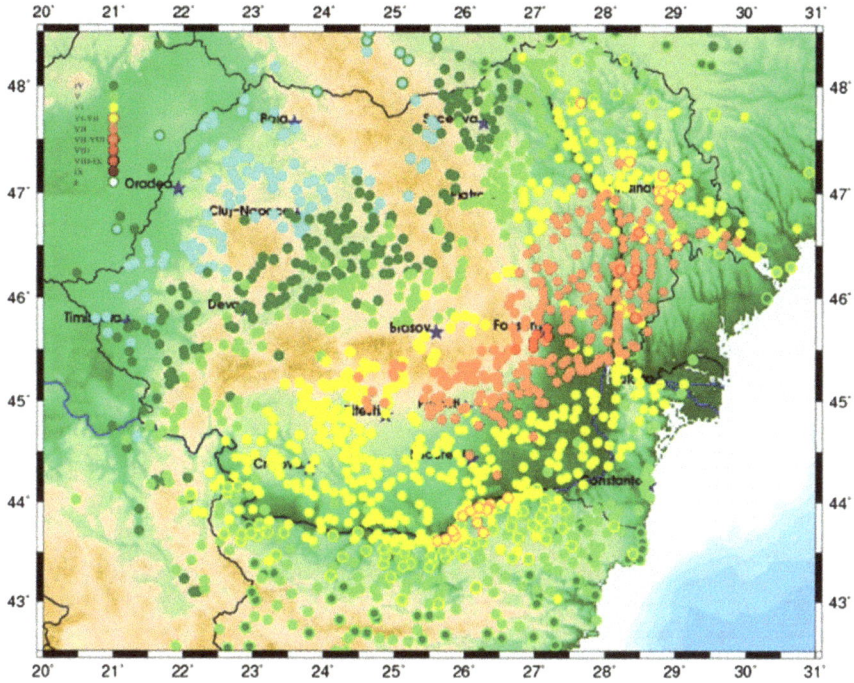

Fig. 5 Intensity map of the event from August 30, 1986

2.1 Macroseismic Intensity Data Points

The macroseismic intensity depends on distance between seismic source and point of observation, on geology, on type of building and how the earthquakes were felt by people. It varies from place to place. The intensity scale of Medvedev et al. (1964) (MSK-64), that was used in this paper is subjective and uses as degrees, numbers ranging from 1 (not felt) to 12 (complete destruction), and are many scales of intensity. It is known that the intensity increases with magnitude and decreases with epicentral distance.

In the graphs from Fig. 7 is show the distributions of historical earthquakes intensity (31st May 1738, 26th October 1802, and 23rd January 1838). Where it can be seen that for the event that take place in 1802 we have more intensity values over VII with even IX–X intensities. For the 1838 event there have been a lot of information that the event was felt in different parts of the country but without description on how it was felt or the associated effects, and that's why we can't attributed intensity values.

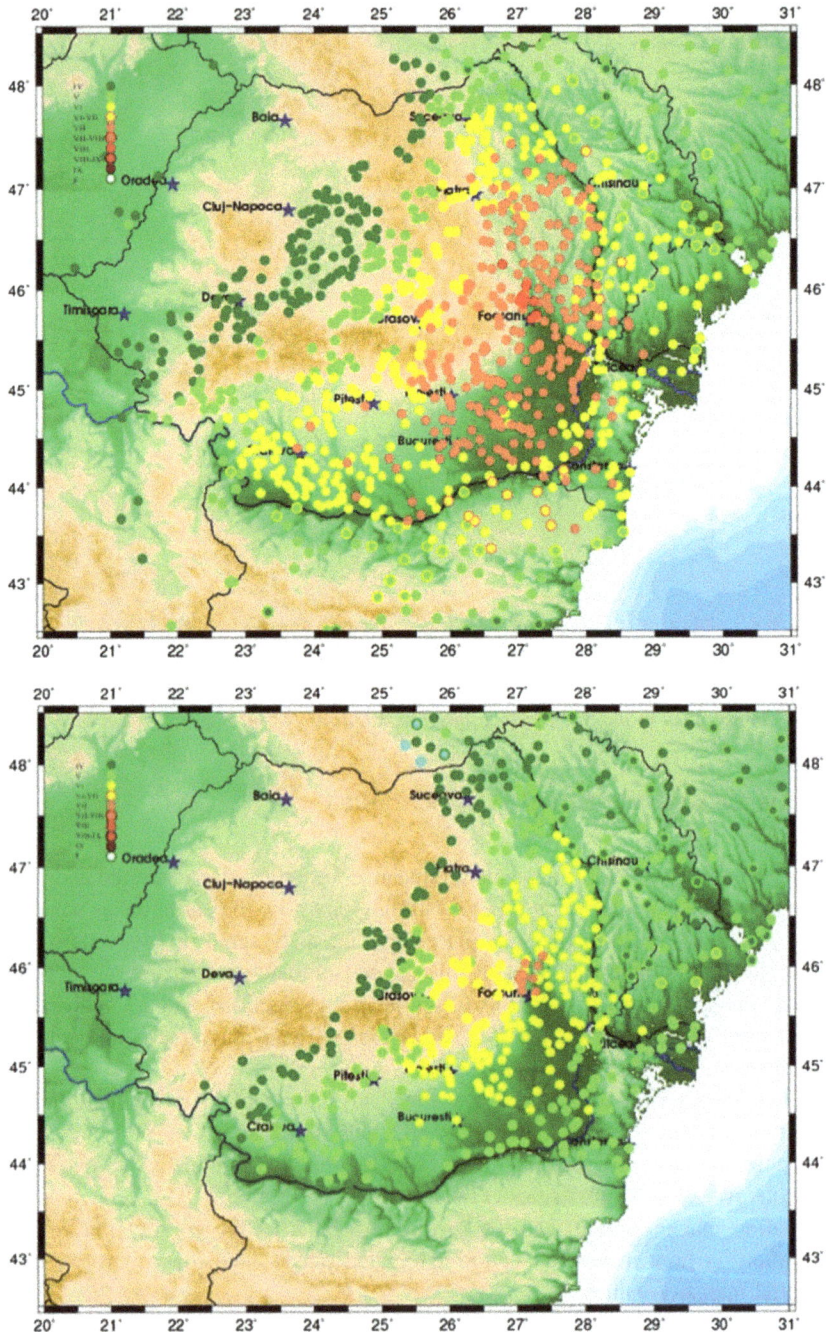

Fig. 6 Intensity map of the events: May 30 (top) and May 31 (bottom), 1990

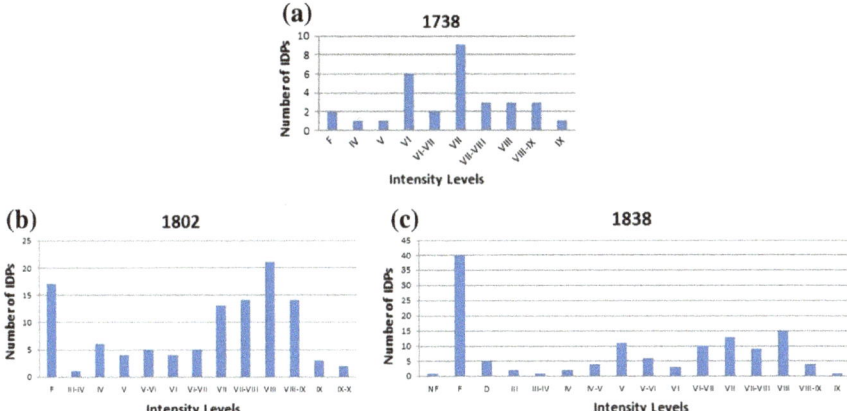

Fig. 7 Number of IDPs' for each intensity level for **a** 1738, **b** 1802, and **c** 1838 earthquakes

2.2 The Attenuation Relations

The macroseismic attenuation laws of Moldovan (2007), Sørensen et al. (2010) and Vacareanu et al. (2015), were used to test which is the best to model the effects distribution of Vrancea intermediate depth earthquakes, and to be implemented in real time shake map.

2.2.1 Moldovan (2007)

The attenuation relation for intermediate Vrancea zone, was determined by Moldovan (2007) (Eq. 2), by using information from four major Vrancea earthquakes, (March 04th, 1977, August 30th, 1986 and May 30th and 31st, 1990). The law proposed has the following form:

$$I = I_0 + n \log \sqrt{1 + \frac{d^2}{h^2}} + c \cdot \alpha \left(\sqrt{h^2 + d^2} - h \right) \quad \text{where } \alpha = 0.0012; \quad (1)$$

I and I_0 are the values of the macroseismic intensities at the point of study and in the respectively in the epicentral n and c are the parameters of the distribution, d is the epicentral distance, h is the depth (in kilometers). This law is based on the division of macroseismic maps into angular sectors equal to 15°, and to determine the parameters for each considered azimuth (Table 2).

Table 2 Parameters of the attenuation relation of Moldovan (2007)

Azimuth	n	c	Azimuth	n	c
0	−3.59516	−2.3	180	−2.36612	−1.73
15	−5.94089	−1.22	195	−2.28276	−1.55
30	−1.69964	−1.5	210	−1.09136	−1.8
45	−1.95074	−1.39	225	−1.355	−1.76
60	−3.25824	−1.4	240	−1.0693	−1.75
75	−5.56363	−1.18	255	−2.13533	−1.7
90	−7.02662	−1.1	270	−4.79691	−1.64
105	−5.48569	−1.39	285	−5.11263	−2.4138
120	−5.402	−1.55	300	−4.76939	− 2.77
135	−3.71915	−1.73	315	−4.76954	−3.09
150	−3.7476	−1.72	330	−6.84275	−2.5968
165	−2.49525	−1.77	345	−3.03096	−3.06

2.2.2   Sorensen et al. (2010)

The attenuation relation for intermediate Vrancea zone, was determined by Sorensen et al. 2010 (Eq. 2), by using information from five major Vrancea earthquakes, (November 11th, 1940, March 04th, 1977, August 30th, 1986 and May 30th and 31st, 1990). The events range was between magnitude M_w = 6.4–7.7 and deep h = 79–150 km.

$$I = c_1 \cdot M_W + c_2 \cdot \log(h) + c_3 - c_4 \cdot \log \sqrt{\frac{R^2 + h^2}{h^2}} - c_5 \cdot \left(\sqrt{R^2 + h^2} - h\right) + c_6 \cdot M_W \cdot dI$$

$$(2)$$

where:

I represent the intensity, M_w is moment magnitude, h–the depth to the centre of the fault plane,

R was defined as the Joyner-Boore distance, the shortest distance to the surface projection of the fault plane, $c1$–$c6$—isotropic parameter;

dl is an empirical regional correction function as a spatial function of longitude λ and latitude θ, combining five anisotropic two-dimension Gaussian-distribution function:

$$dI(\lambda, \theta) = \sum_{j=1}^{5} p_{6,j} \cdot \exp\left(-\left[p_{3,j} \cdot (\lambda - p_{1,j})^2 + 2p_{5,j}(\lambda - p_{1,j})(\theta - p_{2,j}) + p_{4,j} \cdot (\theta - p_{2,j})^2\right]\right)$$

where

- λ—longitude
- θ—latitude
- Pij—parameters.

2.2.3   Vacareanu et al. (2015)

The data used by Vacareanu et al. (2015) to determine the attenuation low (Eq. 3) were taken from Kronrod et al. (2013), given in electronic format, i.e. 9822 de intensity data points (in MSK scale), from six events: November 10, 1940 (Mw = 7.7); March 4, 1977 (Mw = 7.4); August 30, 1986 (Mw = 7.1); May 30, 1990 (Mw = 6.9); May 31, 1990 (Mw = 6.4); and October 27, 2004 (Mw = 6.0). The range of magnitude M_w = 6.0–7.7 and deep h = 87–150 km. The relationship takes into account the geographic position of the site in relation to the curvature area of the Carpathian Mountains.

$$I_{i,j} = c_1 + c_2 \cdot M_{W,i} + c_3 \cdot \log \sqrt{\frac{d_{ij}^2 + h_i^2}{h_i^2}}$$
$$- c_4 \left(1 - ARC_j\right)\left(\sqrt{d_{ij}^2 + h_i^2} - h_i\right) + c_5 ARC_j \left(\sqrt{d_{ij}^2 + h_i^2} - h_i\right) + \eta_i + \varepsilon_{ij} \quad (3)$$

where: i is the earthquake's index, j is the site's index, I is the MSK macroseismic intensity, M_w is the moment magnitude, d is the epicentral distance (in kilometers), and h is the focal depth (in kilometers). The ARC term introduces the location of the site with reference to the Carpathians Mountains arc (ARC = 0 for back-arc site and ARC = 1 for force-arc site), and c_k (k = 1–5) are coefficients obtained from the dataset by regression analysis. η_i and ε_{ij} are the inter-events residuals.

3   Results

In Figs. 8, 9, 10, 11, 12 and 13 are represented the intensities (blue dots) computed with Moldovan 2007 (Eq. 1 and Table 2), Sørensen et al. 2010 (Eq. 2) and Vacareanu et al. 2015 (Eq. 3) attenuation laws. With grey dots are represented the IDPs for all historical and instrumental events. With red dots are represented the differences (DI) between the observed IDPs' and the computed ones. The corresponding scale for the differences is represented on the secondary axis, on the right side of the graphs.

The data represented on graphs are grouped using the event occurrence time (year), then for each event the data were ordered taking into account the observed intensity value and then the epicentral distance. The attenuation laws were tested only for epicentral distances lower than 700 km.

The relationship of Moldovan (2007), (Fig. 9) fits well the observed data from earthquakes with depth between 60 and 100 km, especially at high intensities and distances lower than 600 km. At distances larger than 700 km, the calculated intensities tend to be lower than those observed. The errors increase with the depth of the hypocenter. As a rule, Moldovan relationship gives higher values than the

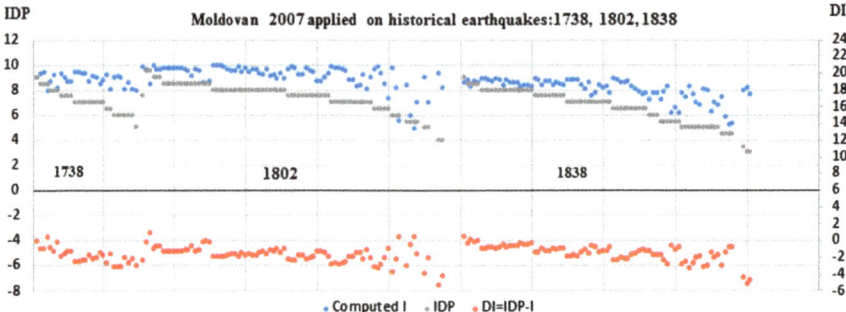

Fig. 8 Testing Moldovan (2007) relationship for the historical intermediate depth earthquakes

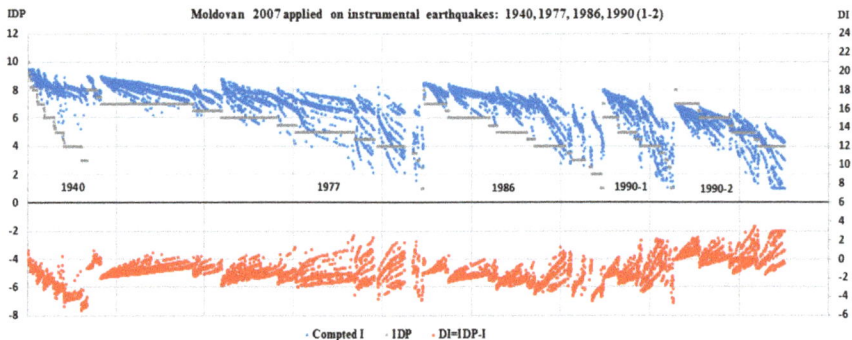

Fig. 9 Testing of Moldovan (2007) relationship for the instrumental intermediate depth earthquakes

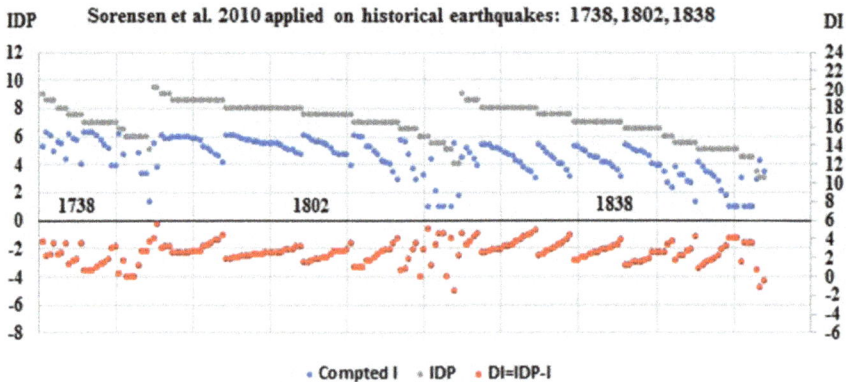

Fig. 10 Testing Sorensen et al. (2010) relationship on the historical intermediate depth earthquakes

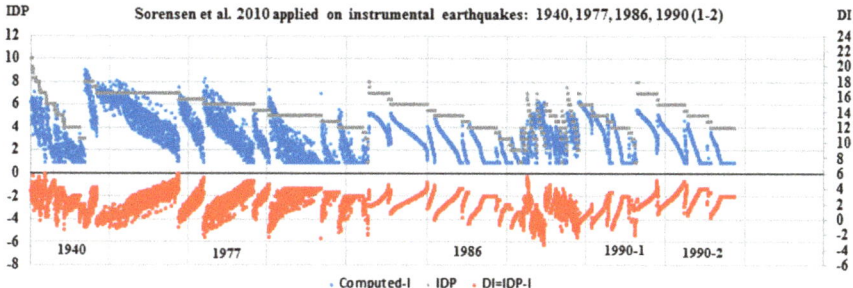

Fig. 11 Testing Sorensen et al. (2010) relationship on the instrumental intermediate depth earthquakes

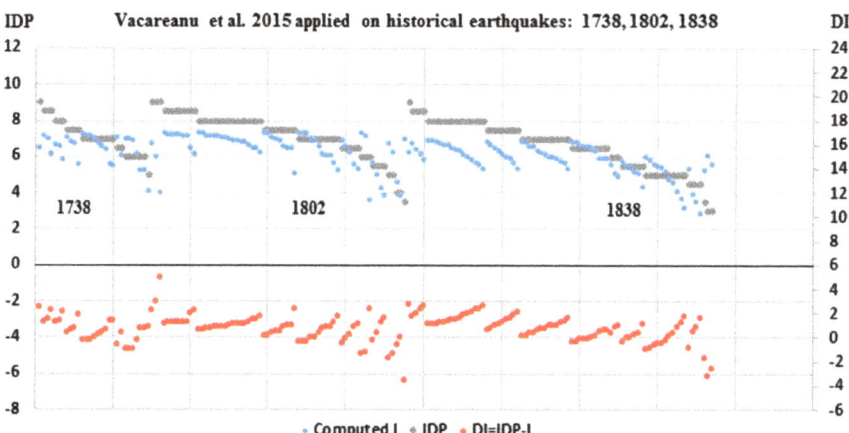

Fig. 12 Testing of Vacareanu et al. (2015) relationship on the historical intermediate depth earthquakes

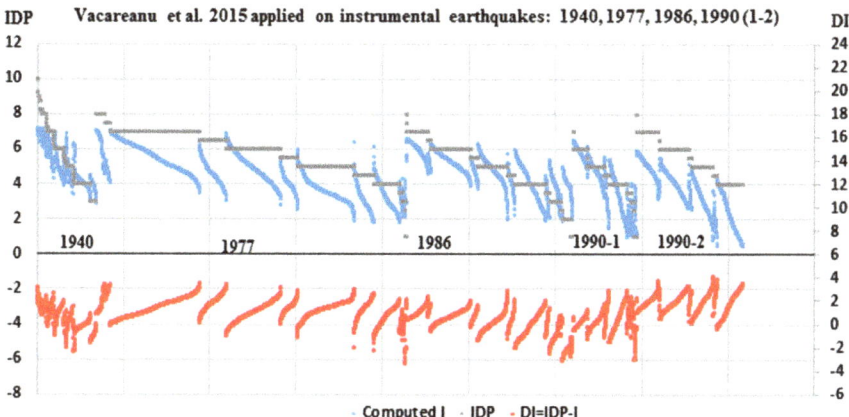

Fig. 13 Testing of Vacareanu et al. (2015) relationship on the instrumental intermediate depth earthquakes

observed ones, with a DI between (−3, 0) degrees of intensity, with very few exceptions where DI is between (0, +2) degrees, especially at larger distances.

The tests (Figs. 10 and 11) demonstrates that the Sørensen et al. 2010 relation is working well for events dipper than 100 km. For Sørensen et al. (2010) relationship we used only parameters for Joyner-Boore distance. For the other two ruptures and epicentral distance are obtained the same graphs. In contrast with Moldovan results, the intensity values obtained with Sørensen relation, are underestimating the observed intensities. DI is mostly positive, having values between (−2, 4) degrees, with higher values at distances larger than 1000 km.

Vacareanu et al. 2015 relationship is fitting very well the historical data and seems to be more appropriate for distant sites as the other two relations. His relation is also underestimating the real observed values, and DI is between (0, +2) for the case of historical earthquakes and (−2, +3) for the instrumental ones.

4 Conclusions

In this paper there have modelled the macroseismic field for all subcrustal Romanian earthquakes for which we had observed macroseismic intensities, using existing attenuation relationships and then we have compared the results with the real maps. The goal was to find the best relation to be used in Shake Map.

The conclusion is that for Vrancea intermediate depth earthquakes Moldovan (2007) attenuation law is good to model the macroseismic field of subcrustal earthquakes produced at depth between 60–100 km, at distances not larger than 600 km, Sørensen et al. (2010) is better for events with depth larger than 100 km and Vacareanu et al. (2015) has to be used at large epicentral distances.

Because the errors for all used models are high enough, raise the necessity to improve the existing relations, in order to lower the errors.

Acknowledgements This paper was partially carried out within Nucleu Program, supported by ANCSI, projects no. PN 16 35 01 06, PN 16 35 01 09, PN 16 35 03 01 and the Partnership in Priority Areas Program—PNII, under MEN-UEFISCDI, DARING Project no. 69/2014, and project: Disaster risk assessment at national level- RO-RISK.

References

Ardeleanu L, Leydecker G, Bonjer K-P, Busche H, Kaiser D, Schmitt T (2005) Probabilistic seismic hazard map for Romania as a basis for a new building code. Nat Hazards Earth Syst Sci 5:679–684

Demetrescu G (1941) Determination provisoire de l' épicentre du trem-blement de terre de Roumanie du 10 Novembre 1940. Observatoire de Bucuresti, Station sismique, p 10

Glavcheva R (1983) Impacts on PR Bulgaria, 178–187. In: Brankov G (ed) Vrancea earthquake in 1977, its after-effects in the People's Republic of Bulgaria, Earthquake Engineering National Committee, Bulgarian Academy of Sciences, Sofia, 428 pp

Grigorova E, Glavtcheva R, Sokerova D (1978) The earthquake on March 4, 1977—some results of seismic observations in Bulgaria. In: Proceedings of the symposium on the analysis of seismicity and on seismic risk, Liblice, October 1977, pp 109–113

Kronrod T, Radulian M, Panza G, Popa M, Paskaleva I, Radovanovich S, Gribovszki K, Sandu I, Pekevski L (2013) Integrated transnational macroseismic data set for the strongest earthquakes of Vrancea (Romania). Tectonophysics. https://doi.org/10.1016/j.tecto.2013.01.019

Leydecker G, Busche H, Bonjer K-P, Schmitt T, Kaiser D, Simeonova S, Solakov D, Ardeleanu L (2008) Probabilistic seismic hazard in terms of intensities for Bulgaria and Romania—updated hazard maps. Nat Hazards Earth Syst Sci 8:1–9

Lungu D, Zaicenco A, CorneaT, van Gelder P (1997) Seismic hazard: recurrence and attenuation of subcrustal (60–170 km) earthquakes. Struct Saf Reliab 7:1525–1532

Marza VI, Pantea AI (1995) Probabilistic estimation of seismic intensity attenuation for Vrancea (Romania) subcrustal sources. In: Proceedings of XXIV General Assembly of the E.S.C., 19–24 Sep 1994, vol III, Athens, Greece, pp 1752–1761

Medvedev SV, Sponheuer W, Karnik V (1964) Neue seismic Skala (Intensity scale of earthquakes), 7. Tagung der Europaischen Seismologischen Kommission vom 24.9. bis 0.9.1962 in Jena, Veroff. Insttite fur Bodedynamik und Erdbebenforschunh in Jena 77, 69-6. Medvedev

Moldovan IA, Enescu BD, Ionescu C (2000) Predicting peak ground horizontal acceleration for vrancea large earthquakes using attenuation relations for moderate shocks. Rom J Phys 45 (9–10):785–800

Moldovan IA (2007) Metode si modele statistice in seismologie, pag 236. Editura Morosan, Bucuresti

Oncescu MC, Marza VI, Rizescu M, Popa M (1999) The Romanian earthquake catalogue between 1984–1997. In: Wenzel F, Lungo D, Novack O (eds) Vrancea earthquakes: tectonics, hazard and risk mitigation, Kluwer Academic Publishers, Dordrecht, pp 43–47

Pantea A (1994) Macroseismic intensity attenuation for crustal sources on Romanian territory and adjacent areas. Nat Hazards 10:65–72

Pantea A, Moldovan IA (2000) Attenuation relationships using macroseismic intensity curves. Part I. Crustal earthquakes from Fagaras and adjacent zones. Rom J Phys 45(7–8):615–631

Radu C, Radulescu D, Sandi H (1990) Some data and considerations on recent strong earthquakes of Romania, cahier technique, 3. AFPS, Paris, pp 19–31

Radu C, Utale A (1989) A new version of the March 4, 1977 Vrancea earthquake (in Romanian), Report CEPS, theme 30.86.3/1989, II, A4, 31–32, Bucharest

Rogozea M (2016) Impactul cutremurelor majore din Romania: trecut, present si viitor. Editura Electra, Bucuresti

Rogozea M, Marmureanu Gr, Radulian M, Toma D (2014) Reevaluation of the macroseismic effects of the 23 January 1838, Vrancea earthquake. Romania Report in Physics 66

Rogozea M, Radulian M, Marmureanu Gh, Mandrecu N, Paulescu D (2013) Large and moderate historical earthquakes of 15th and 16th centuries in Romania reconsidered, Romanian reports in physics, vol 65

Sokolov V, Bonjer KP, Wenzel F, Grecu B, Radulian M (2008) Ground-motion prediction equations for the intermediate depth Vrancea (Romania) earthquakes Bulletin of Earthquake Engineering, vol 6, pp 367–388

Sørensen MB, Stromeyer D, Grunthal G (2008) Estimation of macroseismic intensity—new attenuation and intensity versus ground motion relations for different parts of Europe. In: Proceedings of the 14th world conference on earthquake engineering 12–17 Oct 2008, Beijing, China

Sørensen MB, Stromeyer D, Grunthal G (2010) A macroseismic intensity prediction equation for intermediate depth earthquakes in the Vrancea region, Romania. Soil Dyn Earthq Eng. https://doi.org/10.1016/j.soildyn,2010.05.009

Vacareanu R, Iancovici M, Neagu C, Pavel F (2015) Macroseismic intensity prediction earthquake for Vrancea intermediate-depth seismic source. Hazards, Nat. https://doi.org/10.1007/s11069-015-1994-y

Zsiros T (1996) Macroseismic focal depth and intensity attenuation in the Carpathian region. Acta Geod Geophys Hung 31(1–2):115–112

Historical Earthquakes: New Intensity Data Points Using Complementary Data from Churches and Monasteries

Gheorghe Marmureanu, Radu Vacareanu, Carmen Ortanza Cioflan,
Constantin Ionescu and Dragos Toma-Danila

Abstract The Vrancea seismogenic zone denotes a peculiar source of seismic hazard which represents a major concern in Europe, especially to Romania and neighbouring regions from Bulgaria, Serbia and Republic of Moldova. The strong seismic events that can occur in this area can generate the most destructive effects in Romania and may affect high-risk manmade structures such as nuclear power plants, chemical plants, large dams and pipelines located within a wide area including the Northern zone from the Republic of Bulgaria and the SW of the Moldavia Republic. A major part of the information for determining the design basis earthquakes consists of a complete set of historical earthquake data. Therefore, it is necessary that the available historical records to be collected, extending as far back in time as possible. Most of these historical records will be of descriptive nature, including such information as the number of houses damaged or destroyed, the behaviour of population etc. But from such information a measure of the intensity scale value of each earthquake in modern macroseismic intensity scale values may be determined. During the past project "*Bridging the gap between seismology and earthquake engineering: from the seismicity of Romania towards a refined implementation of seismic action EN1998-1 in earthquake resistant design of buildings (BIGSEES)*", the authors developed the macroseismic intensity map of

G. Marmureanu (✉) · C. O. Cioflan · C. Ionescu · D. Toma-Danila
National Institute for Earth Physics, Calugareni, Nr12, Magurele,
077125 Ilfov, Romania
e-mail: marmur@infp.ro

C. O. Cioflan
e-mail: cioflan@infp.ro

C. Ionescu
e-mail: viorel@infp.ro

D. Toma-Danila
e-mail: toma@infp.ro

R. Vacareanu
Seismic Risk Assessment Research Center, Technical University of Civil Engineering
of Bucharest, Bd. Lacul Tei 122-124, Bucharest, Romania
e-mail: radu.vacareanu@utcb.ro

© Springer International Publishing AG, part of Springer Nature 2018
R. Vacareanu and C. Ionescu (eds.), *Seismic Hazard and Risk Assessment*,
Springer Natural Hazards, https://doi.org/10.1007/978-3-319-74724-8_7

103

Romania by using newly compiled information about the damages experienced by 115 churches and monasteries after 10 strong earthquakes (Mw > 6.9) occurred in Vrancea zone starting with XVth century.

Keywords Historical seismicity · Intensity seismic hazard map Macroseismic intensity

1 Introduction

Historical seismicity is seismicity for which evidence can be found in the historical record of humankind. For pre-instrumental earthquakes, however, this evidence is preeminent. Their existence, interpretation and use are of major interest to seismic hazard analysis, and are playing a key role in defining and litigating the seismic design of many critical facilities in last time. The historical data, consisting of written evidence of the effects of earthquakes on humans, their civilization and the environment, has been used to determine earthquake location and size, the strength of ground motions and its frequency of occurrence in most parts of the world.

Historical records differ greatly in the length, completeness and quality of the earthquake histories they portray. As we go back in time, the minimum size earthquake mentioned in any given history will usually increase. Typically, only those events which led to disastrous consequences are mentioned in ancient records. Additional information about these events would have been of great benefit to present-day seismologists and engineers. Of practical interest is the estimate of the time duration of the earthquake. These details provide the analyst with much information that allows an improved estimation of seismic hazard from future great earthquake in the area.

Seismic activity in Romania is confined to the crust, except the Vrancea zone, where earthquakes with focal depth down to 200 km occur. Figure 1 displays the spatial distribution of earthquake epicentres in Romania and surrounding areas considering also magnitudes and focal depths. The statistical analysis of the earthquake catalogue compiled in the frame of the BIGSEES Project (2017) gives at a mean return interval of 100 years a magnitude Mw 7.5 for Vrancea intermediate-depth seismogenic zone and much lower magnitudes for crustal seismogenic zones: Mw 5.7 for Banat and Danubian zones (located in South-West of Romania) and Mw 5.6 for a very active zone in Central Romania (Fagaras-Campulung).

The strongest event in the documented history and originating in Vrancea area occured on October 26, 1802, having a 7.9 magnitude (Mw) and a depth of 150 km, according to ROMPLUS catalogue (NIEP 2017). It was felt over a large area in Europe: in the Balkans up to Constantinopolis (Istanbul), up to Moscow and Sankt-Petersburg in the North (North-East) and to Mediterranean Sea in the South-West, as stated by Rogozea et al. (2016) and references therein. Actually, a peculiarity of earthquakes occurring at intermediate depths (70–200 km) in Vrancea

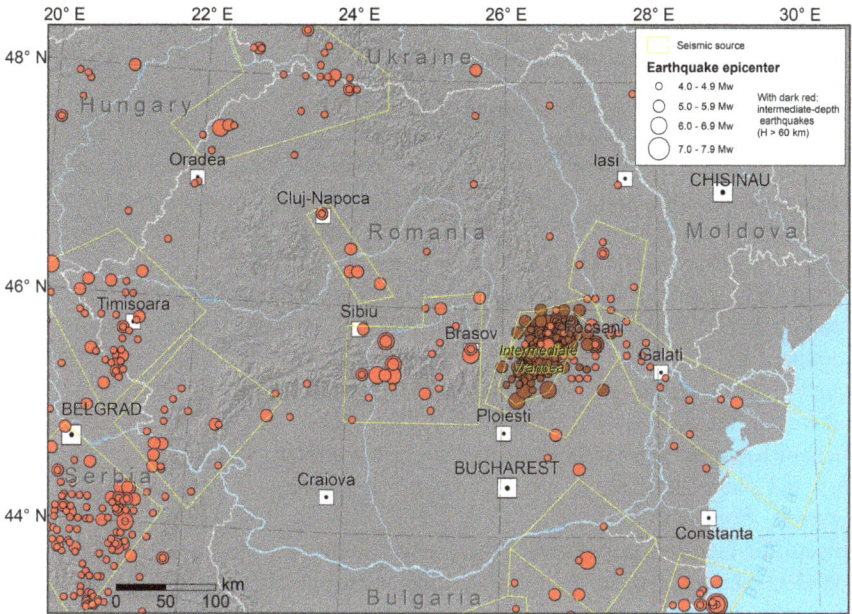

Fig. 1 Seismicity and seimogenic zones contributing to seismic hazard of Romania. *Datasource* BIGSEES Project 2017

zone is the quasi-elliptical shape of the macroseismic field with the major axe oriented along NE–SW direction; events with magnitudes higher then 6 can be felt in the whole Balkan area up to Turkey (in the South) and to Poland (in the North).

During the last century in Vrancea seismogenic zone 4 strong earthquakes (Mw > 7) occured at intermediate depths: October 6, 1908 (Mw 7.1) at 125 km depth, November 10, 1940 (Mw 7.7) at 150 km depth, March 4, 1977 (Mw 7.4) at 94 km and August 31, 1986 (Mw 7.1) at 131 km. In spite of being the most destructive event in the previous century, the 1977 earthquake had numerous precedents in Romania. The historical record here extends to Roman times and includes similar intermediate—depth earthquakes with magnitudes of 7 or greater about 2–3 times per century. Over the past several centuries, earthquakes with likely magnitudes of 7 or greater caused damage in Bucharest in 1681, 1738, 1802, 1829, 1838 and more recently in 1940, 1977.

A major part of the information for determining the design basis earthquakes to important structure like nuclear power plants, bridges etc. is a complete set of historical earthquake data. Therefore, it was necessary that the available historical records be collected, extending as far back in time as possible. Most of these historical records will naturally be of descriptive nature, including such information as the number of houses damaged or destroyed, the behaviour of population etc.; from such information a measure of the intensity scale value of each earthquake in modern macroseismic intensity scale values may be determined.

2 New Intensity Data by Using Complementary Data from Churches and Monasteries

In estimating intensity scale values for historical earthquakes, account should be taken of the fact that information obtained from old chronicles, from churches, monasteries and individuals who may have experienced the event shows a tendency to exaggerate the importance of damage and other phenomena associated with earthquakes There are many data on the historical earthquakes. Many writers and journalists along time made many descriptions on earthquake effects on churches, monasteries etc. Authors collected, centralised and analysed also information from literature. For example, many descriptions of damages and macroseismic effects induced by the earthquake of 1802 ($M_W = 7.9$, $h = 150$ km) are from Mihail Sadoveanu. In the novel "Jderi Brothers" he stated that in many places from Bucharest around Bucharest Cathedral the earth unbind, went out oil, water and sand and many large slits with transversal dimensions between 50 and 150 cm occurred. Considering the effects on the natural environment and taking into account the characteristics of the monasteries (building material, size of construc-tive elements etc.) values of macroseismic intensity (intensity data points, IDP) were associated to the recorded damages. Some examples are discussed below.

After the strong earthquake of 1620 (Mw = 7.5; h = 150 km), Vasile Lupu, prince of Moldova, started Putna (MO/1) monastery rehabilitation, in fact its recovery. The restoration of the Church at Putna Monastery was completed in the year 1662. The large earthquake of 1738 ($M_W = 7.7$; h = 130 km) seriously damaged the church, the walls of the enclosure and towers of the four corners.

Stavropoleos Monastery (MO/76) (44.431789 latitude; 26.098869 longitude; altitude = 79 m) in downtown Bucharest (the word "*stavropoleos*", name in Greek translates as "*cross town*") was build up in 1724. Another period in the building of the Stavropoleos church (1730–1733) extended the altar, built the side apses in stone columns and carved porch. The monastery was strongly affected by the earthquake 1802 ($M_W = 7.9$; h = 150 km). The effects were devastating. Another earthquake took place on January 23, 1838. French consul in Bucharest, noted that "*disasters caused in this city of this event are huge and cannot be evaluated*". The structure of the church and inn was so severely affected that on November 15, 1840, the Department of Internal Affairs is forced to submit an address in which require demolition of the tower Stavropoleos. In 1940, when the church was restored after an absence of decades, the monastery was less affected by the 7.7 magnitude event.

Bogdana Monastery (MO/104) (47.842480 latitude; 25.917498 longitude; alti-tude = 372 m) is located in the heart of historic Bukovina, in Rădăuți, on one of the roads leading to Putna Monastery. Here in Rădăuți, Dragoș Vodă has established one of the royal residences and founder Bogdan I, the first voivode of Moldavia is buried here in wooden church in 1367. Moldova have information to historian Ioan Neculce in 31 mai/June 11, 1738 ($M_W = 7.7$; h = 110 km) … have made a great earthquake, the Golia monastery from Iași fell and many houses and many other monasteries in Focşani; disaster followed by a "great plague in Iasi", at Bacău overturned Răchitoasa Monastery (Corfus I., (1967)—"Notes old", Iaşi, Junimea), Solomon church was destroyed, monastery Bogdana (Rădăuţi) collapsed; also in some places the ground cracked and "went out water and sand".

Comana Monastery, located in the homonymous village on the meadow of Neajlov river, is mentioned in a charter issued by the chancellery of Vlad Ţepeş on September 27, 1461. In 1588, Radu Şerban, the local nobleman, built a new monastery on the old foundation. Between 1699 and 1701, governor Şerban Cantacuzino started the works of restoration of the monastery. In August 1709, hard times came for Comana monastery. The state of the monastery is sporadically recorded in some writings: "…because of the earthquakes, like that of 1701 ($M_w = 7.1$; h = 150 km), and because of lack of care, the monastery is in danger of ruining" (Ghenadie Enăceanu notes in quoted work Catagraphy from 1834, quoted by Maria Valeria Picu). This reflects a sad reality, as at Comana Monastery there were: "a chapel in the wall of monastery and the four towers totally broken down, the walls of the surrounding courtyard were totally broken and cracked" (Archives of the state Bucharest. *Ministry of Public Instruction Fund, Wallachia, File 4/1828, f.3*). In 1854, after the earthquake of 1838 ($M_w = 7.5$), "*demolition and rebuilding*" of the monastery is done, as Cezar Bolliac says. The earthquake of October 26, 1802 ($M_w = 7.9$; h = 150 km) caused great damages to the whole construction, and in 1854 a new restoration began, conducted by architect J. Schlatter, after the earthquake of 1838 ($M_w = 7.5$). The church of monastery was retrofitted after the earthquakes of 1977 ($M_w = 7.4$; h = 94 km) and 1986 ($M_w = 7.1$; h = 131.4 km).

3 Maps with Maxim Macroseismic Intensities Observed Along Time in Romania

Newly collected data regarding seismic damages of 115 monasteries has been synthesized and analyzed through the engineering point of view, considering materials and dimensions of the buildings, towers etc. Some of the resulting intensities presented in Table 1 are completing the known macroseismic field of each event, and data may be used to construct the isoseismic map of the maximum possible Vrancea earthquake. Figures 2, 3, 4, 5, 6 and 7 show the intensities evaluated from the analysis of the damage records at the specific sites.

The estimated intensity values for the 1738, 1802 and 1940 events are consistent with map of the macroseismic intensities expected in the case of a Vrancea "maximum credible" earthquake (Marmureanu et al. 2011) where the shape of isoseismal surfaces are elongated much more towards NE than to the SW. The sites with maximum intensities observed in the last century (IX and X MSK) are shifted towards East with respect of the epicentre area, inside or at the edges of the Focsani depression (a foredeep basin formed in front of the East Carpathians, where sediments are reaching even 14 km depth).

Table 1 Macroseismic intensities values (MSK64 scale) evaluated using historical data from churches and monasteries

Name of monastery (M) or church (C)	Earthquakes									
	1471 $M_W = 7.5$ $H = 110$ km	1516 $M_W = 7.5$ $H = 150$ km	1590 $M_W = 7.3$ $H = 100$ km	1620 $M_W = 7.5$ $H = 150$ km	1701 $M_W = 7.1$ $H = 150$ km	1738 $M_W = 7.7$ $H = 130$ km	1802 $M_W = 7.9$ $H = 150$ km	1838 $M_W = 7.5$ $H = 150$ km	1940 $M_W = 7.7$ $H = 150$ km	1977 $M_W = 7.4$ $H = 94$ km
Putna (M)			IX	IX½		IX½	IX½			
Bogdana (M)						X				
Probota (M)			IX					VIII½		
Moldoviţa (M)		IX½								
Neamţ (M)	IX½								IX½	
Sihăstria (C)										VIII½
Secu (M)										
Sf. Nicolae, Iassy (C)								IX½		
Bistriţa Neamţ (M)		IX½					IX			
Coşula (M)										VIII
Golia (M)				IX½		IX½	VIII½			
Caşin (M)						IX	IX½	IX		
Sf. Trei Ierarhi (M)					IX½		IX½	IX½		IX½
Cetăţuia (M)							IX½			
Hlincea (M)										
Domn. Bârlad (C)							X	IX½	VIII	VIII
Sf. Ilie Bârlad (C)									VIII	
Domn. Vaslui (C)							IX½–X		IX	
Rafaila (M)									IX	
Făstăci (M)									IX	
Sămurcăşeşti (M)								IX½	IX½	IX½
Viforâta (M)				IX½	IX½		X		X	
Sf. Gh. Nou Buc. (M)							IX½	IX½		

(continued)

Table 1 (continued)

Name of monastery (M) or church (C)	Earthquakes									
	1471	1516	1590	1620	1701	1738	1802	1838	1940	1977
	$M_W = 7.5$	$M_W = 7.5$	$M_W = 7.3$	$M_W = 7.5$	$M_W = 7.1$	$M_W = 7.7$	$M_W = 7.9$	$M_W = 7.5$	$M_W = 7.7$	$M_W = 7.4$
	H = 110 km	H = 150 km	H = 100 km	H = 150 km	H = 150 km	H = 130 km	H = 150 km	H = 150 km	H = 150 km	H = 94 km
Antim (M)						IX½	IX½		X	
Icoanei (C)								X	IX	
Stavropoleos (M)							IX½	IX½		
Zlătari (C)					X		X	IX½		IX½
Cărămidarii Sus (C)								X		
Precupeţii Noi (C)										X
Patriarch. Cathedral							IX½		IX½	IX
Radu Vodă (M)								IX½		
Comana (M)		X			X	X	X	IX½		VIII½
Răteşti (M)								IX½		VIII½
Curtea de Argeş (M)			IX½				X	IX½		
Zamfira (N)								X	IX	IX
Brâncoveni (M)				IX½			IX½	IX½	IX½	
Bistriţa-Vâlcea (M)				IX½				IX½		
Arnota (M)					X			IX½		IX
Nucet (M)						X	X	X	IX½	
Dealu (M)						IX		IX	IX½	
Căldăruşani (M)					IX½	IX½				
Sitaru (M)							IX½	IX	IX	VIII
Glavacioc (M)				IX½	IX½		X	IX½	IX	IX
Pissiota (M)									IX½	VIII½
Mamu (M)				X			IX½	IX	IX	IX
Clocociov (M)										VIII½

(continued)

Table 1 (continued)

Name of monastery (M) or church (C)	Earthquakes									
	1471	1516	1590	1620	1701	1738	1802	1838	1940	1977
	$M_W = 7.5$	$M_W = 7.5$	$M_W = 7.3$	$M_W = 7.5$	$M_W = 7.1$	$M_W = 7.7$	$M_W = 7.9$	$M_W = 7.5$	$M_W = 7.7$	$M_W = 7.4$
	H = 110 km	H = 150 km	H = 100 km	H = 150 km	H = 150 km	H = 130 km	H = 150 km	H = 150 km	H = 150 km	H = 94 km
Stelea (M)						IX½	X			
Negru Vodă (M)				X			IX½	X		
Coșuna (M)	IX½							IX½	X	
Domn. Pitești (C)							IX½			
Popânzălești (M)								X		IX½
Pasărea (M)								X		
Plătărești (M)							IX½			
Mărcuța (M)										VIII½
Snagov (M)					X	IX½			IX½	IX½
Gura Motru (M)				X		IX				
Câmpina (C)									VIII½	VIII
Căciulați (C)									VIII½	VIII
Comarnic (C)									VIII	VIII½
Hurezi (M)						IX½				
Govora (M)										VIII
Ocnele Mari (C)								X		
Celic Dere (M)									IX½	
Saon (M)									IX½	
Cocoș (M)									IX	
Topolinița (M)							IX			VIII½
Sfânta Ana (M)										VIII½
Cozia (M)		IX½			IX½			X		

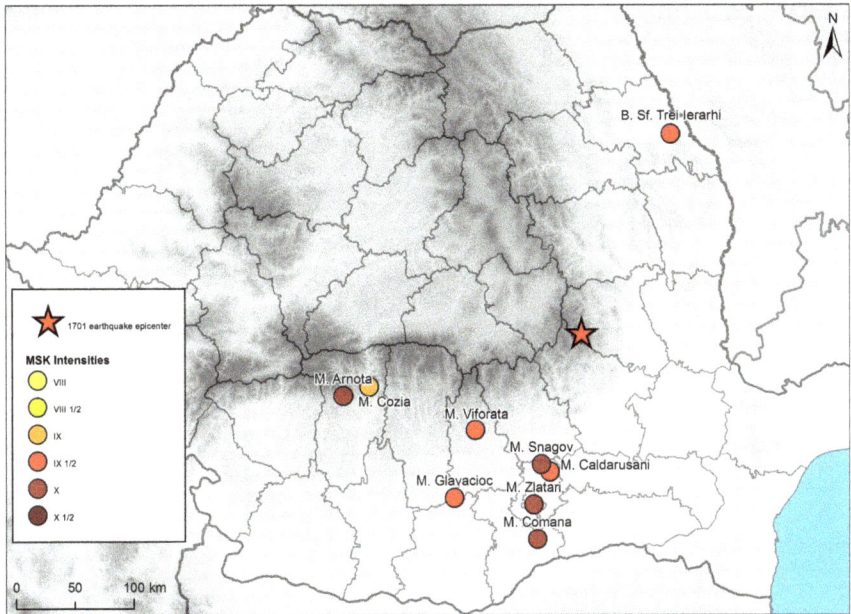

Fig. 2 IDP's evaluated for Vrancea earthquake of June 12, 1701: $M_W = 7.1$ h = 150 km

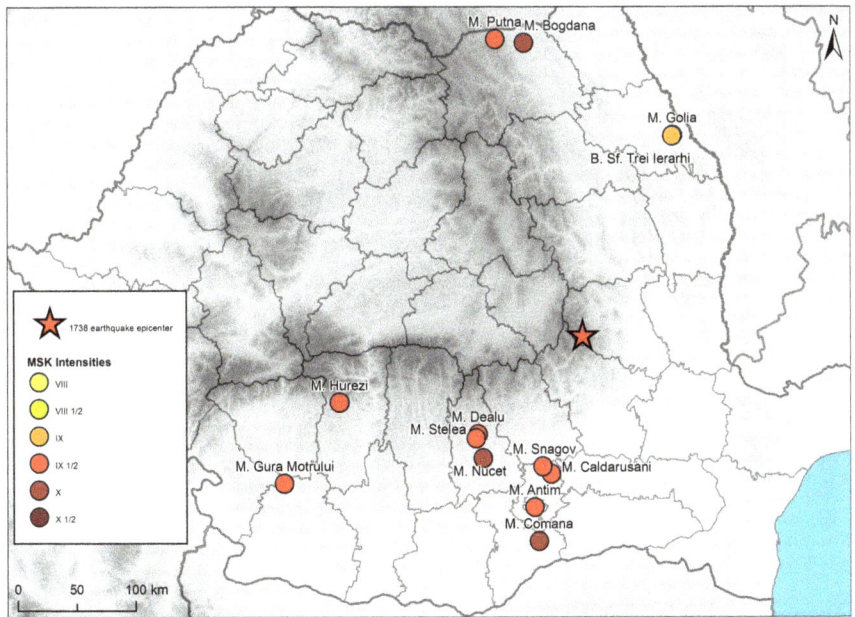

Fig. 3 IDP's evaluated for Vrancea earthquake on June 11, 1738, $M_W = 7.7$

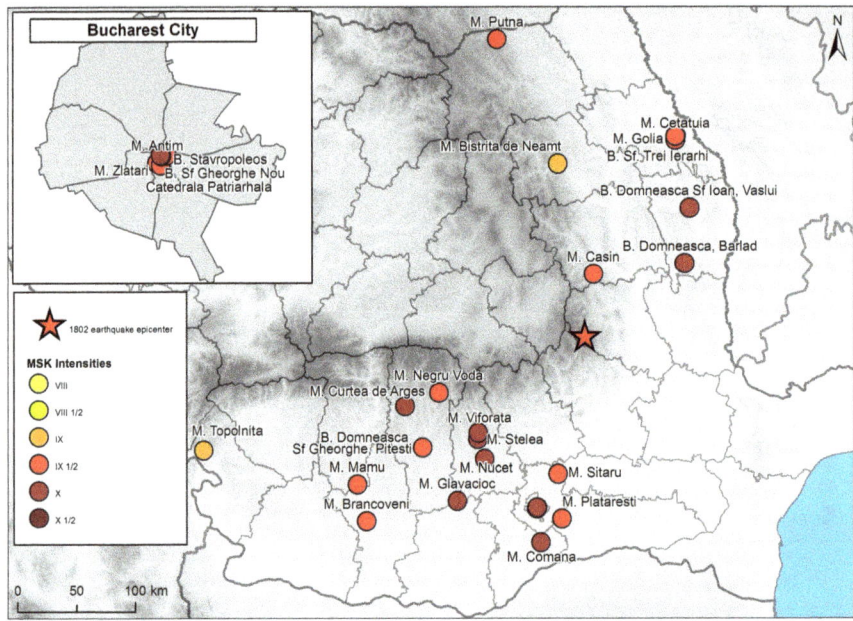

Fig. 4 New IDPs for Vrancea earthquake on October 26, 1802. M_W = 7.9

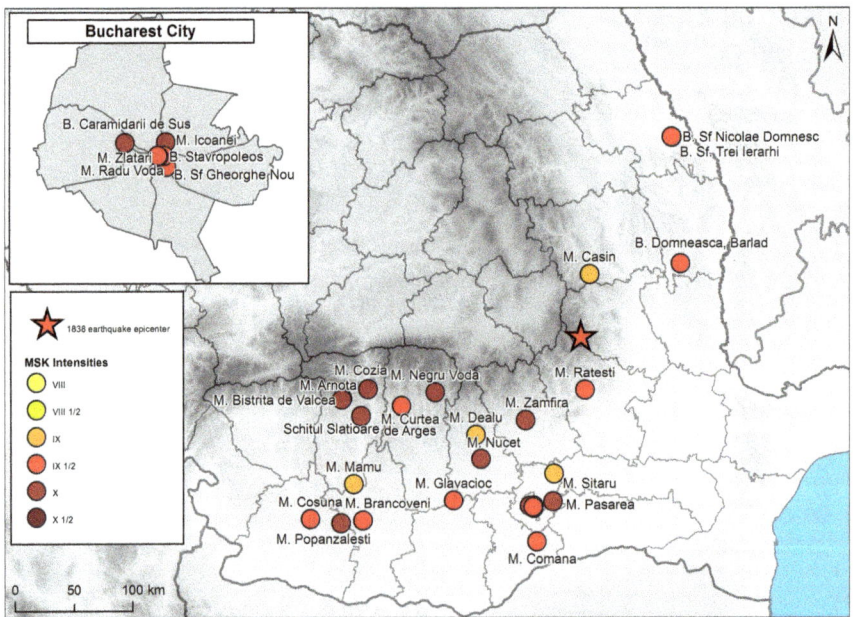

Fig. 5 IDP evaluated for Vrancea earthquake on January 23, 1838. M_W = 7.5

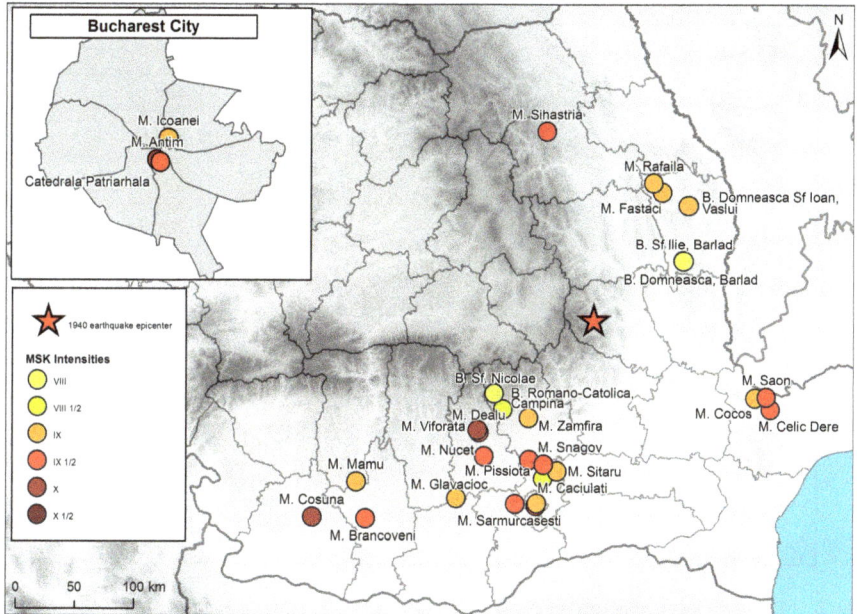

Fig. 6 IDP's evaluated Vrancea earthquake on November 10, 1940. $M_W = 7.7$

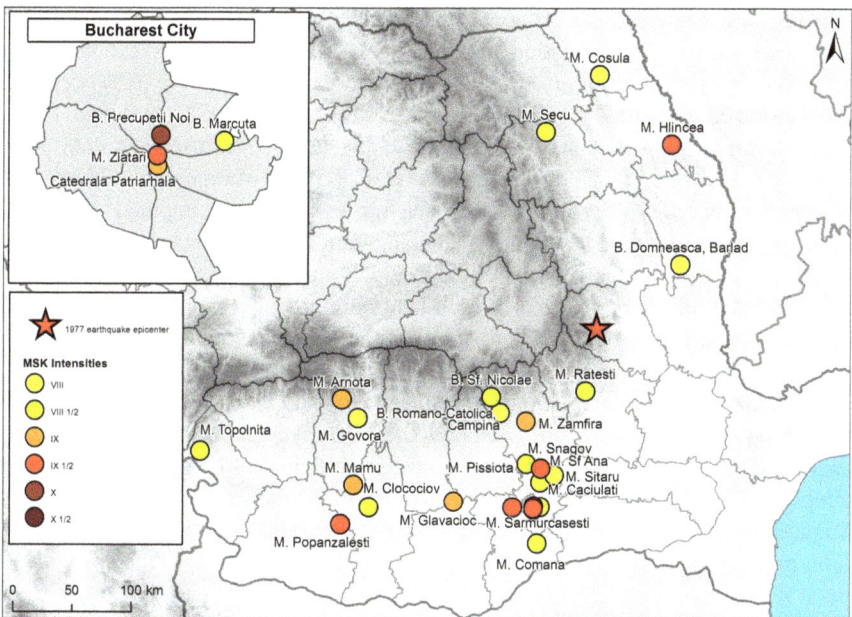

Fig. 7 Re-evaluated IDPs from Vrancea earthquake on March 4, 1977. $M_W = 7.4$

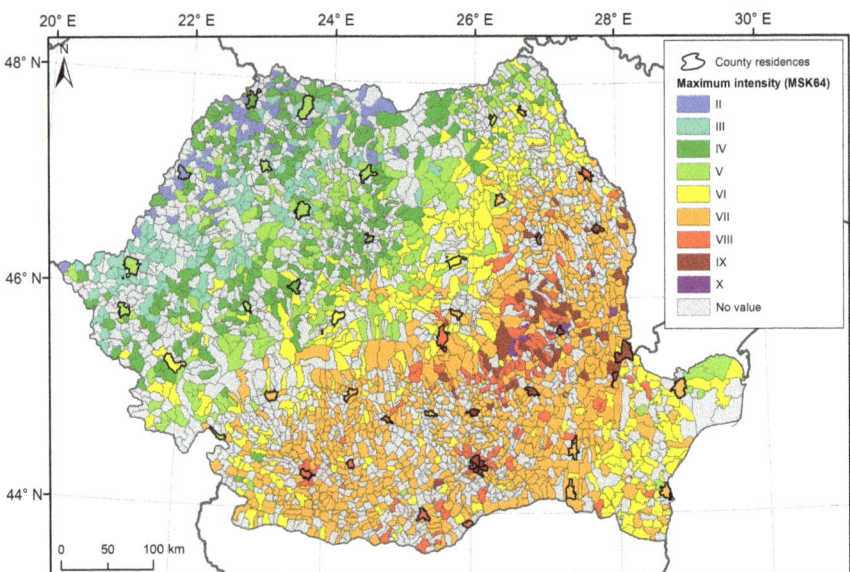

Fig. 8 Maximum observed macroseismic intensities due to Vrancea intermediate-depth earthquakes. New intensity data points using complementary data from churches and monasteries where averaged within each land administrative unit for each earthquake considered in this study

In Fig. 8 is presented the observed macroseismic intensity map of Romania by using newly compiled information about the damages experienced by 115 churches and monasteries after strong intermediate-depth earthquakes (Mw $\geq$ 6.9) occurred in 1471, 1516, 1590, 1620, 1701, 1738, 1802, 1838, 1940, 1977, 1986 and 1990. The new information is compiled together with relevant macroseismic data published by Constantin et al. (2011), Kronrod et al. (2013), Marmureanu (2016) and Rogozea et al. (2014, 2016). For each seismic event were computed the mean values of the intensity data points available within the administrative units, as defined nowadays. For the map presented in Fig. 8 only the maximum value among the earthquake means is reported in each land administrative unit.

Many attempts have been made to correlate the observed intensity with recorded ground motion parameters, as well as the hazard assessment for future events, which require a regression relationship between intensity and strong-ground-motion parameters. Enescu et al. (2001) elaborated an estimation method of the seismic hazard generated by strong and deep Vrancea earthquakes considering only the strong earthquakes produced in the area after 1940 (Table 2). It was observed that all epicentres are situated on a line oriented NE–SW and the corresponding maximum macroseimic intensities I_0 occur on a line which is parallel to the epicentre's line at distance of about 23 km.

Table 2 Strong Vrancea earthquakes ($M_{GR} > 6.5$) after 1935

Date	Time	Φ° N	λ° E	h(km)	M_{GR}	M_W	I_0
Nov. 10, 1940	01:39:07	45.8	26.70	150	7.4	7.7	X
March 4, 1977	19:22:15	45.34	26.30	94	7.2	7.4	IX
August 30, 1986	21:28:37	45.53	26.47	131	7.0	7.1	VIII
May 30, 1990	10:40:06	45.82	26.90	90	6.7	6.9	VIII

4 Concluding Remarks

Strong, intermediate-depth (70–200 km) earthquakes in the Vrancea area (Romania) occur irregularly, but not infrequently. They have caused a high toll of casualties and extensive damage *over the last several centuries*. The three strongest earthquakes (M > 7) that occurred in Vrancea during the last century (1940, $M_W = 7.7$ and h = 150 km; 1977, $M_W = 7.4$ and h = 94 km; 1986, $M_W = 7.1$ and h = 131.4 km) caused large damage over a wide European territory, including distant, long period elements of the built environment. The quake of March 4, 1977, $M_W = 7.4$ caused significant damage in Romania & Bulgaria and was felt up to Central Europe.

This study is of interest to seismologists and other earth scientists who want to gain an understanding of how historical information is used in estimating earthquake hazard *along time of centuries* in *Romania*, engineers who want to know more about the seismic input to be used in design and risk calculations, decision makers and other public officials who want to make decisions, particularly that related to the seating and evaluation of nuclear plants etc. Typically, only those events which led to disastrous consequences are mentioned in historical records, however it is not to be neglected that smaller earthquakes can lead to a more insightful seismic hazard evaluation.

In estimating intensity scale values for historical earthquakes, account should be taken of the fact that information obtained from old chronicles, from churches, monasteries and individuals who may have experienced the event shows a tendency to exaggerate along time the importance of damage and other phenomena associated with earthquakes.

The results on seismic hazard obtained in BIGSEES Project provide a very solid background for the upcoming update of the EN 1998-1 National Annex.

Acknowledgements This study was performed in the frame of the projects BIGSEES (contract 72/2012) and PN 16-35-02-02 (contract 21N/2016). The financial support of SEER project number PN-III-P2-P2.1-PED-2016-1014 is gratefully acknowledged.

References

Constantin AP, Pantea A, Stoica R (2011) Vrancea (Romania) subcrustal earthquakes: historical sources and macroseismic intensity assessment Rom. J Phy 56(5–6):813–826

Enescu D, Marmureanu G, Enescu BD (2001) A procedure for estimating the seismic hazard generated by Vrancea earthquakes and its application. I. "Etalon" earthquake. In: 2nd national conference on earthquake engineering, Bucharest, November 8–9, 2001, pp 1–10, Paper I

Kronrod T, Radulian M, Panza G, Popa M, Paskaleva I, Radovanovich S, Gribovszki K, Sandu I, Pekevski L (2013) Integrated transnational macroseismic data set for the strongest earthquakes of Vrancea (Romania). Tectonophysics 590:1–23

BIGSEES Project (2017). BIGSEES earthquake catalog. http://bigsees.infp.ro/Results.html

Mănăstiri Ortodoxe (MO) No. 1, 76, 104. Edit. De Agostini Hellas SRL; Vouliagments 44-46,16673 Atena; Editor: Petros Kapnistos; www.deagostini.ro

Marmureanu G, Cioflan CO, Mărmureanu A (2011) Intensity seismic hazard map of Romania by probabilistic and (neo)deterministic approaches, linear and nonlinear analyses. Rom Rep Phys 63(1):226–239

Marmureanu G, Cioflan CO, Marmureanu A, Manea EF (2016) Main characteristics of November 10, 1940 strong Vrancea earthquake. In: Vacareanu R, Ionescu C (eds) The 1940 Vrancea earthquake. Issues, insights and lessons learnt. Proceedings of the symposium commemorating 75 years from November 10, 1940 Vrancea earthquake. Springer Natural Hazards Series, Springer International Publishing, pp 72–83. https://doi.org/10.1007/978-3-319-29844-3_30

Marmureanu G (2016) Certainties and uncertainties in seismic hazard and risk assessment of strong Vrancea earthquakes. Romanian Academy Ed., Bucharest

NIEP (2017) Romplus earthquake catalog. http://www.infp.ro/catalog-seismic

Rogozea M, Marmureanu G, Radulian M, Toma D (2014) Reevaluation of macroseismic effects of the 23 January 1838 Vrancea earthquake. Rom Rep Phys 66(2):520–538

Rogozea M, Radulian M, Popa M (2016) Comparison of three major historical earthquakes with three recent earthquakes. In: Vacareanu R, Ionescu C (eds) The 1940 Vrancea earthquake. Issues, insights and lessons learnt. Proceedings of the symposium commemorating 75 Years from November 10, 1940 Vrancea earthquake. Springer Natural Hazards Series, Springer International Publishing, pp 267–283. https://doi.org/10.1007/978-3-319-29844-3_30

Presignal Signature of Radon (Rn^{222}) for Seismic Events

Maria Zoran, Roxana Savastru, Dan Savastru and Doru Mateciuc

Abstract Rock microfracturing in the Earth's crust preceding a seismic rupture may cause local surface deformation fields, rock dislocations, charged particle generation and motion, electrical conductivity changes, radon and other gases emission, fluid diffusion, electrokinetic, piezomagnetic and piezoelectric effects as well as climate fluctuations. Space-time anomalies of radon gas emitted in underground water, soil and near the ground air weeks to days in the epicentral areas can be associated with the strain stress changes that occurred before the occurrence of medium and strong earthquakes. This paper presents some results of continuous monitoring of radon in air near the ground with short term (ten days exposure time) solid state nuclear track detectors (SSNTD) CR-39 at seismic stations Plostina (Vrancea), and Bucharest Magurele, Romania. During 2012–2016 periods, radon concentration anomalies along with meteorological parameters were found to be statistically significant for the seismic events within the moment magnitudes Mw $\geq$ 5.0 and epicentral distances of 15–200 km for the Vrancea source. The frequent registered positive anomalies with constant environmental perturbation indicate the opening and closing of micro cracks within the volume of dilatancy by strain stress energy, result which is very important for short term earthquake prediction.

M. Zoran (✉) · D. Savastru
National Institute of R&D for Optoelectronics, MG5, Magurele, Romania
e-mail: maria@dnt.ro

D. Savastru
e-mail: dsavas@inoe.ro

R. Savastru
National Institute of R&D for Optoelectronics,
MG5, Atomistilor 409, Magurele, Romania
e-mail: rsavas@inoe.ro

D. Mateciuc
National Institute of R&D for Earth Physics, Magurele, Romania
e-mail: dmateciuc@infp.ro

© Springer International Publishing AG, part of Springer Nature 2018
R. Vacareanu and C. Ionescu (eds.), *Seismic Hazard and Risk Assessment*,
Springer Natural Hazards, https://doi.org/10.1007/978-3-319-74724-8_8

117

Keywords Radon anomalies · Solid state nuclear track detectors
Seismic precursors · Vrancea region

1 Introduction

The earthquakes have been classified into five stages (precursor stages I–III; earthquake at stage IV; rapid stress-relief, aftershocks at stage V) with changes in geophysical/geochemical parameters associated with these stages. One of these parameters is radon emission considered as precursor, which increases during stage II, levels off in stage III and returns to normal in stage V. During stage II, microcracks form in the rocks resulting in an increase in the surface area of the rocks, as P-wave velocity decreases and the ground uplifts and tilts. The increased surface area exposes more of the radon to water in rocks and the radon can then be dissolved into it. The dissolved radon gas in water is forced out of the rock during stage II, when the radon may be released and then the microcracks act as escape pathways for the gas. During stage III, P-wave velocity increases, ground uplift and tilt decrease, microcracks stop forming and radon emissions decrease (Crockett et al. 2006). Changes during stage II allow scientists to make short/medium-term seismic predictionsfunction of radon abundance in different areas.

Understanding the earthquake cycle and assessing earthquake hazards is a topic of both increasing potential for scientific advancement and social urgency in active geotectonic areas in the world.

In order to minimize the earthquake hazard risk several attempts have been performed to establish earthquake precursory phenomena using sudden changes in radon emanation, crustal ground deformations, changes in electric resistivity and magnetic properties of the rocks, the level of groundwater table change, climate change fluctuations, etc. (Wakita et al. 1980; Gosh et al. 2009; Cicerone et al. 2009; Barman et al. 2016; Jilani et al. 2017).

Most earthquakes have precursors (defined as changes in the Earth physical-chemical properties that take place prior to an earthquake). Earthquake precursors emerging as a strong earthquake approaches are fuzzily differentiated by the characteristic lead-time between precursor and the strong earthquake: long-term (LT), tens of years; intermediate-term (IT), years, short-term (ST), months; and immediate (Im), days and less. This gradation reflects a more general phenomenon: unfolding of seismicity in a cascade of earthquakes' clusters, culminated by a strong earthquake. Among several seismic precursors, radon has been suggested to be one possible presignal (Zoran et al. 2012a).

Radon (^{222}Rn) is a radioactive gas with a half life of 3.8 days that belongs to the uranium (^{238}U) decay series and distributes roughly universally throughout the Earth's crust. It is a daughter of radium (^{226}Ra) and can be transported effectively from deep layers of Earth to the surface by carrier gases. The great part of the natural radiation activity is not from radon itself, but from the short-lived alpha particle-emitting radon daughters, most notably ^{218}Po (radioactive T1/2 = 3 min), and ^{214}Po (radioactive T1/2 = 0.164 ms), along with beta particles from ^{214}Bi (T1/2 = 19.7 min), where T1/2 is physical half-life. Since radon does not react chemically to another elements and its origin is restricted mostly to land surface, it is used as an earthquake precursor or tracer for location of active seismic faults and investigating of atmospheric circulation (Zoran 1979; Zoran et al. 2012b).

Anomalous radon (Rn^{222}) emissions enhanced by forthcoming earthquakes is considered to be a precursory phenomenon related to an increased geotectonic activity in seismic areas. Rock microfracturing in the Earth's crust preceding a seismic rupture may cause local surface deformation fields, rock dislocations, charged particle generation and motion, electrical conductivity changes, radon and other gases emission, fluid diffusion, electrokinetic, piezomagnetic and piezoelectric effects as well as climate fluctuations. The strain changes occurring within the earth's surface during an earthquake is expected to enhance the radon concentration in soil gas. Space-time anomalies of radon gas emitted in underground water, soil and near the ground air weeks to days in the epicentral areas can be associated with that strain stress alterations that occurred before the occurrence of medium and strong earthquakes (Zoran 2002; Gosh et al. 2011).

Due to the subcrustal earthquakes located at the sharp bend of the Southeast Carpathians, Vrancea zone in Romania, placed at conjunction of four tectonic blocks is considered one of the most seismically active areas in Europe with a high potential of seismic hazard for the neighbouring countries. This paper aims to investigate temporal variations of radon concentration levels in air near or in the ground during 2012–2016 periods in relation with some meteorological parameters by the use of solid state nuclear track detectors (SSNTD) CR-39 for some important seismic events recorded in Vrancea region, Romania.

2 Geological Background and Seismicity of Vrancea Region

The Vrancea zone (Fig. 1) in Romania shows a peculiar source of seismic hazard, which represents a major concern in Europe, especially to neighbouring countries (Bulgaria, Hungary, Serbia, Republic of Moldavia etc.). It is placed at the sharp bend of the Southeast Carpathians being characterized by the occurrence of low moment magnitude (M_w) crustal (0–40 km depth) earthquakes ($M_w < 5.5$ and moderate seismic activity) and intermediate depth (70–200 km) strong earthquakes ($6 \leq M_w \leq 8$) in a narrow epicentral and hypocentral region.

The epicentral area is confined to about 30 km × 70 km, depth range 70 and 180 km within an almost vertical column. Deeper and shallower events have also been recorded but only with small magnitudes. The depth interval of the strong events is bounded by zones of low seismicity between 40 and 60 km and beneath 180 km. Four major events struck within this century: November 10, 1940, 7.7 M_w at depth of 150–180 km; March 4, 1977, 7.5 M_w at depth 90–110 km; Aug. 30,

Fig. 1 Vrancea active zone on geomorphologic map of Romania

1986, 7.2 M_w, depth 130–150 km; May, 30, 1990, 6.9 M_w, depth 70–90 km. The ruptured areas migrated from 150–80 km (1940) to 9–110 km (1977) to 130–150 km (1986) and to 70–90 km (1990) depth.

The depth interval between 110 and 130 km remained unruptured since at least 200 years. This depth interval is a natural candidate for the next strong Vrancea event (Matenco and Bertotti 2000; Raileanu et al. 2005).

Placed at the border of the great East-European Platform, Romanian territory is a region of a complex geological structure dominated by the presence of the Alpine Orogenic Belt of the Carpathian Mountains. This mountains arc has a spectacular change of direction just above the well-known Vrancea seismic region. Subduction of the Black Sea Sub-Plate under the Pannonian Plate produces faulting processes. The fault plane that results is oriented approximately parallel to the Carpathian Bend, i.e. NE–SW. The Peceneaga-Camena Fault, a deep crustal fracture with dextral slip, is considered to be North-Eastern boundary of the Moesian Platform. The Eastern unit of the Moesian Sub-Plate is characterized by a series of principal faults with a North-Western orientation and by a secondary system of faults orientated NE-SW. NW trending crustal fractures are also evidenced East of the Peceneaga-Camena Fault. The Black Sea Sub-Plate would have a NW displacement along the "markers" formed by the Moesian and Eurasian Sub-Plates only provided that we admit of another fault plane, oriented NW-SE, approximately perpendicular to the first (Wenzel et al. 1998; Bala et al. 2003) as can be seen on geologic map of Romania in Fig. 1. Seismic data have been provided by ROMPLUS catalogue (www.infp.ro/romplus) and (www.earthquakes.usgs.gov).

3 Materials and Methods

^{222}Rn radon gas, significant for seismo-tectonic applications is produced from his immediate parent radium (^{226}Ra) in rock grains that contains ^{238}U and its daughters in secular equilibrium. Are known different techniques for radon monitoring: nuclear emulsion, adsorption, solid scintillation, liquid scintillation, gamma spectrometry, beta monitoring, solid state nuclear track detectors (SSNTD), ionization chambers, surface barrier detectors, termoluminiscent detectors, electrets detectors, etc. Solid state nuclear track detectors (SSNTD) allowed integrated measurements of radon concentration by alpha tracks produced by its alpha active descendants. This paper used radon in air and soil monitoring for seismic investigation of Vrancea area.

For radon concentration measurements in surface air at 1 m height above the ground have been employed Radon Analytics SSNTDs type CR-39 detectors mounted in small wooden experimental houses in Vrancea (Plostina) seismic station located on active fault in Vrancea zone as well as in Bucharest-Magurele. Detectors were exposed for consecutive intervals of 10 days during 5 years period (2012–2016). Exposimeters CR-39 provided by Radon Analytics (http://www.radon-analytics.com), Germany used for radon concentrations monitoring in the air near the ground offer an integrating long-term measuring procedure suitable for extensive campaigns in Vrancea area. The measuring range of approx. 15 Bq/m^3 up to over 5.000 Bq/m^3 enables the accurate determination of all radon concentrations usually occurring in air just above the ground in seismic fault zones. The detector itself—a plastic film, SSNTD CR-39 type consisting of a polycarbonate foil (approx. 1 to 10 cm^2 in size, approx. 0.1 to 1 mm thick)—is located inside the exposimeter housing. Air containing radon diffuses through the housing into the measuring chamber; radon decay products, dust and humidity are retained here. In addition, the sufficiently large diffusion resistance prevents the intrusion of thoron (Rn-220; half-life: approx. 55 s). In the measuring chamber radon decays emitting alpha particles into its so called radon daughters. These short-lived decay products are alpha-emitters too. The alpha particles create microscopic tracks in the film, which are chemical etched with the standard method and optically counted in the laboratory. A specific radon concentration and corresponding uncertainty are calculated in Bq/m^3, based on the number of the tracks, the size of the detector and the exposure time (Zoran et al. 2012b). Meteorological data have been provided by National Administration of Meteorology (http://anm.meteoromania.ro/) and seismic stations data.

4 Results

The purpose of this study was to investigate the relationship between radon concentration in the air near the ground and earthquake data, taking into account the meteorological factors in active seismic zone Vrancea in Romania. In this paper we evaluated radon data seasonally, yearly and as long term, to build a relationship with tectonic movements. Radon in air near the ground measurements were taken continuously with CR-39 SSNTDs both in Vrancea (Plostina station) as well as in Bucharest distant location throughout five years (2012–2016) period.

4.1 Seismicity

Regarding seismicity in Vrancea region during 1 January 2012–31 December 2016 investigated period have been recorded hundred events of moment magnitude in the range of $2.0 \leq M_w \leq 5.7$. As Fig. 2 and Table 1 show, during 5 years period have been recorded 56 events of moment magnitude Mw in the range of $4.0 \leq M_w \leq 5.7$ on Richter scale, of which only three earthquakes with $M_w \geq 5.0$. The crustal stress pattern of the SE-Carpathian region is expected to reflect both crustal tectonics as well as deep-seated mantle processes resulting from the Vrancea slab. It seems that the crustal stress pattern of Romania is characterized by small differential horizontal stresses where local stress sources are responsible for the observed heterogeneity of stress orientations and that the subducted slab under Vrancea is only weakly coupled to the crust. The coupling of the subducted Vrancea slab to the crust is estimated to be weak. The seismic gap between 40 and 70 km could be explained as a zone of weakness where no major earthquakes will occur.

The crustal seismicity evidenced an increase of microseismic events at the depths of 67–173 km in Vrancea area (registered at seismic stations Plostina and Vrancioaia,), but also an increase of crustal earthquakes number in the range of 0–30 km depth. Figure 2 presents seismicity evolution (Depth/Moment magnitude M_w at seismic station Plostina, Vrancea zone, during time period 1 January 2012–31 December 2016.

4.2 Radon (Rn^{222}) in Air Near the Ground Concentration Variability

Earthquakes (EQ) are large-scale fracture phenomena arising in the Earth's crust, the occurrence of which is perceived in the form of a sudden violent shaking of the Earth's surface. It is considered that during and before earthquake radon gas is removed by underground rocks or waters from cracks in the Earth's crust just made

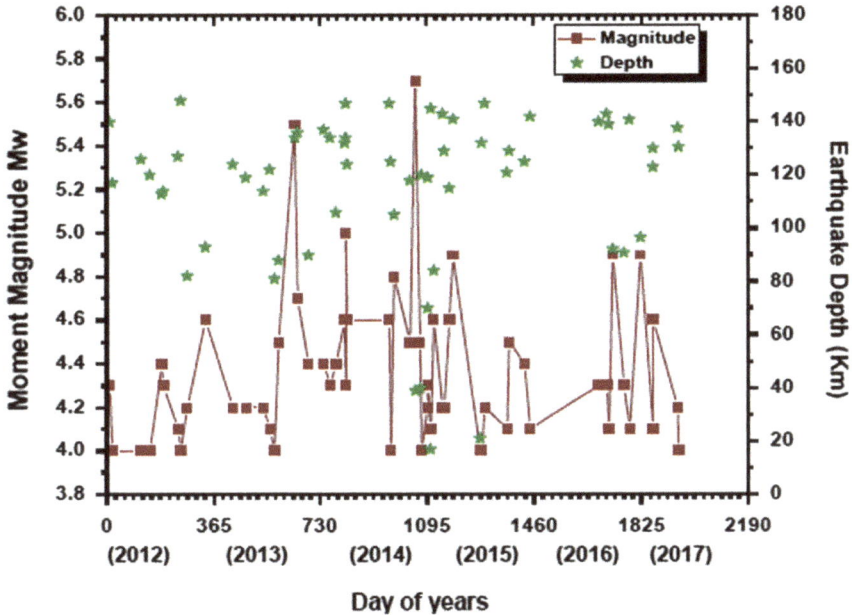

Fig. 2 Seismicity evolution in Vrancea region during 2012–2016 period

in associated deformation processes. The close correlation of changes of radon concentration with time in underground waters, soils, and air above the ground in active geotectonic areas with movements of the Earth's crust before earthquakes is evidence of that. According to an analogous mechanism, the zones of active tectonic breaks are characterized by anomalous radon concentration. Measurement of the radon concentration is widely used in seismic testing areas to study active tectonic breaks and earthquake precursors.

An attempt was made to establish a relationship between the intensity of radon anomaly and size of impending earthquake that may be valid for a specified source distance range from the recording site. Analysis of in situ observations at Plostina in Vrancea test site and Bucharest distant station revealed an increase of air radon concentration just above the ground before seismic events of moment magnitude $4.0 \leq M_w \leq 5.7$ registered during monitoring periods, followed by a sudden decrease (Fig. 3). Higher ^{222}Rn concentrations have been related with three seismic events of moment magnitude Mw ≥ 5.0 recorded during 2012–2016 period.

The average ten days mean radon concentration C_{Rn} in air above the ground measured with CR-39 detectors recorded for five years period in Vrancea area was 528.77 ± 89.19 Bq m^{-3}, and ten days C_{Rn} fluctuations in the range of 71 ± 20 Bq m^{-3} and 2854 ± 300 Bq m^{-3}, while the average ten days mean radon concentration C_{Rn} in air above the ground measured with CR-39 detectors recorded for the same period in Bucharest-Magurele area was 604.02 ± 97.47 Bq m^{-3} and ten days C_{Rn} fluctuations in the range of 71 ± 15 Bq m^{-3} and

Table 1 Earthquake events recorded in Vrancea region during 2012–2016 periods

Earthquake Nr.	Day of years	Data	Moment magnitude M_w	Depth (km)	Earthquake Nr.	Day of years	Data	Moment magnitude M_w	Depth (km)
1	10	10.01.2012	4.3	140	29	984	10.09.2014	4.8	105
2	19	19.01.2012	4	117	30	1038	03.11.2014	4.5	118
3	117	26.04.2012	4	126	31	1057	22.11.2014	5.7	39
4	147	26.05.2012	4	120	32	1072	07.12.2014	4.5	40
5	188	06.07.2012	4.4	113	33	1077	12.12.2014	4	120
6	194	12.07.2012	4.3	114	34	1099	03.01.2015	4.2	70
7	244	31.08.2012	4.1	127	35	1100	04.01.2015	4.3	119
8	253	09.09.2012	4	148	36	1108	12.01.2015	4.1	17
9	275	01.10.2012	4.2	82	37	1110	14.01.2015	4.1	145
10	336	01.12.2012	4.6	93	38	1120	24.01.2015	4.6	84
11	431	06.03.2013	4.2	124	39	1150	21.02.2015	4.2	143
12	477	21.04.2013	4.2	119	40	1156	27.02.2015	4.2	129
13	537	20.06.2013	4.2	114	41	1173	16.03.2015	4.6	115
14	559	12.07.2013	4.1	122	42	1186	29.03.2015	4.9	141
15	575	28.07.2013	4	81	43	1276	29.06.2015	4	21
16	589	11.08.2013	4.5	88	44	1282	05.07.2015	4	132
17	645	06.10.2013	5.5	134	45	1292	15.07.2015	4.2	147
18	654	15.10.2013	4.7	136	46	1368	29.09.2015	4.1	121
19	691	21.11.2013	4.4	90	47	1376	07.10.2015	4.5	129
20	743	12.01.2014	4.4	137	48	1429	29.11.2015	4.4	125
21	765	03.02.2014	4.3	134	49	1448	18.12.2015	4.1	142
22	786	24.02.2014	4.4	106	50	1678	04.08.2016	4.3	140
23	816	26.03.2014	4.6	132	51	1706	01.09.2016	4.3	143

(continued)

Table 1 (continued)

Earthquake Nr.	Day of years	Data	Moment magnitude M_w	Depth (km)	Earthquake Nr.	Day of years	Data	Moment magnitude M_w	Depth (km)
24	819	29.03.2014	4.3	147	52	1713	08.09.2016	4.1	139
25	819	29.03.2014	5	134	53	1728	23.09.2016	4.9	92
26	824	03.04.2014	4.6	124	54	1766	31.10.2016	4.3	90.8
27	967	24.08.2014	4.6	147	55	1785	19.11.2016	4.1	140.8
28	972	29.08.2014	4	125	56	1823	27.12.2016	4.9	96.6

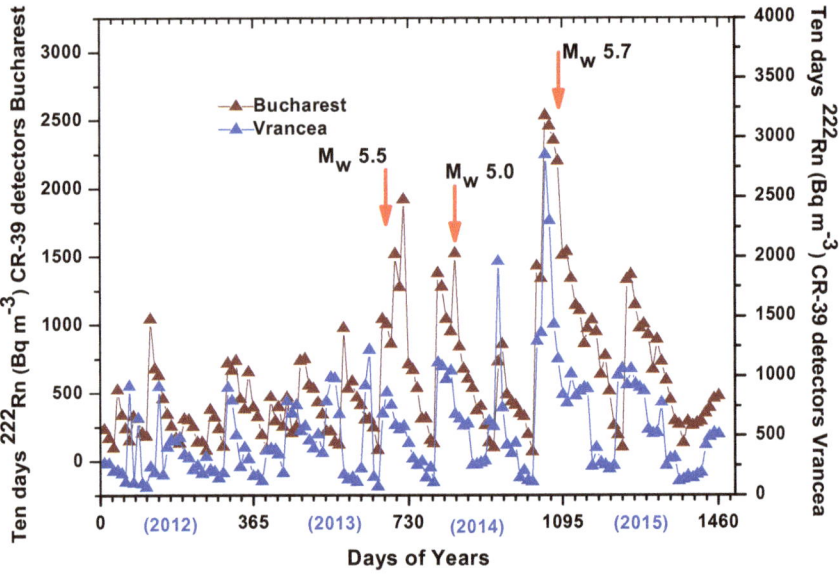

Fig. 3 Temporal evolution of ten days ^{222}Rn concentration recorded by CR-39 SSNTD at Vrancea and Bucharest stations during 2012–2015 period in relation with seismic events of moment magnitude Mw $\geq$ 5.0

2537 $\pm$ 330 Bq m^{-3}. The values of ^{222}Rn in the air near the ground at Plostina and Vrancioaia stations are placed in the close ranges. Changes in the air near the ground radon concentrations may be associated with the changes in radon in soil regime and geological settings in the epicentral and surrounding regions.

At both sites (Vrancea and Bucharest), the radon concentration was low in the summer and high in the winter. It was done the assumption of a ^{222}Rn anomaly when its concentration C_{Rn} is satisfying the following relationship (1):

$$C_{Rn} \geq Mean\ C_{Rn} \pm 2\ s \tag{1}$$

where s is standard deviation. The variation of the radon concentration was examined between the mean values and plus/minus two standard deviations. The standard deviation of the radon measurements was about 10% of the average radon concentration. During the monitoring of radon in air above the ground concentration, numerous anomalies that equal or twice standard deviations were observed. Any increase in radon concentration above this limit was assumed to be ^{222}Rn anomaly. These anomalies usually appeared between a few days or weeks before the moderate/strong earthquakes occurrence. Furthermore, in the data sets for all three sites, anomalous radon changes were observed before and after the three recorded earthquakes (6 October 2013, moment magnitude M$_w$ = 5.5, depth 134 km; 29 March 2014, moment magnitude M$_w$ = 5.0, depth 134 km; 22 November 2014, moment magnitude M$_w$ = 5.7, depth 39 km). Clear radon anomalies (with sharp increase week before and decline after the quakes) have been observed at Vrancea stations in epicentral area as well as in Bucharest before these three moderate recorded earthquakes as well as for other earthquakes of $4.5 \leq M_w \leq 5.0$. The duration of recorded radon anomalies is linearly related to the earthquake preparation radius and inversely to the distance from earthquake focus to the radon measuring site. This observation supports the theory that radon in air above the ground could be used as a presignal of seismic events. As the radon near the ground concentration values taken shortly after that seismic events showed a sudden decrease returning to normal concentrations as before earthquakes, we concluded that these anomalies could be attributed to the seismic activity. This finding is in a good accordance with scientific literature in the world. An attempt was made to establish a relationship between the intensity of radon anomaly and size of impending earthquake that may be valid for a specified source distance range from the recording site. The statistical analysis of measured time series ^{222}Rn concentrations in air near the ground was correlated with moment magnitude of recorded seismic events during measuring periods. The relative changes of radon concentration from its average for a particular period of the year is depending on the magnitude and depth of the impending earthquake, the distance of recording station from the earthquake source, and seasonal changes of meteorological parameters (temperature, pressure, humidity).

Prior to the seismic events the earthquake precursors appear at different distances and heights over the active seismogenic areas. It seems that the earthquake preparation area on the ground can be estimated according to the relation (2):

$$R = 10^{0.43 M_w} \tag{2}$$

where R is the radius of the preparation zone in Km and M_w is the earthquake moment magnitude. For Vrancea earthquake of 22 November 2014 with $M_w = 5.7$, the corresponding calculated radius would be $R \approx 282.5$ km in the latitudinal and longitudinal directions from the epicenter, for earthquake 6 October 2013, moment magnitude $M_w = 5.5$ the corresponding calculated radius would be $R \approx 231.7$ km, while for earthquake 29 March 2014 earthquake, moment magnitude $M_w = 5.0$ the corresponding calculated radius would be $R \approx 141.3$ km in the latitudinal and longitudinal directions from the epicentre. As Bucharest station is located at almost 150 km of Vrancea seismic area, it seems that preparation area of a moderate/strong earthquake will start from about the same distance of Vrancea. Scatter plots of monthly mean of ten days ^{222}Rn activity concentrations in air near the ground presented in relation with main meteorological parameters (air temperature, pressure and wind speed are presented in Fig. 4, which also show a seasonal variation at Vrancea station with a minimum in the late spring and summer, then gradually increases in autumn and winter.

The time-series variation in the radon concentration near the ground in Vrancea zone was found to be affected by atmospheric turbulence which was linked to the variation in the air surface temperature and wind intensity. The relatively positive high correlation ($R^2 = 0.68$) between ^{222}Rn concentrations in air near the ground in

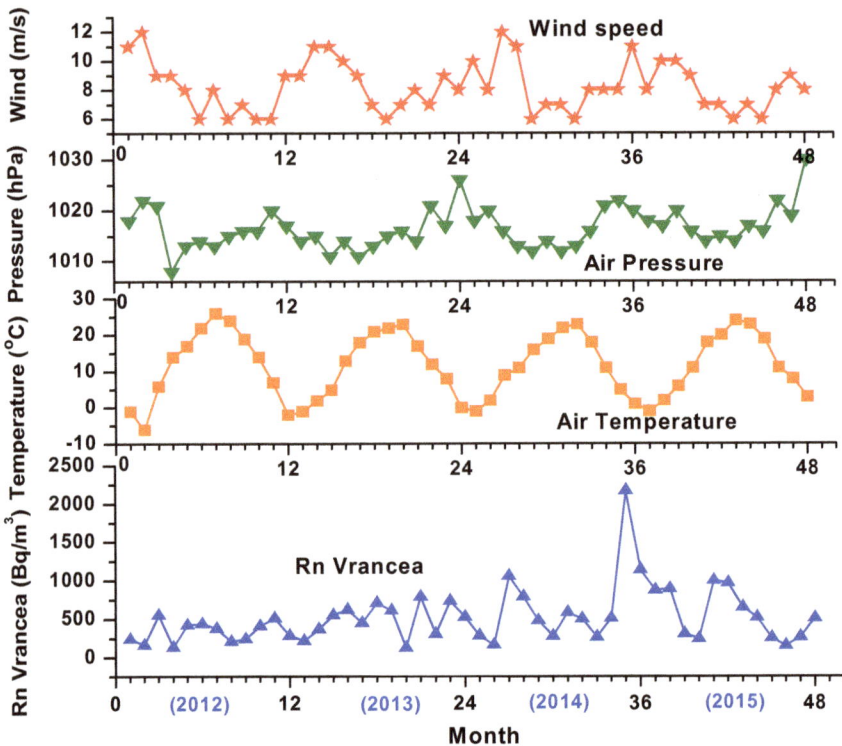

Fig. 4 Temporal evolution of ten days ^{222}Rn concentration recorded by CR-39 SSNTD at Vrancea in relation with meteorological parameters during 2012–2015 period

Vrancea and Bucharest stations suggests that for moderate recorded earthquakes during 2012–2016 period the change of radon in Vrancea corresponds well to the change of the radon concentration in Bucharest. The results are in good accordance to theory proposed by Dobrovolski et al. (1979) who assumed that when stress conditions changes rapidly, the change in radon concentration is proportional to strain change. The increases in the atmospheric radon may reflect the formation of microcracks in the granitic rocks in the geotectonic active area Vrancea in Romania.

Sudden changes in radon near the ground concentration, which clearly exceed the diurnal and seasonal variations occur and are always linked with seismic activities. Decreased gas concentration may indicate compression resulting in reduced fault permeability as is implied by negative peaks following local earthquake swarms. A sudden increase in radon concentrations may indicate an increased fault permeability caused by stress redistribution, giving rise to opening of migration pathways. This implies a repeatedly sudden rise in gas concentration before local earthquake swarms. So, changes in radon emission, or in electrical or magnetic fields, represent a natural amplification of pre-earthquake deformation under special geological conditions. Fluid, gas or electromagnetic measurements might thus detect deformation and strain indicators indirectly at localized sites and with amplitudes related nonlinearly to strain

5 Conclusions

Investigation of radon concentration anomalies in air near the ground in seismic areas is an important issue in the field of earthquake short term forecasting of Vrancea moderate or strong earthquakes in Romania. The recorded radon anomalies presignals of earthquakes during four years monitoring period performed with solid state nuclear track detectors CR-39 (short term-10 days exposure time) suggest that earthquake precursors registered before moderate or strong seismic events are associated with some physical processes in or near the Vrancea earthquake fault zones or its neighbouring. It can be concluded that some of the short-term radon anomalies can be correlated with seismic events, when they had magnitudes of $M_w \geq 5$ or greater. The association between radon concentration anomalies and major tectonic faults in Vrancea region could result from mechanical cracks in the rocks or too slow crack growth determined by local strain of the media. Mostly with several days up to weeks before three moderate recorded earthquakes have been observed anomalous radon in the air near the ground variations both in seismic zone Vrancea as well as in Bucharest-Magurele area located at almost 150 km far of Vrancea. This paper considered also the effects of meteorological parameters (air temperature, pressure, relative humidity, wind intensity and rainfall) on radon in air near the ground concentrations. Understanding the seasonal and weather-related influences on the atmospheric radon concentration variations in geotectonic active area Vrancea is essential for studying the relationship between seismic activity and radon, which might be considered a preseismic crustal strain indicator and an important earthquake surveillance tool for Romania.

Efforts to advance understanding of earthquake prediction in Romania require detailed observation of all phases of the earthquake cycle (pre-, co-, and post-seismic), across multiple fault systems and tectonic environments, with integrated knowledge of all geophysical, geochemical, geological, seismological, geodinamic parameters. Unique "precursory fingerprints of radon" for individual seismic geologic structures are expected to be pointed out as an outcome of target-oriented strategic prediction research. Due to its particular features and geotectonic setting, the Vrancea seismic structure in Romania appears to be an excellent experimental laboratory for prediction research.

Acknowledgements This work was supported by Romanian National Authority for Scientific Research by grant CNDI-UEFISCDI, project number PN-II-PT-PCCA-86/2014 VRAGEO and Program NUCLEU Contract 5N/09.03.2016, under 16 40.01.01 research theme.

References

Bala A, Radulian M, Popescu E (2003) Earthquakes distribution and their focal mechanism in correlation with the active tectonic zones of Romania. J Geodyn 36:129–145

Barman C, Ghose D, Sinha B, Deb A (2016) Detection of earthquake induced radon precursors by Hilbert Huang Transform. J Appl Geophys 133:123–131

Cicerone RD, Ebel JE, Britton J (2009) A systematic compilation of earthquake precursors. Tectonophysics 476:371–396

Crockett RGM, Gillmore GK, Phillips PS, Denman AR, Groves-Kirkby CJ (2006) Radon anomalies preceding earthquakes which occurred in the UK, in summer and autumn 2002. Sci Total Environ 364:138–148

Dobrovolsky IP, Zubkov SI, Miachkin VI (1979) Estimation of the size of earthquake preparation Zones. Pure appl Geophys 117:1025–1029

Gosh D, Deb A, Sengupta R (2009) Anomalous radon emission as precursor of earthquake. J Appl Geophys 69:67–81

Ghosh D, Deb A, Sahoo SR, Haldar S, Sengupta R (2011) Radon as seismic precursor: new data with well water of Jalpaiguri, India. Natural Haz 58(3):877–889

Jaishi HP, Singh S, Tiwari RP, Tiwari RC (2014) Correlation of radon anomalies with seismic events along Matfault in Serchhip District, Mizoram, India. Appl Radiat Isot 86:79–84

Jilani Z, Mehmood T, Alam A, Awais M, Iqbal T (2017) Monitoring and descriptive analysis of radon in relation to seismic activity of Northern Pakistan. J Environ Radioact 172:43–51

Matenco L, Bertotti G (2000) Tertiary tectonic evolution of the external East Carpathians (Romania). Tectonophysics 316:255–286

Raileanu V, Bala A, Hauser F, Prodehl C, Fielitz W (2005) Crustal properties from S-wave and gravity data along a seismic refraction profile in Romania. Tectonophysics 410:251–272

Wakita H, Nakamura Y, Notsu K, Noguchi M, Asada T (1980) Radon anomaly: a possible precursor of the 1978 Izu-Oshimakinkai earthquake. Science 207:882–883

Wenzel F, Achauer U, Enescu D, Kissling E, Russo R, Mocanu V, Musacchio G (1998) Detailed look at final stage of plate break-off is target of study in Romania. Eos, Trans Am Geophys Union 79(48):589

Zoran M (1979) Anomalous high concentrations of radon and its alpha active descendents in the lower atmosphere afterward strong Romanian earthquake from March 4, 1977. In: Cornea C, Radu I (eds) Seismological researches for Romanian earthquakes from March 4, 1977, pp 447–452

Zoran M (2002) Radon in soil variations for Vrancea seismic area. Revue Roumanie de Geophysique 46:111–118

Zoran M, Savastru R, Savastru D (2012a) Radon levels assessment in relation with seismic events in Vrancea region. J Radioanal Nucl Chem 293:655–663

Zoran M, Savastru R, Savastru D, Chitaru C, Baschir L, Tautan M (2012b) Monitoring of radon anomalies in South-Eastern part of Romania for earthquake surveillance. J Radioanal Nucl Chem 293(3):769–781

Abnormal Animal Behavior Prior to the Vrancea (Romania) Major Subcrustal Earthquakes

Angela Petruta Constantin, Iren-Adelina Moldovan
and Raluca Partheniu

Abstract The goal of this paper is to present some observations about the abnormal animal behavior prior to and during some Romanian subcrustal earthquakes. The major Vrancea earthquakes of 4 March 1977 (M_W = 7.4, Imax = IX–X MSK), 30 August 1986 (M_W = 7.1, I_0 = VIII–IX MSK) and 30 May 1990 (M_W = 6.9, I_0 = VIII MSK), were preceded by extensive occurrences of anomalous animal behavior. These data were collected immediately after the earthquakes from the affected areas. Some species of animals became excited, nervous and panicked before and during the earthquakes, such as: dogs (barking, howling and running in panic), cats, birds (hens, geese, turkey hens, ducks, pigeons, parrots), cattle, pigs, mice and rats (came into the houses and have lost their fear), horses, fishes, snakes etc. These strange manifestations of the animals were observed on the entire territory of the country, especially in the extra-Carpathian area. This unusual behavior was noticed within a few seconds to days before the seismic events, but for the most of cases the time of occurrence was within two hours prior to the quakes. We hope that one day the abnormal animal behavior will be documented enough in order to be used as a reliable seismic precursor for the intermediate depth earthquakes.

Keywords Unusual animal behavior · Vrancea intermediate depth earthquakes Macroseismic questionnaires

A. P. Constantin (✉) · I.-A. Moldovan · R. Partheniu
National Institute of Research, Development for Earth Physics,
Magurele, Romania
e-mail: angela@infp.ro

I.-A. Moldovan
e-mail: irenutza_67@yahoo.com

R. Partheniu
e-mail: raluca@infp.ro

© Springer International Publishing AG, part of Springer Nature 2018
R. Vacareanu and C. Ionescu (eds.), *Seismic Hazard and Risk Assessment*,
Springer Natural Hazards, https://doi.org/10.1007/978-3-319-74724-8_9

1   Introduction

For millennia, in high seismic regions, people are reporting strange animal behavior before, during and after significant earthquakes.

Beside unusual animal behavior, those who have survived major earthquakes report also the occurrence of other mysterious phenomena such as: plant behavior, lightning, big noise, strange clouds. It seems that the concept that animals can sense earthquakes before they occur may have originated in ancient Greece in 373 B.C. when rats, weasels, snakes, and centipedes were reported to have moved to safety several days before a destructive earthquake (Quammen 1985; Schaal 1988). In the next centuries many other nations associated the occurrence of the earthquakes with the animals and their abnormal behaviour.

Starting with the 70s, earthquake prediction based on animals premonitory behaviour has been attempted in many countries, in present China and Japan being leaders. As in a number of other countries (Allen et al. 1975, 1976; Evernden 1976, 1980; Rikitake 1976, 1978; Ling-Huang 1978; Nikonov 1981; Buskirk et al. 1981; Tributsch 1982), in the 70s a great interest was also shown by Romanian scientists in biological phenomenon (geophysical, geological, biological) associated to the earthquakes, and primarily in abnormal animal behavior (AAB) prior to the earthquakes, especially after the 1977 Vrancea earthquake, taking as an example the success of the seismic prediction using these abnormal animal manifestations and geophysical measurements in the case of 1975 Haicheng (China) earthquake. In fact, this earthquake is considered the only major earthquake that has been ever predicted (Wang et al. 2006). This way, a review has been made and the information regarding the abnormal biological behaviour prior to, during and after the earthquakes in Romania from historical times has been examined (Radu and Spanoche 1976; Radu et al. 1979b). In that study, the authors have used information about 270 earthquakes occurred between 1471 and 1977, from which 11 were preceded by biological phenomena.

In 1997, in the framework of the Department of the Magnetotelluric and Bio-seismic Studies, a bio-laboratory was founded at the Vrancioaia seismic station for monitoring the abnormal behavior of various species of animals. In the recent years, a project entitled: "Monitoring methods of the animal behavior as a precursor for the short-term prediction of the strong earthquakes" has been developed in this field. The major result of this project consists in the achievement, for the first time, of a database concerning animal behavior prior to earthquakes that is compatible with the seismic data. The goal of this project was to use the animal behaviour as a precursor factor for the seismic prediction of the Vrancea strong earthquakes.

This research field has registered at international level a growing number of papers studying animal behavior in relation to seismic events (i.e. Rikitake 1998; Ikeya 2000, 2004; Wang et al. 2006; Bhargava et al. 2009, 2010, 2011; Hayakawa 2013; Berberich et al. 2013; Freund and Stolc 2013; Apostol et al. 2014 etc.).

This type of AAB was notice during all the major and moderate Vrancea intermediate depth earthquakes occurred in the 20th and 21st centuries in Romania.

The major Vrancea earthquakes of 4 March 1977, 30 August 1986, and 30 May 1990 were preceded by extensive occurrences of anomalous animal behaviour. Some observations about AAB prior to and during the above earthquakes are presented in this paper.

2 Vrancea Subcrustal Earthquakes

2.1 The March 4, 1977 Earthquake

The Vrancea subcrustal earthquake occurred during the evening of 4 March 1977, at 19:21 GMT (21:21 local time), with the magnitude Mw = 7.4 (Romplus catalogue 2017), was one of the strongest in Romania, with the epicentre in the bending area of the South-Eastern Carpathians, a region known as Vrancea seismogenic zone. The epicentral intensity was assigned at I_0 = VIII MSK (Radu et al. 1979a) and maximum intensity assigned after the revision of the macroseismic effects was IX–X MSK (Pantea and Constantin 2013).

The March 1977 earthquake represents the greatest tragedy in the past 40 years in Romania that resulted in major human and property losses. The greatest damages were concentrated in the southern part of the country, namely in urban centers with high population density and constructions, especially in cities such as Bucharest, Craiova, Zimnicea, Turnu Magurele etc. (Figs. 1 and 2). The earthquake caused the death of 1578 people, of which 1424 only in Bucharest. Over 10,000 people were found under the collapsed buildings in the capital of Romania while in the entire country about 11,300 were injured and about 35,000 buildings were severely damaged or collapsed.

2.2 The August 30, 1986 Earthquake

Next major earthquake occurred in Vrancea seismogenic zone, after strong 1940 and 1977 earthquakes (Pantea and Constantin 2011, 2013), was the intermediate depth earthquake from August 30, 1986, at 23:28:37 (local time) at a depth of 133 km and epicentral coordinates: latitude 45.53 N and longitude 26.47E (Radu 1988). The magnitude of the seismic event was Mw = 7.1. The epicentral intensity was assigned at I_0 = VIII–IX MSK (Radu et al. 1987). The main shock was followed by 124 aftershocks. The August 30 seismic event was felt on a large area and it caused serious damage in epicentral area, including the collapse of a church. Two people killed, 558 injured, and about 55,000 homes partially damaged. The largest aftershock (M_{GR} = 4.2) occurred at about 52.5 h after the main shock. The August 30 earthquake was preceded by a magnitude M_{GR} = 5 event on August 16.

Fig. 1 Partial collapse of some buildings in Bucharest after the 1977 earthquake (photo by Ene 1977)

Fig. 2 Partial or total collapse of some buildings [**a** masonry house and **b** church] in Prahova county after the 1977 earthquake (photos by Pantea 1979)

2.3   The May 30, 1990 Earthquake

The last major intermediate depth earthquake occurred in the 20th century was that of May 30, 1990 generated in the NE part of the Vrancea zone, same as those of March 1977 and November 1940. The shock occurred at 13:40:06 (local time) at a depth of 90 km with magnitude Mw = 6.9, I_0 = VIII MSK and epicentral coordinates: Latitude 45.82 N and Longitude 26.90 E (Romplus catalogue 2017). This was followed by a shock with Mw = 6.4 and I_0 = VII MSK, which occurred on May 31 at 03:17:48.6 (local time), with the focus at a depth of 79 km and the epicenter located at latitude 45.83 N and longitude 26.89 E (Romplus catalogue 2017). The May 31, 1990 earthquake can be considered an aftershock of the May 30, 1990 earthquake, but, as well, it can be considered the pair of the main shock, in this case the sequence is a doublet. The seismic event of May 31, 1990 occurred after 13.6 h and it can be considered the second shock in a slow rupture process.

3   Data Collection About Unusual Animal Behaviour

Information about macroseismic effects and evidence of abnormal animal behavior were collected after the seismic events by questioning people from the entire territory of Romania sending out macroseismic questionnaires (MQ). In the questionnaire used for the study of the effects of these earthquakes some questions regarding the animals behavior prior to earthquake are included (Constantin et al. 2016a, b). Two pages of an example of MQ filled after the 1977 earthquake are presented in Fig. 3.

Even if the reported observations were done by non-specialists, most of the information existing in MQs is accurate. In many MQs, the observers particularly described the types of manifestations, the coincidence of facts received from different persons, even pulling attention to animal behavior and debating with other persons before the earthquakes.

As it can be seen in the MQ used also for collecting this type of data, the reports were related to groups or species of animals rather than to individuals (domestic animals in general such as poultry, cats, dogs in cities and villages, cattle, birds in cages, fishes in aquariums etc.).

There were some cases in which the respondents did not notice any changes in animal behavior before the earthquakes. Some animal owners did not observe or did not report any disturbances in animal behavior prior to, during and after the quakes.

Out of the 1650 localities surveyed for March 1977 earthquake, positive information about unusual animal behavior were received from 1560 localities. In the case of the 30 August 1986 from 970 localities, 765 reported anomalous animal behavior. Also for May 30, 1990 earthquake the useful data from 335 localities were collected (705 localities surveyed). The information came from cities, communes and villages. For the 1977 earthquake, out of the 1560 observed evidence,

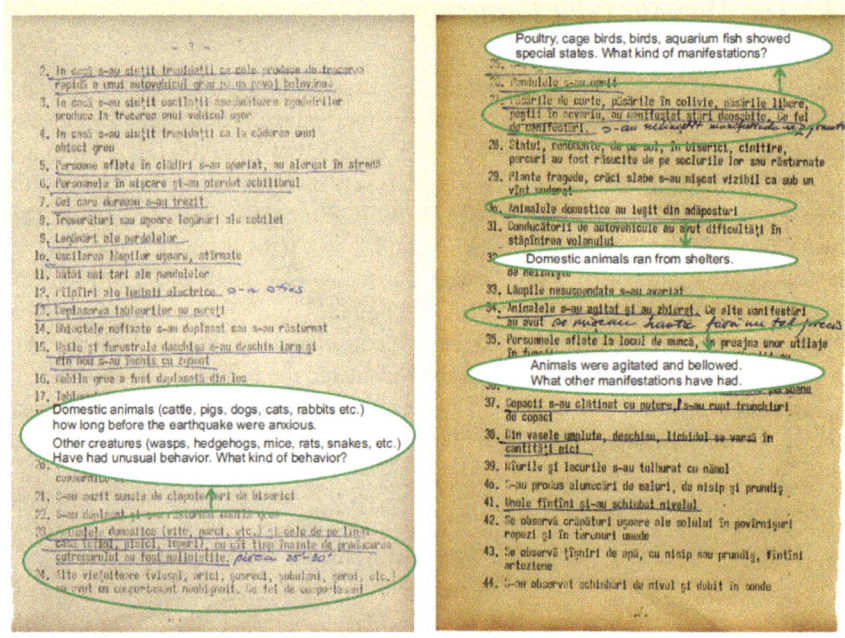

Fig. 3 Questions regarding AAB existing in the macroseismic questionnaire

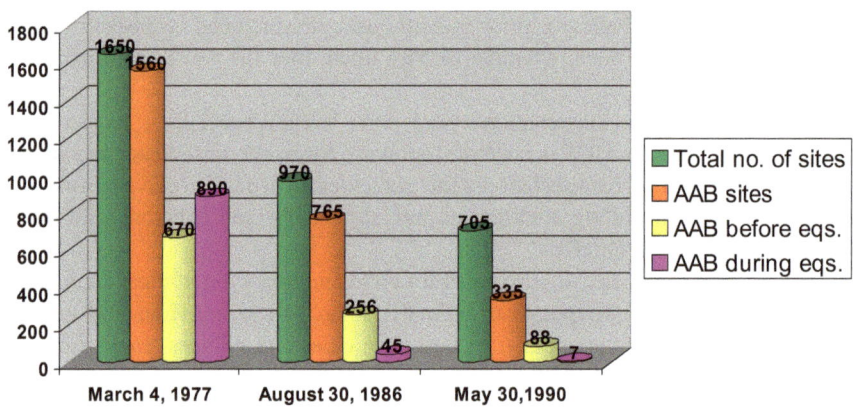

Fig. 4 The data set used for the study of the AAB

670 refer to the manifestations prior to the earthquake, and the rest of 890 during the earthquake (Fig. 4). In the case of 1986 earthquake 256 evidences were recorded before, and 45 during the earthquake. For 1990 earthquake, 88 observations were notice before and 7 during the seismic event. In Fig. 4 the statistics about the collected data used for this study is presented.

4 Abnormal Animal Behavior Before and During the 1977, 1986 and 1990 Earthquakes

The most common and typical manifestations counted as being unusual, were the distinctive signs for the majority of the animals: agitation and fear, the urge to leave the regular habitation place, and a refusal to eat. Some domestic animals indicated aggressive intentions. Various classes and species of animals simply showed their own natural behavior. For example, dogs usually got nervous before the shocks, and barked, howled, whined, or ran. After the earthquakes, some of them ran away. Many scouted for protection in houses and stayed close to people after the events. The most general reports concerning the behavior of dogs was their barking and howling before, during and after the earthquakes. Poultry generally escaped from shelters, with their wings outstretched, squawking or refusing to go back in their henhouses within a few hours before the earthquakes. Cocks crowed, panicked, flapped their wings, and shrieked. Parrots were agitated and showed signs of panic.

Rats and mice were seen running in open spaces or entering or getting out of the houses and having lost their fear. Few isolated reports about the increased number of rats with 1–2 weeks before the quakes were given, and one day before or after the earthquakes rats seem to have vanished.

The manifestation of cats before, during and after the earthquakes point out that they became agitated. Some of them took a run outside or some hid in dark places and somehow behaved abnormally; some disappeared for a few days shortly prior to the shocks. The cats seemed to perceive the shake before people, and crouched in fright or escaped. There were cases when cats were not seen for 2 days after the shocks. Bellowing of the cattle before and at the time of the shocks was generally observed and reported. Moreover, the cattle and horses became unusually uneasy, snapped their bridles and broke loose. In some cases, within minutes before the earthquakes, they were banging against the doors of the shelters, trying to run. The sheep seemed troubled, and bleating restless. Birds flocked together or flew uneven, fishes swam wildly or dive on the bottom of the aquarium, and in some cases the fishes in rivers came up to the water surface and tried to leap out of it.

Observers recorded that animals were scared and escaped during the earthquakes, but in a different way compared to what had been observed in other circumstances. For example, dogs were not howling as they did before and during the earthquakes, they were only barking.

The majority of the reports concerning the unusual manifestations at these three earthquakes came from birds (poultry, pigeons, etc.), dogs, cattle, cats, etc. Among the species with a recorded abnormal behavior, birds, especially hens, geese, turkey hens, ducks, and dogs were more evidenced (Fig. 5).

There were some reports of rats and mice, and few information on the behaviour of fishes from aquariums. Also a few reports about the unusual behavior of wild animals from ZOO (wolves howled, bears had pulled the bars of their cages, herbivores got restless, etc.) were recorded.

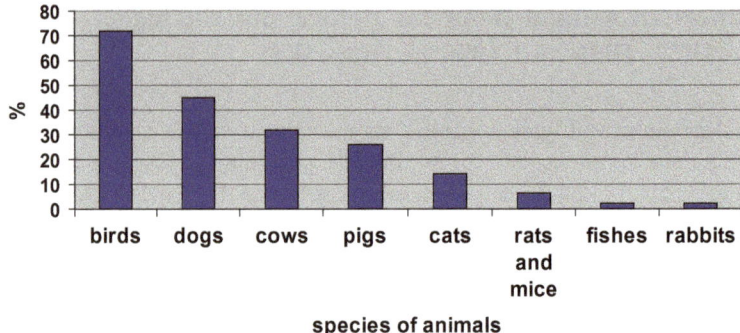

Fig. 5 Species of animals whose unusual behavior was recorded before and during the Vrancea earthquakes

The biggest precursor time period of all was observed in the Piatra Neamt city (Neamt county), near the epicentral zone, where dogs started to howl about one week before the 1977 earthquake.

Majority of the places from where the information was collected are located in the extra-Carpathian area. The helpful observations of unusual animal behavior were recorded hundreds of kilometers to the North, Southwest, and Northwest of the epicentral area. Some reports arrived from far distance, more precisely 400 km away from epicentral zone. The place furthest from the epicenter is situated 475 km away (Uivar, Timis). As Nikonov (1992) said, the premonitory abnormal behavior of a different species of animals was recorded, not only in the epicentral zone and areas with high intensity, but also where the macroseismic intensity was VI or less, possibly even in such places where people did not notice the earthquakes effects.

A few hours and minutes before the earthquakes the unusual animal manifestation became evident and happened on a mass scale (Table 1). Shortly before the shocks (seconds and minutes) all the respondents have noticed the strange animal manifestations and this could have been forecasted by any observer if they possessed proper knowledge of the existence of biological precursors. We need to point out that this strange behavior was revealed by the animals no matter where they lived: in the air, on the ground, in water and underground. The distribution of these observations reported by population for the Vrancea earthquakes studied in this paper, is presented in Figs. 6, 7, and 8.

Table 1 Distribution of animal behavior according to the epicentral distance and the time of occurrence before the Vrancea subcrustal earthquakes

Δ (km)	Time before earthquake				
	Few sec—5 min	10–30 min	1–10 h	1 day	Few days
Epicentral zone	Poultry, dogs, cats, cattle, pigs, mice	Chickens, dogs, cows, cats, pigs	Exotic fish, dogs, poultry, cats, rabbits	Cattle, rats	Rats, mice, dogs, cats
>100 km	Cattle, pigs, dogs, cats, poultry	Poultry, dogs, bird cage, rats, mice, horses, cattle, pigs, rabbits	Poultry, dogs, cattle, cats, rabbits, pigs, fish, horses	Dogs	Dogs
>200 km	Dogs, cage birds, poultry, rats, pigs, cows, sheep, cats, pigeons, rabbits	Dogs, poultry, pigs, cats, rats, hedgehogs	Poultry, sparrows, dogs, cats	Cattle, pigs, dogs, fish	Poultry, dogs, cats
>300 km	Poultry, dogs, cattle, pigs, cats, rabbits	Poultry	Turkeys, pigeons, dogs, cats, wasps, hedgehogs, cattle, pigs, rabbits		Dogs
>400 km	Mice, rats, poultry, dogs	Poultry, dogs, cats	Chickens, turkeys, geese		

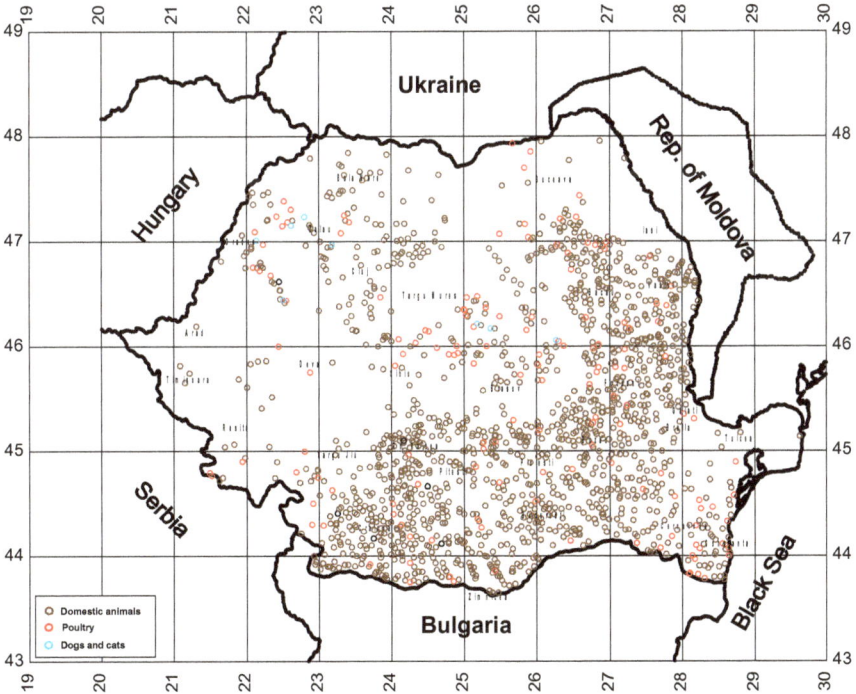

Fig. 6 Sites of recorded unusual animal behaviour before and during the 4 March 1977 Vrancea earthquake

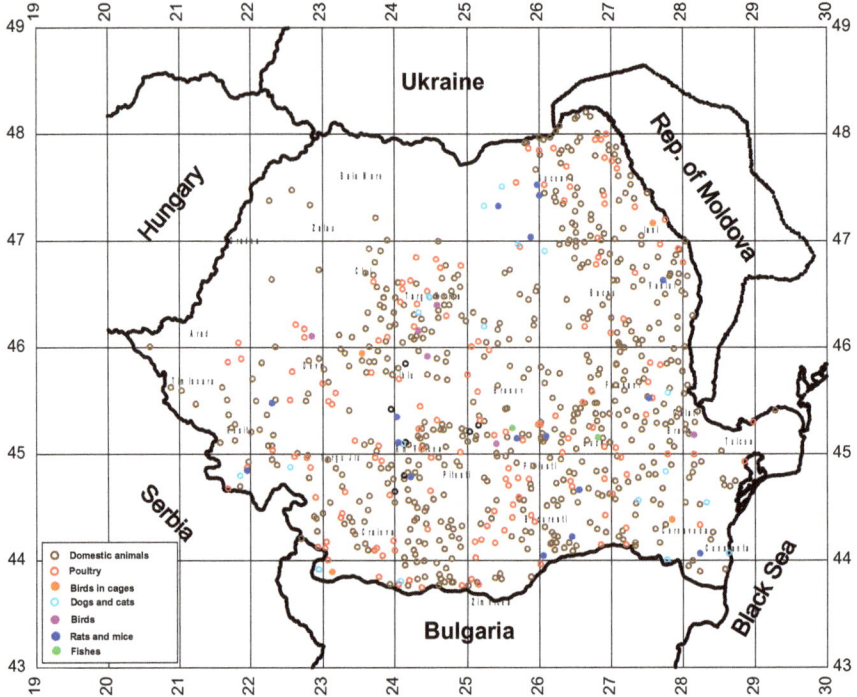

Fig. 7 Sites of recorded unusual animal behaviour before and during the Vrancea earthquake of 30 August 1986

5 Conclusion

The collected evidence of unusual animal behaviour before and during the Vrancea earthquakes can be regarded as sufficient to make a definite conclusion. The main conclusion is that the abnormal animal behaviour before the Vrancea earthquakes was so obvious and occurred on the whole Romania's territory that it would have been impossible to miss the precursors and perhaps the large number of casualties would have been avoided if the population would had been informed about such a precursor.

It is important to emphasize that all the evidences related to abnormal animal behaviour before the Vrancea earthquakes, is no different from those mentioned for other earthquakes in other seismic regions (Nikonov 1981, 1992; Rikitake 1978; Ling-Huang 1978 etc.).

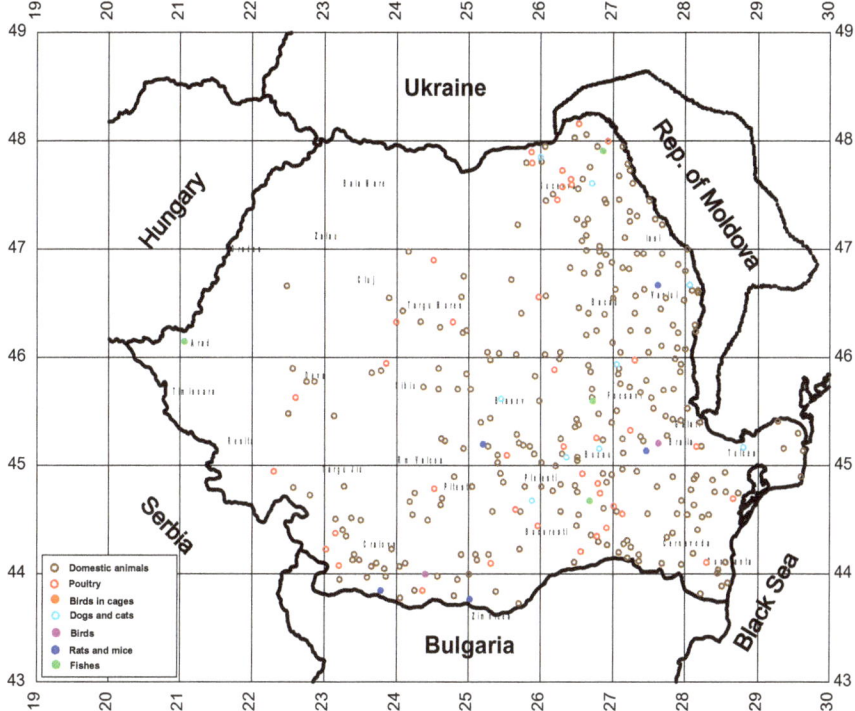

Fig. 8 Sites of recorded unusual animal behaviour before and during the Vrancea earthquake of 30 May 1990

The monitoring of unusual animal behaviour in all the Romanian seismic regions should be continued, in particular within Vrancea region where the intermediate depth earthquake forecasting has been made, it could become a reliable seismic precursor for these events.

Acknowledgements This paper was partially carried out within Nucleu Program, supported by ANCSI, projects no. PN 16 35 01 06, PN 16 35 03 01, PN 16 35 01 12 and the Partnership in Priority Areas Program—PNII, under MEN-UEFISCDI, DARING Project no. 69/2014.

References

Allen CR, Bonilla MG, Brace WF, Bullock M, Clough RW, Hamilton RM, Hofheinz R Jr, Kisslinger C, Knopoff L, Park M, Press F, Raleigh CB, Sykes LR (1975) Earthquake research in China. Eos Trans Am Geophys Union 56:838–879

Allen CR (Chair), Edwards W, Hall WJ, Knopoff L, Raleigh CB, Savit CH, Toksoz MN, Turner RH (1976) Predicting earthquakes: a scientific and technical evaluation—with implications for society. Panel on Earthquake Prediction of the Committee on Seismology,

Assembly of Mathematical and Physical Sciences, National Research Council, US National Academy of Sciences, Washington, DC

Apostol A, Moldovan IA, Constantin AP, Toader VE (2014) Are red wood ants earthquake predictors? In: Proceedings of Second European Conference on Earthquake Engineering an Seismology, 24–29 August, Istanbul, Turkey

Berberich G, Berberich M, Grumpe A, Wöhler C, Schreiber U (2013) Early results of three-year monitoring of red wood ants' behavioral changes and their possible correlation with earthquake events. Animals 3:63–84. https://doi.org/10.3390/ani3010063

Bhargava N, Katiyar VK, Sharma ML, Pradhan P (2009) Earthquake prediction through animal behavior: a review. Indian J Biomech 7–8:159–165

Bhargava N, Pradhan P, Sharma ML, Katiyar VK (2010) Electric charge developed by seismic stress on earthquake sources and its effect on animals. J Int Acad Phys Sci 14(2):205–214, ISSN 0974-9373

Bhargava N, Pradhan P, Katiyar VK, Sharma ML (2011) Unusual animal behavior due to acoustic waves generated by an earthquake. J Int Acad Phys Sci 15(4):435–445. ISSN 0974-9373

Buskirk RE, Frohlich K, Litham GV (1981) Unusual animal behavior before earthquakes: a review of possible sensory mechanisms. Rev Geophy Space Phy 19:247–270

Constantin AP, Partheniu R, Moldovan IA (2016a) Macroseismic intensity distribution of some recent Romanian earthquakes. Rom J Phys 6(5–6):1120–1132

Constantin AP, Moldovan IA, Craiu A, Radulian M, Ionescu C (2016b) Macroseismic intensity investigation of the November 2014, M=5.7, Vrancea (Romania) crustal earthquake. Ann Geophys 59(5):0542. https://doi.org/10.4401/ag-6998

Ene M (1977) Images of the effects of March 4, 1977 earthquake in Bucharest, Contract 30-77-1/ 1977, pp 137–153, Bucharest

Evernden JF (ed) (1976) Abnormal animal behavior prior to earthquakes. U.S. Dept. of Interior Geological Survey, Conference I. Convened under the auspices of the National Earthquake Hazards Reduction Program, USGS, Menlo Park, CA, 23–24, September 1976

Evernden JF (ed) (1980) Abnormal animal behavior prior to earthquakes. In: Proceedings of Conference Menlo Park, Calif.: USGS, 2, 237 p

Freund F, Stolc V (2013) Nature of pre-earthquake phenomena and their effects on living organisms. Animals 3:513–531. https://doi.org/10.3390/ani3020513

Hayakawa M (2013) Possible electromagnetic effects on abnormal animal behavior before an earthquake. Animals 3:19–32. https://doi.org/10.3390/ani3010019

Ikeya M (2004) Earthquakes and animals. World Scientific, Singapore

Ikeya M, Yamanaka C, Mattsuda T, Sasaoka H, Ochiai H, Huang QH, Ohtani N, Komuranani T, Ohta M, Ohno Y, Nakagawa T (2000) Electromagnetic pulses generated by compression of granitic rocks and animal behaviour. Episodes 23:262–265

Nikonov AA (1981) Abnormal animal behaviour as earthquake precursors. "VINITI", Moscow, Dep. 4316–81, 54 (in Russian)

Nikonov AA (1992) Abnormal animal behaviour as a precursor of the 7 December 1988 Spitak, Armenia, earthquake. Nat Hazards 6:1–10

Pantea A (1979) Macroseismic effects of the March 4, 1977 earthquake as observed in Prahova, Dambovita and Arges districts, in the monograph. In: Cornea I, Radu C (eds) Seismological research on March 4, 1977 earthquake, Bucharest, pp 615–641

Pantea A, Constantin AP (2011) Reevaluated macroseismic map of Vrancea (Romania) earthquake occurred on November 10, 1940. Rom J Phys 56(3–4):578–589

Pantea A, Constantin AP (2013) Re-evaluation of the macroseismic effects produced by the March 4, 1977 strong Vrancea earthquake in Romanian territory. Ann Geophys 56(1):37–49. https://doi.org/10.4401/ag-5641

Quammen D (1985) Animals and earthquakes: This World. San Francisco Chronicle, April 21, pp 15–16

Radu C, Spanoche E (1976) Some phenomena related to the Romanian earthquakes. Publ Inst Geophys Pol Acad. Sci A-4(115):163–172

Radu C, Polonic G, Apopei I (1979a) Macroseismic field of the March 4, 1977 Vrancea earthquake, in the monograph. In: Cornea I, Radu C (eds) Seismological research on March 4, 1977 earthquake, Bucharest, pp 345–357

Radu C, Utale A, Galeriu A (1979b) Biological phenomena related to the March 4, 1977 earthquake (in Romanian). In: Cornea I, Radu C (eds) Seismological research on March 4, 1977 earthquake, Bucharest, pp 431–446

Radu C (1988) The Vrancea Romania 30 August 1986 major earthquake: seismological and engineering aspects. Programme and abstracts, 21th Gen. Ass., August 23–27, Sofia, pp 29–30

Radu C, Utale A, Winter V (1987) Cutremurul Vrancean din 30 August 1986: Distributia Intensitatilor Seismice (Text prescurtat). Raport CFPS/CSEN, Tema 30.86.3/1987 (Responsabil—C. Radu), II, A3 ICEFIZ Bucuresti, 25 mai 1987

Rikitake T (1976) Earthquake prediction. Elsevier, Amsterdam, p 357

Rikitake T (1978) Biosystem behavior as an earthquake precursor. Tectonophysics 51:1–7

Rikitake T (1998) The science of macro-anomaly precursory to an earthquake. Kinmiraisha, Nagoya, Japan

Romplus catalogue (2017) Updated to 2017 by the Department of Data Acquisition of the National Institute for Earth Physics (NIEP), Bucharest

Schaal RB (1988) An evaluation of the animal-behavior theory for earthquake prediction. Calif Geol 41(2)

Ling-Huang S (1978) Can animal help to predict earthquakes. Earth Inform Bull US Dept Inter Geol Surv 10(6):231–233

Tributsch H (1982) When the snakes awake, animals and earthquake prediction. The MIT Press, Cambridge, Mass

Wang K, Chen QF, Sun S, Wang A (2006) Predicting the 1975 Haicheng Earthquake. Bull Seismol Soc Am 96(3):757–795. https://doi.org/10.1785/0120050191

Part II
Site Response and Seismic Risk Assessment

Investigation of Local Site Responses at the Bodrum Peninsula, Turkey

Hakan Alcik and Gülüm Tanırcan

Abstract Bodrum Peninsula is situated on the southwest coast of Turkey, near the Aegean Sea coast. The Peninsula extends ∼42 km in the E-W direction and ∼15 km in the N-S direction between the Gulfs of Güllük and Gökova. The Bodrum peninsula with a population over a million in summer season is one the touristic centers of Turkey. The region is also surrounded by numerous active seismic entities such as Ula-Ören Fault Zone, Gökova Graben, eastern part of the Volcanic Arc and Hellenic Arc-Trench System etc. These systems demonstrate high seismic hazard and pose a great threat to settlements in and around the region. Considering the high seismic risk and high population of the peninsula, a strong ground motion monitoring system, consists of five accelerometric stations, was deployed in June 2015 by Boğaziçi University, Kandilli Observatory and Earthquake Research Institute (KOERI), Earthquake Engineering Department. Three out of five stations (B1, B2 and B3) are on alluvium sediments. The rests are on Limestone (B4) and Volcanic rock (B5). Up to now the network has recorded more than 100 earthquakes. Among the dataset, 25 events with magnitudes (Ml) from 3.0 to 5.5 occurred within 200 km epicentral distances were selected for site effect calculation. Predominant frequencies and amplification values of shallow soil layers under the stations were estimated through Horizontal to Vertical Spectral Ratio and Standard Spectral Ratios. The results indicate that (1) predominant frequencies change between 2.1 and 3.5 Hz for soft soils, where it is 5.8 Hz for the reference site, (2) relative amplifications are in the range of 2.3–6.8, and (3) empirically estimated sediments thickness beneath the B1, B2 and B3 stations vary between 35.6 and 64.2 m.

Keywords Earthquakes · Seismic hazard · Predominant frequencies
Site effects

H. Alcik (✉) · G. Tanırcan
Earthquake Engineering Department, Boğaziçi University,
Kandilli Observatory and Earthquake Research Institute, Istanbul, Turkey
e-mail: alcik@boun.edu.tr

© Springer International Publishing AG, part of Springer Nature 2018
R. Vacareanu and C. Ionescu (eds.), *Seismic Hazard and Risk Assessment*,
Springer Natural Hazards, https://doi.org/10.1007/978-3-319-74724-8_10

1 Introduction

The southwest coasts are the most important touristic centers of Turkey. One of these centers is the Bodrum peninsula with a population over one million in summer season. The region is also located in one of the seismically active regions of the southeast Aegean Sea and surrounded by numerous active seismic entities such as Ula-Ören Fault Zone, Gökova Graben (Kalafat and Horasan 2012; Yolsal and Taymaz 2010), Datça faults (Dirik 2007), eastern part of the Volcanic Arc and Hellenic Arc-Trench System (Papadopoulos et al. 2007; Sakkas et al. 2014) (Fig. 1). All of those systems have capability of producing large magnitude earthquakes. Frequent occurrence of historical destructive and instrumental earthquakes clearly demonstrates high seismic hazard in Bodrum and its surrounding area. Among destructive historical earthquakes, 1493 Kos event (Mw = 6.94 ± 0.32) caused complete collapse of the Bodrum district. 1741 (Mw = 7.54 ± 0.30), 1863 (Mw = 7.5 ± 0.30) and 1869 (Mw = 6.77 ± 0.37) are other the important earthquakes (SHEEC 2017, Yolsal and Taymaz 2010). In the instrumental period seismic activity in the Gökova region includes M > 6 earthquakes: 23 April 1933 (Ms = 6.4), 23 May 1941 (Ms = 6.0) and 13 December 1941 (Ms = 6.5) events (Kalafat and Horasan 2012). Among above mentioned faults, Gökova graben system poses larger hazard due to its close proximity to Bodrum region. This fault zone has a potential to produce earthquakes varying in size from M6.9 to M7.8 and annual probability of occurence of a M7+earthquake has been

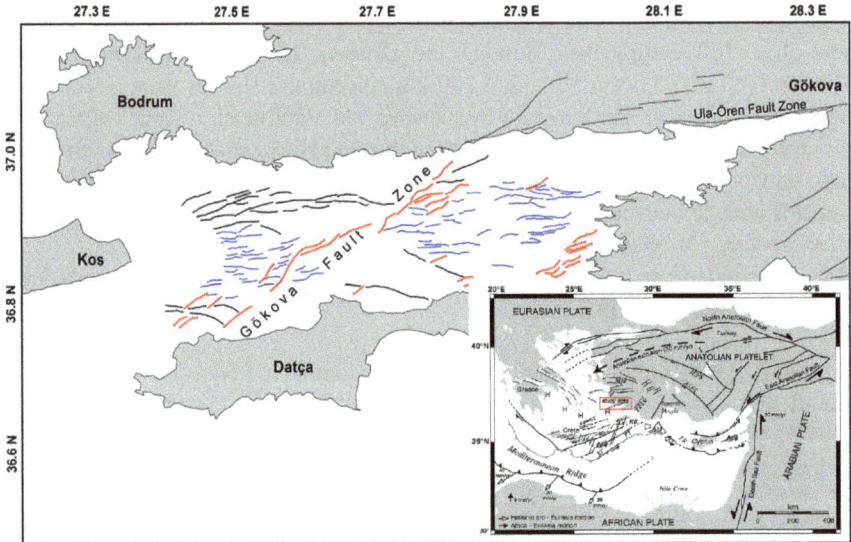

Fig. 1 Active fault map of the Gulf of Gökova (red and black lines: strike-slip and normal faults respectively; blue lines are reverse faults and folds (Iscan et al. 2013). Inset: The geodynamic framework of the Eastern Mediterranean (TenVeen et al. 2009)

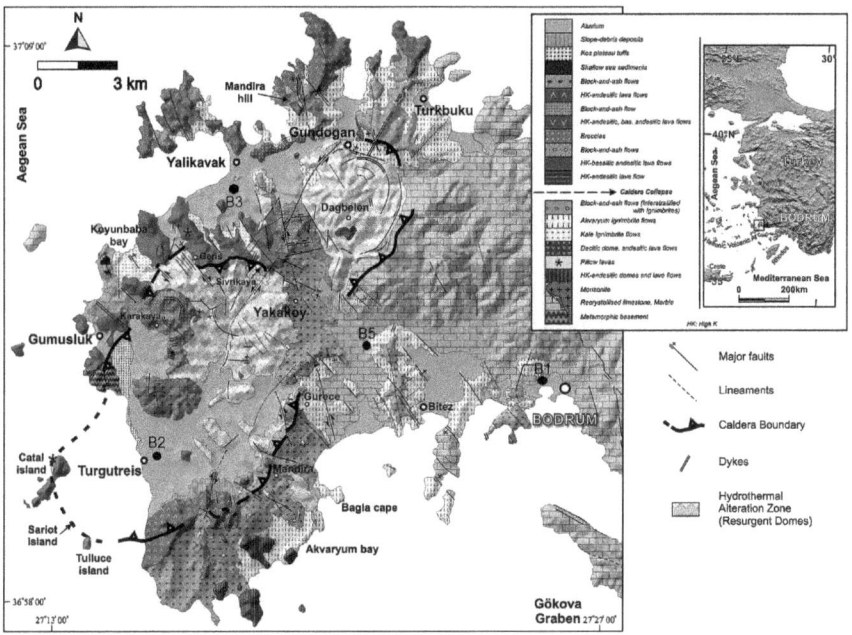

Fig. 2 Distribution of the B-Network stations at the Bodrum peninsula (B1 = Bodrum downtown, B2 = Turgutreis, B3 = Yalikavak, B4 = Yali-Ciftlik, B5 = Ortakent). B4 station is situated outside of the map, on the eastern side of the B1 station. Geological sketch map is taken from Ulusoy et al. (2004)

estimated as 2% (Demircioğlu 2010). The whole region falls into 1st degree seismic zone, being the highest (AFAD 2017) in the seismic zonation map of Turkey.

Despite its high seismicity, only three stations are in service at the region for seismological monitoring purposes. Hence, with the aim of collecting accurate and reliable data for engineering and scientific research purposes a strong ground motion monitoring system, shortly B-network consists of 5 accelerometric stations (Fig. 2), has been set up in 2015 (Alcik et al. 2015).

There are little known about the site effects of the Bodrum Peninsula. During the literature survey, the authors have not come across any site related studies for this particular region. Hence, calculation of the predominant frequencies and amplification values of shallow soil layers by utilizing the B-network seismic data are the main objective of the study.

2 Methods

Two very well-known methods were employed for the calculation; Horizontal-to-Vertical Spectral Ratio (HVSR) and Standard Spectra Ratio (SSR).

2.1 Horizontal-to-Vertical Spectral Ratio (HVSR or H/V) Method

The HVSR technique applied to ambient noise recordings (Nakamura 1989) has been extensively used in recent times to estimate the site effects. The spectral ratio between the horizontal and the vertical components of the recorded motion eliminates the contributions of the Rayleigh waves, but it conserves the effects resulting from the geological structure of the site (Gök and Polat 2012; Lermo and Chavez-Garcia 1994; Nakamura 1989; Yalcinkaya and Alptekin 2005). Although the theoretical basis of the method is controversial, technique has been validated by both simulations and earthquake recordings (i.e., Flores et al. 2013; Lermo and Chavez-Garcia 1993; Parolai et al. 2002). In HVSR method only a 3-components record of one station is required. The transfer function (TF) at the measurement point is calculated by the following equation:

$$TF(f) = Hs(f)/Vs(f) \qquad (1)$$

where $Hs(f)$ is the amplitude spectra of the horizontal components, and $Vs(f)$ is the amplitude spectrum of the vertical component (Dikmen et al. 2013).

2.2 Standard Spectral Ratio (SSR) Method

The SSR technique was first used by Borcherdt (1970) near San Francisco Bay, California. The method involves comparison of pairs of records from nearby stations, one representing a soil (an alluvial) site and the other representing a reference (a rock) site (Gök et al. 2012; Şafak 1997). Distance between the stations must be much smaller than their hypocentral distances, so that the source and path effects on the records are nearly identical (Şafak 1997). Therefore, any differences in the records can be attributed to site effects. This method is required two horizontal component records of two stations and has been to use spectral ratio, the ratio of the Fourier amplitude spectrum of the soil site recording to that of the rock site recording (Borcherdt 1970; Mittal et al. 2013; Şafak 1997). Transfer function (TF) of the measurement points is calculated by the following equation:

$$TF(f) = Hs(f)/HR(f) \qquad (2)$$

where $Hs(f)$ and $HR(f)$ denote the smoothed horizontal component of Fourier amplitude spectrum at the site of interest and reference site respectively (Dikmen et al. 2013). SSR method is considered as the most reliable method in determining the effects of local site conditions (Yalcinkaya and Alptekin 2003).

3   Data Set and Analyses

B-network was put into action on the 2nd June 2015 and till now it is fully operational with five stations. The network consists of 18-bit digitizers and acquisition modules of GeoSIG Limited, CMG-5T accelerometers of Guralp Systems Limited, DC batteries, GPS units and 3G modems. It has been stationed in dense settlements in the Bodrum Peninsula and generally located at grade level in small and medium-sized buildings of the Municipality and the primary schools: Bodrum (B1), Turgutreis (B2), Yalikavak (B3), Ciftlik (B4) and Ortakent (B5). B1, B2 and B3 are on Alluvial sediment, B4 is on Limestone and B5 is on Volcanic rock (Fig. 2; Table 1). These geological explanations were taken from the 1:500,000 scale geology map of Turkey, Denizli sheet (MTA 2012). Soil profiles of the sites are not currently available.

So far, more than 100 earthquakes have been recorded by the network. The dataset used here consists of 25 earthquakes. Local magnitudes range from 3.0 to 5.5 and epicentral distances range from 1 to 200 km with amplitude values of all acceleration records are greater than 1 mg. The information of these earthquakes is listed in Table 2.

The SSR and the HVSR methods were used to determine site effects at the stations. Data processing was done using MATLAB® (http://www.mathworks.com/) software codes in accordance with the techniques mentioned above.

Before computing the spectral ratios, all recorded time-series were visually checked to identify possible inaccurate measurements. The records with signal-to-noise ratio greater than 3 were kept in the analyses. The full-length records were de-trended, baseline corrected and band-pass filtered between 0.05 and 20 Hz. In the analysis of both methods, the same data processing procedure was followed except for data window lengths to be processed. The window lengths were selected 30 s after the S-wave of earthquake accelerograms for SSR method, and 60 s after the P-wave for HVSR method. Selected data were windowed by a 5% cosine taper before performing a Fast Fourier Transform. Each spectrum was smoothed by a Hamming window following Şafak (1997). Then, spectral ratios, and finally, mean spectral ratios for each site were computed (Figs. 3 and 4).

Table 1 Stations information

Station	Latitude (E)—Longitude (N)	Location	Altitude (m)	Geological information
B1	37.037°—27.424°	Bodrum downtown	8	Alluvial
B2	37.007°—27.257°	Turgutreis	4	Alluvial
B3	37.102°—27.293°	Yalikavak	7	Alluvial
B4	37.023°—27.563°	Yali-Ciftlik	87	Limestone
B5	37.047°—27.347°	Ortakent	28	Volcanic rock

Table 2 Earthquake lists used in this study

No	Date and Time (UTC)	Latitude (N)—Longitude (E)	Magnitude (ML)	Depth (km)
1	20150529 08:02:51	36.916°—27.607°	3.9	1.5
2	20150709 10:41:35	36.857°—29.963°	4.3	7.5
3	20150719 20:32:07	37.074°—27.499°	3.7	7.6
4	20150724 09:58:37	36.639°—26.798°	4.7	131.8
5	20150913 02:57:26	37.128°—28.890°	4.5	3.7
6	20151020 19:00:50	39.941°—28.043°	3.8	5.0
7	20151023 18:29:39	37.996°—26.820°	4.0	8.0
8	20151226 11:00:50	35.488°—27.393°	5.4	14.8
9	20160223 21:32:43	37.263°—26.855°	3.4	5.0
10	20160226 18:32:16	37.058°—27.618°	3.8	3.8
11	20160312 15:44:34	36.571°—27.971°	4.1	84.1
12	20160318 19:22:34	36.720°—26.701°	3.5	8.2
13	20160505 07:34:08	36.890°—26.453°	3.8	5.0
14	20160612 09:2250	36.957°—27.948°	3.8	5.0
15	20160614 14:43:57	37.175°—26.850°	3.6	4.1
16	20160719 01:55:42	37.181°—26.851°	4.0	5.9
17	20160813 21:38:03	37.020°—27.391°	3.1	7.3
18	20160906 05:27:26	36.337°—28.166°	4.4	50.9
19	20160909 10:45:09	36.933°—27.577°	3.0	6.6
20	20160927 20:57:07	36.386°—27.585°	5.4	83.3
21	20161017 01:30:30	37.906°—26.891°	4.7	8.3
22	20161105 07:35:48	36.982°—27.386°	3.2	5.0
23	20161220 06:03:43	36.565°—26.927°	5.4	118.8
24	20170125 01:19:33	36.797°—27.657°	4.2	7.9
25	20170125 18:50:52	35.445°—26.448°	5.5	21.7

4 Results

Examining HVSR data, it was found out that predominant frequencies change
between 2.1 and 3.5 Hz for soft soils, where it is 5.8 Hz for B5. Hence for SSR
analyses B5 was selected as the reference site. SSR spectra show that relative
amplifications are in the range of 2.3–6.8 (Table 3). Even though B1, B2 and B3
stations rest on similar geological units, B1 gives the lowest relative amplification.
Since shear wave velocity profiles of the stations are not available, authors can not
make a direct comparison between the profiles, Vs 30 and amplifications but for
further analyses Vs 30 information will be obtained from Bodrum Municipality or
related institutions. In general both HVSR and SSR spectra show similar trend. It is
also possible to estimate the bedrock depth of the soil under stations based on a
previous empirical equation given by Birgören et al. (2009). The equation is,

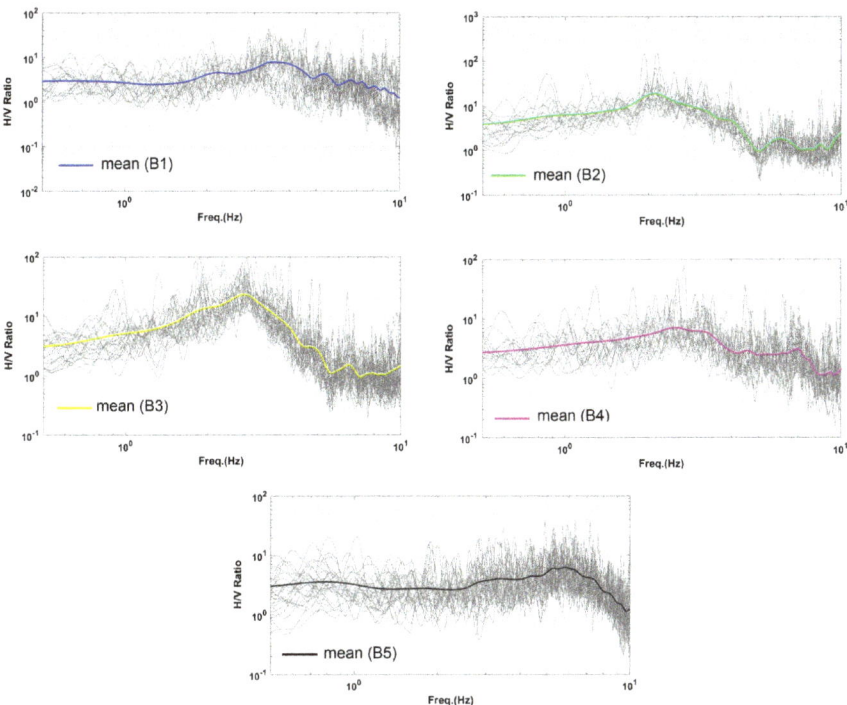

Fig. 3 HVSR results

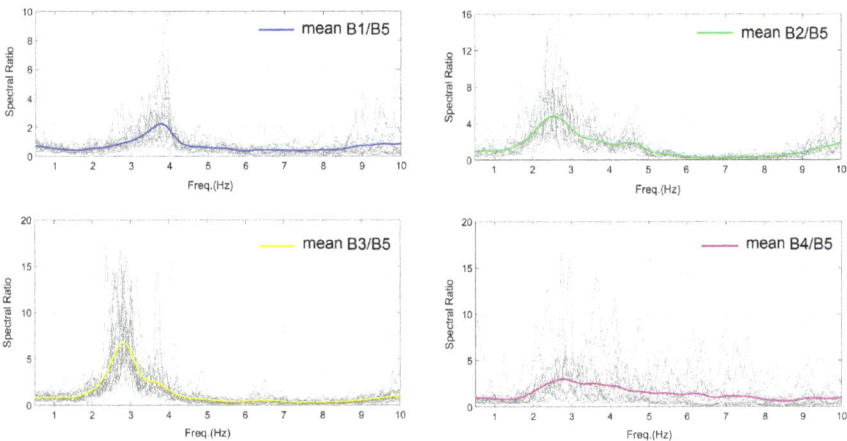

Fig. 4 SSR results

Table 3 HVSR, SSR and H estimations

Station	Predominant frequency (Hz) (HVSR)	Ratio	Soil Amplification (SSR)	Estimated bedrock depth (m) (H)
B1	3.5	B1/B5	2.3	35.6
B2	2.1	B2/B5	4.9	64.2
B3	2.7	B3/B5	6.8	48.0
B4	2.5	B4/B5	3.0	52.5
B5	5.8	–	–	19.9

$$H = 150.99 * f_r^{-1.1531} \tag{3}$$

In this equation f_r denotes resonance frequency ($\sim$ predominant frequency) of a site and H is bedrock depth. Estimated values under B1, B2, B3, B4 and B5 stations are 36, 64, 48, 53 and 20 m, respectively (Table 3). Comparing these findings with the geological map of the region (Fig. 2), it was found out that thickness associated with resonance frequencies are in remarkable agreement with surface geology information. Thickness to bedrock is the highest at Turgutreis valley, the largest alluvial valley of the region and lowest at the Bodrum downtown, the smallest alluvial valley of the region.

However, it is difficult to say that there is a linear relation between the site amplification and the estimated thickness as a result of the maximum amplification was calculated at B3 station, since amplification is also directly related with impedance contrast of the layers.

5 Conclusion

In the present study an attempt has been made to calculate predominant frequencies and amplification values considering local site effects related to the Bodrum peninsula. We estimated site effects using the standard spectral ratio and horizontal-to-vertical ratio techniques. For this purpose, a set of time histories contains 106 acceleration records, totally 25 earthquakes with different magnitudes and epicentral distances were analysed.

The results of analyses indicate that (1) predominant frequencies change between 2.1 and 3.5 Hz for soft soils, where it is 5.8 Hz for the reference site B5; (2) relative amplifications are in the range of 2.3–6.8; and (3) estimated thicknesses of sediments beneath the B1, B2 and B3 stations vary between 35.6 and 64.2 m. The results are believed to motivate research groups to perform further geotechical investigations for the Bodrum Peninsula, in particular, regions with thick soil deposits.

Acknowledgements This study was supported by Boğaziçi University Scientific Research Project Funds (Project #10260/15TP2).

References

Alcik H, Tanırcan G, Korkmaz A (2015) Bodrum strong motion network, Muğla, Turkey. In: American Geophysical Union Fall Meeting. Available via https://agu.confex.com/agu/fm15/webprogram/Paper62410.html. Accessed on 04 Sep 2017

AFAD (2017) Republic of Turkey, Prime Ministry Disaster & Emergency Management Authority. http://www.deprem.gov.tr/en/Category/earthquake-zoning-map-96531. Accessed on 10 Aug 2017

Birgören G, Ozel O, Siyahi B (2009) Bedrock depth mapping of the coast south of Istanbul: comparison of analytical and experimental analyses. Turkish J Earth Sci 18:315–329

Borcherdt RD (1970) Effects of local geology on ground motion near San Francisco Bay. Bull Seism Soc Am 60(1):29–61

Demircioglu BM (2010) Earthquake hazard and risk assessment for Turkey. PhD Dissertation, Boğaziçi University

Dikmen U, Arısoy MO, Akkaya I, Demirci I, Hasancebi N (2013) Yer tepkisinin belirlenmesinde kullanılan yöntemlerin ivme kaydı üzerinde değerlendirilmesi. In: 2nd Turkish earthquake engineering and seismology conference. Available via http://www.tdmd.org.tr/TR/Genel/pdf/TDMSK071.pdf

Dirik K (2007) Neotectonic characteristics and seismicity of the Reşadiye peninsula and surrounding area, Southwest Anatolia. Geol Bull Turkey 50(3):130–149

Flores H, Malischewsky P, Jentzsch G (2013) H/V spectral ratio analysis and Rayleigh modelization in Eastern Thuringia, Germany. Geofisica Internacional 52(4):355–364

Gök E, Kececioglu M, Ceken U, Polat O (2012) IZMIRNET istasyonlarında standart spektral oran yöntemi kullanılarak zemin transfer fonksiyonlarının hesaplanması. Dokuz Eylül Üniversitesi Mühendislik Fakültesi Mühendislik Bilimleri Dergisi 14(41):1–11

Gök E, Polat O (2012) Microtremor HVSR study of site effects in Bursa city (Northern Marmara Region, Turkey). In: D'Amico S (ed) Earthquake research and analysis-new frontiers in seismology

Iscan Y, Tur H, Gökasan E (2013) Morphologic and seismic features of the Gulf of Gökova, SW Anatolia: evidence of strike-slip faulting with compression in the Aegean extensional regime. Geo-Mar Lett 33:31–48

Kalafat D, Horasan G (2012) A seismological view to Gökova region at southwestern Turkey. Int J Phys Sci 7(30):5143–5153

Lermo J, Chavez-Garcia FJ (1993) Site effect evaluation using spectral ratios with only one station. Bull Seism Soc Am 83:1574–1594

Lermo J, Chavez-Garcia FJ (1994) Are microtremors useful in the site Response evaluation? Bull Seism Soc Am 84:1350–1364

Mittal H, Kamal Kumar A, Singh SK (2013) Estimation of site effects in Delhi using standard spectral ratio. Soil Dyn Earthq Eng 50:53–61

MTA (2012) Maden Tetkik Arama/General Directorate of Mineral Research and Exploration, http://yerbilimleri.mta.gov.tr/anasayfa.aspx. Accessed on 10 Aug 2017

Nakamura Y (1989) A method for dynamic characteristic estimations of subsurface using microtremor on the ground surface. RTRI 30(1):25–33

Papadopoulos GA, Daskalaki E, Fokaefs A, Giraleas N (2007) Tsunami hazard in the Eastern Mediterranean: strong earthquakes and tsunamis in the east Hellenic Arc and Trench system. Nat Hazards Earth Syst Sci 7:57–64

Parolai S, Bormann P, Milkereit C (2002) New relationships between Vs, thickness of sediments, and resonance frequency calculated by the H/V ratio of seismic noise for the Cologne area (Germany). Bull Seism Soc Am 92(6):2521–2527

Safak E (1997) Models and methods to characterize site amplification from a pair of records. Earthq Spectra 13(1):97–129

Sakkas V, Novali F, Lagios E, Bellotti F, Vassilopoulou S, Damiat BN, Allievi J (2014) Ground deformation study of Kos Island (SE Greece) based on Squee-SARTM interferometric technique. In: International geoscience and remote sensing symposium (IGARSS 2014) and 35th Canadian Symposium on Remote Sensing (35th CSRS), Québec, Canada, 13–18 July 2014

SHEEC (2017) SHARE European earthquake catalogue, http://www.emidius.eu/SHEEC/catalogue/. Accessed on 10 Aug 2017

TenVeen JH, Boulton SJ, Alcicek MC (2009) From palaeotectonics to neotectonics in the Neotethys realm: the importance of kinematic decoupling and inherited structural grain in SW Anatolia (Turkey). Tectonophysics 473:261–281

Ulusoy I, Cubukcu E, Aydar E, Labazuy P, Gourgaud A, Vincent PM (2004) Volcanic and deformation history of the Bodrum resurgent caldera system (southwestern Turkey). J Volcanol Geotherm Res 136:71–96

Yalcinkaya E, Alptekin O (2003) Dinar'da zemin büyütmesi ve 1 Ekim 1995 depreminde gözlenen hasarla ilişkisi (Site amplification in Dinar and relationship with damage observed on the October 1, 1995 Earthquake). Bull Earth Sciences Application and Research Centre of Hacettepe University 27:1–13

Yalcinkaya E, Alptekin O (2005) Site effect and its relationship to the intensity and damage observed in the June 27, 1998 Adana-Ceyhan earthquake. Pure appl Geophys 162:913–930

Yolsal S, Taymaz T (2010) Gökova Körfezi depremlerinin kaynak parametreleri ve Rodos-Dalaman bölgesinde tsunami riski. İTÜ dergisi/dmühendislik 9(3):53–65

Ground Types for Seismic Design in Romania

Cristian Neagu, Cristian Arion, Alexandru Aldea,
Elena-Andreea Calarasu, Radu Vacareanu and Florin Pavel

Abstract The paper presents an overview of available data concerning ground types for seismic design in Romania. A short overview of a ground information database created during BIGSEES Romanian project is presented. A comparison of shear wave velocity for 19 sites in Bucharest determined by PS logging measurements and by Wald topographic slope method is discussed. The paper reiterates the conclusion of a study regarding the Eurocode soil factor S derived from the Romanian seismic motions. The need for an enlarged database of in situ determined ground condition is underlined, at least at the location of seismic stations. Based on borehole-specific data (geotechnical properties, hydrologic factors) and velocity profiles, evaluations of soil liquefaction potential and related indices were performed by using empirical equations proposed in literature. The application of GIS tools provided a spatial distribution of liquefaction susceptibility of Quaternary alluvial sediments in Bucharest.

Keywords Shear wave velocity $\cdot$ $V_{S,30}$ $\cdot$ Soil conditions $\cdot$ S factor
Liquefaction potential $\cdot$ GIS

C. Neagu (✉) $\cdot$ C. Arion $\cdot$ A. Aldea $\cdot$ F. Pavel
Technical University of Civil Engineering of Bucharest, Bucharest, Romania
e-mail: cristi.neagu@utcb.ro

C. Arion
e-mail: cristi.arion@utcb.ro

A. Aldea
e-mail: alexandru.aldea@utcb.ro

F. Pavel
e-mail: florin.pavel@utcb.ro

E.-A. Calarasu
National Institute for Research and Development in Construction, Urban Planning
and Sustainable Spatial Development URBAN-INCERC, Bucharest, Romania
e-mail: andreea.calarasu@gmail.com

R. Vacareanu
Seismic Risk Assessment Research Center, Technical University of Civil Engineering
of Bucharest, Bucharest, Romania
e-mail: radu.vacareanu@utcb.ro

© Springer International Publishing AG, part of Springer Nature 2018
R. Vacareanu and C. Ionescu (eds.), *Seismic Hazard and Risk Assessment*,
Springer Natural Hazards, https://doi.org/10.1007/978-3-319-74724-8_11

1   Introduction

The evaluation of seismic action for design is under continuous development from early 20th century. Initially, the seismic design action was simply evaluated as a horizontal force defined as a percent of total weight of the building superstructure. Nowadays, the seismic action has a complex definition also involving the consideration of local soil conditions whose importance highlighted by the seismic records since 1970. The famous study of Seed et al. (1974) put in evidence the striking differences between normalized acceleration response spectra on different ground condition and strongly influenced the development of modern seismic design codes.

Seismic design codes such as EN 1998-1, ASCE/SEI 7-10, IBC 2012 and other modern design regulations define local soil conditions based on shear wave velocity averaged on the top 30 m of soil profile, $V_{S,30}$. This parameter was used for the first time by Borcherdt and Glassmoyer (1992) for ground types classification in ASCE seismic design code in USA.

In the Romanian seismic design code P100-1 editions from 1992, 2006 and 2013 due to insufficient data (soil profile and shear wave velocity profile correlated with seismic records), the parameter that defines the site conditions is the control (corner) period T_C. Lungu et al. (1997) proposed a moving average procedure for computing effective peak acceleration, velocity and displacement used for computing T_C values that are the base of T_C national zonation map in the code.

During the national BIGSEES project "Bridging the gap between seismology and earthquake engineering: from the seismicity of Romania towards a refined implementation of Seismic Action EN1998-1 in earthquake resistant design of buildings", a national geological Database was created, putting together all soil information from all three partners in the project: National Institute of Earth Physics —INFP, Technical University of Civil Engineering of Bucharest—UTCB and National Institute for Building Research—URBAN-INCERC.

According to historical and recent data, liquefaction and associated permanent ground deformations cause severe damages on structures and lifelines: Niigata and Great Alaska (1964), San Fernando (1971), Tangshan (1976), Loma Prieta (1989), Great Hanshin (1995), Kocaeli and Chi-Chi (1999), Canterbury (2010), Christchurch and Tohoku (2011). Detailed studies have revealed that the most susceptible to liquefaction are unconsolidated and Quaternary near-surface deposits represented by sandy soils and silty sands, with high degree of saturation, low plasticity and uniform gradation in loose relative density state. A great part of research studies has been especially focused on sandy soils liquefaction (Seed and Idriss 1967; Ishihara and Koga 1981; Ishihara 1985), however the observations of past earthquakes conducted to well-documented case of liquefaction of gravelly soils (Andrus 1994; Lewis et al. 2013), low plasticity silts and clays (Boulanger and Idriss 2004; Chu and Stewart 2004) or reclaimed land (Ishihara et al. 1998; Yasuda et al. 2012).

Several procedures have been proposed for predicting and zoning the liquefaction hazard by combining seismic records, geotechnical, geological and geomorphologic data (Ishihara and Ogawa 1978; Youd and Perkins 1978; Iwasaki et al. 1982; Wakamatsu 1993). The liquefaction resistance of soils is commonly assessed using empirical correlations between soil parameters determined by field investigations such as Standard Penetration Test (Seed and Idriss 1971; Idriss and Boulanger 2010), Cone Penetration Test (Robertson and Campanella 1985; Suzuki et al. 1997; Robertson and Wride 1998; Idriss and Boulanger 2003) and shear wave velocity (V_S) test (Robertson et al. 1992; Andrus and Stokoe 2000; Andrus et al. 2004) or laboratory tests (Tokimatsu and Uchida 1990; Chen et al. 2005). Other contributions based on probabilistic approaches include the work of Liao et al. (1988), Youd and Noble (1997), Juang et al. (2002), Moss et al. (2006), Idriss and Boulanger (2012), Kayen et al. (2013).

As a result of accumulated knowledge on liquefaction occurrence and ground failures, several requirements are specified in the modern building codes when designing new structures on liquefiable soils in earthquake-prone areas. Eurocode 8 Part 5, Sect. 4.1.4 and Annex B contains requirements, detailed guidelines and charts regarding specific analysis for potentially liquefiable soils, included in sub-soil class S2. It is mentioned that a minimum in situ investigation (Standard Penetration Test—SPT or Cone Penetration Test—CPT) and laboratory grain size analysis should be performed. Furthermore, several conditions when the risk of liquefaction on site may be neglected are specified. EC8 provide a limit for the acceleration at the peak ground surface for the occurrence of liquefaction: $PGA \geq 0.15$ g. In Romania, a technical guideline for studying the properties of cohesionless liquefiable soils (P 125-84) consists in some requirements regarding the necessary laboratory and field tests which should be carried out.

2 Romanian Geological Database

A national geological Database (geographically depicted in Fig. 1) was created as an output of BIGSEES project (2012–2016). It contains borehole information (where available) such as depth, compression wave velocity V_P, shear wave velocity V_S, density and stratigraphy. Within the database is also available a list of 55 soil types (such as clay, sand, loess, marl) with density, shear modulus G and damping ratio D values at low shear strain.

The database consists mostly of boreholes drilled in the '70s that have no stratigraphy information but only data about compression wave velocity (V_P). It also includes a smaller number of boreholes with depth varying from 13 to 153 m, most of them in Bucharest area, with more complete information (stratigraphy, V_S, V_P).

In Romania V_S measurements were performed using down-hole method and sometimes by cross-hole method. A campaign of measurements was performed in '70s but data is not available anymore. First modern shear wave and compression wave velocity measurements were performed in 1997 within a UTCB project (Lungu et al. 1999) at INCERC site (the site where 1977 strong earthquake was recorded) by GEOTEC—Institute for Geotechnical and Geophysical Studies. Soon

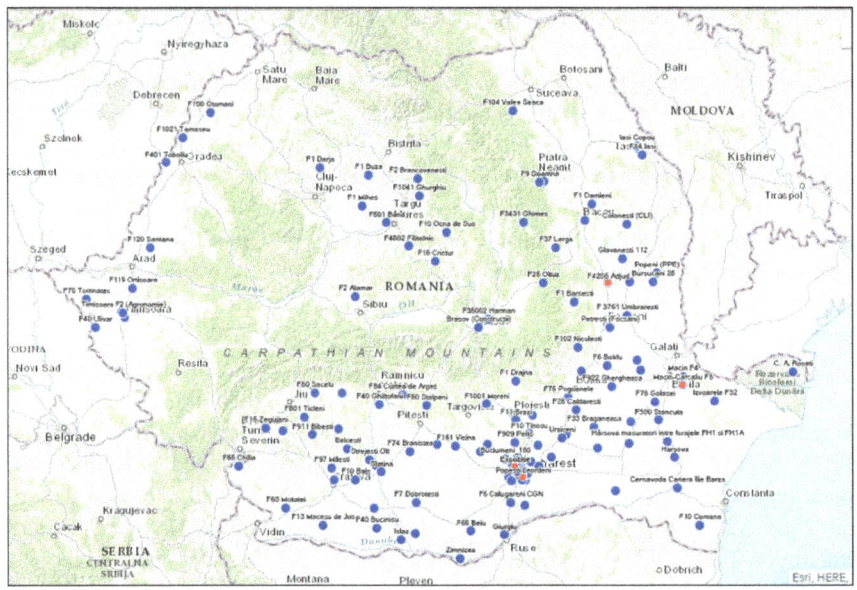

Fig. 1 BIGSEES geological database (http://infp.infp.ro/bigsees/Results.html)

after, two other boreholes were investigated at INCERC site within the German—Romanian research program "Strong Earthquake: A Challenge to Geoscience and Civil Engineering" (financed by German Science Foundation at Karlsruhe University), program which later financed other 7 borehole measurements, Orlowski et al. (2003).

In Bucharest in 2002 ÷ 2008 down-hole measurements were performed at over 20 boreholes in Bucharest region within INFP national and international research projects. In the recent years, sometimes, private design companies are requesting site dependent seismic hazard studies including shear wave velocity measurements, especially in Bucharest.

During the Japan International Cooperation Agency (JICA) Project for Seismic Risk Reduction on Building and Structures in Romania (2002–2008) V_S measurements were continued in Bucharest by National Center for Seismic Risk Reduction (NCSRR) using PS Logging method and equipment produced by Tokyo Soil Research Co. Ltd. Figure 2 presents the 19 boreholes locations in Bucharest where PS logging measurements were performed by NCSRR staff for JICA Project and NATO Science for Peace Project 981882 (2005–2008). Details about these sites are presented in Table 1 (Aldea et al. 2006; Bala et al. 2013; Neagu 2015). Table 1 also presents for comparison (Neagu 2015), the $V_{S,30}$ determined by the topography slope method proposed by Wald et al. (2004).

Wald et al. (2004) and Wald and Allen (2007) proposed a methodology that correlate topographic slope data obtained from 30 arc-sec (SRTM30—Shuttle Radar Topography Mission 30 arc-sec) topographic data recorded in 2000 by space

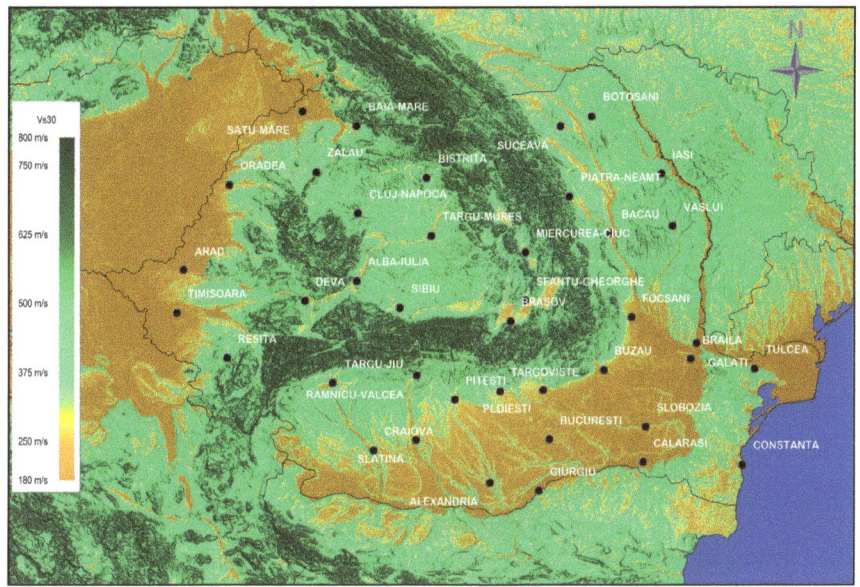

Fig. 2 $V_{S,30}$ map for Romania (data from USGS)

Table 1 Shear wave velocity at 19 sites in Bucharest

No.	Site	Depth (m)	PS Logging measurements		Topographic slope method	EC8 ground class	Difference[a] (%)
			$V_{S,med}$ (m/s)	$V_{S,30}$ (m/s)	$V_{S,30}$ (m/s)		
1	UTCB Tei	78	349	310	271	C	12
2	UTCB Instalații	66	332	288	226		21
3	UTCB Utilaj	34	319	314	236		25
4	Apărarea Civilă	68	323	292	221		24
5	Spitalul Municipal	69	303	245	224		9
6	Piața Victoriei	151	354	284	232		18
7	INCERC (CNRRS)	153	364	270	247		8
8	Primăria București	54	261	219	271		−23
9	Parcul Studențesc Tei	50	320	295	248		16
10	Parcul Bazilescu	50	328	288	209		27
11	Parcul Tineretului	50	304	263	283		−8
12	Parcul Motodrom	50	327	288	273		5
13	Parc Titan	50	339	299	249		17
14	INFP	50	331	316	256		19
15	Federația Română de TIR	50	307	271	285		−5
16	Muzeul de Geologie	50	320	298	214		28
17	Universitatea Ecologică	50	324	286	262		8

(continued)

Table 1 (continued)

No.	Site	Depth (m)	PS Logging measurements		Topographic slope method	EC8 ground class	Difference[a] (%)
			$V_{S,med}$ (m/s)	$V_{S,30}$ (m/s)	$V_{S,30}$ (m/s)		
18	Institutul Astronomic	50	310	273	289		−6
19	EREN	30	288	286	213		26
Average				283	248		12

[a]The differences is between $V_{S,30}$ from PS logging and $V_{S,30}$ from topografic slope method

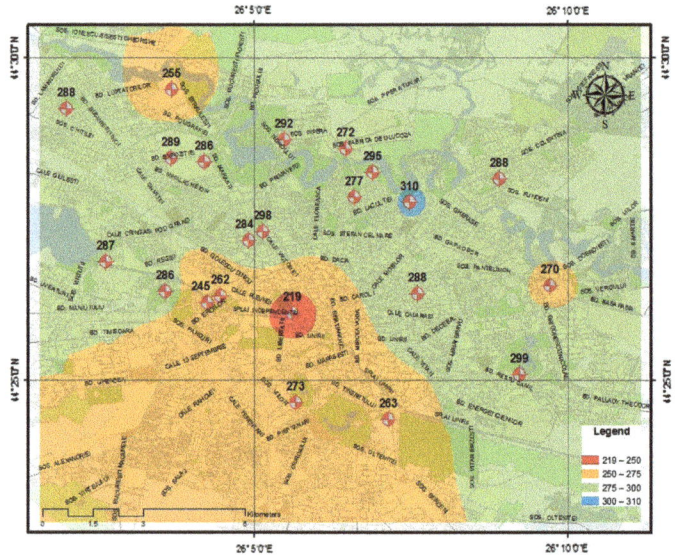

Fig. 3 Bucharest spatial distribution map of $V_{S,30}$ values based on NCSRR measurements (2003–2010)

shuttle Endeavor and $V_{S,30}$ values obtained from in situ measurements from different sites in USA, Australia, Taiwan, Italia, Puerto Rico, New Zeeland and Japan. The results were extrapolated and used to create a global map for $V_{S,30}$ values, available on USGS server. The method was and tested by Lemoine et al. (2012) for 706 sites in Europe and Middle East. Their conclusions reiterate the recommendation of Wald and Allen (2007) for using with caution this method for local or site-specific first order studies.

The Availability of average $V_{S,30}$ values for Romania territory based on Wald topographic slope methodology (Fig. 3) allowed the identification of ground types at seismic stations. This information was used for the development of ground motion prediction equations GMPE which include a parameter of ground condition,

(Vacareanu et al. 2014a, b). Figure 2 presents the $V_{S,30}$ map for Romania region (Neagu 2015) created with data downloaded from USGS website.

One can notice from Table 1 that the differences between borehole measured values and topographical slope estimated values of $V_{S,30}$ vary between −23 and 28% with a 12% mean value. Both methods place all 19 Bucharest sites in ground type "C" according to Eurocode 8 ground type classification (shear wave velocity between 180 and 360 m/s).

In absence of more in situ $V_{S,30}$ measurements, the Wald topographical slope $V_{S,30}$ estimation may be used for ground type classification in regional or national studies.

A GIS-mapping of $V_{S,30}$ values obtained from down-hole measurements performed in Bucharest urban area was developed (Fig. 3). According to EC8 and P100-1/2013, the site classification suggests that all the considered sites correspond to type C (intermediate soil). However, depending on the $V_{S,30}$ values, two subclasses can be observed. The subclass with $V_{S,30}$ values ranging from 250 to 275 m/s in the central part of Bucharest along Dambovita River and southern part, while the subclass with $V_{S,30}$ values ranging from 275 to 300 m/s is corresponding to northern, north-western and eastern part of the city.

This map may supply the up-to-date knowledge for Bucharest sites applicable for engineering and other research purposes. One should taking into account that the accuracy is primarily dependent on the amount and geographic distribution of data.

3 Soil Factor S

Eurocode 8 elastic response spectra is dependent on ground type (five ground types from A—rock to E—soft soil) defined according to $V_{S,30}$. A soil factor S is also considered with different values corresponding to the ground types.

Romanian seismic design code P100-1/2013 defines elastic response spectra for three different site categories which are classified according the corner period T_C through a national zonation map, and no soil factor is considered.

Vacareanu et al. (2014a, b) evaluates the values of the soil factor S using strong ground motions recorded from Vrancea intermediate-depth earthquakes and the procedures described in Rey et al. (2002) and Pitilakis et al. (2012, 2013). The study used a database of 233 strong ground motion (465 horizontal components) recorded during 9 intermediate-depth Vrancea earthquakes with moment magnitude M_W between 5.2 and 7.4. Ground types at seismic station sites were classified according to Eurocode 8 ($V_{S,30}$ determined using Wald methodology) and by corner period T_C in the Romanian seismic code. Tables 2 and 3 present the soil factor

Table 2 Soil factor values for Eurocode 8 ground types classification

Ground type	All earthquakes	$M_W > 6.3$ earthquakes
B	3.59	2.67
C	2.97	2.63

Table 3 Soil factor values for P100-1/2013 site classification	Site classification	All earthquakes	$M_W > 6.3$ earthquakes
	$0.7\ s < T_C \leq 1.0\ s$	2.05	1.03
	$T_C > 1.0\ s$	1.95	0.94

S values evaluated for both ground types classifications (by Eurocode 8 and by P100-1/2013).

Since in Table 3 for strong earthquakes the computed values are nearly 1, therefore there is no need for S factor. The results showed that the actual definition of spectral shapes from Romanian seismic design code with equal values of maximum amplification factor for all site classes is appropriate.

Measured V_S profiles are recommended for local and site-specific studies. However, one should consider that even measured soil properties are not free of uncertainties and should be regarded as best estimates.

4 Liquefaction in Bucharest

The earthquake-induced liquefaction in Bucharest during 1977 Vrancea strong earthquake ($M_W = 7.5$) occurred within a limited area of the old river bank of Dambovita River, in Quaternary alluvial sandy deposits in the form of sand boils, without reported damages on structures (Ishihara and Perlea 1984; Youd 1977). For the liquefaction occurrence zonation of this area, geotechnical data were collected from more than 100 boreholes. The simplified zonation procedure was based on the presence of saturated sands up to 15 m depth, with a thickness larger than 3 m, covered by thin non-liquefiable layers. Extensive field measurements based on borings and penetration tests were carried out for assessing the liquefaction potential of two sites near the Dambovita River (Ishihara and Perlea 1984). A maximum horizontal acceleration of about 0.2 g similar to the one recorded at INCERC station in 1977 and the ground water level at 1 m depth from the ground surface were considered. Empirical correlations to estimate the cyclic strength value of sandy layers on the basis of N-value obtained from SPT were used for computing the safety factor against liquefaction. Recent studies have been carried out for assessing the liquefaction potential of sites based on the results from situ investigations using the correlations given in existing procedures (Hannich et al. 2007; Arion et al. 2007, 2015).

4.1 Assessment of Liquefaction Potential in Bucharest

In the present study three methods for evaluation of soil liquefaction potential were applied.

The first method evaluates the liquefaction potential in terms of safety factor against liquefaction, F_s, defined by Ishihara (1985) using the following equation:

$$F_s = \frac{CRR}{CSR} \tag{1}$$

when $F_s \leq 1.0$, the liquefaction is predicted to occur, while if $F_s > 1.0$ no liquefaction can occur.

The cyclic resistance ratio CRR is computed using the Andrus and Stokoe (1997) equations:

$$CRR = a\left(\frac{V_{s1}}{100}\right)^2 + b\left(\frac{1}{V_{s1}^* - V_{s1}} - \frac{1}{V_{s1}^*}\right) \tag{2}$$

where: V_{S1}—overburden-stress corrected shear wave velocity; a, b—curve fitting parameters with $a = 0.022$, $b = 2.8$, V_{S1}^*—limiting upper value of V_{S1} for liquefaction occurrence ($V_{S1}^* = 200$–215 m/s).

For correcting V_S to a reference overburden stress, the Robertson et al. (1992) equation was used:

$$V_{s1} = V_s\left(\frac{P_a}{\sigma'_{v0}}\right)^{0.25} \tag{3}$$

where: V_S—measured shear-wave velocity (m/s), P_a—reference stress of 100 kPa, σ'_{v0}—initial effective vertical stress (kPa).

The value of CRR was adjusted for the magnitude of the earthquake under consideration, with a magnitude scaling factor, MSF:

$$CRR = CRR_{M=7.5} \cdot MSF \cdot K_\sigma \tag{4}$$

where: $CRR_{7.5}$—cyclic resistance ratio with reference to an earthquake of magnitude of $M_W = 7.5$, MSF—magnitude scaling factor with $MSF = 1$ for the earthquake with $M_W = 7.5$ (Youd and Noble 1997), K_σ—overburden stress correction factor.

The cyclic stress ratio CSR generated by an earthquake is determined from the peak horizontal ground acceleration by considering the weighing factor to calculate the equivalent uniform stress cycles required to generate the same pore water pressure during an earthquake with 0.65 value according to the simplified equation suggested by Seed and Idriss (1971):

$$CSR = 0.65 \cdot \frac{a_{max}}{g} \cdot \frac{\sigma_{vo}}{\sigma'_{vo}} \cdot r_d \tag{5}$$

where: g—acceleration of gravity, σ_{vo} and σ'_{vo}—total vertical overburden stress and effective vertical overburden stress at the depth of interest, a_{max}—peak horizontal ground acceleration, r_d—depth-dependent stress reduction factor (Youd and Noble 1997).

The second method to quantify the liquefaction potential is in terms of liquefaction probability (P_L) based on the correlation with the safety factor proposed by Juang et al. (2002) and the classification of liquefaction likelihood followed the classes proposed by Chen and Juang (2000):

$$P_L = \frac{1}{1 + (F_S/0.73)^{3.4}} \tag{6}$$

The third method is the evaluation of liquefaction potential index (*LPI*). The index is computed by combining depth, thickness and the safety factor against liquefaction of soil layers for predicting the liquefaction potential at the ground surface at the site of interest in relation to earthquake magnitude and peak value of ground acceleration. For the analysis, the layers up to a depth of 20 m of the soil profile were considered. Using the expression proposed by Luna and Frost (1998), the liquefaction potential index is calculating:

$$LPI = \sum_{i=1}^{n} w_i F_i H_i \tag{7}$$

where: w_i—weighting factor ($w_i = 10 - 0.5z_i$, with z_i—depth of *i*-th layer), H_i—thickness of soil layers; n_i—number of layers; F_i—liquefaction severity of *i*-th layer ($F_i = 0$ for $F_{si} \geq 1.0$, $F_i = 1 - F_{si}$ for $F_{si} < 1.0$); F_{si}—safety factor for *i*-th layer.

4.2 GIS Mapping of Liquefaction Susceptibility Using V_S Data

The liquefaction parameters corresponding to saturated cohesionless layers are computed and presented in Table 4. Two earthquake magnitudes ($M_W = 7.5$ and $M_W = 7.7$, as in case of 1977 and 1940 Vrancea earthquakes) were considered.

Table 4 Example of parameters used in liquefaction analysis

Depth (m)	R_d	Overburden pressure		Fines content (%)	Measured V_S value (m/s)	Corrected V_S value (m/s)	Cyclic stress ratio (CSR)	Cyclic resistance ratio ($CRR_{7.5}$)	Safety factor
3	0.982	49.5	49.5	5	140	167	0.191		n.a
11	0.882	185.5	146.27	7	220	200	0.218	0.16	0.75
18	0.779	302.5	194.62	5	300	254	0.236	0.16	0.69
28	0.658	467.5	261.55	5	360	283	0.229	0.22	0.97
35	0.618	587.5	312.9	12	390	293	0.226	0.24	1.07
50	0.618	857.5	435.8	12	400	277	0.237	0.21	0.89

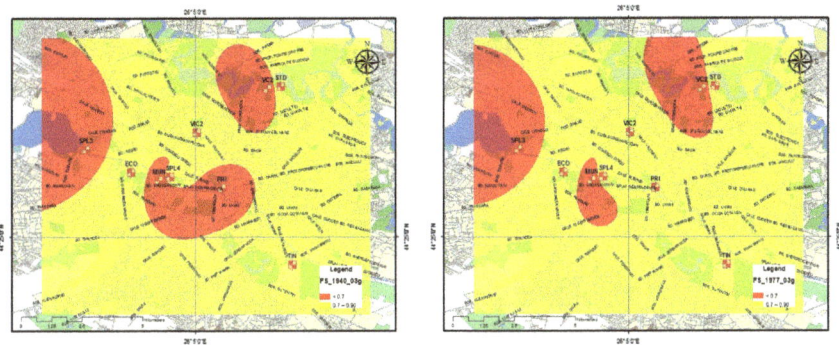

Fig. 4 Spatial distribution of liquefaction safety factor (*Fs*)

Peak horizontal ground acceleration (*PGA*) of 0.3 g for a mean recurrence interval of earthquakes of 225 years is also used as input seismic data according to the specifications of P100-1/2013, Romanian seismic code.

Soil liquefaction potential in terms of safety factor (*Fs*), probability of liquefaction (*PL*) and liquefaction potential index (*LPI*) are determined. Maps of the spatial distribution of these values are generated using GIS tools. It should be noted that there are some limitations of the present liquefaction analysis consisting in the following issues: (i) use of V_S field data which involves small-strain measurement, poor accuracy near the ground surface and influence of vibration sources, (ii) independent option in selecting the empirical correlation used in liquefaction analysis, (iii) calculation of *LPI* values for layers with a depth up to 20 m of soil profile, (iv) the simplified definition of seismic actions lead to conservative estimation of liquefaction parameters.

The spatial distribution maps illustrate a concentration of lower *Fs* < 0.7 values in the central part of Bucharest along Dambovita River and northern part, near Colentina River, that corresponds to very high potential for liquefaction (*LPI* > 15). Moreover, the other sites indicate the occurrence of liquefaction (*Fs* < 1) and high potential for liquefaction (5 < *LPI* < 15) if the scenario conditions are similar. In case of liquefaction probability, *PL* values show that liquefaction or non-liquefaction is equally likely for the most part of sites, except for two locations near Dambovita and Colentina River, where the liquefaction is very likely (Figs. 4, 5 and 6).

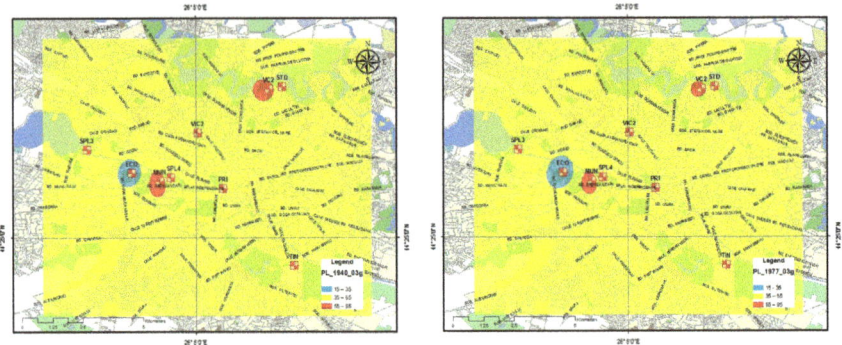

Fig. 5 Spatial distribution of liquefaction probability (*PL*)

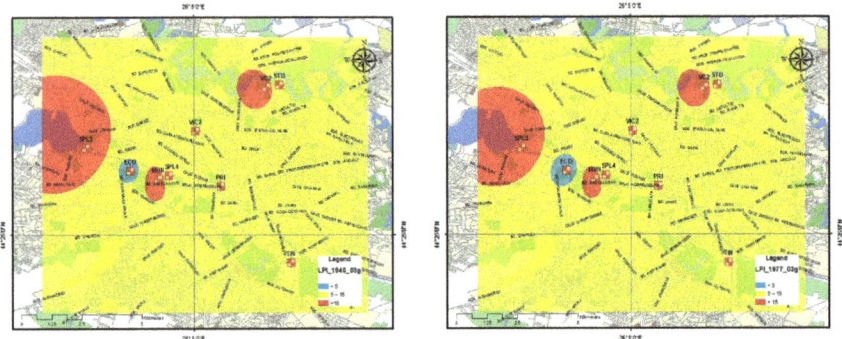

Fig. 6 Spatial distribution of liquefaction potential index (*LPI*)

5 Conclusions

The classification of soil/site categories for seismic design is generally based on shear wave velocity profiles. In Romania, except for Bucharest region, measured V_S data are not available.

The $V_{S,30}$ values derived from slope topography (as proposed by Wald et al. 2004) can be used at regional scale with caution. V_S values derived from SPT and CPT correlations also involve uncertainties, and such tests are possible only over limited depths and generally for medium to soft soils. Other methods such as ambient vibrations (array microtremor techniques), multichannel analysis of surface waves (*MASW*), Multi-Offset-Vertical-Seismic-Profiling method (*MOVSP*), etc. may also be used for the estimation of a velocity profile.

The control period of response spectra T_C can be used as a more direct parameter for differentiating spectral shapes for different site conditions. In order to improve the definition of seismic action for design and to reduce the uncertainties of seismic zonation maps, Romania needs a national strategy and action plan for instrumental investigation of soil characteristics (soil stratigraphy and velocity profile) and the development of its seismic network. Ideally, the soil condition of each seismic station with records should be identified through in situ techniques.

Starting from historical information on liquefaction occurrence during strong Vrancea earthquakes and also the existence of triggering factors related to ground conditions, completed with 1D modelling of site-specific data, the liquefaction susceptibility at different sites in Bucharest is determined. This study reveals a high probability of liquefaction for the sites near Dambovita old riverbed, with factor of safety (F_S < 1). The assessment of liquefaction potential based on V_S data lead to locate the saturated liquefiable deposits at various depths ranging from 8 m up to 17 m. Liquefaction potential index (*LPI*) values show different levels of liquefaction severity in Bucharest area, with no significant difference between the data sets corresponding to the selected scenarios of earthquake magnitudes. The generated GIS maps indicate high susceptibility of soils to liquefaction with *LPI* = 5–15 and very high susceptibility (with *LPI* > 15) mainly in the central part of the city near Dambovita River, linked to large thickness of alluvial layers and shallow ground water levels.

The liquefaction potential GIS spatial distribution using data from different evaluation methods may be a useful analytical tool for future earthquake induced damage.

Acknowledgements Authors would like to acknowledge Japan International Cooperation Agency (JICA) for the equipment donated within JICA Project in Romania, for the training periods in Japan of NCSRR research staff and for dispatching Japanese specialists for short or long period in Romania. We kindly acknowledge the support of Building Research Institute (BRI) and Tokyo Soil Research, Japan.

The results presented in this paper were partially obtained, within JICA Project for Seismic Risk Reduction on Building and Structures in Romania (2002–2007), NATO Science for Peace Project (2005–2008) and BIGSEES—Bridging the gap between seismology and earthquake engineering: from the seismicity of Romania towards a refined implementation of Seismic Action EN1998-1 in earthquake resistant design of buildings (2012–2016).

References

Aldea A, Yamanaka H, Negulescu C, Kashima T, Radoi R, Kazama H, Calarasu E (2006) Extensive seismic instrumentation and geophysical investigations for site-response studies in Bucharest, Romania. In: ESG 2006 third international symposium on the effects of surface geology on seismic motion, Grenoble, France, Paper Number: 69, 10 p, CD-ROM

Andrus RD (1994) In situ characterization of gravelly soils that liquefied in the 1983 Borah peak earthquake. Ph.D. dissertation, University of Texas

Andrus RD, Stokoe KH (1997) Liquefaction resistance of soils from shear-wave velocity. In: Proceedings of NCEER workshop on evaluation of liquefaction resistance of soils, national centre for earthquake engineering research, Buffalo, pp 89–128

Andrus RD, Stokoe KH (2000) Liquefaction resistance based on shear-wave velocity. J Geotech Eng ASCE 126(11):1015–1025

Andrus RD, Stokoe KH, Juang CH (2004) Guide for shear-wave-based liquefaction potential evaluation. Earthq Spectra 20(2):285–305

Arion C, Tamura M, Calarasu E, Neagu C (2007) Geotechnical in situ investigation used for seismic design of buildings. In: 4th International conference on earthquake geotechnical engineering, paper no. 1349

Arion C, Calarasu E, Neagu C (2015) Evaluation of bucharest soil liquefaction potential. Math Model Civil Eng 11(1):5–12

ASCE/SEI 7-10 (2010) Minimum design loads for buildings and other structures. American Society of Civil Engineers, USA

Bala A, Arion C, Aldea A (2013) In situ borehole measurements and laboratory measurements as primary tools for the assessment of the seismic site effects. Rom Rep Phys 65(1):285–298

Borcherdt RD, Glassmoyer G (1992) On the characteristics of local geology and their influence on ground motions generated by the Loma Prieta earthquake in the San Francisco Bay region, California. Bull Seismol Soc Am 82:603–641

Boulanger RW, Idriss IM (2004) Evaluating the potential for liquefaction or cyclic failure of silts and clays. Rep. No. UCD/CGM-04/01, Center for Geotechnical Modeling, Department of Civil and Environmental Engineering, University of California, Davis, California

Chen CJ, Juang CH (2000) Calibration of SPT and CPT based liquefaction evaluation methods, innovations applications in geotechnical site characterization. Geotechnical special publication, ASCE, New York, vol 97, pp 49–64

Chen YM, Ke H, Chen RP (2005) Correlation of shear wave velocity with liquefaction resistance based on laboratory tests. Soil Dyn Earthq Eng 25(6):461–469

Chu DB, Stewart JP (2004) Documentation of soil conditions at liquefaction and non-liquefaction sites from 1999 Chi-Chi Taiwan earthquake. Soil Dyn Earthq Eng 24(9–10):647–657

Hannich D, Hoetzl H, Ehret D, Huber G, Danchiv A, Bretotean M (2007) Liquefaction probability in Bucharest and influencing factors. In: International symposium on strong Vrancea earthquakes and risk mitigation, 4–6 Oct 2007, Bucharest, Romania, pp 205–221

http://infp.infp.ro/bigsees/default.htm

IBC 2012 (2011) International building code. International Building Council, USA

Idriss IM, Boulanger RW (2003) Relating $K\alpha$ and $K\sigma$ to SPT blow count and to CPT tip resistance for use in evaluating liquefaction potential. In: Proceedings of the 2003 dam safety conference, ASDSO, 7–10 Sept, Minneapolis, MN

Idriss IM, Boulanger RW (2010) SPT-based liquefaction triggering procedures. Report UCD/CGM-10/02, Department of Civil and Environmental Engineering, University of California, Davis, CA, 259 p

Idriss IM, Boulanger RW (2012) Probabilistic standard penetration test-based liquefaction—triggering procedure. J Geotech Geoenviron Eng ASCE 138(10):1185–1195

Ishihara K (1985) Stability of natural deposits during earthquakes. In: Proceedings of 11th international conference on soil mechanics and foundation engineering, vol 1, pp 321–376

Ishihara K, Koga Y (1981) Case studies of liquefaction in the 1964 Niigata earthquake. Soil Found 21(3):33–52

Ishihara K, Ogawa K (1978) Liquefaction susceptibility map of downtown Tokyo. In: Proceedings of the 2nd international conference on microzonation, San Francisco, vol 2, pp 897–910

Ishihara K, Perlea V (1984) Liquefaction-associated ground damage during the Vrancea earthquake of March 4, 1977. Soils Found 24(1):90–112

Ishihara K, Cubrinovski M, Nonaka T (1998) Characterization of undrained behaviour of soils in the reclaimed area of Kobe. Soils Found 2:33–46

Iwasaki T, Tokida K, Tatsuoka F, Watanabe S, Yasuda S, Sato H (1982) Microzonation of soil liquefaction potential using simplified methods. In: Proceedings of 3rd international conference on microzonation, Seattle, vol 3, pp 1319–1330

Juang CH, Jiang TJ, Andrus RD (2002) Assessing probability-based methods for liquefaction potential evaluation. J Geotech Geoenviron Eng 128(7):580–589

Kayen R, Moss RES, Thompson EM, Seed RB, Cetin KO, Der Kiureghian A, Tanaka Y, Tokimatsu K (2013) Shear-wave velocity-based probabilistic and deterministic assessment of seismic soil liquefaction potential. J Geotech Geoenviron Eng 139(3)

Lemoine A, Douglas J, Cotton F (2012) Testing the applicability of correlations between topographic slope and VS30 for Europe. BSSA 102(6):2585–2599

Lewis MR, Arango I, Stokoe KH (2013) Liquefaction resistance of gravelly soils. In: Proceedings 7th international conference on case histories in geotechnical engineering, Chicago

Liao SSC, Veneziano D, Whitman RV (1988) Regression models for evaluating liquefaction probability. J Geotech Eng ASCE 114(4):389–411

Luna R, Frost JD (1998) Spatial liquefaction analysis system. J Comput Civil Eng 12:48–56

Lungu D, Cornea T, Aldea A, Zaicenco A (1997) Basic representation of seismic action. Design of structures in seismic zones: Eurocode 8—worked examples. In: Lungu D, Mazzolani F, Savidis S (eds) TEMPUS PHARE CM Project 01198: implementing of structural Eurocodes in Romanian civil engineering standards. Bridgeman Ltd., Timisoara, pp 1–60

Lungu D, Aldea A, Moldoveanu T, Ciugudean V, Stefanica M (1999) Near-surface geology and dynamic properties of soil layers in Bucharest, Vrancea earthquakes: tectonics, hazard and risk mitigation. Springer, Netherlands, pp 137–148

Moss RES, Seed RB, Kayen RE, Stewart JP, Der Kiureghian A, Cetin KO (2006) CPT-based probabilistic and deterministic assessment of in situ seismic soil liquefaction potential. J Geotech Geoenviron Eng ASCE 132(8):1032–1051

Neagu C (2015) Local soil condition and nonlinear soil response influence on design seismic action. Ph.D. Thesis, Technical University of Civil Engineering of Bucharest (in Romanian)

Orlowski D, Witte C, Loske B (2003) Execution and evaluation of seismic measurements in Bucharest by the Multi-Offset-Vertical-Seismic-Profiling method (MOVSP) (in German), Internal report, DMT, Mines & More Division, Essen

Pitilakis K, Riga E, Anastasiadis A (2012) Design spectra and amplification factors for Eurocode 8. Bull Earthq Eng 10:1377–1400

Pitilakis K, Riga E, Anastasiadis A (2013) New code site classification, amplification factors and normalized response spectra on a worldwide ground-motion database. Bull Earthq Eng. https://doi.org/10.1007/s10518-013-9429-4

Rey J, Faccioli E, Bommer J (2002) Derivation of design soil coefficients (S) and response spectral shapes for EC 8 using the European strong-motion database. J Seismol 6:547–555

Robertson PK, Campanella RG (1985) Liquefaction potential of sands using the cone penetration test. J Geotech Div ASCE 22(3):277– 286

Robertson PK, Wride CE (1998) Evaluating cyclic liquefaction potential using the cone penetration Test. Can Geotech J 35(3):442–459

Robertson PK, Woeller DJ, Finn WDL (1992) Seismic cone penetration test for evaluating liquefaction potential under cyclic loading. Can Geotech J 29:(4)686–695

Seed HB, Idriss IM (1967) Analysis of liquefaction: Niigata earthquake. ASCE 93(SM3):83–108

Seed HB, Idriss IM (1971) Simplified procedure for evaluating soil liquefaction potential. J Soil Mech Found Div ASCE 97(SM9):1249–1273

Seed HB, Ugas C, Lysmer J (1974) Site-dependent spectra for earthquake-resistant design, earthquake. Engineering Research Center, University of California, Berkley, EERC, pp 74–12

Suzuki Y, Koyamada K, Tokimatsu K (1997) Prediction of liquefaction resistance based on CPT tip resistance and sleeve friction. In Proceedings, 14th International Conference on Soil Mechanics and Foundation Engineering, Hamburg, Germany, Vol 1, pp 603–06

Tokimatsu K, Uchida A (1990) Correlation between liquefaction resistance and shear wave velocity. Soils Found Tokyo 30(2):33–42

Vacareanu R, Mărmureanu G, Pavel F, Neagu C, Cioflan CA, Aldea A (2014a) Analysis of soil factor S using strong ground motions from Vrancea subcrustal seismic source. Rom Rep Phys 66(3)

Vacareanu R, Radulian M, Iancovici M, Pavel F, Neagu C (2014b) Fore-arc and back-arc ground motion prediction model for Vrancea intermediate-depth seismic source. In: Proceedings of the second European conference on earthquake engineering and seismology (2ECEES), Istanbul, Aug 24–29 2014, Paper no. 484

Wakamatsu K (1993) History of soil liquefaction in japan and assessment of liquefaction potential based on geomorphology. Ph.D. thesis, Waseda University, Tokyo

Wald DJ, Allen TI (2007) Topographic slope as a proxy for seismic site conditions and amplification. Bull Seism Soc Am 97(5):1379–1395

Wald DJ, Earle PS, Quitoriano V (2004) Topographic slope as a proxy for seismic site correction and amplification. EOS Trans AGU 85(47):F1424

Yasuda S, Harada K, Ishikawa K, Kanemaru Y (2012) Characteristics of liquefaction in Tokyo Bay area by the 2011 Great East Japan earthquake. Soils Found 52(5):793–810

Youd TL (1977) Reconnaissance report of geotechnical observations for the 4 March 1977 Romanian earthquake. Extract from EERI report

Youd JL, Perkins DM (1978) Mapping of liquefaction induced ground failure potential. J Geotech Eng Div ASCE 104:433–446

Youd TL, Noble SK (1997) Liquefaction criteria based on probabilistic analyses, in NCEER Workshop on evaluation of liquefaction resistance of soils. National Center for Earthquake Engineering Research. Technical report NCEER-97-0022, pp 201–216

Ground Motion Intensity Versus Ground Motion Kinematics. Exploring Various Intensity Measures

Horea Sandi

Abstract The interest for this topic was raised by the occurrence of the destructive Vrancea earthquake of 1977.03.04, when the first strong motion record of Romania was obtained at the Building Research Institute (INCERC) Bucharest. The attempt to assess the ground motion intensity on the basis of instrumental criteria specified by the Medvedev-Sponheuer-Karnik (MSK) intensity scale failed, due to strongly divergent results obtained by means of using the peak ground acceleration (*PGA*) and peak ground velocity (*PGV*) criteria. This failure was due to the fact that the inflexible MSK criteria referred to relied on the implicit non-realistic assumption that all ground motions are characterized by an invariable velocity/acceleration corner period of 0.5 s. A flexible *Spectral Intensity Assessment System* (*SIAS*) relying on ground motion accelerographic information was developed. This makes it possible to estimate, according to needs, global intensity, frequency related intensity, intensity averaged upon a spectral band etc. Alternative basic kinematic ground motion characteristics for global intensities and for frequency related intensities were introduced. Correlation analysis performed revealed quite strong correlations between results of using alternative criteria. Characteristic parameters were calibrated and alternative calibrations are discussed. Some illustrative cases are presented. Comments and recommendations are finally presented.

Keywords Intensity · Instrumental intensity · Frequency related intensity Intensity spectrum

1 Introduction

A need to relate intensity to kinematic characteristics of ground motion was felt already long ago, at a time when neither instrumental data on strong motion, nor appropriate instruments were available. A well-known attempt at providing a

H. Sandi (✉)
Romanian Academy of Technical Sciences, Bucharest, Romania
e-mail: horeasandi@yahoo.com

© Springer International Publishing AG, part of Springer Nature 2018   173
R. Vacareanu and C. Ionescu (eds.), *Seismic Hazard and Risk Assessment*,
Springer Natural Hazards, https://doi.org/10.1007/978-3-319-74724-8_12

Table 1 Average values of kinematic parameters according to the MSK 1976 scale

MSK intensity	PGA (cm/s^2)	PGV (m/s)	$PS_M D$ (mm)
VI	50	4	2
VII	100	8	4
VIII	200	16	8
IX	400	32	16

solution was given in the frame of the MSK scale (Medvedev 1977). These quantifications were rather close to conventional, reduced, engineering design values. According to the most recent version of the instrumental criteria of that scale, the average values for *PGA* (peak ground acceleration), *PGV* (peak ground velocity) and $PS_M D$ (peak displacement of Medvedev's seismoscope, having a natural period of 0.5 s and a logarithmic decrement of 0.5), corresponding to the intensity degrees VI to IX, were as in Table 1.

The examination of this table puts to evidence that:

– the values adopted build geometric progressions (common ratio: 2.0);
– the values adopted correspond to a standard response spectrum shape (more precisely, a velocity/acceleration corner period of 0.5 s, as adopted in (Medvedev 1962), on the basis of examination of response spectra for Californian strong motion records).

The starting point of the studies presented was the occurrence of the destructive Vrancea earthquake of 1977.03.04. The first strong ground motion acceleration record of Romania was then obtained at INCERC Bucharest. In the aftermath of the event, the author was asked by a highly placed government official to assess the ground motion intensity in agreement with the instrumental criteria of the MSK scale, as provided in the Romanian standard (IRS 1971). The attempt to fulfill this request failed, because a huge gap, of two intensity degrees, occurred between the outcomes of using alternatively the *PGA* and the *PGV* criteria prescribed in the standard. It turned out that it is necessary to critically reconsider the concept of ground motion intensity. It turned out that this was due to the assumption on which the standard implicitly relied, namely that the dynamic factor of the motion corresponds to the function $\beta(T)$,

$$\beta(T) = \beta_0 \quad (T \leq 0.5 \text{ s}) \tag{1.a}$$

$$\beta(T) = \beta_0 \times 0.5/T \quad (T > 0.5 \text{ s}) \tag{1.b}$$

On the contrary, experience shows that response spectra of actual motions reveal a wide manifold of kinds of response spectra that should be taken into account. For instance, the response spectra of the event referred to show maximum spectral values for a period $T \approx 1.5$ s.

Following developments are intended to offer a solution to the task of assessing ground motion intensity on the basis of instrumental information, by bridging the gap between:

- the traditional seismological approach [as revealed by the two intensity scales most recently endorsed by the European Seismological Commission, namely Medvedev (1977) and Grünthal (1998)], which are blind to the consideration of spectral characteristics; and
- the use of knowhow of structural dynamics that makes it possible to adopt a much stronger approach, oriented towards an in depth analysis of the features of ground motion.

2 Analytical Background

2.1 General

In order to assess intensity on the basis of instrumental data, the first step is to make available an analytical way to convert some of the kinematic ground motion characteristics into a kind of intensity measure. Several ways to do this (Sandi 1986; Sandi and Floricel 1998) are presented subsequently. According to the needs of the performed ground motion analysis, it is desirable to alternatively consider features like *global intensity*, J_X, or *frequency related intensity*, $j_x(\varphi)$, or *intensity averaged upon a frequency band*, $j_x^\sim(\varphi', \varphi''')$, where the argument φ is measured in Hz. A common way adopted in order to assess *global intensity*, J_X, was to use the expression (2.a) or, in order to assess *frequency related intensity*, $j_x(\varphi)$, to use the expression (2.b). The various intensity characteristics dealt with are defined as functions of some kinematic characteristics of the ground motion, denoted Q_X or $q_x(\varphi)$, defined subsequently. The subscripts X or x are replaced by other ones, depending upon the concrete measures used. All of the functions Q_X, $q_x(\varphi)$ have a physical dimension $<L^2 \times T^{-3}>$, while the various homologous intensity measures J_X, $j_x(\varphi)$ etc. derived on their basis, are non-dimensional. The variable subscripts X and x are to be replaced by other ones, according to Table 2. The general expressions defining the basic intensity measures J_X, $j_x(\varphi)$ are

$$J_X = \log_b Q_X + J_{X0} \qquad (2.a)$$

$$j_x(\varphi) = \log_b q_x(\varphi) + j_{x0} \qquad (2.b)$$

(the expression of $j_x^\sim(\varphi', \varphi''')$ is similar to (2.b), with the same free term j_{x0}).

The value of the subscript b was initially adopted as $b = 4$, in order to correspond to the product of the geometric ratios $\rho_a = \rho_v = 2$, of the kinematic criteria *PSA* and *PSV* of the MSK scale (Medvedev 1977), respectively, while the free term J_{X0} had to be calibrated such as to obtain a best possible correlation coefficient

Table 2 System of instrumental criteria for intensity assessment (SIAS)

Name	Symbols used for intensities: * global; ** related to a frequency φ; *** averaged upon a frequency interval (φ', φ'')			Source of definition/comments
	*	**	***	
Spectrum based intensities	J_S	$j_s(\varphi)$	$\tilde{j_s}(\varphi', \varphi'')$	Product of maxima (with respect to φ) of linear response spectra for absolute accelerations and for absolute velocities Product of linear response spectra for absolute accelerations and for absolute velocities
Intensities based on Arias' type integral	J_A	$j_d(\varphi)$	$\tilde{j_d}(\varphi', \varphi'')$	Integral of square of acceleration of ground (for J_A). Integral of square of acceleration of pendulum of natural frequency φ(for $j_d(\varphi)$). Both extensible to tensorial definitions; averaging rules specified
Intensities based on integral of fourier transforms	$J_F(\equiv J_A)$	$j_f(\varphi)$	$\tilde{j_f}(\varphi', \varphi'')$	Integral of Fourier image of acceleration (for J_F). Integral of square of Fourier images (for $j_d(\varphi)$). Both extensible to tensorial definitions; averaging rules specified

between the values corresponding to different criteria Q_X, $q_x(\varphi)$ etc. The subscript X, corresponding to global intensities, and the subscript x, corresponding to frequency related intensities, had to be replaced by some different subscripts respectively, corresponding to the alternative types of criteria (see Table 2).

It is possible to meet situations in which averaging intensities is of interest. This operation can be related to averaging upon spectral intervals and/or upon directions of motion. The spectral averaging is to be performed, for parameters specific to intensities $j_x(\varphi)$, by means of the relation between homologous entities, $\tilde{q_x}(\varphi', \varphi'')$, and $q_x(\varphi)$,

$$\tilde{q_x}(\varphi', \varphi'') = [1/\ln(\varphi''/\varphi')] \int_{\varphi}\backslash^{\varphi''} q_s(\varphi)\mathrm{d}\varphi/\varphi \qquad (3)$$

and/or averaging upon two orthogonal, horizontal, directions, by means of the relations

$$Q_{X,1,2} = (Q_{X,1} + Q_{X,2})/2 \qquad (4.a)$$

$$q_{x,1,2}(\varphi) = [q_{x,1}(\varphi) + q_{x,2}(\varphi)]/2 \qquad (4.b)$$

The relation (4.b) can be applied also to intensities averaged upon a time interval (3).

2.2 Global Intensities

<1> A first idea to look for a measure of an unique, or global, measure of intensity was suggested by the way which was adopted in some cases in order to specify design spectra for industrial equipment, which led to the use of peak spectral acceleration and peak spectral velocity,

$$Q_S = (PSA/2.5) \times (PSV/2.5) \tag{5}$$

Note here that the corner frequency T_S is given by the expression

$$T_S = (2\pi) \times PSV/PSA \tag{6}$$

<2> Another idea, suggested by the Arias definition of intensity (Arias 1970), starts from the integral of the square of ground acceleration, $a(t)$,

$$Q_A = \int [a(t)]^2 dt \tag{7}$$

<3> A quite similar starting point relies on the use of the Fourier transform of the accelerogram,

$$a^{(\varphi)}(\varphi) = \int_{-\infty}^{\infty} \exp(-2i\pi\varphi t)a(t)dt \tag{8}$$

of which the integral of the square (argument φ: Hz)

$$Q_\varphi = \int \left[|a^{(\varphi)}(\varphi)|\right]^2 d\varphi \tag{9}$$

is used (note here that, due to analytical reasons, $Q_\varphi \equiv Q_A/2$).

2.3 Intensities Related to Frequency (Intensity Spectra)

The intensities depending on frequency can be considered, in analytical terms, to be homologous to the global intensities referred to previously.

<4> In order to define intensities homologous to the version <1>, the motion of a pendulum of undamped frequency φ and of 0.05 critical damping is used, for which the product of peak values of response spectra for the absolute acceleration $\max_t |w(t, \varphi, 0.05)|$, and the absolute velocity, $\max_t |v(t, \varphi, 0.05)|$ respectively, are considered:

$$q_s(\varphi) = \max_t |w(t, \varphi, 0.05)| \times \max_t |v(t, \varphi, 0.05)| \tag{10}$$

<5> In order to define intensities homologous to the version <2>, the acceleration $a^{(\varphi)}(t, \varphi, 0.05)$ of the motion of a pendulum of undamped natural frequency φ and of 0.05 critical damping, for which the integral

$$q_d(\varphi) = \int \left[a^{(\varphi)}(t, \varphi, 0.05) \right]^2 dt \tag{11}$$

is used.

<6> In order to define intensities homologous to the version <3>, the function

$$q_f(\varphi) = \varphi \left| a^{(\varphi)}(\varphi) \right|^2 \tag{12}$$

is adopted.

2.4 Discretized Intensity Spectra

The intensities $j_x(\varphi)$ previously defined may be referred to as continuous intensity spectra. An alternative approach is to use discrete intensity spectra $j_x^{\sim}(\varphi', \varphi'')$, related to frequency intervals (φ', φ''). It is desirable to organize the sequence of limits <φ', φ''> as a geometric sequence. As an example, the discrete, average intensity spectra $j_x^{\sim}(\varphi', \varphi'')$ used subsequently, correspond to a geometric ratio $\varphi''/\varphi' = 2$ (or 6 dB). More concretely, the sequence of limits adopted is <0.25, 0.5, ... 16.0 Hz>, where a *common central value* is 2.0 Hz. The sequence of values φ referred to includes practically the most important domain of accelerograms. Some illustrative examples of intensity spectra are presented in the next section.

2.5 Correlation Analysis. Calibrations Derived

It is interesting to perform an analysis of the correlation between the various kinds of intensities referred to. A high value of correlation coefficients means a tendency to stability of the results of use of the system dealt with and thus a promising credibility of the results. A correlation analysis was performed using the strong motion instrumental results obtained in Romania during the earthquakes of 1986.08.30, 1990.05.30 and 1990.05.31 (more than one hundred accelerograms). Correlation analysis is practically feasible for discretized intensity spectra and so it was performed. The outcome of analysis is presented in Tables 3, 4, 5 and 6.

Table 3 Correlation coefficients (upper triangle) and standard deviations (lower triangle) for global intensities and for average intensities upon spectral band (0.25, 16.0 Hz)

	J_S	J_A	$\tilde{j}_s$(0.25, 16.0 Hz)	$\tilde{j}_d$(0.25, 16.0 Hz)	$\tilde{j}_f$(0.25, 16.0 Hz)
J_S	*	0.94 ... 0.98	0.96 ... 0.98	0.94 ... 0.97	0.93 ... 0.97
J_A	0.14 ... 0.18	*	0.93 ... 0.98	1.00	0.99 ... 1.00
$\tilde{j}_s$(0.25, 16.0 Hz)	0.12 ... 0.14	0.15 ... 0.23	*	0.93 ... 0.98	0.92 ... 0.97
$\tilde{j}_d$(0.25, 16.0 Hz)	0.14 ... 0.17	0.02 ... 0.03	0.15 ... 0.23	*	0.99 ... 1.00
$\tilde{j}_f$(0.25, 16.0 Hz)	0.15 ... 0.17	0.04 ... 0.05	0.16 ... 0.23	0.04 ... 0.05	*

Table 4 Correlation coefficients for various spectral bands

(φ', φ''), Hz	$\tilde{j}_{sq} \leftrightarrow \tilde{j}_{dq}$	$\tilde{j}_{sq} \leftrightarrow \tilde{j}_{fq}$	$\tilde{j}_{dq} \leftrightarrow \tilde{j}_{fq}$
(0.25, 0.5)	0.96 ... 0.98	0.95 ... 0.98	0.98 ... 1.00
(0.5, 1.0)	0.96 ... 0.98	0.94 ... 0.99	0.99 ... 1.00
(1.0, 2.0)	0.94 ... 0.98	0.92 ... 0.98	0.99 ... 1.00
(2.0, 4.0)	0.92 ... 0.98	0.86 ... 0.96	0.98 ... 0.99
(4.0, 8.0)	0.91 ... 0.96	0.82 ... 0.86	0.95 ... 0.97
(8.0, 16.0)	0.84 ... 0.95	0.52 ... 0.78	0.78 ... 0.88

Table 5 Differences between the conversion constants for various intensity variants

	J_{S0}	J_{A0}	$\tilde{j}_{s0}$	$\tilde{j}_{d0}$	$\tilde{j}_{f0}$
J_{S0}	*	−1.26 ... −1.22	−0.31 ... −0.22	−2.27 ... −2.20	−1.06 ... −1.01
J_{A0}	1.25	*	0.93 ... 1.00	−1.00 ... −0.98	0.21 ... 0.22
$\tilde{j}_{s0}$	0.30	−0.95	*	−1.98 ... −1.93	−0.79 ... −0.72
$\tilde{j}_{d0}$	2.25	1.00	1.95	*	1.19 ... 1.22
$\tilde{j}_{f0}$	1.05	−0.20	0.75	−1.20	*

Table 6 Calibrations adopted for the constants J_{X0} and j_{x0}

Parameter	J_{S0}	J_{A0}	j_{s0}	j_{d0}	j_{f0}
Calibration	8.00	6.75	7.70	5.75	6.95

The correlation coefficients and the standard deviations are given in Table 3 for the alternative intensity definitions used. Further on, the correlation coefficients are given in Table 4 for alternative spectral bands of intensity spectra.

Note also that a comparison of macroseismic assessment of intensities with the outcome of instrumental estimates for several cases was encouraging (Sandi 1986).

In order to make the presented system usable, it is necessary to calibrate the logarithm basis b and the conversion constants J_{X0} și j_{x0}. For the parameter b the value $b = 4$ has been adopted initially. This corresponds to the geometric ratios ρ_a

and ρ_v adopted for acceleration and velocity in the MSK scale. Note that the value of b is equal to the product of geometric ratios ρ_a and ρ_v of acceleration and velocity corresponding to the various intensity values referred to in the scale, $b = \rho_a \times \rho_v = 2 \times 2 = 4$.

Table 5 presents the differences between the conversion constants J_{X0} and j_{x0} of relations (2.a), (2.b). In the upper triangle one can see the ranges of intervals, while in the lower triangle one can see the differences adopted, which correspond to the values rounded up to multiples of 0.05. The calibrations adopted for the constants J_{X0} and j_{x0} are given in Table 6.

2.6 Calibration of Characteristic Parameters—A Statistical Approach

The calibration presented previously cannot be stated to be a final one. The wealth of macroseismic and instrumental information which became gradually available made it possible to develop statistical studies on the relationships between macroseismic intensity and kinematic parameters (Aptikaev 2005). The wealth of data used was considerable, as shown in Table 7. The results obtained stood at the basis of the specification of instrumental criteria adopted in the frame of the draft new Russian Macroseismic Scale, RMS-04 (Aptikaev 2006).

For each case a macroseismic estimate (ranging from intensity II to intensity IX) and instrumental data on PGA, PGV, PGD, P, (where $P = PGA \times PGV$), were available. The empirical relations determined on a statistical basis are [with some updating with respect to (Aptikaev 2005)]: for peak ground accelerations, "A"; for peak ground velocities, "V"; for peak ground displacements "D"; and for peak wave kinematic power, "P" respectively:

$$\lg A (\equiv PGA), \text{cm/s}^2 = -0.75 + 0.4I \pm 0.39(0.25)$$
$$\text{(correlation coefficient: } 0.84) \tag{13}$$

$$\lg V (\equiv PGV), \text{cm/s} = -2.23 + 0.47I \pm 0.33(0.20)$$
$$\text{(correlation coefficient: } 0.84) \tag{14}$$

$$\lg D (\equiv PGD), \text{cm} = -4.26 + 0.68I \pm 0.65(0.33)$$
$$\text{(correlation coefficient: } 0.81) \tag{15}$$

Table 7 Number of records of various intensities used for statistical analysis (Aptikaev 2006)

Intensity	2	3	4	5	6	7	8	9
No. of records	75	75	172	391	353	212	178	84

$$\lg P, \text{cm}^2/\text{s}^3 = -2.22 + 0.87I \pm 0.49(0.41)$$
$$\text{(correlation coefficient: } 0.89)$$

$$(16)$$

It turned out that the values of geometric ratios were about $\rho_a = 2.51 \approx 2.5$ for *PGA* and $\rho_v = 2.95 \approx 3.0$ for *PGV*, $4.79 \approx 4.8$ for *PGD* and $7.41 \approx 7.5$ for *P*. This should imply a value of *b* of about $2.5 \times 3.0 \approx 7.5$. This means, among other, that the velocity /acceleration corner period T_c should tend to increase (as known) with increasing intensity. In parentheses are given values for intensities $J > 6$.

The fact that the factor 0.47 of relation (14) is higher than the homologous factor 0.40 of relation (13), while the factor 0.68 of relation (15) is higher than the homologous factor 0.47 of relation (14), correspond to a rather well known trend of increase of dominant oscillation periods of ground motion with increasing intensity (this trend was quite systematically observed, on the basis of instrumental data obtained at a same location during different earthquakes, in Romania too). These results, which correspond to reality, are in direct contradiction with the features of the MSK scale criteria, which relied on the assumption of fixed corner periods, irrespective of intensity.

Looking at the values of kinematic parameters derived on the basis of previous relations, it turns out that adopting these assumptions, one obtains reasonable values even for lowest intensities, for which the assumption of fixed values of $\rho_a = \rho_v = 2.0$ for a jump of one intensity unit did no longer work. Therefore, it appears to be reasonable to adopt alternative rates, referred to further on. In case the rounded up values suggested are accepted, the result would be a value $b = 7.5$, which would make it possible to cover in a satisfactory manner an extensive interval of intensities, going e.g. downwards up to intensity 2.

3 Illustrative Cases

Some significant cases, where strong motion accelerograms were obtained, are presented subsequently. The most significant aspect revealed is represented by the relationship between response spectra and discrete intensity spectra. The eight cases dealt with (Table 8) differ mainly by the spectral locations of the domains of their most significant amplitudes.

The case <1> of the table corresponds to the classical, reference, motion record of 1940.05.18 at El Centro, Calfornia.

The case <2> of the table corresponds to the motion with long dominant periods, recorded on 1985.09.19 at the site of the Segretería Comunicaciones y Transportes, of Mexico City, where the existence of a very soft upper geological package of about 40 m thickness was at the origin of a motion with a long dominant period, exceeding 2 s.

Table 8 Illustrative ground motion characteristics

No.	General data on event and on record site	Characteristics of ground motion along two orthogonal horizontal directions			Magnitudes and global intensities
		Accelerograms	Response spectra for absolute accelerations (5% critical damping)	Intensity spectra: $\tilde{i_s}(\varphi', \varphi'')$, red, and $\tilde{i_d}(\varphi', \varphi'')$, blue, averaged upon 6 dB intervals	
1	*Source:* USA, Ca, Imperial Valley 1940.05.18 *φ, λ:* 32.73, −115.45 *Record site:* El Centro *Code:* 1940ELC *φ, λ:* 39.72, −115.55				M_S: 7.0 I_S: 8.01 (8.18, 7.80) I_{SI}: 8.34 (8.52, 8.09) I_A: 8.43 (8.55, 8.30) I_{DI}: 8.44 (8.56, 8.31)
2	*Source:* N. A./Cocos Plates (Mexico) 1985.09.19 *φ, λ:* 18.08, −102.94 *Record site:* Mex. City SCT *Code:* 1985SCT *φ, λ:* 19.43, −99.13				M_S: 8.1 I_S: 8.92 (8.55, 9.16) I_{SI}: 8.42 (7.90, 8.72) I_A: 8.40 (8.03, 8.64) I_{DI}: 8.47 (8.09, 8.72)

(continued)

Table 8 (continued)

No.	General data on event and on record site	Characteristics of ground motion along two orthogonal horizontal directions			Magnitudes and global intensities
		Accelerograms	Response spectra for absolute accelerations (5% critical damping)	Intensity spectra: $\tilde{i}_s^{\sim}(\varphi', \varphi'')$, red, and $\tilde{i}_d^{\sim}(\varphi', \varphi'')$, blue, averaged upon 6 dB intervals	
3	*Source:* Romania, VSZ φ, λ: 45.34, 26.30 1977.03.04 *Record site:* Bucharest/ INCERC *Code:* 7711INC φ, λ: 44.44, 26.16				M_{G-R}: 7.2 I_S: 8.09 (7.64, 8.37) I_sI: 8.37 (7.97, 8.63) I_A: 7.85 (7.67, 7.99) I_DI: 7.89 (7.67, 8.04)
4	*Source:* Romania, VSZ φ, λ: 45.53, 26.47 1986.08.30 *Record site:* Bucharest/ INCERC *Code:* 8611INC φ, λ: 44.44, 26.16				M_{G-R}: 7.0 I_S: 6.68 (6.48, 6.84) I_sI: 6.86 (6.73, 6.96) I_A: 6.65 (6.61, 6.68) I_DI: 6.68 (6.63, 6.72)

(continued)

Table 8 (continued)

No.	General data on event and on record site	Characteristics of ground motion along two orthogonal horizontal directions			Magnitudes and global intensities
		Accelerograms	Response spectra for absolute accelerations (5% critical damping)	Intensity spectra: $i_s^\sim (\varphi', \varphi'')$, red, and $i_d^\sim (\varphi', \varphi'')$, blue, averaged upon 6 dB intervals	
5	*Source:* Romania, VSZ φ, λ: 45.82, 26.90 1990.05.30 *Record site:* Bucharest/ INCERC *Code:* 901INC1 φ, λ: 44.44, 26.16				M_{G-R}: 6.7 I_S: 6.33 (6.55, 6.02) I_SI: 6.61 (6.88, 6.16) I_A: 6.45 (6.62, 6.22) I_DI: 6.46 (6.64, 6.21)
6	*Source:* Romania, VSZ φ, λ: 45.53, 26.47 1986.08.30 *Record site:* Cernavodă/ Town Hall *Code:* 861CVD1				M_{G-R}: 7.0 I_S: 5.87 (6.05, 5.64) I_SI: 5.67 (5.79, 5.53) I_A: 6.46 (6.56, 6.35) I_DI: 6.48 (6.57, 6.37)

(continued)

Table 8 (continued)

No.	General data on event and on record site	Characteristics of ground motion along two orthogonal horizontal directions			Magnitudes and global intensities
		Accelerograms	Response spectra for absolute accelerations (5% critical damping)	Intensity spectra: $i_s^{\sim}(\varphi', \varphi'')$, red, and $i_d^{\sim}(\varphi', \varphi'')$, blue, averaged upon 6 dB intervals	
7	φ, λ: 45.23, 28.03 *Source:* Romania, VSZ φ, λ: 45.82, 26.90 1990.05.30 *Record site:* Cernavodă/ Town Hall *Code:* 901CVD1 φ, λ: 45.23, 28.03				M_{G-R}: 6.7 I_S: 6.93 (6.92, 6.93) I_SI: 6.75 (6.81, 6.69) I_A: 6.93 (6.95, 6.90) I_DI: 6.95 (6.98, 6.92)
8	*Source:* Romania, VSZ φ, λ: 45.82, 26.90 1990.05.30 *Record site:*				M_{G-R}: 6.7 I_S: 6.93 (6.92, 6.93) I_SI: 6.75 (6.81, 6.69) I_A: 6.93 (6.95, 6.90) I_DI: 6.95 (6.98, 6.92)

(continued)

Table 8 (continued)

No.	General data on event and on record site	Characteristics of ground motion along two orthogonal horizontal directions			Magnitudes and global intensities
		Accelerograms	Response spectra for absolute accelerations (5% critical damping)	Intensity spectra: $\tilde{i_s}\,(\varphi',\,\varphi'')$, red, and $\tilde{i_d}\,(\varphi',\,\varphi'')$, blue, averaged upon 6 dB intervals	
	Cernavodă/ Town Hall *Code:* 901CVD1 φ,λ: 45.23, 28.03				

Note indices used

L (longitudinal): direction X of record

T (transversal): direction Y of record

I (one): averaging of frequency dependent intensities over the frequency interval (0.25, 16.0 Hz)

(no index): global intensities

Units used for kinematic parameters: m, m/s, m/s^2

Abscissa scale (periods) for response spectra and intensity spectra: logarithmic

Ordinate scale: natural

The triad <3>, <4>, <5> of the table corresponds to the sequence of strong motion accelerograms recorded at the site of INCERC (Building Research Institute, Bucharest), where the records obtained on 1977.03.01, 1986.08.30 and 1990.05.30 presented a sequence with quite strong spectral differences. The differences referred to between the characteristics of successive motions were presented simultaneously by the results of response spectra of locations inside the City of Bucharest and even by the spectra at some sites located outside Bucharest (Sandi and Borcia 2010). Note that the influence of local geological conditions presented a quite low and weak influence, since the geological profiles at the corresponding sites did not present strong contrasts of S-wave propagation velocity at relatively small depths.

The triad <6>, <7>, <8> of the table corresponds to the case of the sequence of three records of 1986.08.30, 1990.05.30 and 1990.05.31 at the site of the Cernavodă Town Hall, for which there was a stability of the dominant period, of about 0.4 s, while the local geological conditions were characterized by a strong contrast of S-wave propagation velocity at a depth in the range of 20 m.

Some comments on the results presented: the discrete intensity spectra as functions of oscillation period, are characterized, as the response spectra as well, by a significant variation; note also that there is not a detailed proportionality of the variation as functions of period. In case one wants to compare the level of significance of the two kinds of spectra, it turns out that intensity spectra reflect definitely better than response spectra the destructive capability of ground motions. Spectral domains, for which the ordinates of intensity spectra are 9.0 or more, correspond to zones of disastrous earthquake effects.

The case <1> corresponds to a relatively uniform destructive power of ground motion, which nevertheless has a maximum of severity in the spectral band (0.1, 1.0 s). The case <2> is characterized by a strong non-uniformity of the destructive power that appears especially in the spectral band of periods longer than 1 s. Indeed, the cases of collapse occurred especially for such situations. The case <3> reveals a stronger destructive power especially for periods around 1 s, which corresponds to relatively tall (around 10 story tall) buildings, during the 1977.03.04 event. The cases <4> and <5> corresponded to non-destructive events, in which the intensities were lower than 8.0, (the most severe effects occurring for periods shorter than 1 s). The cases <6>, <7> and <8> are characterized by a quite strong stability of the spectral band for which the intensities were maximum.

4 Final Considerations

It may be stated that the system briefly presented, SIAS, offers a quite flexible instrument of ground motion characterization that makes it possible to provide information ranging from the approach specific to macroseismic analysis up to a more detailed approach, as required by engineering activities.

The instrumental data collected during earthquakes provide a rich source of information, but this information is rigorously valid just for the locations of instruments. On the contrary, the macroseismic analysis makes it possible, in favorable cases, to get a holistic view on the features of ground motion along certain areas. This fact makes it possible to develop ways of analysis that could help to investigate some specific aspects of seismic hazard for various systems at risk.

The use of SIAS provides more accurate information than the traditional macroseismic analysis of intensity. On the other hand, the instrumental information provided by strong motion networks of the type currently at hand is bound to be limited in the future too. It turns out that it is desirable to collect after earthquakes instrumental, as well as macroseismic data and to finally combine in an appropriate way the data at hand.

The intensity scales of traditional type [like MSK (Medvedev 1977) or (Grünthal 1998)] do not provide appropriate tools for a spectral characterization of ground motion. This, in spite of the fact that the comments attached to the latter scale referred to the recognition of the capability of instrumental data to fully characterize a ground motion recorded.

Note also that the new Russian scale (ШИЗ 2013) provides an instrumental criterion relying on the product $(PSA \times PSV)$, which is quite similar to J_S.

Besides the possibility to use, in the frame of the system, various ways of averaging with respect to oscillation frequency, averaging with respect to direction of oscillation is also feasible if of interest.

Acknowledgements The author is deeply indebted to his former colleagues of INCERC, Ioan Sorin Borcia and Ion Floricel, for their competent cooperation in performing a vast amount of computer work, as required by the data processing for a considerable number of cases.

References

Aptikaev F (2005) Instrumental seismic intensity scale. In: Proceedings of symposium on the 40-th anniversary of IZIIS, Skopje

Aptikaev F (2006) Project of Russian seismic intensity scale RIS-04. In: Proceedings of first European conference on earthquake engineering and seismology, Geneva, Switzerland, paper No. 1291

Arias A (1970) A measure of earthquake intensity. In: Hansen RJ (ed) Seismic design for nuclear power plants. The MIT Press, Cambridge

Grünthal G (1998) European macroseismic scale 1998, vol 15. Cahiers du Centre Européen de Géodynamique et Séismologie, Luxembourg

IRS Scara de intensități seismice. STAS 3684-71

Medvedev SV (1962) Inzhenernaya seismologia. GIFML, Moscow, Russia

Medvedev SV (1977) Seismic intensity scale MSK-76. Publ Inst Géophys Pol Ac Sc, A - 6. Warsaw

Sandi H (1986) An engineer's approach to the scaling of ground motion intensities. In: Proceedings of 8-th European conference on earthquake engineering, Lisbon, Portugal

Sandi H, Borcia IS (2011) Intensity spectra versus response spectra. Basic concepts and applications. PAGEOPH topical volume on advanced seismic hazard assessments (online edition 2010), printed in Pure Appl Geophys 168(1):261

Sandi H, Floricel I (1998) Some alternative instrumental intensity measures of ground motion severity. In: Proceedings of 11th European conference on earthquake engineering, Paris, France

ШИЗ (2013) ШКАЛА ИНТЕНСИВНОСТИ ЗЕМЛЕТРЯСЕНИЙ (Scale of earthquake intensity). Federal Agency of Technical Regulation and Metrology, Moscow, Russia

On the Ground Motions Spatial Correlation for Vrancea Intermediate-Depth Earthquakes

Ionut Craciun, Radu Vacareanu, Florin Pavel and Veronica Coliba

Abstract The spatial correlation of ground motions is a subject extensively anal-ysed in the literature, with many studies obtaining spatial correlation models for different datasets, using multiple ground motion prediction equations for various ground motion parameters, being a necessary tool in assessing the seismic risk of building portfolios or spatially distributed systems. A subject tackled mostly for shallow earthquakes datasets, in recent months, two studies considering a database consisting of strong ground motions generated by earthquakes originating from Vrancea intermediate-depth seismic source have been developed. The differences between the two studies consist of the approach in obtaining the correlation model, directly evaluating the correlation coefficients and using the semivariogram approach. A comparison of the two studies is made, resulting in different decay ratios for the same ground motion parameter, the main reasons being the restrains encountered in the first methodology and the subtraction of some ground motion records in the second study. An investigation regarding the influence of the dataset is performed by developing a correlation model for an adjusted dataset using the direct approach. Comparisons with other available models are performed, revealing higher correlation values and more gradual decays for the two studies discussed here, which is mainly caused by the different seismo-tectonic context of the Vrancea intermediate-depth source.

Keyword Correlation coefficient · Models · Semivariogram · Comparison

I. Craciun (✉) · R. Vacareanu · F. Pavel · V. Coliba
Seismic Risk Assessment Research Center, Technical University
of Civil Engineering of Bucharest, Bucharest, Romania
e-mail: ionut.craciun@utcb.ro

R. Vacareanu
e-mail: radu.vacareanu@utcb.ro

F. Pavel
e-mail: florin.pavel@utcb.ro

V. Coliba
e-mail: veronica.coliba@utcb.ro

© Springer International Publishing AG, part of Springer Nature 2018
R. Vacareanu and C. Ionescu (eds.), *Seismic Hazard and Risk Assessment*,
Springer Natural Hazards, https://doi.org/10.1007/978-3-319-74724-8_13

1   Introduction

Probabilistic Seismic Hazard Assessment (PSHA) originates in 1968 (Cornell 1968), being currently used on an international level for the seismic hazard analyses. An important step in PSHA is considering the uncertainties, which can be classified in aleatory and epistemic uncertainties. The epistemic uncertainties derive from the incomplete knowledge of a model and the aleatory uncertainties are related to the probabilistic nature of the ground motion parameters. Aleatory uncertainties are incorporated in PSHA through the standard deviation of the data resulted by using the ground motion prediction equations (Sokolov et al. 2010), and can be separated in inter-event variability (between earthquakes) and intra-event (within earthquake) variability.

The spatial correlation concept demonstrates that ground motion amplitudes for two given sites are correlated, the correlation coefficient depending on the separation distance between the two sites. A spatial correlation model is required in order to assess the seismic hazard/risk of spatially distributed systems (lifelines) and regionally located building assets (portfolios), given that the traditional PSHA does not offer any information about simultaneous ground motions at different sites.

Because the aleatory uncertainties/variability introduced through the use of GMPEs can be classified in inter-event (between earthquakes) variability and intra-event (within earthquake) variability, one can define inter-event, intra-event and total correlation coefficients. While the inter-event correlation is considered to be constant for a given earthquake, which means that all ground motions from that earthquake have some characteristics that separate them from other records, as described in Cimellaro et al. (2011), the intra-event correlation describes the correlation between the ground motions recorded in different sites for a given earthquake, and the total correlation is a combination of the two. Various relations for obtaining inter-event, intra-event and total correlation coefficients are available in literature, being a commonly used tool to study the ground motion spatial correlation.

The ground motion spatial correlation has been tackled in many studies beginning with the end of the 20th century. While the early studies focused on proving its existence and importance (Kawakami and Sharma 1999; Wang and Takada 2005; Boore et al. 2003), modern studies focused on the seismic hazard assessment using ground correlation models (Goda and Hong 2008b; Goda and Atkinson 2009, 2010), or on seismic risk assessment of spatially distributed systems (Sokolov and Wenzel 2011a, b; Esposito et al. 2014; Goda and Hong 2008a). With regard to the chosen ground motion parameter, correlation models were developed for peak ground velocity (Sokolov and Wenzel 2013), peak ground acceleration (Boore et al. 2003; Wang and Takada 2005), spectral response accelerations (Goda and Atkinson 2009, 2010) or Arias Intensity (Foulser-Piggott and Goda 2015).

Various datasets have been used for developing correlation models, including records from: earthquakes in Taiwan (e.g. Sokolov et al. 2010; Goda and Hong 2008b; Wang and Takada 2005), Californian earthquakes (e.g. Goda and Hong

2008b; Boore et al. 2003; Hong et al. 2009), earthquakes in Japan (various net-works were used in Goda and Atkinson 2009, 2010), earthquakes in the Istanbul region in Wagener et al. (2016), the European Strong-Motion Database (ESD) and the Italian Accelerometric Archive (ITACA) in Esposito and Iervolino (2011, 2012).

Prior to two recent studies (Vacareanu et al. 2017; Pavel and Vacareanu 2017), no information was known concerning ground motion spatial correlation for a dataset consisting of earthquakes generated by the Vrancea intermediate-depth seismic source. Given that in Vacareanu et al. (2017) the correlation coefficients are evaluated directly and Pavel and Vacareanu (2017) use the semivariogram approach, the methodology of obtaining the correlation model being the main difference between the two studies, the main scope of the present study is to perform a comparison between the results obtained in the two studies, developing a correlation model using the first approach for a dataset with similar characteristics to the one used in Pavel and Vacareanu (2017), as well as a comparison with other models available in the literature. To simplify, the model presented in Vacareanu et al. (2017) will also be referred hereinafter as the first model and the model developed in Pavel and Vacareanu (2017) as the second model.

2 Description of the Two Correlation Models

2.1 The Correlation Model Developed in Vacareanu et al. (2017)

A spatial correlation model was developed in Vacareanu et al. (2017) using a database consisting of earthquakes originating from Vrancea intermediate-depth seismic source, for the peak ground acceleration (PGA) and spectral response accelerations at various periods. Results are obtained for both the geometric mean of the two horizontal components and the random horizontal component of the ground motion parameter (in the present study, only the former will be discussed, as in Pavel and Vacareanu (2017) the results are given only in these terms). The present sub-chapter presents the main characteristics of the ground motion correlation model developed in Vacareanu et al. (2017).

The general current form of a ground motion prediction equation GMPE is as follows:

$$\ln Y_{ij}(T_n) = \ln \bar{Y}_{ij}(T_n) + \eta_i(T_n) + \varepsilon_{ij}(T_n) \tag{1}$$

where $Y_{ij}(T_n)$ is the ground motion parameter at the natural vibration T_n for site j, during earthquake i, $\ln \bar{Y}_{ij}(T_n)$ is the mean value of the natural logarithm of the ground motion parameter, $\eta_i(T_n)$ represents the inter-event residual, with zero mean and standard deviation $\sigma_\eta(T_n)$, and $\varepsilon_{ij}(T_n)$ represents the intra-event residual, with zero mean and standard deviation $\sigma_\varepsilon(T_n)$.

The total correlation coefficient was obtained as in Goda and Hong (2008b):

$$\rho_T(\Delta, T_n) = 1 - \frac{\sigma_d^2(\Delta, T_n)}{2\sigma_T^2(T_n)} \tag{2}$$

where $\sigma_d^2(\Delta, T_n)$ is the variance between the residual differences (Boore et al. 2003).

The intra-event correlation coefficient is obtained using an expression given in Goda and Hong (2008b):

$$\rho_\varepsilon(\Delta, T_n) = 1 - \frac{\sigma_d^2(\Delta, T_n)}{2\sigma_\varepsilon^2(T_n)} \tag{3}$$

The inter-event correlation coefficient is given in Wesson and Perkins (2001):

$$\rho_\eta(T_n) = \frac{\sigma_\eta^2(T_n)}{\sigma_T^2(T_n)} = \frac{\sigma_\eta^2(T_n)}{\sigma_\eta^2(T_n) + \sigma_\varepsilon^2(T_n)} \tag{4}$$

A database consisting of 10 earthquakes with $M_W > 5$ and focal depths of $87 \div 154$ km originating from the Vrancea intermediate-depth seismic source was chosen to develop a correlation model. The database was developed for the BIGSEES national research project (http://infp.infp.ro/bigsees/default.htm), from the seismic networks of Romania: INFP (National Institute for Earth Physics), INCERC (Building Research Institute), CNRRS (National Centre for Seismic Risk Reduction) and GEOTEC (Institute for Geotechnical and Geophysical Studies).

The GMPE developed in Vacareanu et al. (2015) was chosen in order to perform the analysis, the main reasons including: its modern form, its applicability to the selected database, considering the site conditions in the regression analysis, taking into account the differences in attenuation between fore-arc and back-arc regions and separation of the intra- and inter-event components of the GMPE's standard deviation.

After the residual pairs have been determined and paired according to their separation distance, and after obtaining the variances given in Eq. (3), the empirical correlation coefficients are computed accordingly. Using a bin size of separation distances of 5 km, out of 11,404 residual pairs, only 2669 pairs were used in the analysis, up to a separation distance of 100 km, resulting in a minimum number of pairs per bin of 59.

Like in other studies (Goda and Hong 2008b; Goda and Atkinson 2009, 2010), an exponential functional form was chosen for fitting the empirical correlation coefficients previously determined:

Table 1 Values of the parameter α for the correlation model developed in Vacareanu et al. (2017)

Period (s)	$\alpha(T_n)$
T = 0.0	0.218
T = 0.2	0.267
T = 0.7	0.158
T = 1.0	0.115
T = 2.0	0.126
T = 3.0	0.152

$$\rho_\varepsilon(\Delta, T_n) = \exp\left(-\alpha(T_n)\Delta^{\beta(T_n)}\right) \qquad (5)$$

where $\alpha(T_n)$ and $\beta(T_n)$ are the model parameters; $\beta(T_n)$ was considered equal to 0.5, spectral period independent, and $\alpha(T_n)$ is determined through nonlinear regression; its values for six spectral periods are presented in Table 1.

2.2   The Correlation Model Developed in Pavel and Vacareanu (2017)

The database used in Pavel and Vacareanu (2017) consists of eight seismic events from the database used in Vacareanu et al. (2017). The main differences with respect to the previous study are the use of only Ground type B and C records, as defined in EN 1998-1 (CEN 2004), and only the records located in the south–eastern areas of Romania (without taking into account the intra-Carpathian records). The reasons for the two differences are strictly related to the chosen ground motion prediction equation, the same as in the first case (Vacareanu et al. 2015): the intra-Carpathian records were eliminated due to the different attenuation in this area, as discussed in Vacareanu et al. (2015), and the reason for keeping only the ground type B and C records was that the GMPE functional form explicitly offers regression coefficients for the two ground types. Also, because the ground motion prediction equation uses the geometric mean of the two horizontal components of the ground motion parameter, the results are only presented in these terms.

The main steps in the procedure used in Pavel and Vacareanu (2017) are: obtaining the empirical semivariograms, selecting a fitting functional (there are many options in the literature: exponential, spheric, gaussian etc.), choosing the exponential model to perform the analysis and obtaining the model parameters.

The experimental semivariograms are obtained based on the normalized intra-event residuals, and can be determined using the following relation (Clark and Harper 2000):

$$\breve{\gamma}(h) = \frac{1}{2N(h)} \sum_{N(h)} [\varepsilon(u+h) - \varepsilon(u)]^2 \tag{6}$$

where $N(h)$ is the number of residual pairs separated by the separation distance h that were considered in the analysis, and $\varepsilon(u+h)$ and $\varepsilon(h)$ are the normalized intra-event residuals obtained for the separation distance h.

As in other studies that used the semivariogram approach to obtain a spatial correlation model (Esposito and Iervolino 2011, 2012; Jayaram and Baker 2009 etc.), the correlation coefficient can be obtained as:

$$\breve{\gamma}(h) = 1 - \breve{\rho}(h) \tag{7}$$

The exponential model (exponential functional form) adopted in Pavel and Vacareanu (2017) is presented in literature as (e.g., Clark and Harper 2000):

$$\gamma(h) = a\left[1 - \exp\left(\frac{-3h}{b}\right)\right] \tag{8}$$

where a and b represent the sill (considered equal to one, in this case), and the range of the semivariogram, respectively.

The parameter b was obtained through the two-stage least-squares method, similarly to the methodology presented in Esposito and Iervolino (2012). By using a bin size of 4 km, which led to a minimum number of data pairs per bin of 30, empirical values of the b parameter were determined for eight spectral period values.

The results obtained showed the fastest decay of the spatial correlation for period values of 0.2, 0.4 and 0.7 s and the slowest for long periods, 1.4 and 2.0 s, and a slower decay than other studies, which will be discussed in chapter 5.

3 Comparison Between the Two Correlation Models

Chapter 3 aims to compare the two correlation models and the procedure influence on the final results. The two correlation models are presented in Figs. 1, 2, 3, 4, 5 and 6, for *PGA*, *PSA* at $T = 0.2$, 0.7, 1.0, 2.0 and 3.0 s respectively.

The model developed in Vacareanu et al. (2017) shows a slower decay than that developed in Pavel and Vacareanu (2017). Also, while the first model offers a correlation coefficient of about $0.1 \div 0.35$ at a separation distance of 100 km, the second offers values of $0.05 \div 0.15$ at the same distance. The model developed in Pavel and Vacareanu (2017) presents higher values of the correlation coefficient

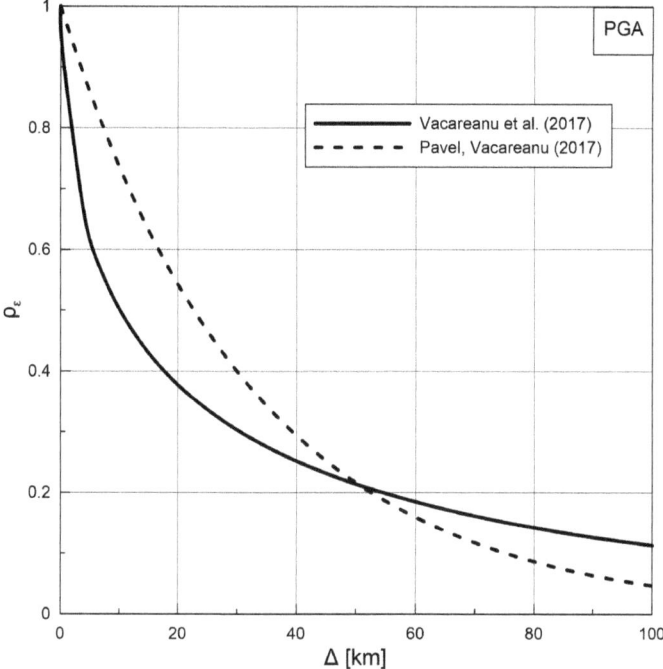

Fig. 1 Comparison between the two models for *PGA*

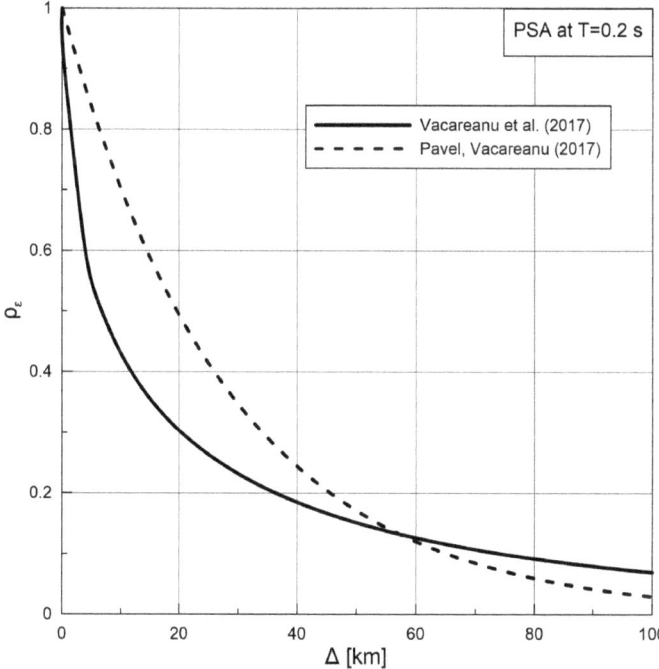

Fig. 2 Comparison between the two models for *PSA* at *T* = 0.2 s

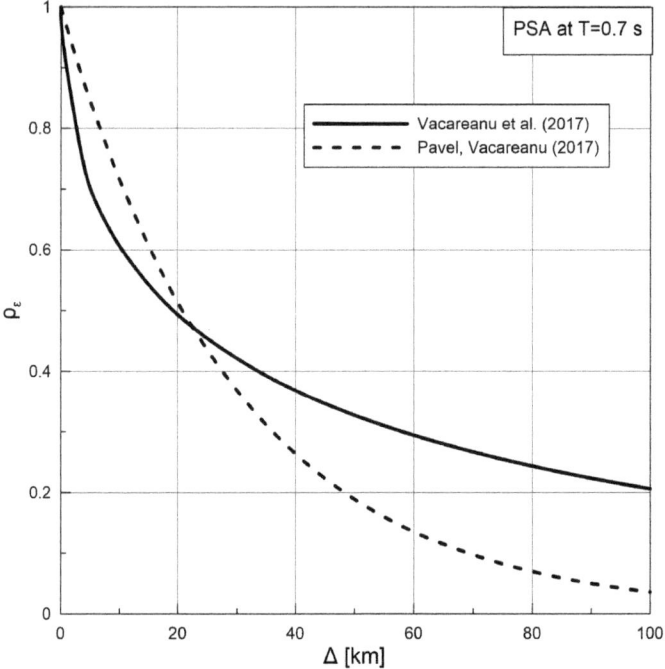

Fig. 3 Comparison between the two models for *PSA* at *T* = 0.7 s

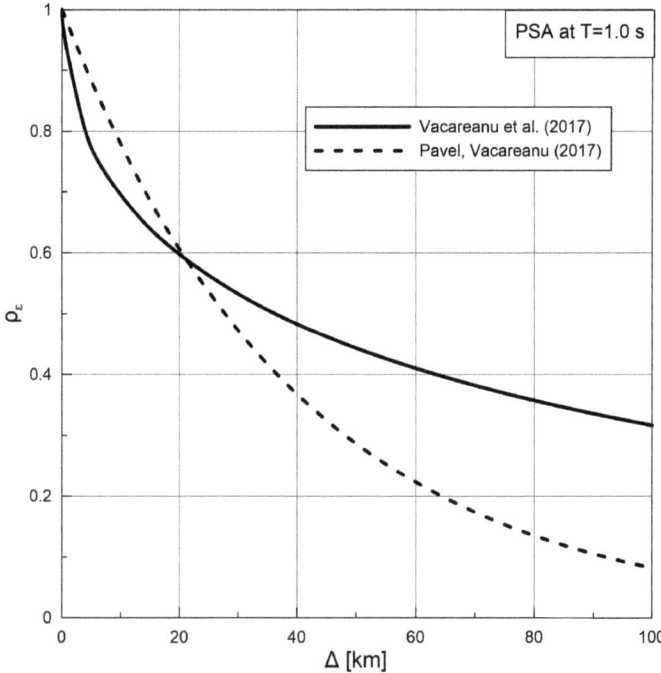

Fig. 4 Comparison between the two models for *PSA* at *T* = 1.0 s

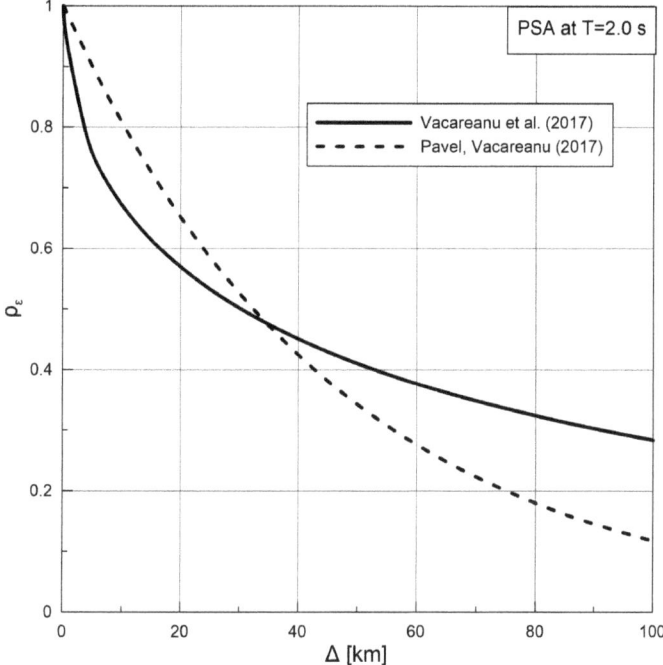

Fig. 5 Comparison between the two models for *PSA* at *T* = 2.0 s

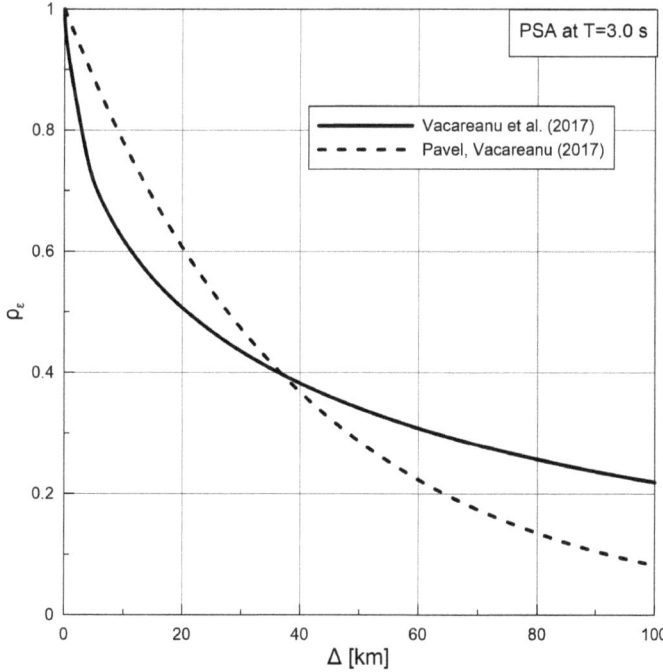

Fig. 6 Comparison between the two models for *PSA* at *T* = 3.0 s

than the first model for distances up to 60 km (for short periods, like 0, 0.1 s) and up to 35 km (for long periods, like 1, 3 s). It can be said that the model developed in Vacareanu et al. (2017) offers higher values of the correlation coefficient at long distances and the model developed in Pavel and Vacareanu (2017) for relatively short distances.

The reason for the lower decay visible in Vacareanu et al. (2017) can be caused by the different methodology of obtaining the correlation coefficients and especially to the differences in the used records. The main reason for the different decays are considered to stem from not considering in the analysis the records from the intra-Carpathian region and using only the records obtained on ground type B and C. Also, the procedure for obtaining the correlation coefficients for the Vacareanu et al. (2017) model, by using Eq. (3), and the relatively small number of data available, especially at short distances, lead to a limitation of available data in order to obtain the correlation coefficients, which can represent another reason for the different decays.

4   Developing a New Correlation Model

This chapter is focused on developing a correlation model for the Vrancea intermediate-depth database used in the first model (Vacareanu et al. 2017) adjusted so that it follows the assumptions made in the second model, using the direct approach (directly evaluating the correlation coefficients). In other words, only the Ground type B and C records, as defined in EN 1998-1 (CEN 2004), and only the records located in the south-eastern areas of Romania (without using the intra-Carpathian records) are used in the analysis. The database is therefore almost identical to that of the second model, which is slightly smaller because it only uses 8 seismic events. The reason for this is studying the influence of the dataset, especially through the exclusion of the above-mentioned records, which was presented in the previous chapter as a possible reason for the different decay ratios of the two models.

The methodology used to obtain the correlation model is identical to the one used in Vacareanu et al. (2017). Like in the two correlation models, the GMPE presented in Vacareanu et al. (2015) was used to develop a correlation model for the peak ground acceleration and spectral accelerations at periods varying from 0.1 to 3.0 s, consistent with the GMPE's parameter. The results are performed only for the geometric mean of the two horizontal components of *PGA* and *PSA* respectively, the comparison between the three models being possible only in these terms, as the Pavel and Vacareanu (2017) model results are obtained for the geometric mean of the two horizontal components of the ground motion parameter.

The bin size is chosen to be equal to 5 km, resulting in a minimum number of pairs per bin of 56. The histograms of the intra-event residual pairs with regard to distance, for a bin width of 5 km is presented in Fig. 7.

Out of 11,404 residual pairs, only 2316 pairs were used in the analysis (from 2669 pairs used in the first model, resulting in a reduction of 353 data pairs), until a separation distance of 100 km. The number of records, the number of available

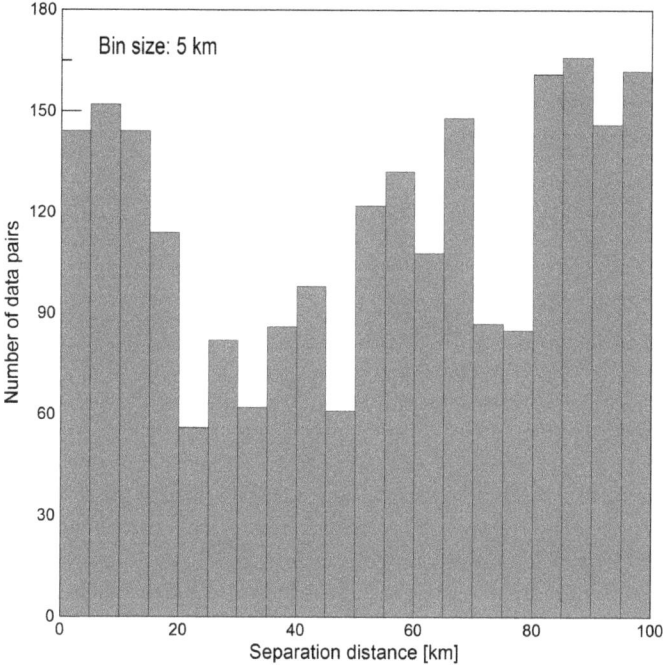

Fig. 7 Histogram of the number of intra-event residual pairs for the adjusted correlation model

Table 2 Number of available and used residual pairs

Event number	Date	Number of records	Number of available intra-event residual pairs	Number of used intra-event residual pairs	
				Vacareanu et al. (2017)	Present study
1	04.03.1977	2	1	0	0
2	30.08.1986	40	780	199	173
3	30.05.1990	52	1326	268	237
4	31.05.1990	36	630	144	127
5	28.04.1999	25	300	99	67
6	27.10.2004	66	2145	717	661
7	14.05.2005	40	780	245	219
8	18.06.2005	37	666	229	206
9	25.04.2009	46	1035	242	210
10	06.10.2013	87	3741	526	416

residual pairs and the number of used residual pairs (for the model presented in Vacareanu et al. (2017) and the model developed in the present chapter) for every seismic event are presented in Table 2.

Table 3 Values of the parameter α for the developed correlation model

Period (s)	α(T_n)
T = 0.0	0.182
T = 0.2	0.184
T = 0.7	0.159
T = 1.0	0.137
T = 2.0	0.162
T = 3.0	0.170

After the correlation coefficients are computed using Eq. 3, an exponential functional form identical to that presented in Vacareanu et al. (2017) is selected for fitting the empirical correlation coefficients. The values of the parameter α for six spectral periods are presented in Table 3 (similarly, the parameter β is considered equal to 0.5).

The developed correlation model and the correlation models presented in Vacareanu et al. (2017), Pavel and Vacareanu (2017) are presented comparatively in Figs. 8, 9, 10 and 11, for *PGA*, *PSA* at *T* = 0.7, 1.0, and 2.0 s respectively.

From analysing Figs. 8, 9, 10 and 11, the main observation is that the model developed in the present paper resembles the model presented in Vacareanu et al.

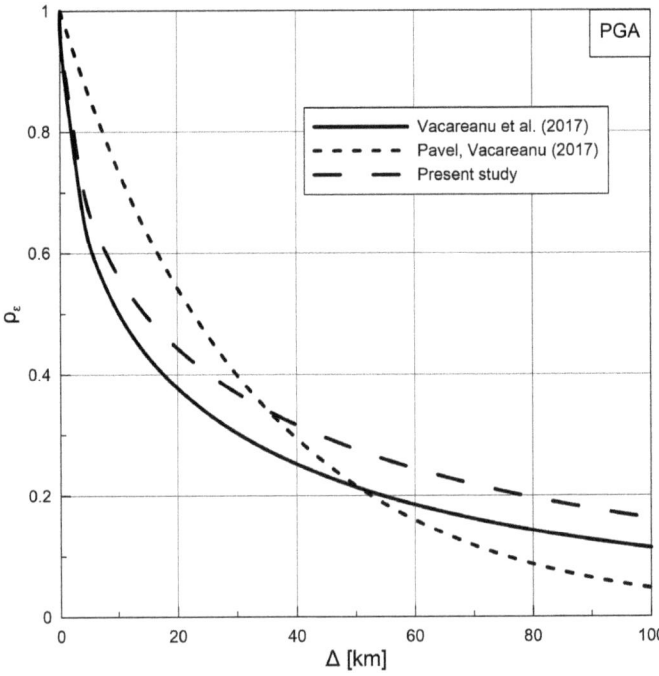

Fig. 8 Comparison between the two models and the developed model for *PGA*

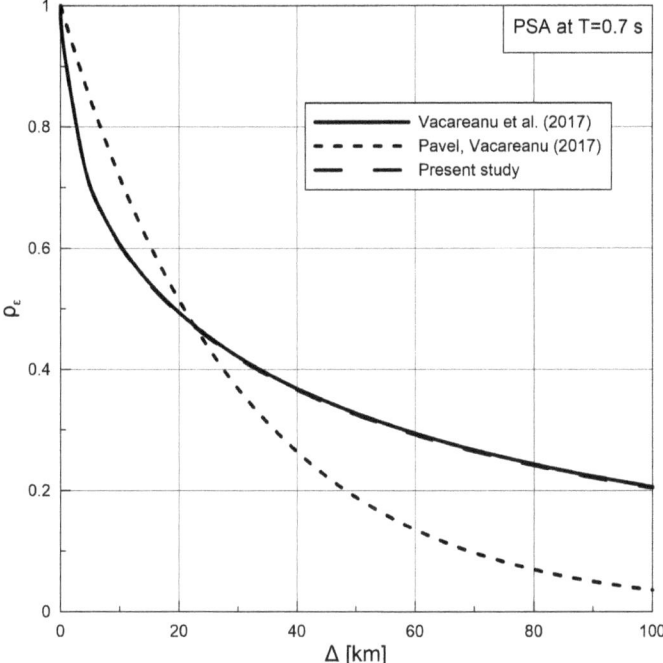

Fig. 9 Comparison between the two models and the developed model for *PSA* at *T* = 0.7 s

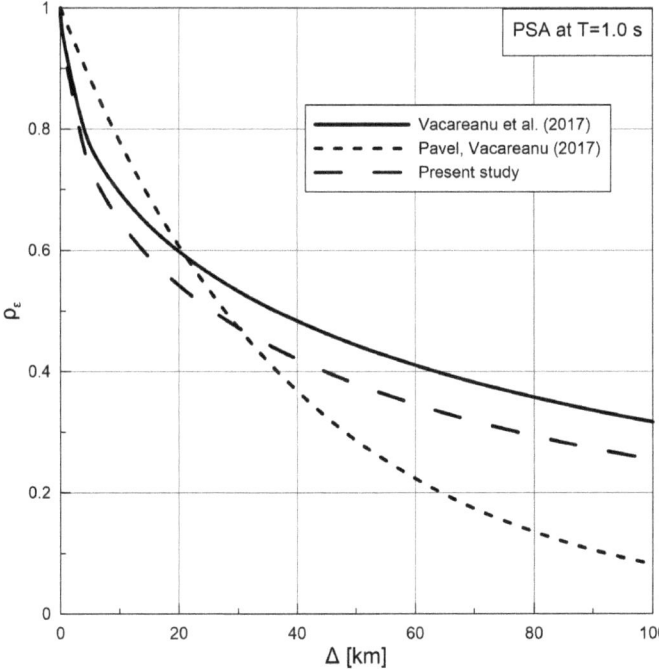

Fig. 10 Comparison between the two models and the developed model for *PSA* at *T* = 1.0 s

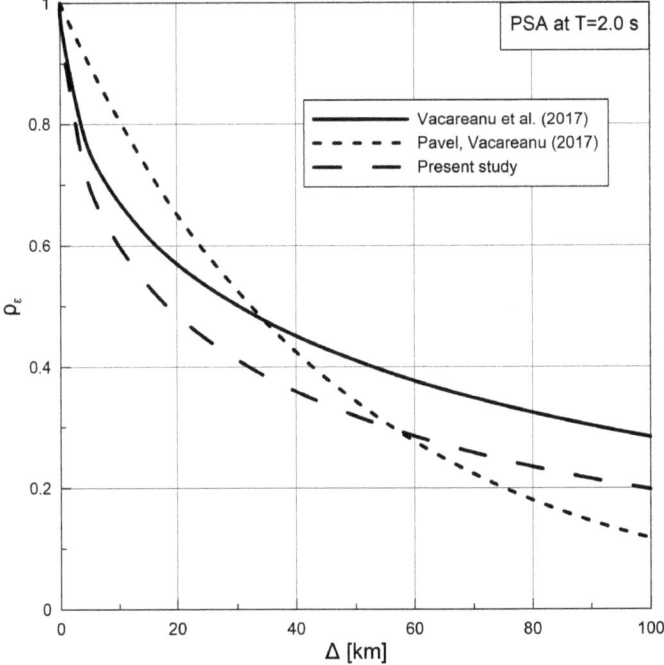

Fig. 11 Comparison between the two models and the developed model for *PSA* at *T* = 2.0 s

(2017). The model offers smaller correlation values than those presented by the Pavel and Vacareanu (2017) model for distances depending on the spectral value (e.g. until approximately 35 km for *PGA*).

One interesting aspect is that for short periods (e.g. for *T* = 0, 0.2 s), the developed model displays higher correlation values than those offered by the Vacareanu et al. (2017) model, until a spectral period value of 0.7 s, where the α parameter values are almost equal, which translates into almost overlapping functions, as visible in Fig. 9. For all spectral values greater than 0.7 s, the correlation values of the developed model are lower than those presented by the Vacareanu et al. (2017) model, as seen in Figs. 10 and 11. This implies an interchange regarding the higher correlation between the developed model and the Vacareanu et al. (2017) model at around a spectral period value of 0.7 s.

After analysing Figs. 8, 9, 10 and 11, no clear conclusions are identified regarding the impact of excluding Ground type A and D and intra-carpathian records on the correlation model, the results showing a similarity to the model presented in Vacareanu et al. (2017). However, the exclusion clearly affects the correlation model, with visible differences between the two models developed with the same procedure.

5 Comparison with Other Studies Available in the Literature

In order to compare the two correlation models with other results, we used the following models available in the literature: Wang and Takada (2005), Goda and Atkinson (2010), Goda and Hong (2008b), Esposito and Iervolino (2012). The datasets consist of the following: ground motion records of Californian and the Chi-Chi earthquakes treated separately (for comparison only the Californian database model was used) in Goda and Hong (2008b), Japanese ground motion records (in the K-NET, KiK-net and SK-net networks) in Goda and Atkinson (2010), Japanese and the Chi-Chi earthquakes in Wang and Takada (2005) and two different datasets, ITACA (Italian Accelerometric Archive) and ESD (European Strong-Motion Database), the present comparison being base only the ITACA results, in Esposito and Iervolino (2012). In Figs. 12 and 13, comparisons between the two Vrancea models correlation models and the above-mentioned models are presented for *PGA* and *PSA* at *T* = 1.0 s respectively.

Regardless of the differences between the two models for Vrancea intermediate-depth earthquakes, both offer higher correlation values and slower decay than most studies available in the literature. The correlation model Vacareanu

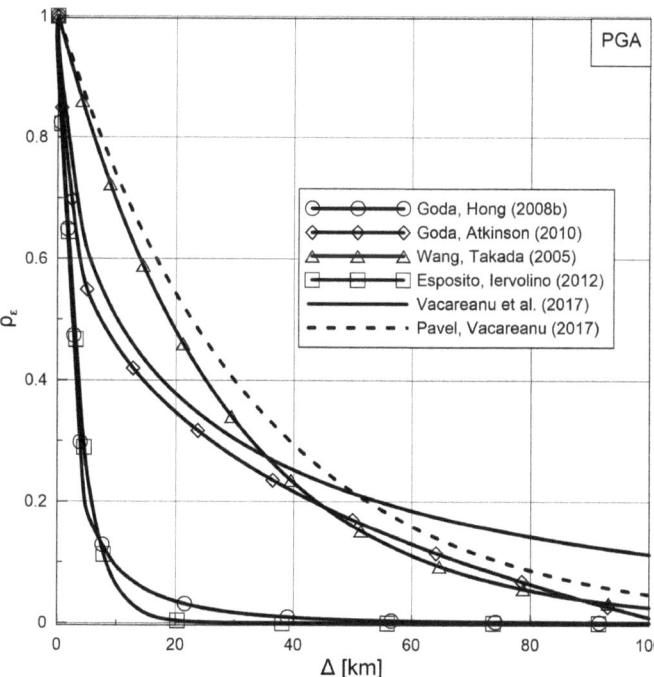

Fig. 12 Comparison between the two studies and other studies available in the literature for *PGA*

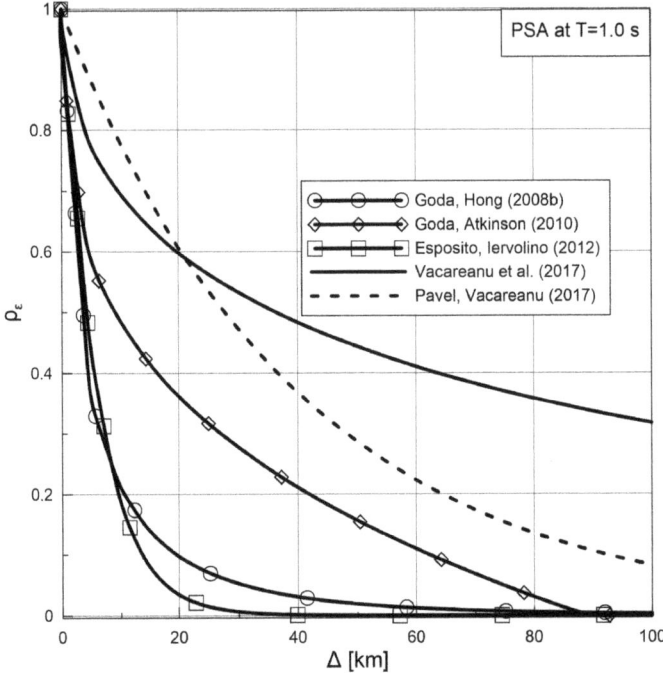

Fig. 13 Comparison between the two studies and other studies available in the literature for *PSA* at *T* = 1.0 s

et al. (2017) developed for *PGA* displays one of the slowest decay, similar to the correlation models developed for the Japanese earthquakes. It also offers the highest values of the correlation coefficient for long distances and relatively high values for short distances.

The similarity between the model developed in Vacareanu et al. (2017) and the one developed in Goda and Atkinson (2010) and that of Pavel and Vacareanu (2017), Wang and Takada (2005) is noticed for *PGA*. While for distances up to 40 ÷ 50 km, Vacareanu et al. (2017) model offers comparative results to that of Goda and Atkinson (2010), for distances above that interval, the Japanese model is more similar to the model developed in Pavel and Vacareanu (2017). Also, both models display much higher correlation values than the models presented in Esposito and Iervolino (2012), Goda and Hong (2008b).

For both Vrancea models, a faster decay is noticed for longer periods (at a separation distance of 100 km, the Vacareanu et al. (2017) model offers a correlation coefficient value of about 0.15 for *PGA*, and of approximately 0.35 for *PSA* at *T* = 1.0 s; the second model displays a correlation coefficient value of about 0.05 for *PGA*, and of about 0.10 for *PSA* at *T* = 1.0 s).

The high correlation obtained by both Vrancea models, especially in comparison with other available models, can be caused by the different mechanism of the earthquakes generated by the Vrancea intermediate-depth seismic source. The studies used for comparison in the present paper are based on databases that consist mostly of shallow earthquakes or of earthquakes with focal depths much smaller than those analysed in the two studies. Regional peculiarities, local geology, wave propagation effects or the ground motions frequency content may represent other reasons for the high correlation displayed in the two models.

6 Conclusions

The objective of the present study was to perform comparisons between two recent correlation models developed using a database of ground motions generated by Vrancea intermediate-depth earthquakes (Vacareanu et al. 2017; Pavel and Vacareanu 2017), comparisons between the two previously-mentioned studies and other models available in the literature and the development of an adjusted correlation model based on the dataset assumptions made in the second model with the procedure described in the first model. Differences in the spatial correlation decay between the two models were observed, which can be caused by the different procedure of obtaining the correlation coefficients and the restrains encountered in the first model methodology, and by the employment of fewer ground motions in the latter case. The impact of excluding records made in the second model was evaluated by developing an adjusted correlation model, which showed a visible impact of these assumptions on the results. Comparisons between the two models and other available models were performed, the former resulting in higher correlation coefficient values and a slower decay than most studies chosen to perform the comparison. The effects are mainly caused by the different mechanism of Vrancea intermediate-depth earthquakes, the studies chosen as comparison using mostly ground motions generated by shallow earthquakes. The results presented in the two Vrancea studies represent another reason for the necessary development of the national seismic networks, in order to obtain a dense array of seismic recorders, but also of the necessity for further development of the spatial correlation models using databases with similar characteristics to the Vrancea intermediate-depth earthquakes.

Acknowledgements The results are obtained within the CoBPEE (Community Based Performance Earthquake Engineering) research project, financed by the Romanian National Authority for Scientific Research and Innovation, CNCS—UEFISCDI, project number PN-II-RU-TE-2014-4-0697. This support is gratefully acknowledged.

References

Boore D, Gibbs J, Joyner W, Tinsley J, Ponti D (2003) Estimated ground motion from the 1994 Northridge, California, earthquake at the site of the interstate 10 and La Cienega Boulevard Bridge Collapse, West Los Angeles, California. Bull Seismol Soc Am 93(6):2737–2751

CEN (2004) Eurocode 8: design of structures for earthquake resistance, Part 1: general rules, seismic actions and rules for buildings. EN 1998–1:2004. Brussels, Belgium

Cimellaro GP, Stefano A, Reinhorn AM (2011) Intra-event spatial correlation of ground motion using L'Aquila earthquake ground motion data. In: 3rd ECCOMAS thematic conference on computational methods in structural dynamics and earthquake engineering, Corfu, Greece, pp 3091–3108

Clark I, Harper W (2000) Practical geostatistics. Ecosse North America, Columbus, Ohio

Cornell CA (1968) Engineering seismic risk analysis. Bull Seismol Soc Am 58:1583–1606

Esposito E, Iervolino I, d'Onofrio A, Santo A (2014) Simulation-based seismic risk assessment of gas distribution networks. Comput Aided Civil Infrastruct Eng 00:1–16

Esposito S, Iervolino I (2011) PGA and PGV spatial correlation models based on European multievent datasets. Bull Seismol Soc Am 101(5):2532–2541

Esposito S, Iervolino I (2012) Spatial correlation of spectral acceleration in European data. Bull Seismol Soc Am 102(6):2781–2788

Foulser-Piggott R, Goda K (2015) Ground-motion prediction models for arias intensity and cumulative absolute velocity for Japanese earthquakes considering single-station sigma and within-event spatial correlation. Bull Seismol Soc Am 105(4):1903–1918

Goda K, Atkinson G (2009) Probabilistic characterization of spatially correlated response spectra for earthquakes in Japan. Bull Seismol Soc Am 99(5):3003–3020

Goda K, Atkinson G (2010) Intraevent spatial correlation of ground-motion parameters using SK-net data. Bull Seismol Soc Am 100(6):3055–3067

Goda K, Hong HP (2008a) Estimation of seismic loss for spatially distributed buildings. Earthq Spectra 24(4):889–910

Goda K, Hong HP (2008b) Spatial correlation of peak ground motions and response spectra. Bull Seismol Soc Am 98(1):354–365

Hong HP, Zhang Y, Goda K (2009) Effect of spatial correlation on estimated ground-motion prediction equations. Bull Seismol Soc Am 99(2A):928–934

Jayaram N, Baker JW (2009) Correlation model for spatially distributed ground-motion intensities. Earthq Eng Struct Dyn 38(15):1687–1708

Kawakami H, Sharma S (1999) Statistical study of spatial variation of response spectrum using free field records of dense strong motion arrays. Earthq Eng Struct Dynam 28:1273–1294

Pavel F, Vacareanu R (2017) Spatial correlation of ground motions from Vrancea (Romania) intermediate-depth earthquakes. Bulletin of the Seismological Society of America 107(1), https://doi.org/10.1785/0120160095

Sokolov V, Wenzel F (2011a) Influence of spatial correlation of strong ground motion on uncertainty in earthquake loss estimation. Earthq Eng Struct Dynam 40(9):993–1009

Sokolov V, Wenzel F (2011b) Influence of ground-motion correlation on probabilistic assessments of seismic hazard and loss: sensitivity analysis. Bull Earthq Eng 9(5):1339–1360

Sokolov V, Wenzel F (2013) Further analysis of the influence of site conditions and earthquake magnitude on ground-motion within-earthquake correlation: analysis of PGA and PGV data from the K-NET and the KiK-net (Japan) networks. Bull Earthq Eng 11(6):1909–1926

Sokolov V, Wenzel F, Jean WY, Wen KL (2010) Uncertainty and spatial correlation of earthquake ground motion in Taiwan. Terr Atmos Oceanic Sci (TAO) 21(6):905–921

Vacareanu R, Pavel F, Craciun I, Aldea A, Calotescu I (2017) Correlation models for strong ground motions from Vrancea intermediate-depth seismic source. In: 16th World Conference on Earthquake Engineering, Santiago, Chile, January 9th to 13th 2017. Paper no. 2026

Vacareanu R, Radulian M, Iancovici M, Pavel F, Neagu C (2015) Fore-arc and back-arc ground motion prediction model for Vrancea intermediate depth seismic source. J Earthq Eng 19: 535–562

Wagener T, Goda K, Erdik M, Daniell J, Wenzel F (2016) A spatial correlation model of peak ground acceleration and response spectra based on data of the Istanbul earthquake rapid response and early warning system. Soil Dyn Earthq Eng 85:166–178

Wang M, Takada T (2005) Macrospatial correlation model of seismic ground motions. Earthq Spectra 21(4):1137–1156

Wesson RL, Perkins DM (2001) Spatial correlation of probabilistic earthquake ground motion and loss. Bull Seismol Soc Am 91(6):1498–1515

Risk Targeting in Seismic Design Codes: The State of the Art, Outstanding Issues and Possible Paths Forward

John Douglas and Athanasios Gkimprixis

Abstract Over the past decade there have been various studies on the development of seismic design maps using the principle of "risk-targeting". The basis of these studies is the calculation of the seismic risk by convolution of a seismic hazard curve for a given location (derived using probabilistic seismic hazard analysis) with a fragility curve for a code-designed structure (ideally derived from structural modelling). The ground-motion level that the structure is designed for is chosen so that the structure has a pre-defined probability of achieving a certain performance level (e.g. non-collapse). At present, seismic design maps developed using this approach are only widely applied in practice in the US but studies have also been conducted on a national basis for France, Romania, Canada and Indonesia, as well as for the whole of Europe using the European Seismic Hazard Model. This short article presents a review of the state of the art of this technique, highlighting efforts to constrain better some of the input parameters. In addition, we discuss the difficulties of applying this method in practice as well as possible paths forward, including an empirical method to estimate an upper bound for the acceptable collapse and yield risk.

Keywords Seismic hazard · Earthquake engineering · Fragility curves
Risk targeting · Design · Acceptable risk

J. Douglas (✉) · A. Gkimprixis
Department of Civil and Environmental Engineering,
University of Strathclyde, Glasgow, UK
e-mail: john.douglas@strath.ac.uk

A. Gkimprixis
e-mail: athanasios.gkimprixis@strath.ac.uk

© Springer International Publishing AG, part of Springer Nature 2018
R. Vacareanu and C. Ionescu (eds.), *Seismic Hazard and Risk Assessment*,
Springer Natural Hazards, https://doi.org/10.1007/978-3-319-74724-8_14

211

1 Introduction

Current seismic building codes (e.g. Eurocode 8), based on results from a probabilistic seismic hazard analysis (PSHA), generally adopt a constant hazard approach
to define the ground motions used for design. In other words, the peak ground
acceleration (PGA, or other intensity measure, IM, e.g. spectral acceleration) used
for design in one location has the same probability of being exceeded in a given
year as the design PGA in another location. Often this annual probability is
1/475 = 0.0021 (equivalent to 10% in 50 years or a return period of 475 years
assuming a Poisson process). Ten years ago, Luco et al. (2007) proposed a new
approach that targets a constant risk level across a territory. This has three principal
advantages over the use of design levels defined in the traditional way: transparency, a uniform risk level across a territory and the ability to compare (and
ideally control) risk for different types of hazard (e.g. earthquake and wind). It does
come, however, with the disadvantage of making more choices explicit, rather than
implicitly assumed because of convention (e.g. the choice of 475 years as the
design return period).

The procedure of Luco et al. (2007), although often using different input
parameters (see below), has been applied to France (Douglas et al. 2013), Romania
(Vacareanu et al. 2017), Indonesia (SNI 2012), Canada (Allen et al. 2015) and at a
European scale (Silva et al. 2016), as well as forming the basis of the current US
seismic design code (ASCE 2010). Despite its numerous attractions (see above) and
the fact that it is a relatively simple procedure to implement, there are a number of
outstanding issues. For example, Douglas et al. (2013) note that the collapse
probabilities targeted by Luco et al. (2007) appear to be at least an order of magnitude too high when compared with observed damage in previous earthquakes
(also see Sect. 3.1 of this article).

The next section presents an overview of the risk targeting approach and discusses previous choices of the critical input parameters. Section 3 highlights the
outstanding problems and some potential solutions, which are currently being
investigated by the authors.

2 Method and Required Inputs to Risk Targeting

The risk of collapse (or other level of structural damage) of a building at a given site
from earthquake shaking can be estimated by convolving the seismic hazard curve,
expressing the probability of different levels of ground motion, with the fragility
curve, expressing the probability of collapse given these ground motions (e.g.
Kennedy 2011). This so-called "risk integral" forms the basis of the risk-targeting
approach. For this approach there needs to be a link between the design acceleration
and the fragility curve used to compute the risk of collapse. For a standard fragility
curve based on the lognormal distribution, a single point on this curve (if the

standard deviation is fixed) is required to define the building's fragility completely. A convenient choice is to use the design IM and the corresponding probability of a building attaining the considered damage state when subjected to that IM.

The general procedure for finding the design value for the considered IM (e.g. PGA) is shown in Fig. 1. The key input parameters, using the nomenclature of Douglas et al. (2013), are: β (the standard deviation of the fragility curve assuming a lognormal distribution), X (the probability of collapse at the design IM) and Y (the targeted annual probability of collapse). Seismic design codes generally do not report these values and hence assessing them has been the focus of considerable efforts over the past decade (see following section).

There is a closed-form solution for the risk given a hazard curve expressed as a power law and a lognormal fragility curve (e.g. Kennedy 2011), which can be used to understand the influence of different parameters on the design IM. For typical hazard curves, however, an iterative technique is required to determine the design IM (Fig. 1). We have found that a bisection method, which bounds the targeted risk from above and below until convergence to a given tolerance, is the best approach. To compute the convolution, numerical integration of the derivative of the fragility curve with the hazard curve using the trapezium rule works well.

The seismic hazard curve used within the calculation needs to be defined down to potentially very low probabilities of exceedance because the probabilities of collapse (or another damage state) defined by the fragility function are often far from one for large accelerations. This means that the hazard curve may need extrapolation, for which the power-law expression (based on the IMs for the smallest calculated probabilities) works well for most examples tested to date. We validated our algorithm by comparing our results with those from the Risk-Targeted Ground Motion Calculator available on the USGS website (https://earthquake.usgs. gov/hazards/designmaps/rtgm.php).

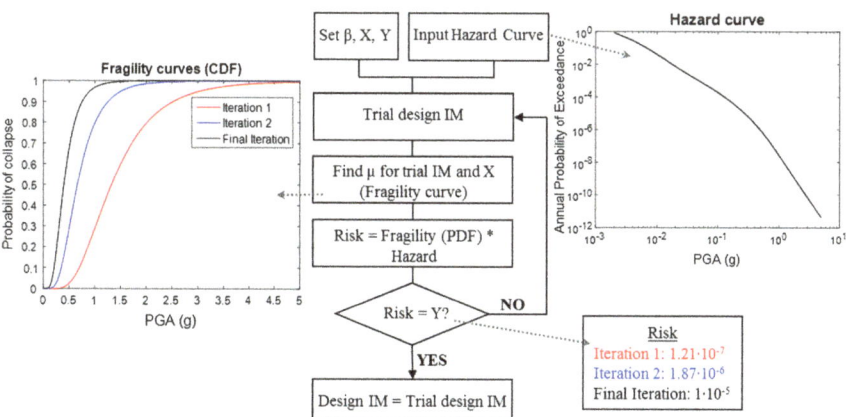

Fig. 1 Flowchart of the method to find risk-targeted design parameters. μ is the mean of the fragility curve assuming a lognormal distribution

2.1 The State of the Art

A summary from the literature of the three key inputs to the procedure is given in this section (we are assuming that the seismic hazard curve has been correctly defined by a recent PSHA for a wide range of probabilities of exceedance and that lognormal fragility curves are used).

ASCE (2005) provides design criteria for nuclear power plants and critical facilities based on a risk-targeting framework. The targeted risk ranges from 10^{-5} to 10^{-4}, while for lower design categories values up to 10^{-3} are given in Braverman et al. (2007). To comply with the code, a nuclear plant must have a smaller than 1% chance of unacceptable performance for the design IM and less than 10% for 1.5 times the design IM.

Kennedy (2011) comments on the ASCE (2005) approach and implements it for 28 US nuclear plants. With a β in the range 0.3–0.6, the Seismic Core Damage Frequencies (SCDFs) are between 6×10^{-7} and 6×10^{-6}. Based on this, the United States Nuclear Regulatory Commission (2007) has set a target SCDF of 10^{-5}.

Luco et al. (2007) use generic fragility curves with $\beta = 0.8$. They assume a probability of collapse of 10% under the ground motion corresponding to a 2475-year return period. The ATC-63 project initially proposed this value after analysing buildings designed with ASCE 7-05. Luco et al. (2007) find that the 2003 NEHRP design ground motions led to non-uniform risk across US territory. By targeting the average collapse probability in 50 years, which they found to be around 1%, Luco et al. (2007) calculated new design ground motions with ratios of 0.7–1.15 of the 2003 NEHRP design ground motions.

The results of that report were considered for the ASCE 7-10 and FEMA (2009) NEHRP Provisions. The latter propose the convolution of a hazard curve with a fragility curve with $\beta = 0.8$ (decreased to 0.6 in ASCE 7-10), X = 0.1 and a target probability of collapse of 1% in 50 years. Liel et al. (2015) comment on ASCE 7-10 and suggest modifications to consider subduction earthquakes and near-fault effects. With the same X and β as in the regulation, the probability of collapse in 50 years varied from 0.21 to 0.62%, for areas affected by subduction earthquakes, while at near-fault sites, the risk was much higher, reaching in some cases 6%.

Allen et al. (2015) comment on the implementation of risk-targeting in Canada for future versions of its National Building Code (NBCC). For comparison, ground motion values derived from the risk-targeting approach of Luco et al. (2007) are divided with those proposed in 2015 NBCC (for a 2%-in-50-year exceedance probability). The resulting risk coefficients for spectral accelerations at 1.0 s are between 0.84 and 1.00 in southwest Canada.

In Goulet et al. (2007) the collapse probabilities for a four-storey reinforced-concrete (RC) frame under the 2%-in-50-year ground motion were calculated as between 0 and 2%; after considering structural uncertainties the values increased to 2–7%. They also found a variation of Y between 0.1×10^{-4} and 0.5×10^{-4} and between 0.4×10^{-4} and 1.4×10^{-4}, when they considered structural uncertainties.

Fajfar and Dolšek (2012) examined a three-storey RC building with no provisions for earthquake resistance. They found an annual probability of collapse equal to 0.65×10^{-2}. When designed with Eurocode 8 (CEN 2004b), this risk reduced to 2.22×10^{-4} and to 2.7×10^{-4} when they accounted for epistemic uncertainties.

Ramirez et al. (2012) examine 30 buildings designed using the 2003 International Building Code (IBC) together with the ASCE provisions. They find the probabilities of collapse at the design PGA to be in the range 0.4–4.2% for these buildings.

Douglas et al. (2013) try to avoid significant changes in the existing design ground motions for France, which correspond to a 475-year return period. Considering previous studies, as well as the results of some sensitivity analyses, they finally choose $\beta = 0.5$ and $X = Y = 10^{-5}$. Under these assumptions, the estimated risk-targeted design PGAs are not very different from the values proposed by the current French code.

Ulrich et al. (2014a) present fragility curves for three-storey RC buildings designed according to Eurocodes 2 and 8 (CEN 2004a, b) for different levels of design PGA, a_g. They find yielding probabilities between 0.14 (for $a_g = 0.07$ g) and 0.85 (for $a_g = 0.3$ g) and in the range 1.7×10^{-7} ($a_g = 0.07$ g) to 1.0×10^{-5} ($a_g = 0.3$ g) for the probabilities of collapse at the design PGA.

In Martins et al. (2015) fragility curves are also derived, firstly considering the spectral acceleration at the fundamental period of vibration $S_a(T_1)$. This results in β ranging from 0.35 to 0.45, which increased up to 0.8 when they used PGA instead. For a three-storey RC building, they found the X to be equal to 5.20×10^{-3} and 2.21×10^{-2} for design accelerations of 0.2 and 0.4 g, respectively, which change to 3.95×10^{-3} and 5.57×10^{-2} for a five-storey RC building. Then, using $S_a(T_1)$ as the IM, the annual probabilities of collapse for the three-storey buildings were 9.50×10^{-5}, 1.67×10^{-5} and 1.07×10^{-5}, and for the five-storey buildings 1.78×10^{-4}, 7.34×10^{-5} and 2.97×10^{-5}, for design accelerations of 0.0, 0.2 and 0.4 g, respectively.

Likewise, the annual rates of collapse for the Italian territory are investigated by Iervolino et al. (2017). The current version of the Italian seismic code was used for the design of regular three-, six-, and nine-story RC buildings for two soil conditions. The results of their analyses indicate that annual collapse risk for code-conforming RC structures is between 10^{-5} and 2.02×10^{-4}. For the case of one-storey precast RC structures (also designed with the seismic code) Y values are found between 10^{-5} and 6.35×10^{-3}. The reader is referred to RELUIS and EUCENTRE (2014–2018) for details of this ongoing project.

Silva et al. (2016) study the fragility curves derived by the European project Syner-G and find an average value of $\beta = 0.5$, which they consider as a lower bound. Assuming a 10% probability of collapse at the 2475-year ground motion and the hazard curves derived by the European project SHARE (www.share-eu.org), for two values of β (0.6 and 0.8) they find a probability of collapse at the 475-year design acceleration ranging from 10^{-3} to 10^{-2}. Choosing then a value of $X = 10^{-3}$ and $Y = 5.0 \times 10^{-5}$ (by relating the risk of collapse to human losses), they propose new design maps for Europe.

Following a different approach, Tsang and Wenzel (2016) firstly define an acceptable fatality risk. Based on a literature review they choose a value of 10^{-6} for the acceptable annual fatality rate. They then estimate the corresponding limit for the targeted annual risk of complete structural damage as roughly 10^{-4}. For a different limit state, the annual risk of "Real collapse" falls within the range 6×10^{-6} to 8×10^{-6}.

Tsang et al. (2017) investigate buildings with precast RC columns designed using the risk-targeted ground motions (MCE_R) proposed in IBC-2012 (ICC IBC 2012) and ASCE/SEI 7-10. Theoretically, a 10% probability of collapse under the MCE_R in 50 years is expected. They estimate, however, probabilities that are in every case lower than 10% (a maximum of 6.1% was found). In addition, the average annual collapse risk was estimated as 2.5×10^{-6} (with a maximum of 1.6×10^{-4}), while the regulation imposes 2×10^{-4}. Also, the value of 0.25% in a design life of 50 years (5×10^{-5} annually), as proposed in Silva et al. (2016), was in some cases exceeded. Judd and Charney (2014) state that "the assumed ASCE 7-10 fragility curve is conservative" and that "the conditional probability of collapse may exceed 10%".

In Vacareanu et al. (2017) the proposed values of Luco et al. (2007) and Silva et al. (2016) are tested for Romania. Compared to previous uniform hazard maps, the risk distribution changes significantly when targeting uniform risk. Vacareanu et al. (2017) calculate the ratios between the design PGA for a mean return period of 475 years and those resulting from risk targeting for $Y = 2 \times 10^{-4}$, $\beta = 0.8$ and $X = 0.1$ and 0.001. For the lower X, the ratios were below 0.6 whilst for the higher X they were larger than 1.0. Finally, considering two mean return periods, 475 and 2475 years, new risk maps are derived using $X = 0.1$ and 0.001. Based on the distribution of risk over the country, they conclude that using $X = 0.001$, as proposed in Silva et al. (2016), leads to a less realistic distribution of Y.

Following US practice, the philosophy of risk targeting has been followed by the new Indonesian Earthquake Resistance Building Code SNI 1726-2012 (SNI 2012). By setting a target of 1% for the probability of collapse in 50 years, this code maps risk-targeted spectral response accelerations for 0.2 and 1.0 s. The generic fragility curves follow a lognormal distribution with $X = 0.1$. Considering the material properties and human-related parameters representative of the broader area of Indonesia, Sengara et al. (2016) report values of β around 0.7, which is adopted by SNI 1726-2012.

3 Outstanding Issues and Possible Paths Forward

As can be seen by Sect. 2 there is a rapidly increasing number of studies that have attempted to apply the risk-targeting approach to different areas or sought to constrain the various inputs upon which this technique relies. Despite these many studies a number of problems remain, which we highlight in this section along with a brief discussion of potential solutions.

As noted above, when evaluating the risk integral, ground motions for very low probabilities of exceedance sometimes need to be estimated. A power-law extrapolation appears to work well for most examples we have studied to date. Further research on the technicalities of the calculations, however, needs to be conducted to define a stable procedure that works for all possible inputs.

Targeting a non-zero value of collapse risk accepts that some buildings will collapse (potentially leading to human casualties) in earthquakes, even when they are designed in compliance with the building code. From a moral point of view this is problematic and it leads to the difficulty of trying to define what risk is "acceptable". One potential solution to this is to estimate an upper bound on the risk that has been "accepted" historically based on the levels of observed damage in previous earthquakes, as attempted by Labbé (2010). This "risk" is an upper bound, as generally after every damaging earthquake the population lament the damage that occurred. In the following Sect. (3.1), some preliminary assessments using this technique are presented. Another potential solution to this problem is to target a damage state that is less severe than collapse, for example structural yielding. Targeting this limit state is less morally problematic and the level of acceptable risk in this context could be defined using, for example, cost-benefit analysis based on the cost of reducing the risk further (perhaps adopting an ALARP "as low as reasonably practicable" philosophy). This choice also has other benefits, e.g. it is generally easier to assess in numerical modelling when a structure yields rather than when it collapses.

Once the targeted risk is chosen, unless great changes to the accelerations currently used for design are accepted by practicing engineers, the probability of collapse at the design acceleration is automatically implied, as shown by Fig. 3 of Douglas et al. (2013). Here there is a trilemma: any two out of the three input parameters, design IM, X and Y, can be chosen independently but not all three.

To make this point clearer, we can use the work of Žižmond and Dolšek (2017). In their article, they follow a risk-targeting approach for the derivation of the q factor for use in force-based design using Eurocode 8 (CEN 2004b). In Eq. 4 of that paper an analytical formula is presented for the median acceleration of the risk-targeted lognormal fragility curve, μ. Adapting it to our nomenclature and recalling that our design acceleration a_g implies a probability of collapse of X, we have the following equations, which show that the various input parameters are connected:

$$\mu = \left(\frac{k_0 \cdot e^{\frac{k^2 \cdot \beta^2}{2}}}{Y} \right)^{\frac{1}{k}} \tag{1}$$

$$a_g = \text{lognormal}^{-1}(X, \mu, \beta) \tag{2}$$

where a_g is the risk-targeted design acceleration and k_0, k are parameters that define the idealized hazard curve. The solution to this potential problem is to check all three values are physically reasonable. This trilemma may be the reason for the apparently high target collapse probabilities (Y) used by Luco et al. (2007), as they were forced to adopt them once X had been defined and they did not want the design IM to change greatly from the previous code.

There is a need to derive fragility curves for a wide variety of code-designed structures with different geometries and materials. Previous studies have adopted generic fragility curves that scale constantly with the design acceleration so that the iterative procedure used to converge to the targeted risk is simple. As shown by Fig. 3 of Ulrich et al. (2014b), however, this desired feature appears not to be true following current design practice, as they were not created with risk targeting in mind. It is possible, however, to generate a suite of fragility curves for all potential design accelerations and then to use the appropriate ones when iterating to find the actual design acceleration for a location.

Two issues concerning the used fragility functions are: what value of β to use and, indeed, whether the lognormal distribution should continue to be used at all? Fragility curves that are used within the risk targeting calculations can imply very strong buildings (e.g. Table 1 of Douglas et al. 2013) as well as non-negligible chances of collapse for very low ground accelerations. This is a consequence of the high values of β that need to be used to account for different types of structures of varying geometries. Two solutions to this problem are: (a) adopt a different functional form for the fragility curve that equals zero for low ground accelerations and unity for very high accelerations, or (b) move from generic fragility curves for all types and geometries to a curve covering only a small set of structures. The second of these changes would mean a change of philosophy of design codes from being associated with a single map giving the design accelerations for all structures to potentially many maps giving design accelerations for different structural types and geometries. This additional complexity, however, appears to be necessary if the risk-targeting approach is to imply physically realistic buildings and levels of risk.

3.1 Assessing Risk Using Earthquake Damage Databases

In this section, we use field observations of building damage from recent earthquakes and an empirical Fermi-type approach to estimate an upper bound on the "acceptable" probability of collapse (Y) discussed above. In particular, the number of RC buildings that collapsed due to earthquakes during recent time periods in Italy and Greece are used to estimate the observed annual collapse rate.

The technique relies on the ratio of the total number of observed cases of a given damage level in a given period of time to the total number of buildings that could have been affected by earthquakes, which is then normalized to give an annual rate. Estimating both of these totals for a given country requires: (a) post-earthquake field mission reports that provide the number of cases of damage and (b) building census data indicating how many structures were present when the earthquake occurred.

There are a number of difficulties and uncertainties in applying this approach. For example, in some reports the number of collapses of RC buildings is combined with the number of collapses of other structural types, e.g. masonry. In others only a general "damage" category is used without specifying, for example, the number of collapses or partial collapses. Sometimes the focus of post-earthquake field missions is on estimating the number of casualties and not on collecting accurate building damage statistics. Also damage records may be incomplete, either by not including all damaging earthquakes during a time period or by not totalling all incidences of damage for a given earthquake (e.g. damage within cities but not rural areas). For example, Colombi et al. (2008), when deriving empirical vulnerability curves from Italian data, state that roughly half of the available observations cannot be used because of a lack of information on the structural type and level of damage. In addition, when combining data from different databases it is necessary to assume equivalence between various damage scales (or convert between scales). Finally (and probably most important), the sample sizes available are small, both in terms of the number of collapsed buildings and the small number of possible earthquake scenarios (locations and magnitudes) that have been sampled during the time period considered, which is relatively short given the recurrence interval of large earthquakes. Therefore, the annual collapse rates estimated here should be considered very rough.

We conduct a preliminary analysis for Italy and Greece, for which the earthquake damage databases are roughly complete in recent years and easily accessible, and for a combined "total or partial collapse" damage state for RC buildings. It would be relatively simple to conduct such an analysis to other developed countries, such as the USA and Japan.

The Cambridge Earthquake Impact Database (CEQID) (http://www.ceqid.org/CEQID/Home.aspx) reports the number of RC building collapses in Italian earthquakes from 1980 to 2009 inclusive (30 years) as: Irpinia 1980 (58 buildings), Eastern Sicily 1990 (3), Umbria-Marche 1997 (49), Umbria-Marche 1998 (0), Pollino 1998 (9), Molise 2002 (29) and L'Aquila 2009 (57), giving a total of 205 total or partial collapses. Colombi et al. (2008) gives an estimate of 842 collapses for almost the same period range, thereby demonstrating large uncertainty in this total.

Next it is necessary to estimate the total number of RC buildings in Italy. EPISCOPE (2014) estimates that 24.7% of Italian residential buildings are of RC and Istat (2011, http://dati-censimentopopolazione.istat.it) states that there were 14,452,680 buildings in Italy of which 12,187,698 are residential. Therefore, it can be estimated that there are roughly 3.5 million RC buildings in total of which about 3 million are residential. We assume that this total has not changed over the 30 years covered by the collapse database, which given the slow population growth of Italy could be thought a reasonable assumption.

Dividing the total number of collapsed buildings by the building population and the number of years gives an estimate of the collapse rate. Using the two estimates of 205 and 842 in 30 and 23 years, respectively, and 3 or 3.5 million buildings (depending on whether the database covers all buildings or only residential) gives annual collapse rates of between 2×10^{-6} and 1×10^{-5}, which are much lower than some of the probabilities assumed in previous risk-targeting exercises (e.g. Luco et al. 2007).

For the period 1978–2003 inclusive the following RC building collapses due to earthquakes in Greece have been reported: Thessaloniki 1978 (4 buildings) (CEQID), Gulf of Corinth 1981 (15) (Carydis et al. 1982), Kalamata 1986 (5), Aigion 1995 (1), Athens 1999 (69) (CEQID) and Lefkada 2003 (0) (Karababa and Pomonis 2011), giving a total of 94 total or partial collapses. Using information from the Greek statistical service (http://www.statistics.gr) and the TABULA project (www.building-typology.eu) for the total number of RC buildings in Greece, and assuming that the building total remained constant, leads to estimates of the collapse rate between 1×10^{-6} and 2×10^{-6}, again much lower than some assumed probabilities for the targeted risk.

The accuracy of these estimates could be improved by using a more complete database both for the numerator (the damage) and the denominator (the number of buildings at risk) and by extending the time period covered so that more potential earthquake scenarios are considered. Nevertheless, such an analysis is constrained by the long recurrence intervals of damaging earthquakes, which means that the assessed rates will always be associated with large uncertainties. In addition, variations due to building type (e.g. masonry versus reinforced concrete), location (e.g. developed versus less-developed countries) and damage level (e.g. collapse versus yield) could be studied as well.

The same procedure was repeated for the same two countries considering a different damage state. A broader class was defined to represent 'yield', as considered by Ulrich et al. (2014a, b), which includes the total number of RC buildings that are neither not-damaged nor collapsed (partially or in total). The same limitations discussed before are also the case here. In addition, complete data were available for a shorter period of 18 years for Greece (but still 30 years for Italy). With these assumptions, the annual percentage of yield is found equal to 3×10^{-5} for Italy and 1×10^{-4} for Greece. Apart from providing useful constraints on the inputs to risk-targeting this type of analysis could have other benefits as well. Firstly, it would provide an observational-based assessment of earthquake risk that can be compared to estimates from computer modelling, which require a large number of inputs and can be opaque to decision makers. If there are large differences in these estimates it may indicate that the computer models require calibration. Secondly, these estimates will enable differences in earthquake risk levels between countries to be judged more easily and used to understand what the impact of reducing risk could be in terms of the average number of buildings that could collapse per year, for example.

4 Conclusions

Crowley et al. (2013) propose that 'Risk-targeted seismic design actions' should be considered for future versions of the Eurocodes. A call echoed by Formichi et al. (2016) in a report on the background and application of Eurocodes, published on behalf of the European Commission. This report also proposes that changes made in other international seismic design codes should be considered when updating the Eurocodes. It is, therefore, clear that application of the risk-targeting approach is being seriously discussed in Europe. In consequence, additional research effort to this end would provide valuable input for the development of risk-targeted design maps for new buildings in Europe.

In this brief article, we have highlighted the critical issues that we believe need to be solved before the risk-targeting approach for the development of seismic design codes can be employed in practice. Some of these (e.g. development of appropriate fragility functions) solely require engineering calculations but others (e.g. choice of the acceptable level of risk) need input from other domains, including decision makers. In the coming years we plan to tackle these issues, particularly with respect to future Eurocodes.

Acknowledgements The second author of this article is undertaking a PhD funded by a University of Strathclyde "Engineering The Future" studentship, for which we are grateful. We thank Florin Pavel for sharing the submitted version of Vacareanu et al. (2017). We thank Roberto Paolucci for discussions concerning empirical estimates of collapse risk.

References

Allen TI, Adams J, Halchuk S (2015) The seismic hazard model for Canada: past, present and future. In: Proceedings of the tenth pacific conference on earthquake engineering building an earthquake-resilient pacific, Sydney, Australia (Paper number 100)

ASCE (2005) Seismic design criteria for structures, systems and components in nuclear facilities. American Society of Civil Engineers, ASCE Standard 43-05

ASCE (2010) Minimum design loads for buildings and other structures. ASCE Standard 7-10. American Society of Civil Engineers, Reston, VA

Braverman JI, Xu J, Ellingwood BR, Costantino CJ, Morante RJ, Hofmayer CH (2007) Evaluation of the seismic design criteria in ASCE/SEI standard 43-05 for application to nuclear power plants. Agencywide Documents Access and Management System (ADAMS)—USNRC

Carydis PG, Tilford NR, Brandow GE, Jirsa JO (1982) The central Greece earthquakes of February–March 1981. Report No. CETS-CND-018 of the Earthquake Engineering Research Institute, Berkeley, CA, National Academy Press, Washington, DC

CEN (2004a) EN 1992-1-1:2004 Eurocode 2: design of concrete structures—Part 1-1: general rules and rules for buildings. European Committee for Standardization, Brussels

CEN (2004b) EN 1998-1:2004 Eurocode 8: design of structures for earthquake resistance—Part 1: general rules, seismic actions and rules for buildings. European Committee for Standardization, Brussels

Colombi M, Borzi B, Crowley H, Onida M, Meroni F, Pinho R (2008) Deriving vulnerability curves using Italian earthquake damage data. Bull Earthq Eng 6(3):485–504. https://doi.org/10. 1007/s10518-008-9073-6

Crowley H, Weatherill G, Pinho R (2013) Suggestions for updates to the European seismic design regulations. SHARE Deliverable 2.6. Universita degli Studi di Pavia

Douglas J, Ulrich T, Negulescu C (2013) Risk-targeted seismic design maps for mainland France. Nat Hazards 65(3):1999–2013

EPISCOPE (2014) Inclusion of new buildings in residential building typologies: steps towards NZEBs exemplified for different European countries. EPISCOPE Synthesis Report no. 1 (Deliverable D2.4). Contract N°: IEE/12/695/SI2.644739. ISBN 978-3-941140-42-4

Fajfar P, Dolšek M (2012) A practice-oriented estimation of the failure probability of building structures. Earthq Eng Struct Dynam 41:531–547

FEMA (2009) NEHRP recommended seismic provisions for new buildings and other structures (FEMA P750). Federal Emergency Management Agency

Formichi P, Danciu L, Akkar S, Kale O, Malakatas N, Croce P, Nikolov D, Gocheva A, Luechinger P, Fardis M, Yakut A, Apostolska R, Sousa ML, Dimova S, Pinto A (2016) Eurocodes: background and applications: elaboration of maps for climatic and seismic actions for structural design with the Eurocodes. EUR, https://doi.org/10.2788/534912

Goulet C, Haselton C, Mitrani-Reiser J, Beck J, Deierlein G, Porter K, Stewart J (2007) Evaluation of the seismic performance of a code-conforming reinforced-concrete frame building—from seismic hazard to collapse safety and economic losses. Earthq Eng Struct Dynam 36: 1973–1997

ICC IBC (2003) International building code (IBC), international code council (ICC), IL, USA

ICC IBC (2012) International building code (IBC), international code council (ICC), IL, USA

Iervolino I, Spillatura A, Bazzurro P (2017) RINTC project: Assessing the (implicit) seismic risk of code-conforming structures in Italy. In: Papadrakakis M, Fragiadakis M (eds) COMPDYN 2017, 6th ECCOMAS thematic conference on computational methods in structural dynamics and earthquake engineering, Rhodes, Greece

Judd JP, Charney FA (2014) Earthquake risk analysis of structures. In: Proceedings of the 9th international conference on structural dynamics, EURODYN 2014, pp 2929–2938

Karababa FS, Pomonis A (2011) Damage data analysis and vulnerability estimation following the August 14, 2003 Lefkada Island, Greece, earthquake. Bull Earthq Eng 9:1015–1046. https:// doi.org/10.1007/s10518-010-9231-5

Kennedy RP (2011) Performance-goal based (risk informed) approach for establishing the SSE site specific response spectrum for future nuclear power plants. Nucl Eng Des 241:648–656

Labbé PB (2010) PSHA outputs versus historical seismicity: example of France. In: Proceedings of fourteenth European conference on earthquake engineering

Liel AB, Luco N, Raghunandan M, Champion CP (2015) Modifications to risk-targeted seismic design maps for subduction and near-fault hazards. In: 12th international conference on applications of statistics and probability in civil engineering, ICASP12, Vancouver, Canada

Luco N, Ellingwood BR, Hamburger RO, Hooper JD, Kimball JK, Kircher CA (2007) Risk-targeted versus current seismic design maps for the conterminous United States. In: SEAOC 2007 convention proceedings

Martins L, Silva V, Crowley H, Bazzurro P, Marques M (2015) Investigation of structural fragility for risk-targeted hazard assessment. In: 12th international conference on applications of statistics and probability in civil engineering, ICASP12, Vancouver, Canada

Ramirez CM, Liel AB, Mitrani-Reiser J, Haselton CB, Spear AD, Steiner J, Deierlein GG, Miranda E (2012) Expected earthquake damage and repair costs in reinforced concrete frame buildings. Earthq Eng Struct Dynam 41:1455–1475

RELUIS and EUCENTRE (2014–2018) Research project DPC—RELUIS/EUCENTRE, technical report and deliverables

Sengara IW, Sidi ID, Mulia A, Muhammad A, Daniel H (2016) Development of risk coefficient for input to new Indonesian seismic building codes. J Eng Technol Sci 48(1):49–65

Silva V, Crowley H, Bazzurro P (2016) Exploring risk-targeted hazard maps for Europe. Earthq Spectra 32(2):1165–1186

SNI (2012) Tata cara perencanaan ketahanan gempa untuk struktur bangunan gedung dan non gedung. Badan Standardisasi Nasional, ICS 91.120.25; 91.080.01. Report 1726:2012 (In Indonesian)

Tsang HH, Wenzel F (2016) Setting structural safety requirement for controlling earthquake mortality risk. Saf Sci 86:174–183

Tsang HH, Lumantarna E, Lam NTK, Wilson JL, Gad E (2017) Annualised collapse risk of soft-storey building with precast RC columns in Australia. Advancements and Challenges, Mechanics of Structures and Materials, pp 1681–1686

Ulrich T, Negulescu C, Douglas J (2014a) Fragility curves for risk-targeted seismic design maps. Bull Earthq Eng 12(4):1479–1491

Ulrich T, Douglas J, Negulescu C (2014b) Seismic risk maps for Eurocode-8 designed buildings. In: Proceedings of the second European conference on earthquake engineering and seismology

United States Nuclear Regulatory Commission (2007) A performance-based approach to define the site specific earthquake ground motion. Tech report 1:208

Vacareanu R, Pavel F, Craciun I, Coliba V, Arion C, Aldea A, Neagu C (2017) Risk-targeted maps for Romania. J Seismolog. https://doi.org/10.1007/s10950-017-9713-x

Žižmond J, Dolšek M (2017) The formulation of risk-targeted behaviour factor and its application to reinforced concrete buildings. In: Proceedings of the 16th world conference on earthquake engineering, Santiago, Chile, 9–13 January (Paper no. 1659)

Earthquake Risk Assessment for Seismic Safety and Sustainability

Alik Ismail-Zadeh

Abstract An interaction of three major risk components (seismic hazards, vulnerability and exposure) are analysed here, and preventive measures to mitigate disasters are discussed. The importance of action-oriented research on earthquake risk reduction co-produced with multiple stakeholders, including engineers and policymakers, is analysed. This importance is evidenced by the increasing vulnerability and exposure of society to risk in many earthquake-prone regions and by the need for cross-cutting actions in policy and practice related to sustainability.

Keywords Earthquake · Hazard · Vulnerability · Exposure

1 Earthquake Risk Analysis

Earthquake risk can be determined as the probability of harmful consequences or expected losses and damages due to an earthquake resulting from interactions between earthquake hazards, physical and social vulnerability, and exposure. Conventionally, earthquake risk is expressed quantitatively by the convolution of these three parameters (e.g. Kantorovich et al. 1973). Seismic hazard as a component of risk assessment is based on the study of the features of seismic wave excitation at the source, seismic wave propagation, and site effect in the studied region. It has been analysed in the accompanying chapter "Earthquake hazard modelling and forecasting for disaster risk reduction" of this book. Assuming that an earthquake happened, exposure and vulnerability are the key determinants of earthquake risk and the main drivers of disaster losses. Changes in any of these alter the risk calculus by increasing or reducing the impacts of disaster risk on affected communities, regions, or countries. Exposure is the location of people, assets, and infrastructure in hazard-prone areas that could be affected, while vulnerability is the

A. Ismail-Zadeh (✉)
Karlsruhe Institute of Technology, Institute of Applied Geosciences,
Adenauerring 20b, 76137 Karlsruhe, Germany
e-mail: alik.ismail-zadeh@kit.edu

© Springer International Publishing AG, part of Springer Nature 2018

225

R. Vacareanu and C. Ionescu (eds.), *Seismic Hazard and Risk Assessment*,
Springer Natural Hazards, https://doi.org/10.1007/978-3-319-74724-8_15

degree of susceptibility or sensitivity of people, assets, and infrastructure to suffer damages (UNISDR 2013). There is temporal and spatial variability in exposure and vulnerability patterns from local to global scales. These geographic patterns are unevenly distributed across the globe and lead to the disproportionate impacts of disasters, especially in disadvantaged communities, regions, or countries. Also, vulnerability depends on an individual's susceptibility; namely, populations may be less vulnerable or more resilient in the face of disasters because of former exposure to disasters that helped to gain strength, or cultural traditions that improve collective action (Ismail-Zadeh and Cutter 2015).

Not every earthquake results in loss of lives, properties and infrastructure, but only that occurring in a close vicinity to vulnerable places. For example, a strong earthquake in the plate interior may lead to disaster if it strikes near a town, and it will turn to become a disaster in the case of high physical and social vulnerability of the town to earthquakes. The following example is illustrative. The energy released by the 2010 Chile M8.8 earthquake was by a factor of about 500 higher than that by the 2010 Haiti M7.0 earthquake. However, the death toll showed the inverse proportionality: several hundred people lost their lives (primarily because of the triggered tsunami) in the case of the Chile earthquake versus several hundred thousand lives in the case of the Haiti earthquake. Disasters happen mainly because of the "unwillingness of some local authorities to invest in resistant construction due to various reasons including irresponsibility, ignorance, corruption, the perceived requirement to balance the need for costs versus the increased costs of implementation, local politics, funding availability and other urgent and more politically competitive needs" (Ismail-Zadeh et al. 2017).

As a tectonic stress/energy release through earthquakes cannot be stopped and as exposure increases with economic development, a major element in disaster risk management is the reduction of vulnerability. High vulnerability is associated with environmental degradation, rapid and unplanned urbanization in hazardous areas, failures of governance to reduce vulnerability, and the scarcity of livelihood options for the poor. Countries more effectively manage disaster risk if they include considerations of disaster risk in national development and sector plans, translating these plans into actions targeting vulnerable areas and groups (IPCC 2012).

A rapid growth of population, intensive civil and industrial building, land and water instabilities, and the lack of public awareness regarding hazards and risks contribute to the increase of vulnerability of big cities. For example, Babayev et al. (2010) assessed an earthquake risk in Baku (Azerbaijan) based on the convolution of scenario-based seismic hazards, vulnerability (due to building construction fragility, population patterns, the gross domestic product per capita, and landslide's occurrence), and exposure of infrastructure and critical facilities. Figure 1 presents the integrated earthquake risk assessment for three earthquake scenarios. One of remarkable results of this assessment was the fact that the western-central part of the city was exposed to the highest risk independent on the earthquake scenarios, i.e. seismic risk does not depend much on the magnitude and the epicentral distance of the earthquakes employed for the assessment. This is because of physical vulnerabilities characterized by low quality of building constructions, high density of population, and significant exposed values in this part of the city (Babayev et al. 2010).

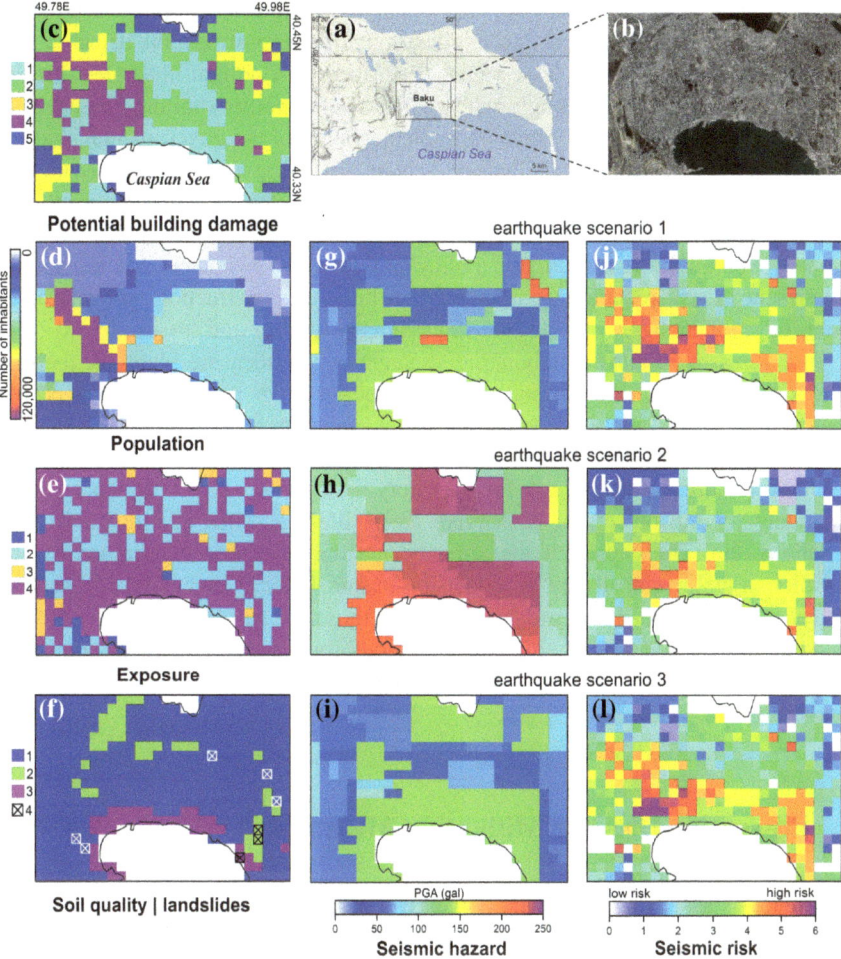

Fig. 1 Earthquake risk for Baku, the capital city of Azerbaijan. **a** Topography of the Absheron peninsula and the location of Baku (the studied region is marked by a solid rectangular). **b** Satellite image of the studied region. **c** Distribution of building constructions: high (type 1), good (type 2) good to poor (type 3), and poor (type 4) quality. Construction free areas are marked by 5. **d** Distribution of urban population. **e** Distribution of exposure. 1: low-value objects (park, lake, and other natural objects); 2: moderate-value objects (main road, square, and their intersections); 3: high-value objects (culture-valued objects); and 4: very-high-value objects (residential objects, hospital, school, commercial, governmental and state buildings, nation-valued object, important communication system). **f** Soil types. 1: Quaternary sand, clay, intrusions of limestone, and sandy clay; the Upper Tertiary limestone and sandstone (minor amplification of shaking). 2: Quaternary sand, gravel-pebble, and limestone with clay intrusions (moderate amplification of shaking). 3: water-saturated sand, clay and rubble (significant amplification of shaking). 4: landslide area. Computed surface peak ground acceleration (seismic hazard, **g–i**) and seismic risk (**j–l**) for three earthquake scenarios. Coastlines are marked by black curves. Modified after Babayev et al. (2010)

Many parts of the world are still vulnerable to earthquakes despite the progress made in earthquake research and engineering for the past several decades. For example, an earthquake risk assessment performed for Portugal illustrates that the highest risk is associated strongly with physical vulnerability (building damage) to earthquakes (Baker 2013). Despite an implementation of the governmental strategy for earthquake risk reduction (Ghafory-Ashtiany 2014), existing physical vulnerability in Iran is a source for potential losses due to an earthquake occurrence. Meanwhile, Japan is an example of a successful and effective implementation of the national earthquake risk reduction policy providing an evidence that substantial investment in risk reduction, particularly, in building construction and reinforcement, pays off in significant decline in earthquake impact and losses.

Risk assessment allows elaborating strategic countermeasure plans for the disaster risk mitigation. An estimation of earthquake risks may facilitate a proper choice in a wide variety of safety measures, ranging from building codes and insurance to establishment of rescue-and-relief resources. Most of the practical problems require estimating risk for a territory as a whole, and within this territory separately for the objects of each type: areas, lifelines, sites of vulnerable constructions, etc. The choice of the territory and the objects is determined by the jurisdiction and responsibility of a decision-maker. Difficulties in decision making are related to uncertainties in data, especially those related to social and physical vulnerabilities and exposure, imperfect methods for hazard assessment, and limitations in using mathematical tools for carrying out the historical analysis and forecasting.

2 Preventive Disaster Mitigation Measures

Risk analysis assists in optimizing preventive mitigation measures to reduce losses from catastrophic disasters. "If about 5 to 10% of the funds, necessary for recovery and rehabilitation after a disaster, would be spent to mitigate an anticipated earthquake, it could in effect save lives, constructions, and other resources" (Ismail-Zadeh and Takeuchi 2007). A large investment is made at the stage of a response to a big earthquake disaster, and the investment decreases until the next large earthquake, especially at the final stage of risk management related to prediction and preparedness. I call this cycle the "seismic-illogical cycle" (Ismail-Zadeh 2010) because it characterized by a decrease of funding toward the next big earthquake, whereas a seismic cycle shows an increase of tectonic stresses toward an earthquake (stress drop) (see Fig. 2). To implement preventive disaster mitigation strategy and move toward seismic safety, the "cycle" should be broken.

However, the investment to avoid potential earthquake losses tends not to be easily accepted in political decision-making (compared to investments to gain immediate positive benefits). It is because the benefit of preventing losses is not easily visible for the terms of a presidential and/or governmental power (normally 4 to maximum 8 years), while the positive benefit is obvious and can easily be agreed

Fig. 2 Seismo-illogical cycle
in seismic risk and earthquake
disaster management.
Modified after Ismail-Zadeh
(2010)

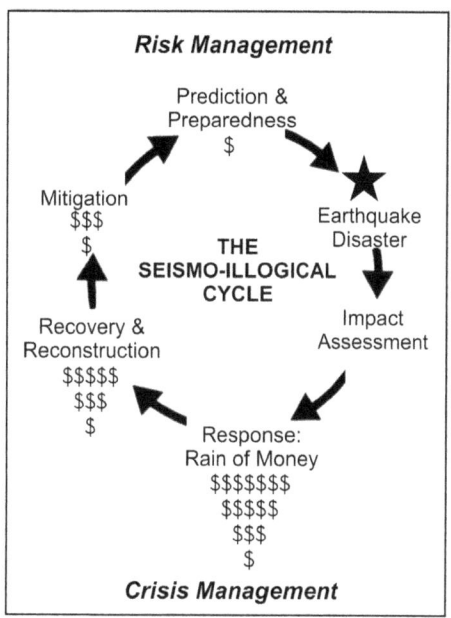

by people. The investment in disaster preparedness is a matter of political decision. Politicians assisted by scientists and engineers could inform the public on major (vital but costly) regional, national or international disaster mitigation project proposals and learn their opinion on the proposals. For example, Ismail-Zadeh and Takeuchi (2007) suggested drawing the public opinion via the contingent valuation method by Cummings et al. (1986) to quantify the benefits of non-marketed goods so that they could be entered directly into cost-benefit calculations.

Preparedness and awareness are important factors in preventive measures to mitigate disasters. The level of preparedness for the 2004 Indian Ocean tsunami disaster in many countries turned out to be extremely low, partly because it happened without any warning. But even if a warning had been sent to the appropriate local authorities of the countries, it is unlikely that it would have been delivered to the public in a timely manner. It is also unlikely that people would have responded to it in appropriate manner, since most of them did not believe that such a disaster could affect them (Ismail-Zadeh and Takeuchi 2007). In Japan, where the preparedness and awareness of people regarding earthquakes and tsunamis are the highest in the world, the initial data on the earthquake of 11 March 2011 sent to the Tohoku coastal region underestimated the height of tsunami waves. In a response, some people did not evacuate their building and did not move to safer places, because they considered that available sea walls (in some places up to 10 m) would protect their houses and lives. Unfortunately, the tsunami waves were much higher than expected, and the sea walls were incapable to prevent the great inundation. Another important issue revealed by a sociological study is that a frequent warning issue with overestimated heights of tsunami waves plays a negative role, as the coastal community begins to disbelieve tsunami alarms (Ando et al. 2011). Emergency management practices should always be planned and exercised well before a disaster occurs. Human adjustment measures to live with risk are important. Such measures include hazard maps, education and awareness raising, warning, evacuation, sea-wall patrol, emergency protection, periodical exercises of evacuation, etc. Such measures apply not only to earthquakes, tsunamis or floods but also to other geological and hydro-meteorological hazards and associated disasters.

Scientists knew about historical devastating earthquakes and tsunamis, such as those happened in the Indian and Pacific Ocean regions or earthquakes in Himalayas-Tibet, in Haiti, and in the Japanese islands. What happened in Indonesia and surrounding countries in 2004 (during the Indian Ocean Earthquake and tsunami events) should have not happened. Despite the knowledge about tsunamis and earthquakes in the region have been available as well as the vulnerability of population and coastal exposure were analysed, scientists could make much more to convince local governments to be prepared in advance. But it was/is a challenging task for scientists to convince their governments to invest in preparedness in advance of a catastrophe; especially it is a difficult task in economically less developed countries, where corruption (Ambraseys and Bilham 2011), ignorance and irresponsibility are sometimes flourishing.

Moreover, scientists made many wise alerts and proposals related to potential earthquake-prone areas. But it seems that their alerts were submerged under other matters like political and financial affairs, environmental issues and other social affairs. Particularly, geophysicists and geodesists alerted about a potential strong earthquake near the capital city of Haiti. They wrote "... the Enriquillo fault in Haiti is currently capable of a Mw7.2 earthquake if the entire elastic strain accumulated since the last major earthquake was released in a single event today" (Manaker et al. 2008). Seismologists and tsunami researchers knew about the 869 Jogan earthquake (off Miyagi, Japan), which produced unusually large tsunamis and took up 1000 people in the Sendai plain (Satake et al. 2007). "Without having the scientific awareness raised, no political and governmental actions are possible. Here there is a large room for geoscientists to take responsibility" (Ismail-Zadeh and Takeuchi 2007).

It is well known that it is not an earthquake that kill people, but collapsing buildings. After the 1755 Great Lisbon earthquake and tsunami, German philosopher Immanuel Kant wrote: "If humans are building on inflammable material, over a short time the whole splendour of their edifices will be falling down by shaking" (Kant 1756; Fuchs 2009). To clarify this sentence, I would mention that in the 18th century scientists believed that an earthquake generates fires, and hence is the Kant's remark about 'inflammable material'. But the meaning of the message is clear: "Build houses safer to withstand an earthquake!" or according to the 2015 Sendai Framework for Disaster Risk Reduction, "Build Back Better" (Sendai Framework 2015). In more than 260 years, the words by Kant are still quite relevant, especially when we think about the L'Aquila (in 2009), Haiti (in 2010), and other disasters caused by strong (but not large or great) earthquakes.

3 Disaster Risk Reduction and Implications for Sustainability

A disaster triggered by an earthquake is a "serious disruption of the normal functioning of a community/society causes widespread human, economic, and/or environmental losses, which exceed the ability of the affected community/society to cope using its own resources" (UN 2017). Disaster risk reduction should be aimed at reducing existing disaster risks, preventing new risks, and managing residual risks, all of which contribute to strengthening resilience and therefore to the achievement of sustainable development. A sustainable society is one which progresses in its development while equitably meeting its present needs and not compromising the ability of future generations to develop and meet their own needs (UNGA 1987). At the United Nations World Conference on Disaster Risk Reduction in Sendai, Japan, in 2015, a new international framework for disaster risk reduction for 2015–2030 was adopted. The framework calls on governments (i) to understand disaster risk; (ii) to strengthen risk governance to manage risks across all sectors; (iii) to invest in risk-reduction measures that promote resilience; and (iv) to enhance disaster preparedness and responses so that nations "Build Back Better" in recovery (Sendai Framework 2015).

Several global Sustainable Development Goals (SDGs) accepted also in 2015 by states of the United Nations are directly related to disaster risk reduction. Namely, they refer to "building the resilience of those in vulnerable situations and to reducing their exposure and vulnerability to climate-related extreme events; to strengthening capacity for adaptation to climate change, extreme weather, drought, flooding and other disasters; to significant reduction of the number of deaths and the number of people affected, and to substantial decreasing the direct economic losses relative to global gross domestic product caused by disasters, including water-related disasters, with a focus on protecting the poor and people in vulnerable situations; to developing and implementation, in line with the Sendai Framework for Disaster Risk Reduction 2015–2030, holistic disaster risk management at all levels; and to strengthening resilience and adaptive capacity to climate-related hazards and natural disasters in all countries" (UN 2015).

How can disaster risk research, including understanding of natural hazards, exposure and vulnerability as well as disasters as such, be leveraged to support the Sendai Framework and to reach the global SDGs? Although science is well-recognized in the priorities of the Sendai Framework and the other political documents, there is a need to enhance the scientific basis for the policy agenda (Cutter et al. 2015; Ismail-Zadeh and Cutter 2015) and to place scientific evidence in policy-making, although it is "neither straightforward nor guaranteed" (Gluckman 2016). To undertake this, a science-based approach to understanding and assessment of disaster risks at all levels should be developed. Risk assessments would provide a clear and unambiguous scientific view on the current state of knowledge in disaster risk, the potential socio-economic impacts of natural hazards, and the ways to reduce significant human and economic losses. Such a scientific view should present various knowledge inputs including local knowledge, business and various other group inputs (Ismail-Zadeh et al. 2017).

According to the Budapest Manifesto (Beer and Ismail-Zadeh 2003), scientists can contribute to decision-making through a risk management framework with which to examine technical and social issues related to sustainability, namely: anticipating disaster risks caused by natural events through wide-spread consultation; determining concerns by using risk assessment techniques for various scenarios; identifying the consequences by systematically cataloguing hazards; undertaking calculations with appropriate models; evaluating the certainties, uncertainties, and the probabilities involved in the calculations of the vulnerability and of the exposure; comparing with criteria to assess the need for further action; determining and acting on options to control, mitigate and adapt to the risk; communicating the results to those who need to know; promoting and guiding monitoring systems to collect, assimilate and archive data relevant to the determination of sustainability and risk, now and in the future; integrating the knowledge and understanding from all relevant disciplines to provide society with the tools to review the sustainability and the risks of proposed policies and plans. Though rational scientific methods hold the promise of an improved science of risk and sustainability, it must be remembered that the priorities for analyses are likely to be heavily influenced by the public and political agenda of the day. This means that implementation of risk management to achieve sustainability can be achieved only through an interaction of theory and praxis (Beer and Ismail-Zadeh 2003).

Disaster risk reduction is essential in achieving sustainable development goals and in building resilience to natural (e.g., earthquake) hazards. This is because sustainability and disaster risk are complementary to the extent that policies and plans are sought to increase sustainability and reduce risk. Disasters undermine poverty eradication and peaceful development, link to unsustainable growth, magnify impacts for small developing countries, impact significantly sustainability of cities, become quite often concatenated, and affect disproportionately women, children, old people and disabled (IRDR 2013).

4 Transdisciplinary Research as a Challenging Problem in Risk Reduction

A way of integration and co-production in disaster risk research is through transdisciplinary studies aiming at in-depth investigations using a system analysis and at recommendations for actions to reduce risks and to improve resilience of society. Such transdisciplinary research offers a practice- and policy-oriented knowledge to reduce disaster risk and disasters (Hirsch Hadorn et al. 2008). "Transdisciplinary research assumes that scientists of different disciplines work together to contribute their unique expertise to the research outside their own discipline. They address a common problem and try to understand the complexities of the entire problem rather than its parts only. To achieve a common goal, scientists exchange data and information, share resources, create conceptual, phenomenological, theoretical, and methodological innovations, integrate disciplines, and move beyond discipline-specific approaches … The co-production of knowledge did not privilege one perspective over the other, but did provide models for connecting theory to observed empirical data in an effort to predict results and solve some of the wicked problems facing society. Divergent perspectives … are the hallmark of transdisciplinary knowledge" (Ismail-Zadeh et al. 2017).

Although a flow chart of co-designed and co-produced work is much complicated, it is presented here using a simple analogy (a meal preparation in a restaurant; Fig. 3). Scientists provide a "menu" of the knowledge available to help for decision making; policymakers and society express their need and order a "meal" from the scientific "menu". A limited budget imposes significant limitations on the willingness of policymakers to pay for disaster reduction due to extreme natural events (Fig. 3a). Scientists and engineers work together with other stakeholders ("cook the meal") keeping in mind the principal aim—to assist policymakers in reducing disaster risks at local, national, regional, and global levels (Fig. 3b). The "cooked meal", that is, new knowledge, risk assessments, and recommendations, is utilized by preventive measures to mitigate disaster risks (Fig. 3c). Such an approach may offer a practice- and policy-oriented knowledge to reduce potential disasters and makes the knowledge gained to be useful, usable and used (Boaz and Hayden 2002).

Despite evolving knowledge in earthquake disaster risks and disaster risk reduction efforts, the society is far from seismic safety, that is, the conditions, which

Fig. 3 A cartoon showing a flow-chart of co-designed and co-produced work on disaster risk reduction (individual images are from dreamstime.com)

make the life protected from risk of death/injury or property losses caused by earthquakes. A reduction of earthquake disaster risks and safety building require an integrated approach coupling natural science, engineering, and social science. The development of a transdisciplinary, team-based, approach is one key way of ensuring that the respective contributions of all disciplines and stakeholders are incorporated into a risk management strategy (Ismail-Zadeh et al. 2017). This requires arduous efforts to tackle the challenging problems of extreme natural hazards and disaster risks.

One of the complications is that scientific community and other stakeholders still live on isolated islands in the 'sea of ignorance'. Consider the geoscience community: seismologists have only recently started to communicate with geodesists on problems of impending large earthquakes; a link between seismologists and earthquake engineers exists but it is still weak; electromagnetic and seismological communities have a little communication (there exists a high barrier between theoretician/statisticians and experimentalists); the line of examples can be continued. When it concerns communication between natural and social sciences, scientists and policy makers, scientists and insurance representatives, the situation becomes even worse. The communication is improving, although quite slowly. All stakeholders dealing with disaster risk research should be properly bridged and strongly linked to each other by means of joint research and actions using a language understandable by each other. An entire great network should be developed to connect the isolated islands in a 'sea of mutual understanding' (Ismail-Zadeh 2014).

5 Conclusion

Disaster losses due to seismic events will continue to impose a great challenge to sustainability unless challenging problems in disaster science discussed here are resolved and implemented in social and political actions. Governments increasingly recognize that the reduction of disaster risks is vital for sustainability (Sendai Framework 2015; UN 2015), and that disaster risk is a cross-cutting issue, requiring action across multiple sectors. "Disaster risk reduction should be based on firm scientific knowledge, vast information/data, and the systematic development and application of policies, strategies and practices to minimize vulnerabilities and disaster risks throughout a society. This will result in avoiding (prevention) or in limiting (mitigation and preparedness) adverse impact of hazards, within the broad context of sustainable development" (Ismail-Zadeh and Cutter 2015). Still greater efforts are required to ensure that integrated (co-engaged, produced and managed) science is better understood and communicated. Science-based approach to risk reduction can help the governments and the society in mitigating and finally preventing disasters.

Let us hope that our descendants will tell one day: 'Scientists in the 21st century … believed that natural events, which they called hazards, lead in many cases to tragedies in families and result in severe losses of lives and properties. They did not know well how to minimize or, as today, to eliminate disasters. We know it now (in the 22nd century). But we should thank them anyway that they thought about us and tried their best to reduce disasters and create a better future for us' (Showstack 2015).

Acknowledgements The author acknowledges a support from the German Science Foundation (DFG grant IS-203/4-1).

References

Ambraseys N, Bilham R (2011) Corruption kills. Nature 469:153–155

Ando M, Ishida M, Hayashi Y, Mizuki C (2011) Interviews with survivors of Tohoku earthquake provide insights into fatality rate. EOS Trans AGU 92(46):411. https://doi.org/10.1029/2011EO460005

Babayev G, Ismail-Zadeh A, Le Mouël J-L (2010) Scenario-based earthquake hazard and risk assessment for Baku (Azerbaijan). Nat Hazard Earth Syst Sci 10:2697–2712

Baker J (2013) Seismology: quake catcher. Nature 498:290–292

Beer T, Ismail-Zadeh A (eds) (2003) Risk science and sustainability. Kluwer Academic Publishers, Dordrecht

Boaz A, Hayden C (2002) Pro-active evaluators: enabling research to be useful, usable and used. Evaluation 8:440–453

Cummings RG, Brookshire DS, Schulze WD et al (eds) (1986) Valuing environmental goods: an assessment of the contingent valuation method. Rowman & Allanheld, Totowa

Cutter S, Ismail-Zadeh A, Alcántara-Ayala I, Altan O, Baker DN, Briceño S, Gupta H, Holloway A, Johnston D, McBean GA, Ogawa Y, Paton D, Porio E, Silbereisen RK, Takeuchi K, Valsecchi GB, Vogel C, Wu G (2015) Pool knowledge to stem losses from disasters. Nature 522:277–279

Fuchs K (2009) The great earthquakes of Lisbon 1755 and Aceh 2004 shook the world. Seismologists' societal responsibility. In: Mendes-Victor LA et al (eds) The 1755 Lisbon earthquake: revisited. Geotechnical, geological, and earthquake engineering, vol 7. Springer, Dordrecht, pp 43–64

Ghafory-Ashtiany M (2014) Earthquake risk and risk reduction capacity building in Iran. In: Ismail-Zadeh A, Urrutia Fucugauchi J, Kijko A, Takeuchi K, Zaliapin I (eds) Extreme natural hazards, disaster risks and societal implications. Cambridge University Press, Cambridge, pp 267–278

Gluckman P (2016) The science–policy interface. Science 353:969. https://doi.org/10.1126/science.aai8837

Hirsch Hadorn G, Hoffmann-Riem H, Biber-Klemm S et al (eds) (2008) Handbook of transdisciplinary research. Springer, Dordrecht

IPCC (2012) Managing the risks of extreme events and disasters to advance climate change adaptation. In: Field CB, Barros V, Stocker TF et al. (eds) A special report of working groups I and II of the intergovernmental panel on climate change. Cambridge University Press, Cambridge

IRDR (2013) Issue brief: disaster risk reduction and sustainable development. Integrated Research on Disaster Risk, Beijing

Ismail-Zadeh A (2010) Computational geodynamics as a component of comprehensive seismic hazards analysis. In: Beer T (ed) Geophysical hazards: minimizing risk and maximizing awareness. Springer, Amsterdam, pp 161–178

Ismail-Zadeh A (2014) Extreme seismic events: from basic science to disaster risk mitigation. In: Ismail-Zadeh A, Urrutia Fucugauchi J, Kijko A, Takeuchi K, Zaliapin I (eds) Extreme natural hazards, disaster risks and societal implications. Cambridge University Press, Cambridge, pp 47–60

Ismail-Zadeh A, Cutter S (eds) (2015) Disaster risks research and assessment to promote risk reduction and management. ICSU-ISSC, Paris

Ismail-Zadeh A, Takeuchi K (2007) Preventive disaster management of extreme natural events. Nat Haz 42:459–467

Ismail-Zadeh A, Cutter SL, Takeuchi K, Paton D (2017) Forging a paradigm shift in disaster science. Nat Haz 86(2):969–988

Kant I (1756) Geschichte und Naturbeschreibung der merkwürdigsten Vorfälle des Erdbebens welches an dem Ende des 1755 sten Jahres einen großen Theil der Erde erschüttert hat. Königsberg

Kantorovich L, Keilis-Borok VI, Molchan G (1973) Seismic risk and principles of seismic zoning. In: Keilis-Borok VI (ed) Computational and statistical methods for interpretation of seismic data. Nauka, Moscow, pp 3–20 (in Russian)

Manaker DM, Calais E, Freed AM, Ali ST, Przybylski P, Mattioli G, Jansma P, Prepetit C, de Chabalier JB (2008) Interseismic plate coupling and strain partitioning in the Northeastern Caribbean. Geophys J Int 174:889–903

Satake K, Sawai Y, Shishikura M, Okamura Y, Namegaya Y, Yamaki S (2007) Tsunami source of the unusual AD 869 earthquake off Miyagi, Japan, inferred from tsunami deposits and numerical simulation of inundation. In: Abstracts of the American Geophysical Union, Fall Meeting 2007. Abstract T31G-03

Sendai Framework (2015) Sendai framework for disaster risk reduction 2015–2030. http://www.unisdr.org/we/inform/publications/43291. Accessed 28 Aug 2017

Showstack R (2015) Geoscientists: focus more on societal concerns. Eos Trans AGU 96(17). https://doi.org/10.1029/2015eo034063. https://eos.org/articles/geoscientists-focus-more-on-societal-concerns. Accessed 23 Aug 2017

UN (2015) Transforming our world: the 2030 agenda for sustainable development. A/RES/70/1. United Nations, New York

UN (2017) Report of the open-ended intergovernmental expert working group on indicators and terminology relating to disaster risk reduction. http://www.preventionweb.net/files/50683_oiewgreportenglish.pdf. Accessed 28 Aug 2017

UNGA (1987) United Nations General Assembly Report of the World Commission on Environment and Development: Our Common Future. UN Headquarters, New York. http://www.un-documents.net/wced-ocf.htm. Accessed 28 Aug 2017

UNISDR (2013) The global assessment report on disaster risk reduction (GAR2013). UN Office for Disaster Risk Reduction, Geneva

Real-Time Safety Assessment of Disaster Management Facilities Against Earthquakes

Saito Taiki

Abstract One of the most crucial issues in times of earthquake disaster is securing the functions of disaster management facilities such as city halls, hospitals and fire stations. However, in the case of past earthquake damage, there are many problems that important structures cannot be used after the earthquake. In order to solve this problem, it is necessary to predict earthquake damage beforehand and take countermeasures such as seismic retrofitting. Also, when an earthquake actually occurs, it is necessary to analyse the degree of damage of the building as soon as possible and diagnose whether it is safe or not by aftershocks. In this paper, seismic simulations are conducted on the east and the west buildings of Toyohashi City Hall using the earthquake ground motion waveform for the future Nankai Trough earthquake. As the result, it was found that the buildings have sufficient earthquake resistance. Additionally, IT strong-motion seismometers were installed in the city hall to continuously monitor vibrations. If an earthquake strikes, the vibration data obtained by the seismometers are sent to the computer through the Internet to simulate the building. Then, the results of simulation are sent to the building owner to assess whether the building is safe for aftershock. By automating this process, real-time safety assessment is possible.

Keywords Safety assessment · Earthquake · Disaster management facility
Seismic monitoring · Simulation

1 Introduction

The Kumamoto earthquake occurred on April 14, 2016 with Magnitude 6.5 at Kumamoto Prefecture, Japan, causing direct loss of 50 people and extensive damage to the buildings and houses. Several local city offices such as the Mashiro

S. Taiki (✉)
Department of Architecture and Civil Engineering,
Toyohashi University of Technology, Aichi, Japan
e-mail: tsaito@ace.tut.ac.jp

© Springer International Publishing AG, part of Springer Nature 2018
R. Vacareanu and C. Ionescu (eds.), *Seismic Hazard and Risk Assessment*,
Springer Natural Hazards, https://doi.org/10.1007/978-3-319-74724-8_16

town office and Otsu town office near the epicentre were severely damaged and became unusable. Also, the Uto City Hall which was constructed in 1965 completely collapsed and municipality works had to be done in the temporary tent outside of the building. Also, many gymnasiums suffered damage such as the falling of ceiling panels and could not be used as evacuation shelters. In this way, the disaster management facilities that should function normally in the event of a disaster were unable to fulfil their roles (Fig. 1). The same problems can be pointed out in the previous earthquake disasters, too.

The objective of this study is to develop a real-time earthquake monitoring system of disaster management facilities. The computer simulation technique developed by the author is applied to assess the damage of the building immediately after the earthquake using the acceleration data transferred from the sensors installed inside the building. The simulation model is also used to estimate the potential damage due to the future earthquakes.

2 Concept of Earthquake Response Monitoring System

Figure 2 shows the concept of earthquake response monitoring system. IT strong-motion seismometers are installed in the target building to continuously monitor its vibration. Every five minutes, the vibration data obtained by the seismometers are sent to the Internet cloud storage and available anytime from the computers in Toyohashi University of Technology via the Internet for the analysis tools to run simulations. The system is designed to select the data exceeding a certain threshold of acceleration. If an earthquake strikes, the results of the simulations will be used to assess whether it is safe to continue using the building.

(a) Otsu town office   (b) Mashiki town office   (c) Mashiki town gymnasium   (d) Uto city hall

Fig. 1 Damage of disaster management facilities due to the 2016 Kumamoto earthquake

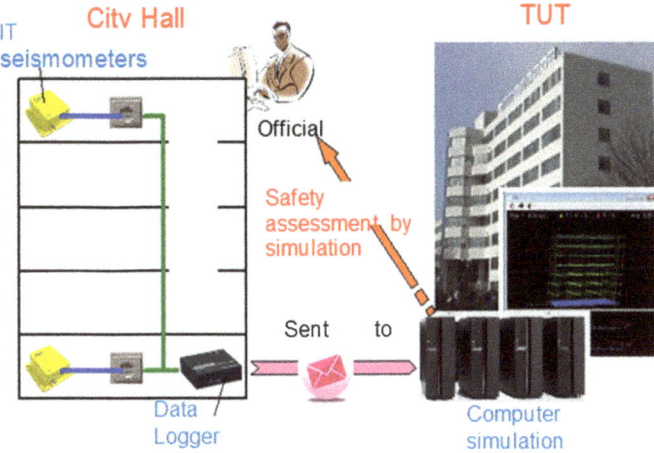

Fig. 2 Concept of earthquake response monitoring system

3   Earthquake Response Monitoring System of Toyohashi City Hall

3.1   Outline of the Building

The Toyohashi City Hall is consisted of the West Building (Fig. 3a) and the East Building (Fig. 3b). The West Building has eight upper stories (one story rooftop) and one story basement. The building height is 38.3 m and the plan is 58.2 m in X direction and 25.6 m in Y direction. Two shear wall cores are in both sides of the longitudinal directions, and are connected by the rooftop truss beam made of steel reinforced concrete. The East Building has thirteen upper stories and three stories basement. The building height is 55.55 m and the plan is 52.9 m length in X direction and 43.5 m length in Y direction. The building has 12.8 m × 13.6 m of atrium in the middle of upper story. The columns are steel reinforced concrete and the beams and bracings are steel.

The frame analysis model of each building is created using the earthquake response analysis software, STERA_3D developed by the author (Saito 2017). STERA_3D is an integrated software for seismic analysis of buildings with various structures (reinforced concrete, steel, masonry, base isolation, response control, etc.) in 3-D space. STERA_3D has a visual interface as shown in Fig. 4 to create building models and show the results easily and rapidly.

The Toyohashi city hall buildings are modelled by STERA_3D as follows. The in-plane deformation of the floor and the deformation of beam-column joint are assumed to be rigid. The beam is modelled as a line element with nonlinear flexural springs at both ends and a nonlinear shear spring at the center. The degrading tri-linear slip model is used for the flexural hysteresis. The column and the wall are

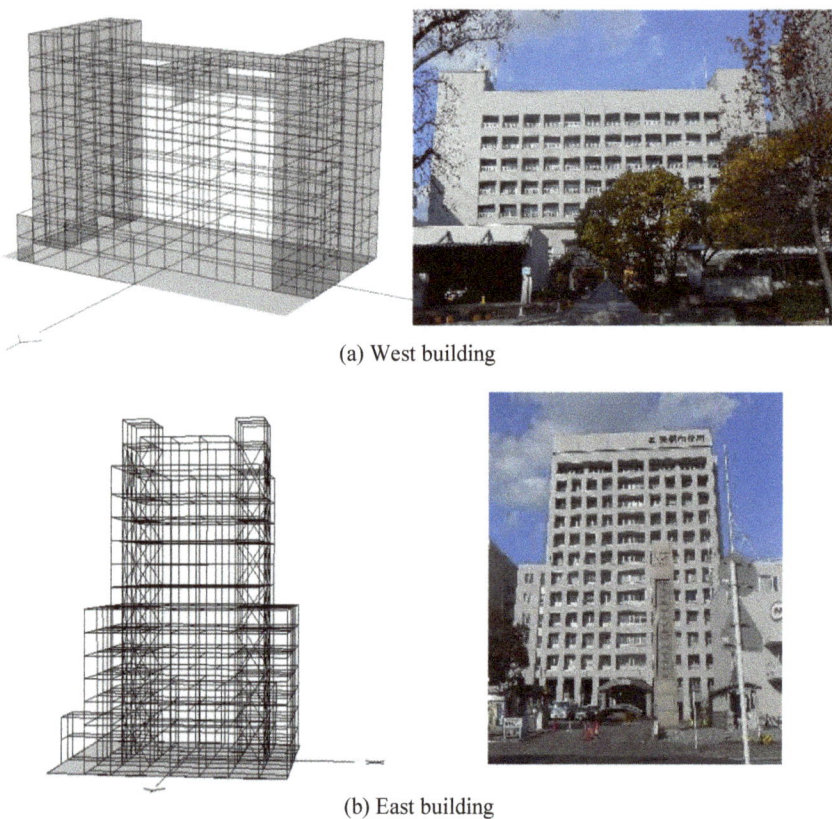

(a) West building

(b) East building

Fig. 3 Toyohashi City Hall buildings and STERA_3D models

modelled in a similar manner, while nonlinear interaction between axial force and moment is expressed using axial springs of concrete and steel arranged in the sections at both ends (so called MS-model) and the nonlinear shear characteristics are modeled by the nonlinear shear springs (Fig. 5). The damping is assumed as an initial stiffness proportional type with 4% damping factor for West building and 2% for East building. Figure 3 shows 3D frame models of the Toyohashi city hall buildings created by STERA_3D.

3.2 Vibration Characteristics of the Building

The first natural periods of the buildings are compared between the micro-tremor observation of the real buildings and the simulation results by STERA_3D as

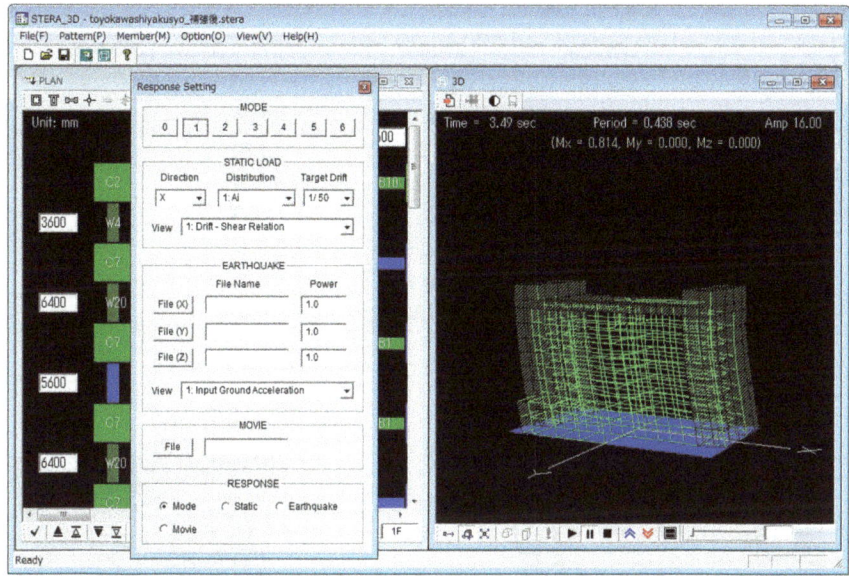

Fig. 4 Interface of STERA_3D

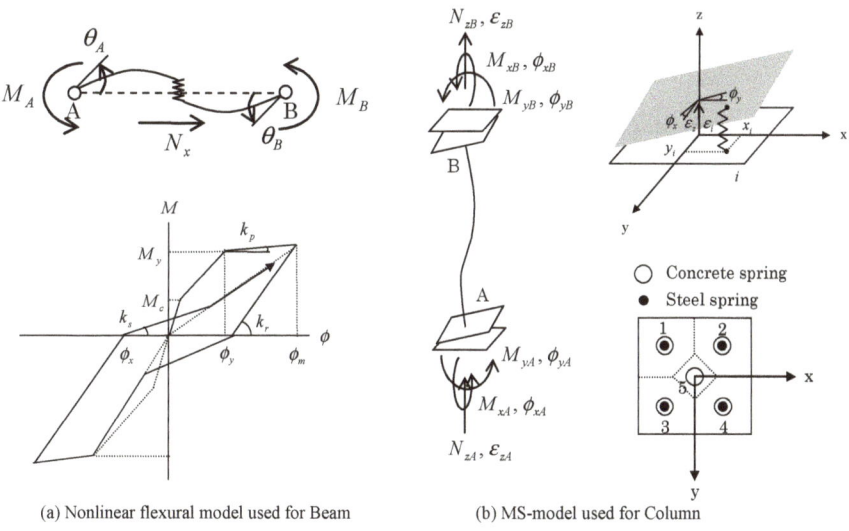

(a) Nonlinear flexural model used for Beam   (b) MS-model used for Column

Fig. 5 Nonlinear member models used in STERA_3D

Table 1 Comparison of the first natural periods

Building	Direction	Method	First natural period (s)
West building	EW (X)	Observation	1.04
		Simulation	1.05
	NS (Y)	Observation	1.01
		Simulation	1.07
East building	EW (X)	Observation	0.43
		Simulation	0.44
	NS (Y)	Observation	0.40
		Simulation	0.42

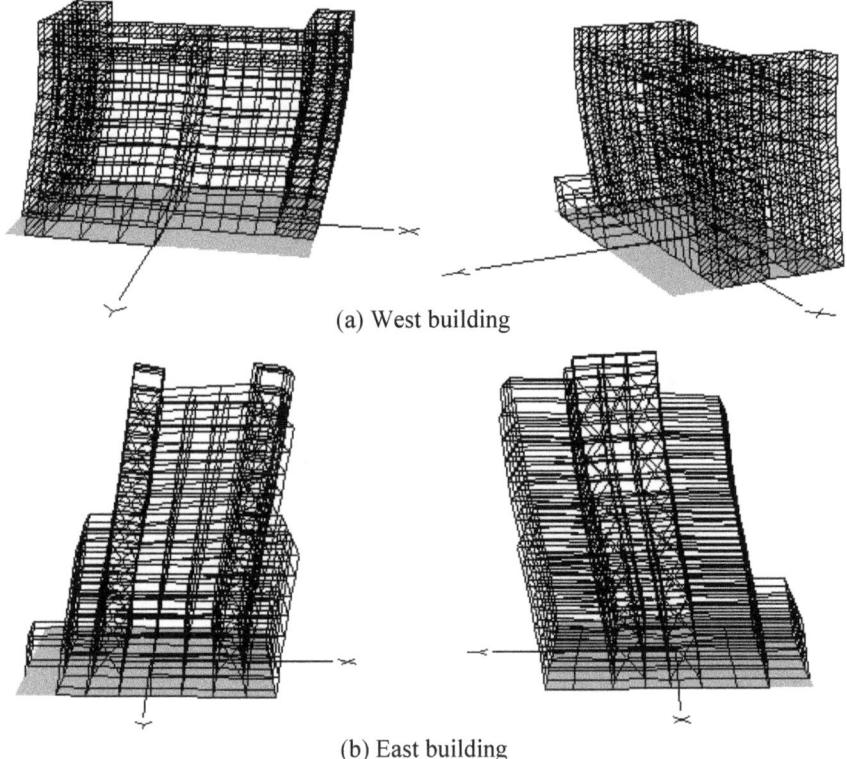

(a) West building

(b) East building

Fig. 6 The first vibration mode shapes of the buildings in each direction

shown in Table 1. It is seen that the both natural periods are close each other. Therefore, the models created by STERA_3D are consistent with the real buildings. Figure 6 is the first vibration mode shapes of the West building and the East building obtained by STERA_3D.

3.3 Monitoring System Using IT Seismometers

The observation network using the IT strong-motion seismometers (SU501, manufactured by Hakusan Co., Ltd) was installed in the city hall. This seismometer is a servo-type acceleration sensor easy to be installed inside EPS (electric pipe space) which operates using power supply from the LAN cable. The measurement range of the sensor is ±4 G and the resolution is 0.0006 gal. Three seismometers were installed on the 1st story basement (W_B1F) and the 8th story floor (W_8F) of the West building, and 13th story floor (E_13F) of the East building and constant observation has started from November 2016. The installation positions of the seismometers are shown in Fig. 7. The text data of observation records is sent to the cloud on the Internet in every 5 min.

Figure 8 shows the acceleration records observed by the IT seismometers during the earthquake (M4.8) occurred on southern part of Wakayama Prefecture on 19 November 2016. The acceleration response of the basement of the building was amplified at the upper floor of the building. Especially, in the EW direction of the East building, the maximum acceleration was 18.76 gal at the 13th floor which is about seven times larger than that of basement.

The acceleration of the top story of each building is compared between the response analysis of STERA_3D model and the observation data of the IT seismometers. The acceleration records at the basement are used for input data of STERA_3D. The green line on Fig. 9 shows the acceleration response calculated by STERA_3D and the red line shows the observed acceleration records by the IT seismometers. For the East building, the analysis results match well with the observation records. For the West building, they are generally correspondent each other, however the analysis results tend to overestimate after 50 s in EW direction.

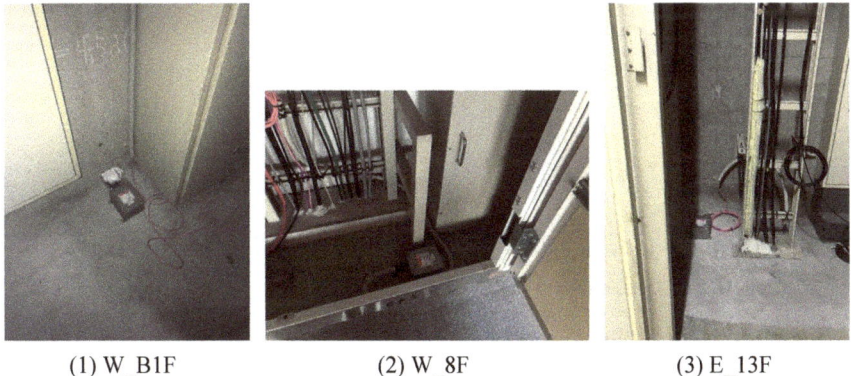

(1) W_B1F   (2) W_8F   (3) E_13F

Fig. 7 Locations of IT strong-motion seismometers

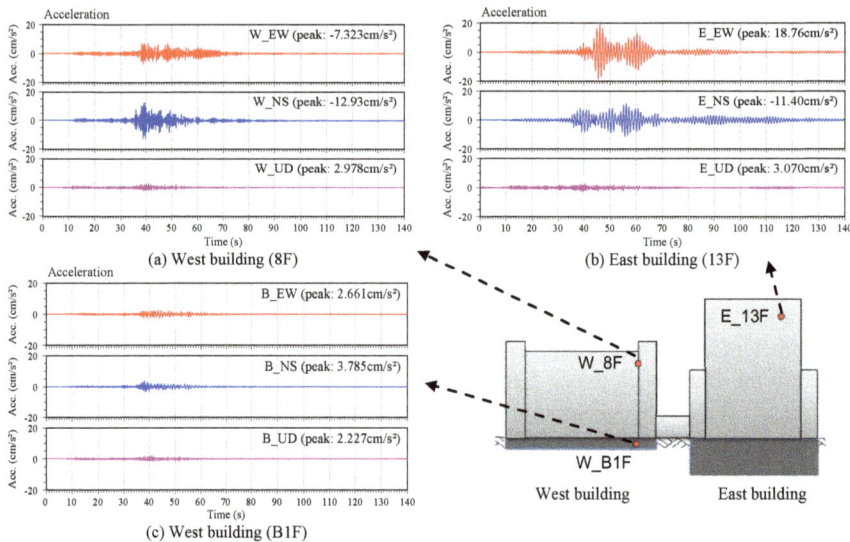

Fig. 8 Observation records at the earthquake on 19 November 2016

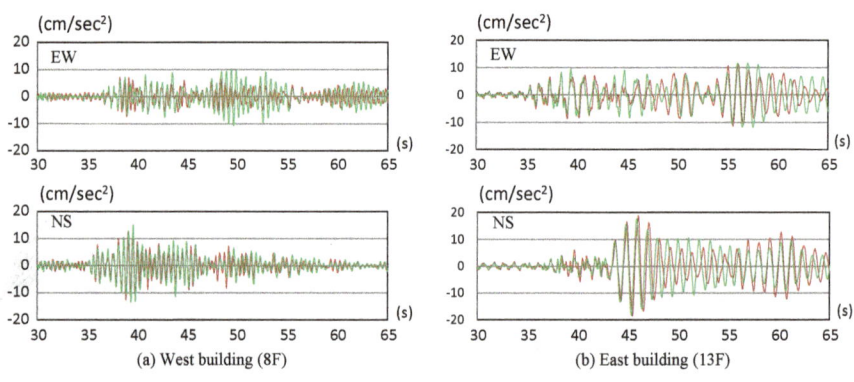

Fig. 9 Comparison of top story accelerations (———— analysis ———— observation)

4   Safety Assessment of Toyohashi City Hall Under Nankai Trough Earthquake

Toyohashi city situated above the Nankai Trough is estimated to have a 70% chance of experiencing an earthquake of magnitude 6–7 on the seismic intensity scale within the next 30 years. The Ministry of Land, Infrastructure, Transport and Tourism created seismic motion waveforms of various places expected in the case

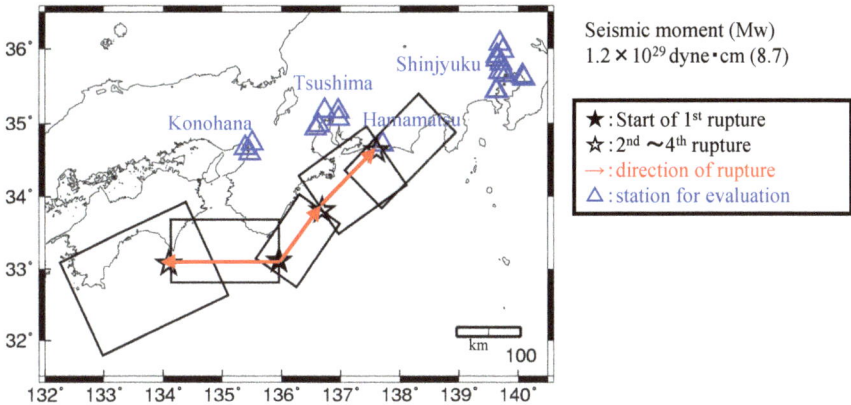

Fig. 10 Rupture sequence of faults assumed for the Nankai Trough earthquake

of the Nankai Trough massive earthquake with the aim of verifying the seismic safety of buildings (Okawa et al. 2012). The rupture sequence of the Nankai Trough faults is indicated in Fig. 10.

Figure 11 shows the simulated acceleration waves at Hamamatsu city for the average estimation (AV) and the average plus one standard deviation (SD). Figure 12 shows the pseudo velocity response spectra with 5% damping factor for the simulated. The spectrum exceeded the velocity of 80 cm/s which is the Level 2 (safety limit) earthquake in the notification of the Building Standard Law of Japan at the engineering bedrock. Since Toyohashi city is close to Hamamatsu city, the simulated ground motions at Hamamatsu city are used to evaluate the safety of the Toyohashi City Hall.

Figure 13 shows the maximum floor acceleration and the maximum story drift angle calculated for EW and NS directions of West and East buildings. The black line shows the results under the average earthquake (AV) and the grey line shows the results under the average plus standard deviation earthquake (AV + SD) at Hamamatsu city. The story drifts of West buildings exceed the safety limit (1/100) at lower stories, whereas those of East buildings maintain less than the safety limit.

Figure 14 shows the locations of yield hinges of West building during the earthquake response analysis by STERA_3D. The yellow part in the figure is a lightly damaged part and the red part is a severe destruction part. In order to clearly show the state of deformation, the deformation is magnified 32 times and displayed. In case of the West building in Fig. 14, although the damage of the whole building is slight, there are severe destruction points in the columns and beams of the lower stories in the middle of the building.

Figure 15 shows the damage parts of East building. Also the deformation is magnified 32 times and displayed. Although minor damage is caused by many members, it can be seen that there are no serious destruction points and the building can be continuously used after the earthquake. On the other hand, from Fig. 13,

(a) AV: Average

(b) SD: AV + standard deviation

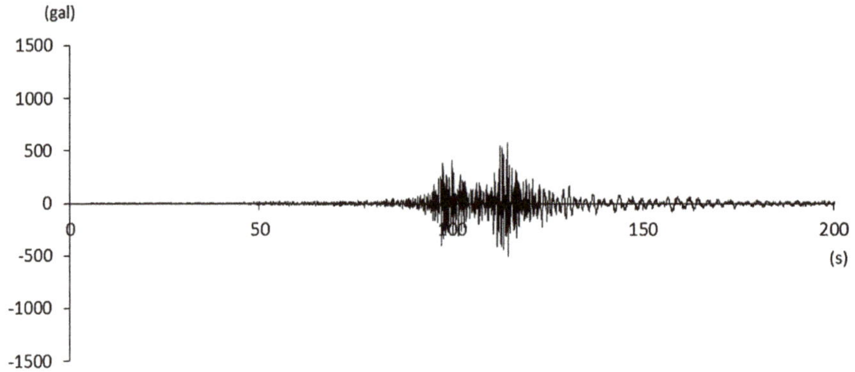

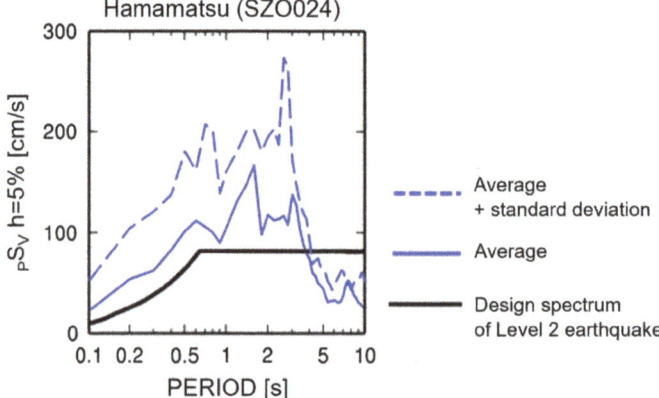

Fig. 11 Acceleration waves simulated at Hamamatsu city (AV: average, SD: AV + standard deviation)

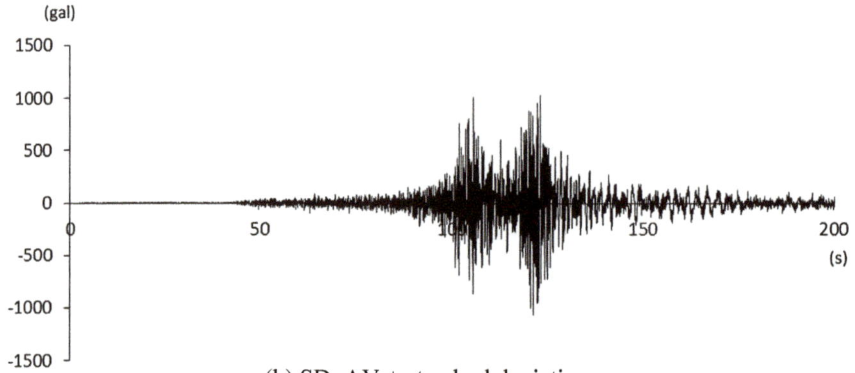

Fig. 12 Pseudo velocity response spectra with 5% damping factor at Hamamatsu city

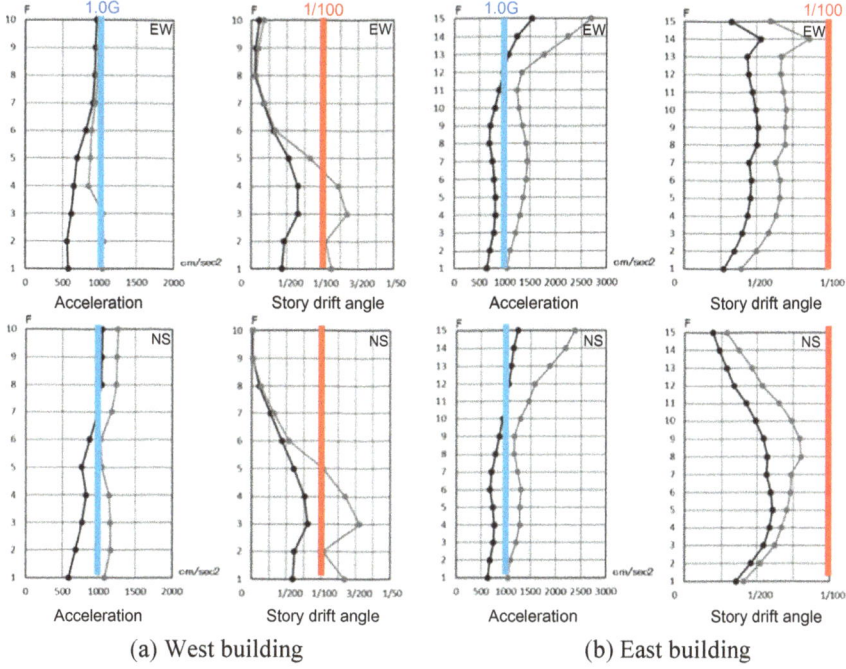

(a) West building   (b) East building

Fig. 13 The maximum responses of city hall under the Nankai Trough earthquake (——●—— AV ———●——— AV + SD)

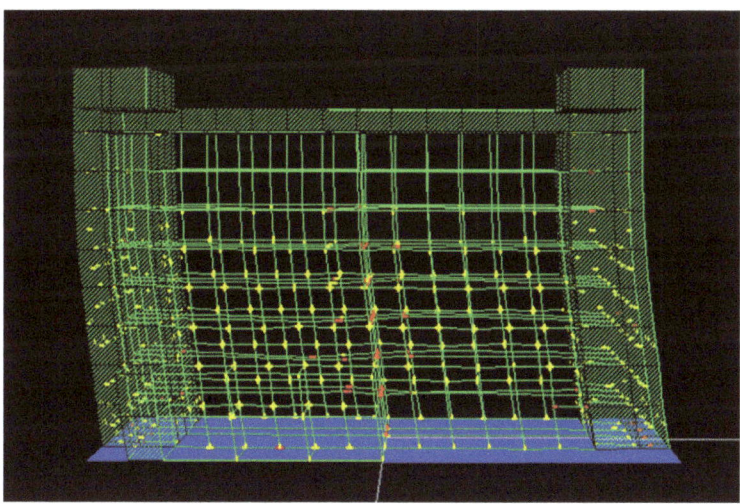

Fig. 14 Damage distribution of STERA_3D model of West building

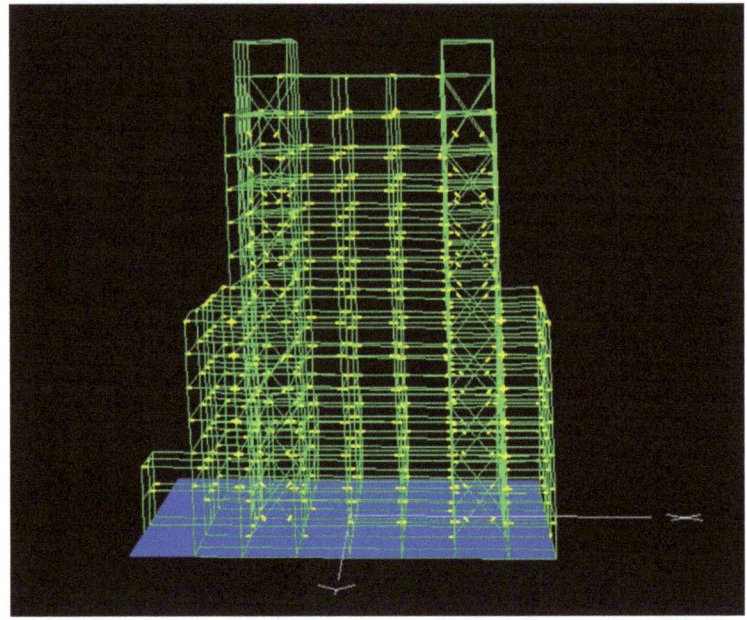

Fig. 15 Damage distribution of STERA_3D model of East building

the maximum floor accelerations of East building become far larger than 1.0 G in upper stories which may cause damage such as overturning of the furniture or falling of the ceiling.

5 Regional Network of Real-Time Safety Assessment of Disaster Management Facilities

In collaboration with local governments in the East Mikawa area such as Toyohashi City, we have started creating a regional network of real-time safety assessment of disaster management facilities (government buildings, fire departments, schools, etc.) in the area. The following actions are in progress:

(1) We set up IT strong motion seismometers to the facilities to monitor the vibration of the buildings anytime anywhere through the Internet connection.
(2) We give a lecture to building managers or disaster prevention personnel on how to use the earthquake response analysis software "STERA_3D" so that they can construct a structural analyse of the building to assess the damage by themselves.
(3) We conduct a simulation analysis of the buildings subjected to the future Nankai Trough Earthquake and visualize the building damage using CG to

Fig. 16 Regional network of real-time safety assessment of disaster management facilities

appeal the necessity of seismic retrofitting of the buildings and improve the awareness of disaster prevention.

(4) When an earthquake occurred, simulation analysis of the buildings will be carried out based on the observation records from IT strong motion seismometers to clarify damaged parts and report to the building managers. We will automate the process as much as possible (Fig. 16).

6   Conclusions

This paper explains the newly installed earthquake monitoring system at the Toyohashi City Hall combined with the computer simulation techniques developed by the author which enables to assess the detail damage of the building immediately after the earthquake by using the observed records from the Internet. Also the simulation model is used to estimate the potential damage due to the future Nankai trough earthquake. This system with monitoring and simulation techniques can be used for both post-event and pre-event measures.

It is planned to create a regional network with this system installing IT seismometers and creating STERA_3D models for the disaster management facilities in the region so that at the moment of earthquake it enables to identify the damage status of disaster management facilities in the region and support the decision of immediate action by the local governor.

Acknowledgements This study is partly supported by Toyohashi City and KAKENHI Grant Number 16H03143.

References

Cabinet Office, "About the damage situation relating to the earthquake with seismic source in the Kumamoto district of Kumamoto prefecture (as of November 14th, 2016)", http://www.bousai. go.jp/updates/h280414jishin/

Okawa I, Satoh T, Sato T, Nishikawa T (2012) An Empirical Evaluation of Long-Period Earthquake Motion for Building Design. Proceedings of the 15th World Conference of Earthquake Engineering, IAEE, Lisbon, Portugal

Saito T (2017) Making buildings safer by pre-emptively visualizing earthquake damage, Feature Story, TUT Research No. 8. https://www.tut.ac.jp/english/newsletter/contents/2017/08/ features/features.html

Saito T. STERA_3D, www.rc.ace.tut.ac.jp/saito/sofware-e.html

Seismic Risk Assessment of Romania

Cristian Arion, Florin Pavel, Radu Vacareanu, Cristian Neagu,
Mihail Iancovici, Viorel Popa and Ionuț Damian

Abstract This paper summarizes the UTCB results for the "National Risk Assessment—RO RISK" project. Within the RO-RISK project, coordinated by the General Inspectorate for Emergency Situations, the first nation-wide assessment of all types of natural risks was performed in 2016. The work was supported by collaboration of disaster reduction experts and earthquake risk modelling specialists from INFP and URBAN-INCERC. The seismic risk assessment was performed for the entire country, at the most detailed resolution available, which is the administrative-territorial unit. For each building typology, four limit states were considered in order to generate fragility curves. Each limit state is associated with a loss percentage, in order to generate vulnerability curves. The assessment shows that among the 10 analysed hazards, the seismic hazard produces the largest impact at country level, 75% of the population and 45% of the vital networks are exposed to moderate and high earthquake risk and Romania's capital Bucharest, is highly exposed to earthquakes.

Keywords Buildings · Population · Networks · Hazard · Earthquake scenario
Losses

1 Introduction

The RO-RISK project coordinated by the General Inspectorate for Emergency Situations (IGSU 2016) begun in March 2016 and aimed at making the first risk assessment at country level for Romania. A total of 10 hazards were taken into consideration and their impact was evaluated at country level by teams of experts from various institutions. The evaluation of seismic risk, which is the focus of this

C. Arion (✉) · F. Pavel · R. Vacareanu · C. Neagu · M. Iancovici
V. Popa · I. Damian
Seismic Risk Assessment Research Center, Technical University of Civil Engineering
of Bucharest, Bucharest, Romania
e-mail: Cristian.arion@utcb.ro

© Springer International Publishing AG, part of Springer Nature 2018   251
R. Vacareanu and C. Ionescu (eds.), *Seismic Hazard and Risk Assessment*,
Springer Natural Hazards, https://doi.org/10.1007/978-3-319-74724-8_17

paper was performed by a team of experts from the Technical University of Civil Engineering of Bucharest (UTCB), National Institute for Earth Physics (INCDFP) and from URBAN-INCERC.

Due to its location, Romania is associated with two contributions to seismic hazard: (i) The major contribution, which comes from subcrustal (intermediate depth) Vrancea seismic zone; (ii) The usual crustal contribution, which comes from various, less active and less intensive, shallow seismic zones distributed over the all territory of the country (the strongest earthquakes in 20th century from Timisoara source were the July 12th, 1991).

The Vrancea region, located where the Carpathians Mountains Arch bends, at about 135 ± 35 km epicentral distance from Bucharest, is a source of subcrustal seismic activity. Vrancea intermediate depth source (depth of focus from 60 to 170 km) induces the largest level of seismic hazard on more than 2/3 of Romania territory but also on large areas in Republic of Moldova, Bulgaria and Ukraine. According to the 20th century seismicity, the epicentral Vrancea area is confined to a rectangle of 40×80 km^2 having the long axis oriented N45E and being centered at about 45.6° Lat. N and 26.6° Long. E. Vrancea earthquakes produced significant damage in built areas and triggered earthquake induced phenomena like liquefaction, landslides, etc.

The most powerful Vrancea earthquake was, probably, the Oct. 26th, 1802 event (Gutenberg-Richter magnitude $M_{G-R} \geq 7.5$). It was felt on a surface over 2,000,000 km^2 (Popescu 1941). Strong earthquakes followed the 1802 event on Nov. 26th, 1829 and on Jan. 23rd, 1838 (both with maximum seismic intensity 8–9).

During the 20th century, Bucharest was threatened by 4 strong Vrancea events: Nov. 10th, 1940 (moment magnitude $M_w = 7.7$, focal depth h = 150 km), March 4th, 1977 ($M_w = 7.4$, h = 109 km), Aug. 30th, 1986 ($M_w = 7.1$, h = 133 km) and May 30/31, 1990 ($M_w = 6.9/6.4$, h = 91/79 km).

In the Table 1, an overview of the most important Romanian earthquakes is presented. Data on their impact, casualties and degree are provided.

2 Seismic Hazard Scenarios

The seismic risk assessment was performed for the entire country, at the most detailed resolution available, which is the administrative-territorial unit. A total of 3186 units were considered in the analysis. The assessment of seismic risk was performed for hazard five scenarios, three of which were common to all the hazards' assessment. The other two scenarios, required a more detailed analysis and were defined individually. The first three seismic hazard scenarios, considered the effect of all the seismic sources which can affect the territory of Romania. The seismic sources that can affect the territory of Romania are defined in Fig. 1. For each mean return period considered in the assessment, namely $MRI = 10$ years, $MRI = 100$ years and $MRI = 1000$ years, a classic probabilistic seismic hazard assessment (PSHA) was performed. In the case of the two deterministic earthquake

Table 1 Major earthquakes on Romanian territory in the XXth century

	Date	Time	M_w	Casualties	Building affected, economical losses
1	November 10th 1940	03:39	7.7	593 deaths (140 in Bucharest) 1271 injured (300 in Bucharest)	Low rise buildings seriously damaged The tallest reinforced concrete building in Bucharest collapsed
2	March 4th 1977	21:21	7.4	1578 deaths (1424 in Bucharest) 11,321 injured (7598 in Bucharest)	– 156,000 apartments in urban zones and 21,500 rural houses destroyed or very seriously damaged; – 366,000 apartments in urban zones and 117,000 rural houses to be repaired; – destroyed 374 kindergartens, nurseries, primary and secondary schools and badly damaged 1992 others. – destroyed six university buildings and damaged 60 others – destroyed 11 hospitals and damaged 228 others hospitals and 220 polyclinics (health care centers) – destroyed or damaged almost 400 cultural institutions such as theatre's and museums – damaged 763 factories. US$2.048 billion equivalent loss
3	August 30th 1986	23:28	7.1	8 deaths 317 injured	
4	May 30th 1990	12:40	6.9	9 deaths 296 injured	
5	July 12th 1991	12:42	5.6	2 deaths 30 injured	5000 rural houses in Banloc, hundreds to thousands of homeless in Timiş County

scenarios, the peak ground accelerations were computed using two ground motion models (median amplitudes), namely the Vacareanu et al. (2015) model for the Vrancea intermediate-depth seismic scenario and the Cauzzi et al. (2015) for the Banat crustal scenario. The most recent delineation of these seismic sources was performed within the BIGSEES national research project and can be found in the paper of Pavel et al. (2016).

The mean return periods of the first three hazard scenarios was taken as 10, 100 and 1000 years. The remaining two scenarios were defined based on their likely impact on urban settlements. The first scenario was a moment magnitude $M_W = 8.1$ earthquake originating in the Vrancea intermediate-depth seismic source (focal depth = 90 km). The $M_W = 8.1$ scenario corresponds to a mean return interval of 1000 years and can be considered as a "worst-case" scenario due to the combination of focal depth and magnitude. The last considered earthquake scenario was crustal event with a $M_W = 5.8$ generated in the Banat seismic source near the city of Timisoara (the magnitude corresponds to a mean return period of 100 years).

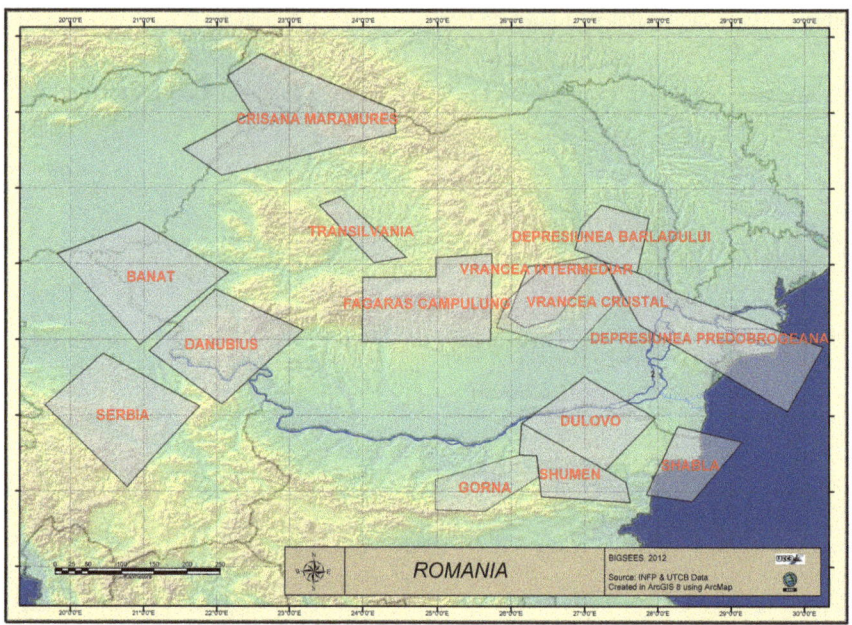

Fig. 1 Seismic sources which can affect the territory of Romania (defined within BIGSEES project)

The selection of the considered earthquake scenarios was based on an exposure analysis for buildings and various infrastructures.

Following a probabilistic hazard assessment methodology the seismic hazard map of Romania for 1000 years mean recurrence interval is presented in Fig. 2. As seen in Fig. 2, the entire eastern part, some areas in the center, the southern and south-western parts of Romania are exposed to a high level of seismic hazard. The results in terms of peak ground accelerations at administrative-territorial unit level are used subsequently in the seismic risk assessment.

3   Buildings, Population and Economic Exposure

The evaluation of exposure includes the quantity and the distribution of people and built environment, and the type of activities they support. Exposure is a very necessary component of risk and, of course, the larger the exposure, the greater the risk. In a general sense, the city's built environment exposure intends to describe when, where and how fast structures have been built and it is assessed in terms of number, size, geographical distribution and value.

Population exposure was characterised by the number and geographical distribution of all city residents, Fig. 3. Exposure data is available from the latest census data, 2011, Table 2. The occupancy exposure was evaluated by the number of

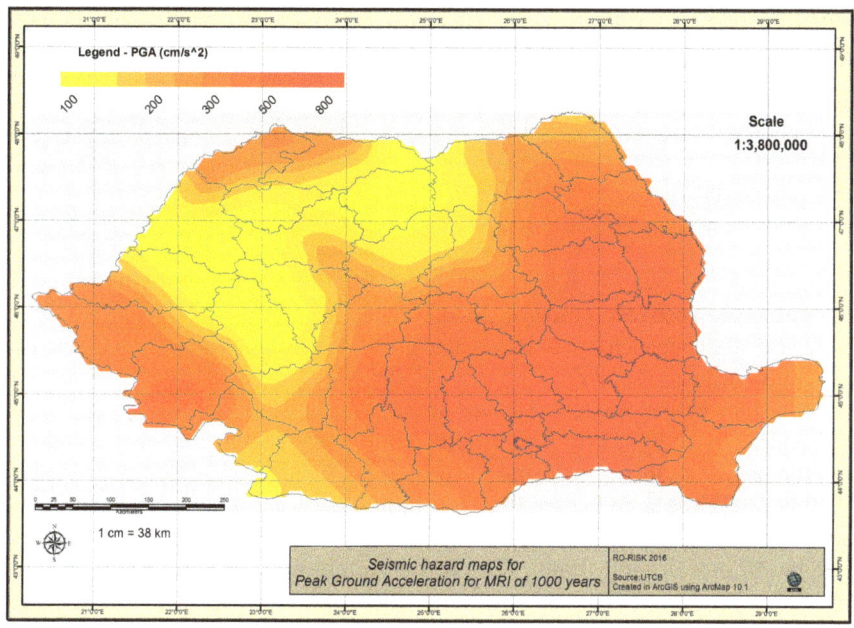

Fig. 2 Seismic hazard map of Romania for 1000 years mean recurrence interval

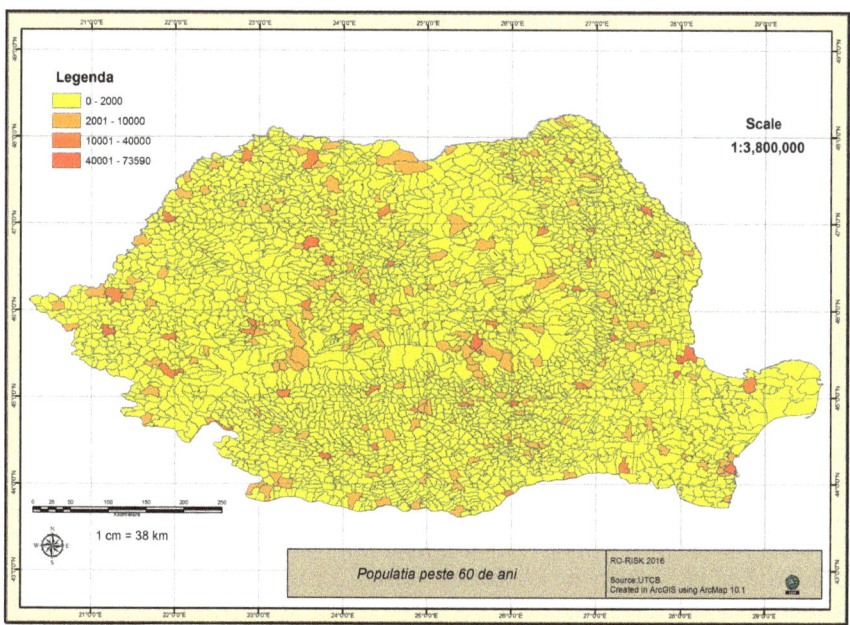

Fig. 3 Distribution of population aged more than 60 years

Table 2 Romania official information from the National Institute of Statistics

	1992 census	2002 census	2011 census
Population	23,286,794	22,628,665	20,121,641
No of buildings	4,482,119	4,837,215	5,341,908
Housing units	7,666,181	8,111,391	8,723,699
GDP (current US$ Billions)	25.12	46.18	185.36

persons expected to be in the building during an earthquake. In terms of post-earthquake recovery, certain types of occupancies are vital to the public needs. These special occupancies are clearly identified as critical emergency services: fire and police stations, hospitals, communication centres, etc.

The economic exposure is reflected in GDP at purchaser's prices (the sum of gross value added by all resident producers in the economy), and was correlated with the regional/national economic importance of each areas within the country.

Romania has the largest gas market in Central Europe and was the first to utilise the natural gas for industrial purposes. The market penetration for natural gas reached high levels in the early 1980s due to a government policy driven to achieve self-sufficiency. The National Company for the Natural Gas Transmission "TRANSGAZ" S.A. Medias has as main object of activity the transmission, dispatching, international gas transit, and the research—project works in the natural gas transmission field, and assures access to the National Transmission System (*NTS*), without discrimination, by any economic agent requesting it. "TRANSGAZ" S.A. is the technical operator of the *NTS*, responsible for the good functioning of the system under quality, safety, economic efficiency and environmental protection conditions.

The natural gas transport system is under high-pressure regime, over 6 Bars, made up of trunk pipelines, as well as of all the installation, equipment and due facilities, that ensure the taking over of natural gas extracted from operation perimeters or of that from the import and its transport for to be delivered to the distributors, direct consumers, export and/or storage. The *NTS* comprises, as follows: 11,000 km of main pipelines with diameters ranging from 150 to 800 mm; 5 gas compressor stations on the pipeline routes; over 600 gas regulating—metering —delivery stations; over 700 cathodic protection stations for the gas transmission pipelines; telecommunication system, Fig. 4.

In 1958 the *Romanian Power System* was created through the interconnection of the local subsystems. In 2000 the *National Electricity Company* was split in into separate legal entities: *Termoelectrica SA* for electricity and heat generation using fossil fuel, *Hidroelectrica SA* for electricity generation from hydro plants, *Electrica SA* for electricity distribution and supply and *Transelectrica SA* (National Power Grid Company) for electricity transmission, power system operation and dispatching, Fig. 5. Also *Transelectrica SA* provides the Romanian Power System control and ensures electricity exchanges with other power systems according to the contacts and technical regulations in force and sets up system services and provides electricity cross-border exchanges.

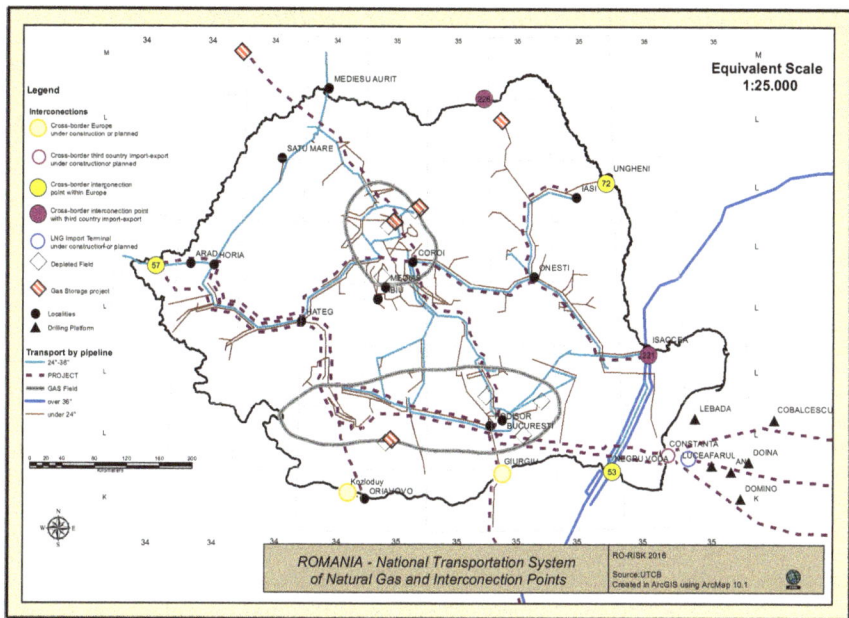

Fig. 4 The natural gas transport system

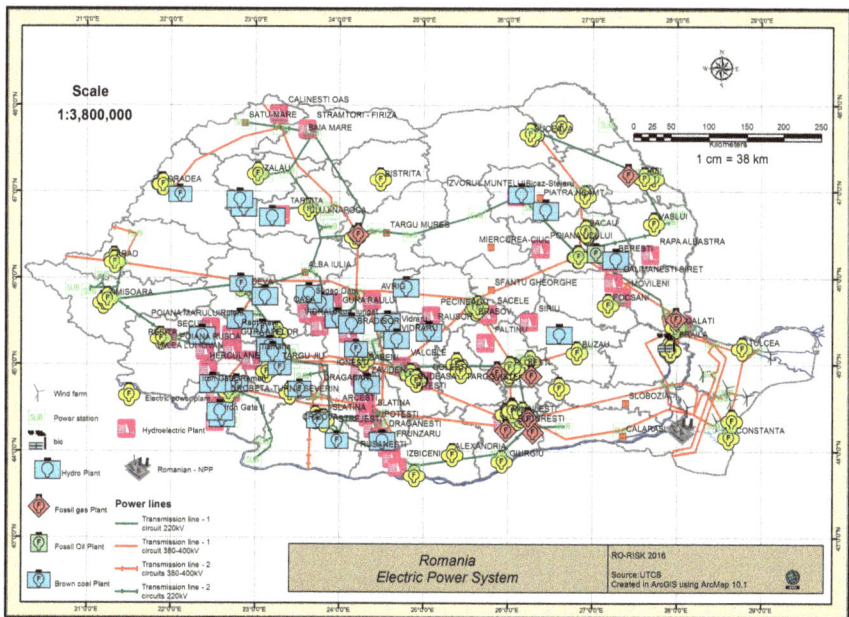

Fig. 5 Electric power system of Romania

3.1 Building Typology and Codes for Earthquake Resistance of Structures, 1940–2017

The buildings with similar structural systems and behaviour characteristics were grouped into a set of pre-defined building classes. According to our criterion of classification, the periods of buildings construction were classified according to the period of validity of Romanian seismic codes: before 1945, 1945–1963, 1964–1970, 1971–1977, 1978–1990, after 1990. This criterion of classification of buildings was later adopted by the Ministry of Public Works and Urban Planning (Governmental Order Nr.6173/NN/26.09.1997).

The quality of seismic design incorporated (Lungu et al. 2000a, b) into existing building stock is modelled by four categories of buildings (levels of seismic design, generations of major developments): *Pre-code* (Gravitational, designed until 1963), *Low-code* (Brittle behaviour, designed between 1963 and 1978), *Moderate-code* (Limited ductility behaviour, designed between 1978 and 1992) and *High-code* (Ductile behaviour, designed after 1992), Table 3.

The purpose of a building typology is to group buildings with similar damage/loss characteristics into a set of pre-defined building classes. We develop a list of parameters (building inventory mapping schemes) required for vulnerability modelling. The building typology matrix has been developed to provide an ability to differentiate between buildings with substantially different damage and loss characteristics. The following primary parameters affecting building damage and loss characteristics were given consideration in developing the building typology matrix: (i) Basic structural system and (ii) Earthquake resistant design level.

The inventories of residential buildings are estimated at lower LAU—local administrative units–level (LAU level 2, formerly NUTS—Nomenclature of Territorial Units for Statistics–level 5)—based on the Housing Census data (2011) available from the National Institute of Statistics (NIS). The LAU level 2 corresponding to Romania contains 3186 units.

Table 3 Classification of codes for earthquake resistant design of buildings in Romania (1940–2017)

Period		Code for earthquake resistance of structures
PC, Pre-code, before 1963	Prior to the 1940 earthquake and Prior to the 1963 code	P.I.—1941 I—1945
LC, Low-code, 1963–1977	Inspired by the Russian seismic practice	P 13-63 P 13-70
MC, Moderate-code, 1977–1990	After the great 1977 earthquake	P 100-78 P 100-81
HC, High-code, after 1990	After the 1986 and the 1990 earthquakes	P 100-90 P 100–92 P 100-1/2006 P 100-1/2013

Table 4 Inventory of building typologies available from the latest census data, 2011

	Population	Housing units	Build area (sqm)	Buildings	
				Romania	Bucharest (%)
RC_PC_LR	724,741	306358	16,603,935	247,324	7.7
RC_PC_MR	114,283	61,165	2,550,190	2605	36.6
RC_PC_HR	28,446	16,799	727,713	273	71.1
RC_LC_LR	840,196	318,907	17,248,189	289,749	3.1
RC_LC_MR	1,699,758	813,170	30,423,504	19,122	8.6
RC_LC_HR	737,437	363,506	14,445,659	4328	45.7
RC_MC_LR	343,886	118,089	7,088,388	96,444	1.7
RC_MC_MR	2,093,137	930,007	40,065,898	29,741	3.7
RC_MC_HR	745,451	328,803	14,961,168	5132	50.2
RC_HC_LR	701,660	274,784	21,174,192	261,264	3.7
RC_HC_MR	128,791	72,112	3,686,716	3711	21.0
RC_HC_HR	68,255	43,386	2,459,026	727	42.0
Total	8,226,041	3,647,086	171,434,578	960,420	5.1

The assignment of each building to one of the categories significantly influences the expected performance of buildings during and after an earthquake. We validate the inventories utilizing a combination of sources, including geo-tagged photographs taken from ground surveys and the internet.

Inventory and approximate classification of building stock as function of period of construction and number of storeys were prepared, Table 4, and is finally mapped as in examples of the Fig. 6. For the classification of reinforced concrete structures (RC) we used the HAZUS (1997) height range methodology (Low-rise: 1–3 stories, h ≤ 9 m; Mid-rise: 4–7stories, 9 m < h ≤ 21 m; High-rise: 8+ stories, h > 21 m). HAZUS is a geographic information system-based natural hazard analysis tool developed and freely distributed by the Federal Emergency Management Agency (FEMA), firstly version released in 1997 as HAZUS97. One may note that from the total number of high rise buildings, in Bucharest are located more than 50% of the total.

Damage and loss prediction models can then be developed for model building types which represent the average characteristics of the total population of buildings within each class.

4 Fragility Curves

The vulnerability assessment for RC buildings was performed using data from the most recent census from 2011. The RC buildings (RC frames and RC structural walls) were divided into four categories as a function of the level of seismic design

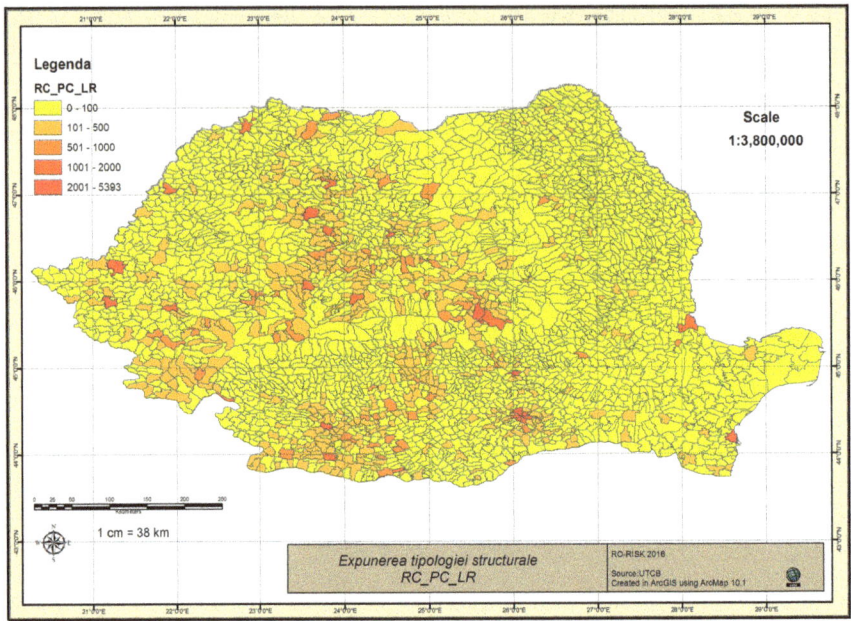

Fig. 6 Distribution of RC_PC_LR type buildings (low rise + pre code + RC)

code, each of them further sub-divided into categories according to the building height. Consequently, by combining the structural system, seismic code level and height regime, a total of 24 structural typologies were defined. Moreover, for each structural typology, the seismic capacity and fragility was defined based on the level of seismic action and if available, soil conditions. The evolution of seismic codes and of their associated zonations was considered for each administrative-territorial unit. Considering structure height, design period and location, a number of 132 different structures resulted. The pairs of capacity and fragility curves were generated for all the administrative-territorial units which cover the territory of Romania using a set of MATLAB (2009) procedures based on the presented assumptions and the calculation parameters presented in Table 5, elaborated specifically for this project.

The procedure proposed to generate fragility curves is based on HAZUS (2003) methodology which calculates the values of lognormal standard deviation and median top story displacement associated with a certain limit state. Fragility curves were generated based on dynamic nonlinear analysis of equivalent single degree of freedom (SDOF) for an accelerogram set. The uncertainties considered were the seismic action, the capacity of the buildings (CU) and the threshold displacement of a limit state (DSU).

Structural strength and limit angular drift for a limit state are considered random variables having a lognormal distribution.

Table 5 Calculation parameters to calculate fragility and vulnerability curves

Structural type	Design code	ε	q_ovs	I*/ m*	β	β*_CU	β_DSU	D_rS [rad]	D_rM [rad]	D_rE [rad]	D_rC [rad]
RCF	PC	0.85	1.5	1.35	2	0.3	0.4	0.005	0.01	0.015	0.02
	LC	0.85	1.5	1.35	1.8	0.25	0.4	0.005	0.01	0.02	0.03
	MC	0.85	1.5	1.35	1.6	0.25	0.4	0.005	0.01	0.025	0.04
	HC	0.85	1.5	1.35	1.5	0.25	0.4	0.005	0.01	0.05	0.08
RCW	PC	0.65	2	1.25	1.25	0.3	0.4	0.005	0.0075	0.012	0.015
	LC	0.65	2	1.25	1.25	0.25	0.4	0.005	0.0075	0.015	0.02
	MC	0.65	2	1.25	1.25	0.25	0.4	0.005	0.0075	0.02	0.03
	HC	0.65	2	1.25	1.25	0.25	0.4	0.005	0.0075	0.04	0.055

*Notations from literature

The seismic demand was defined at administrative-territorial unit level by anchoring the resulting peak ground acceleration to the spectral shape defined as a function of soil conditions defined in the current Romanian seismic design code P100-1/2013 (2013).

The structural performance was then evaluated using the well-known N2 method (Fajfar 2000) in which the inelastic demand spectrum is obtained from the elastic design spectrum by using ductility based reduction factors (in this case, the factors proposed by Vidic et al. 1994). For each structural typology and for each earthquake scenario, the expected structural behaviour is described using five damage states: no damage, slight damage, moderate damage, extensive damage and complete damage. The complete damage state is not necessarily linked to structural collapse as only a fraction of the buildings reaching this limit state actually collapses.

5 Results

The losses were evaluated using the HAZUS (2012) model. The results were obtained by aggregating the data from UTCB which dealt with RC buildings and the URBAN-INCERC contribution that used a different approach for the seismic risk assessment of masonry and adobe buildings. Figure 7 show the number of affected buildings and Fig. 8 show the number of affected people for probabilistic earthquake scenarios, with *MRI* = 1000 years.

The economic impact of the earthquake scenarios, which was evaluated, exceeds 25 bill. € for the *MRI* = 1000 years' scenario. The number of affected people (all degrees of severity of the injuries) exceeds 100,000 for the *MRI* = 100 years' and Vrancea M_W = 8.1 scenario, while in the case of the *MRI* = 1000 years' scenario, the number is three times larger. Consequently, based on the likely impact and the occurrence probability, the seismic risk results were collected in the risk matrix at country level. Figure 9 displays the risk matrix for Romania and from this matrix it is clear that the earthquake has the largest impact both in terms of economic and social impact.

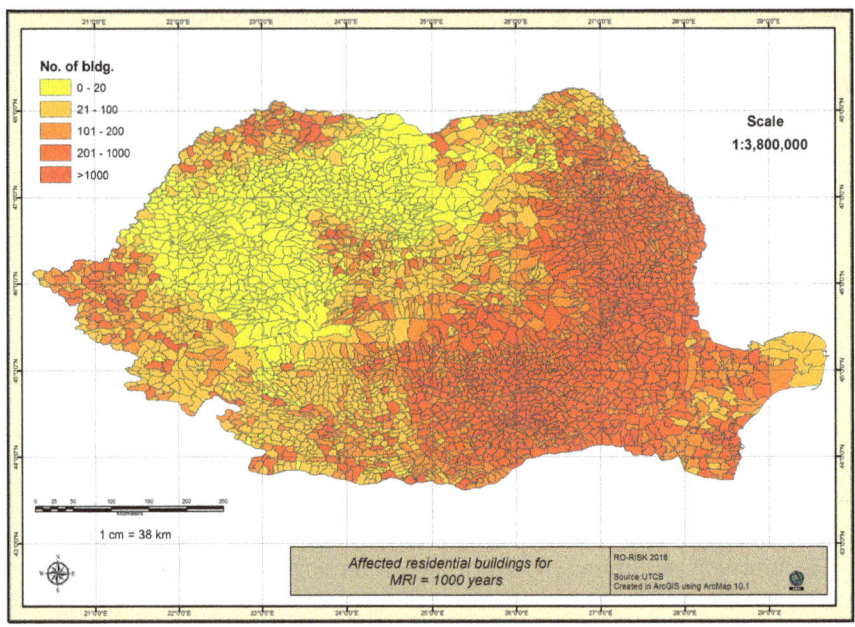

Fig. 7 Number of affected buildings for the earthquake scenario with *MRI* = 1000 years

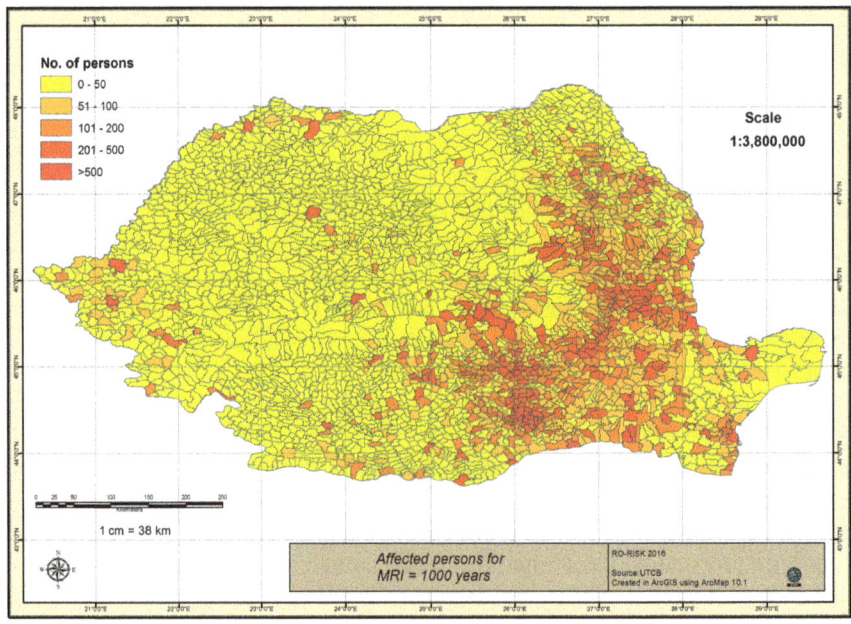

Fig. 8 Number of affected people for the earthquake scenario with *MRI* = 1000 years

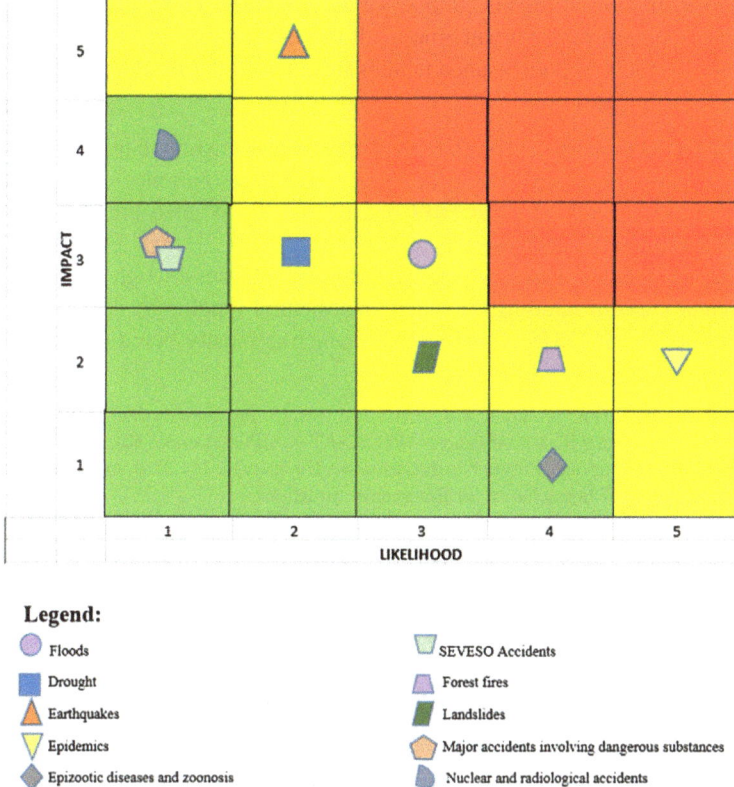

Fig. 9 Risk matrix for Romania (https://www.igsu.ro/index.php?pagina=analiza_riscuri)

6 Conclusions

The risk assessment performed within the RO-RISK project can be viewed as a first step towards a resilience-based analysis that will provide additional key information regarding the risk management policies and strategies.

In this paper, some of the main results of the seismic risk assessment performed within RO-RISK project are discussed. The results show widespread damage and large number of affected people, especially for the two *MRI* = 1000 years' earthquake scenarios (both probabilistic and deterministic).

According to the assessment developed, 75% of the population and 45% of the vital networks are exposed to the risk of an earthquake. One of the main concerns is that Romania's capital, Bucharest, is highly exposed and where the last major earthquake has caused many casualties and damage. Moreover, the entire eastern part, some areas in the center, and the southern part of Romania are close to the epicenter.

Moreover, the impact of the earthquake hazard as shown by its position on the risk matrix is the most important among all the ten analysed risks.

Among the most important uncertainties linked with the results obtained in this study, one could mention:

- The use of building inventory data at territorial-administrative level which in the case of large cities can be considered as a coarse resolution;
- The buildings built after the 2011 census and up to 2016 are not taken into consideration in the analyses;
- The use of statistical data which might not be relevant enough for assigning a particular class of buildings to a certain structural typology;
- The use of the HAZUS casualty model which is mostly based on US data.

Acknowledgements This research was performed within the framework of the RO-RISK research project "National Risk Assessment—RO RISK"—(SIPOCA code: 30), co-financed under EFS through the Operational Program Administrative Capacity 2014–2020 and under the coordination of the General Inspectorate for Emergency Situations.

References

Cauzzi C, Faccioli E, Vanini M, Bianchini A (2015) Updated predictive equations for broadband (0.01–10 s) horizontal response spectra and peak ground motions, based on a global dataset of digital acceleration records. Bull Earthq Eng 13(6):1587–1612

ENTSOG AISBL (2017) European network of transmission system, operators for gas. www.entsog.eu

Fajfar P (2000) A nonlinear analysis method for performance-based seismic design. Earthquake Spectra 16:573–592

Federal Emergency Management Agency (2012) Multi-hazard loss estimation methodology. Earthquake model—HAZUS MH 2.1. Technical manual, Washington, USA

General Inspectorate for Emergency Situations (IGSU) (2016) Country report, 5.1 Conditionality Romania 2016, 78p

HAZUS (1997) Earthquake loss estimation methodology, Technical Manual, Prepared by National Institute of Building Sciences for Federal Emergency Management Agency, NIBS Doc. 5201

INS, National Institute of Statistics. www.insse.ro

Lungu D, Arion C, Baur M, Aldea A (2000a) Vulnerability of existing building stock in Bucharest. In: 6ICSZ sixth international conference on seismic zonation, Palm Springs, California, USA, Nov 12–15, pp 837–846

Lungu D, Aldea A, Arion C (2000b) Engineering, state & insurance efforts for reduction of seismic risk in Romania. In: Proceedings of the 12th WCEE, Auckland, New Zealand, Jan/Feb

Pavel F, Vacareanu R, Douglas J, Radulian M, Cioflan CO, Bărbat A (2016) An updated probabilistic seismic hazard assessment for Romania and comparison with the approach and outcomes of the SHARE project. Pure Appl Geophys 173(6):1881–1905

Popescu IG (1941) Etude comparative sur quelques tremblements de terre de Roumanie, du type ducelui du 10 novembre 1940. Comptes Rendus des Seances de L'Academie des Sciences de Roumanie 5(3). Mai-Juin 1941, Cartea Romaneasca

P100-1/2013 (2013) Code for seismic design—Part I—Design prescriptions for buildings. Ministry of Regional Development and Public Administration, Bucharest, Romania

Vacareanu R, Radulian M, Iancovici M, Pavel F, Neagu C (2015) Fore-arc and back-arc ground motion prediction model for Vrancea intermediate depth seismic source. J Earthquake Eng 19:535–562

Vidic T, Fajfar P, Fischinger M (1994) Consistent inelastic design spectra: strength and displacement. Earthquake Eng Struct Dynam 23:507–521

Comparison of Seismic Risk Results for Bucharest, Romania

Florin Pavel, Radu Vacareanu and Ileana Calotescu

Abstract This paper highlights some of the results regarding the seismic risk and resilience analysis for the residential building stock of Bucharest, obtained within the framework of the COBPEE research project financed by the Romanian National Authority for Scientific Research and Innovation and which was completed in September 2017. The seismic risk metrics are also compared with those from two other recent studies with the same focus. In addition, the issue of economic feasibility of pre-earthquake strengthening of high-rise RC buildings in Bucharest is also analyzed in this study.

Keywords COBPEE · Vulnerability · Structural typology · Seismic damage
Seismic losses

1 Introduction

Bucharest, the capital city of Romania has a population of around two million inhabitants according to the 2011 census and accounts for about a quarter of the Gross Domestic Product (GDP) of Romania. The residential building stock consists of around 130,000 buildings with more than 840,000 dwellings. In our opinion, Bucharest has one of the highest seismic risks in Europe because of a vulnerable building stock (most of the current buildings were built before the large Vrancea 1977 earthquake) and because of its proximity to one of the most active seismic source in Europe, namely the Vrancea intermediate-depth seismic source.

F. Pavel (✉) · R. Vacareanu · I. Calotescu
Seismic Risk Assessment Research Center, Technical University of Civil Engineering
of Bucharest, Bucharest, Romania
e-mail: florin.pavel@utcb.ro

R. Vacareanu
e-mail: radu.vacareanu@utcb.ro

I. Calotescu
e-mail: ileana.calotescu@utcb.ro

© Springer International Publishing AG, part of Springer Nature 2018
R. Vacareanu and C. Ionescu (eds.), *Seismic Hazard and Risk Assessment*,
Springer Natural Hazards, https://doi.org/10.1007/978-3-319-74724-8_18

This seismic source has produced the largest intermediate-depth earthquake produced in Europe in the XXth century, namely the November 10, 1940 event (moment magnitude $M_W = 7.7$). Moreover, nine earthquakes with $M_W \geq 7.0$ were produced in the same seismic source within the past 200 years. The last such event with $M_W \geq 7.0$ occurred in August 1986 and since then three additional seismic events with $M_W \geq 6.0$ struck Romania on May 30, 1990 ($M_W = 6.9$), May 31, 1990 ($M_W = 6.4$) and October 27, 2004 ($M_W = 6.0$).

COBPEE (Community Based Performance Earthquake Engineering) was a nationally funded research project which started in 2015 and finished in September 2017 and aimed at evaluating the level of awareness and preparedness of the population for a hypothetical major seismic event, the level of expected damage of residential buildings, as well the level of implication of the population in the aftermath of a seismic event. A large-scale survey (a total number of 1000 responses) was conducted throughout 2016 in order to evaluate the opinion of the Bucharest residents with regard to the above-stated issues. More details regarding the survey can be found on the project website (http://cobpee.utcb.ro/) and in the papers of Calotescu et al. (2016) and Calotescu and Pavel (2017).

In addition, seismic risk analyses were performed within the framework of COBPEE project in order to have a correct image regarding the seismic damage and losses which are to be expected in the case of Vrancea intermediate-depth earthquake scenarios. Pavel and Vacareanu (2016) and Pavel et al. (2017b) have evaluated the seismic damage and losses using deterministic Vrancea intermediate-depth earthquake scenarios. In the study of Pavel et al. (2017a), the seismic damage is assessed for a Monte-Carlo simulated earthquake catalogue for the Vrancea intermediate-depth seismic source. The seismic risk was evaluated using the well-known HAZUS (2012) methodology in both the studies of Pavel and Vacareanu (2016) and Pavel et al. (2017b). In the research of Pavel et al. (2017b), the seismic risk was assessed by applying the mascroseismic method developed by Lagomarsino and Giovinazzi (2006). The paper of Pavel and Vacareanu (2016) also provides valuable information regarding the seismic resilience analysis for the residential building stock of Bucharest. In this paper, some of the most important results regarding the seismic risk and resilience analyses performed within COBPEE project are presented and discussed in the light of other studies with the same topic.

2 Evaluation of Seismic Hazard for Bucharest

The recent seismic risk studies of Pavel and Vacareanu (2016) and Pavel et al. (2017a, b) have assessed the seismic damage and losses for various Vrancea intermediate-depth earthquake scenarios. Only the Vrancea intermediate-depth seismic source was considered in the analysis, since this source is the main contributor to the seismic hazard of Bucharest, as well as that of southern and eastern

Fig. 1 Seismic hazard disaggregation for peak ground acceleration for Bucharest

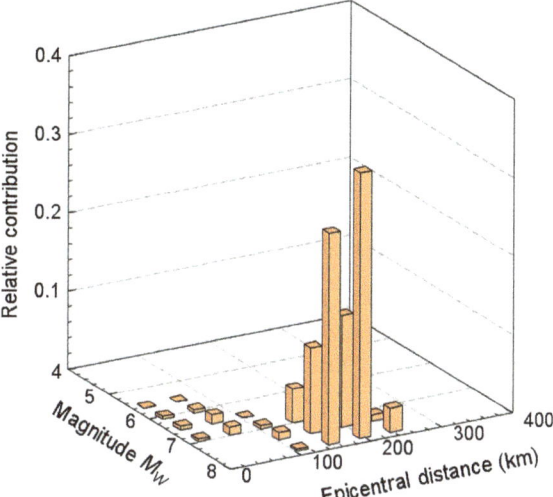

Romania. The seismic hazard disaggregation for peak ground acceleration (*PGA*) is shown in Fig. 1. The disaggregation chart is based on the seismic hazard study of Pavel et al. (2016).

Figure 2 shows the mean normalized acceleration response spectra from the ground motions recorded in Bucharest area during the five most important earthquakes produced by the Vrancea intermediate-depth seismic source in the past 40 years, namely the seismic events of March 4, 1977 (moment magnitude $M_W = 7.4$ and focal depth $h = 94$ km), August 30, 1986 ($M_W = 7.1$, $h = 131$ km), May 30, 1990 ($M_W = 6.9$, $h = 91$ km), May 31, 1990 ($M_W = 6.4$, $h = 87$ km) and October 27, 2004 ($M_W = 6.0$, $h = 105$ km). It is noticeable the fact that the significant long-period spectral amplifications occur only in the case of the largest magnitude Vrancea seismic events.

Fig. 2 Mean normalized acceleration response spectra for ground motions recorded in Bucharest area during past intermediate-depth Vrancea earthquakes

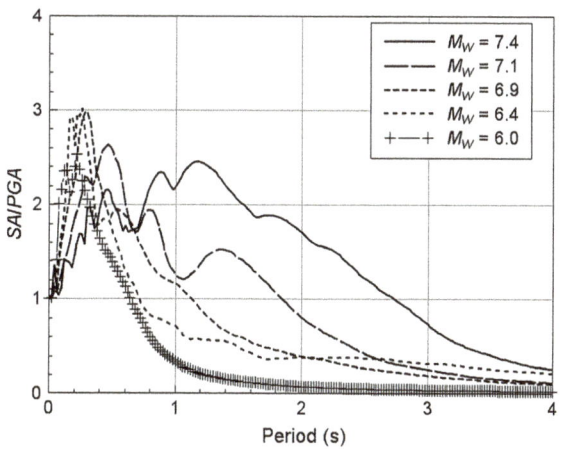

3   Characteristics of the Residential Building Stock in Bucharest

As previously mentioned, the residential building stock of Bucharest comprises of more than 130,000 buildings, made of various materials and having various structural systems and which house around two million inhabitants divided into six City Districts. The main characteristics of the existing residential building stock can be described as follows (Pavel and Vacareanu 2016; Pavel et al. 2017a, d):

- More than 60% of the number of existing buildings have been built prior to the Vrancea 1977 earthquake using a response spectrum which did not take into account long-period amplifications observed during large magnitude Vrancea seismic events;
- More than 60% of the buildings built before 1945 can be found in Districts 1 and 2;
- Around 5% of the number of residential buildings have more than nine stories in height, the majority being in Districts 2, 3 and 6;
- The relative proportion of buildings with more than nine stories in height is 1.4% in District 1, 5.2% in District 2, 8.3% in District 3, 6.3% in District 4, 2.7% in District 5 and 10.2% in District 6;
- Among the buildings with more than 11 stories in height, almost 75% can be found in Districts 2, 3 and 6;
- The largest numbers of residential dwelling units can be found in Districts 2, 3 and 6;
- The mean building area is in the range 192–584 m^2, while the mean number of building occupants is between 6.5 and 31.7. The smallest values correspond to District 1, while the largest ones correspond to District 6;
- More than 50% of the population of Bucharest inhabits buildings with more than nine stories in height. In absolute terms around 6000 buildings house almost one million inhabitants. The only Districts in which the percentage of the population is less than 50% are Districts 1 and 5;
- Around 20% of the population of Bucharest lives in five-storey buildings (around 4000 buildings), most of them built from precast concrete.

Based on the above-mentioned observations, one can conclude that there is a significant difference with regard to the characteristics of the residential buildings stock among the six City Districts. Some Districts, namely 1 and 5 have mainly low-rise buildings, while the majority of high-rise buildings is concentrated in Districts 2, 3 and 6. In addition, one has to notice the fact that around three quarters of the population inhabits roughly 7.5% of the entire residential building stock of Bucharest (five-storey buildings or buildings with more than nine stories).

4 Assessment of Ground Motion Amplitudes

The ground motion amplitudes in Bucharest area were determined in the studies of Pavel and Vacareanu (2016) and Pavel et al. (2017a, b) by using the ground motion model of Vacareanu et al. (2015a) which was specifically developed for the Vrancea intermediate-depth seismic source. The ground motion amplitudes were originally evaluated for a Monte-Carlo simulated earthquake catalog for the Vrancea intermediate-depth seismic source proposed in the study of Pavel et al. (2017a). The other two studies of Pavel and Vacareanu (2016) and Pavel et al. (2017b) use deterministic Vrancea intermediate-depth earthquake scenarios defined in terms of magnitude, focal depth and source-to-site distance. In order to take into account the ground motion variability at city level, a spatial correlation model developed by Pavel and Vacareanu (2017a) was applied in order to generate random ground acceleration fields. In the studies of Pavel and Vacareanu (2016) and Pavel et al. (2017a), the resulting peak ground accelerations were then used as input for anchoring the design response spectrum from the Romanian seismic design code. In the second study of Pavel et al. (2017b), the resulting peak ground accelerations were transformed into macroseismic intensities using the conversion relation proposed by Vacareanu et al. (2015b).

5 Seismic Risk Analysis

The characteristics of the residential building stock in Bucharest were taken from the most recent census performed in 2011. From the point of view of structural systems, the residential buildings were divided into seven categories: two for reinforced concrete structures (RC1—reinforced concrete frame structures and RC2—reinforced concrete shear walls structures, either cast in place or precast), three for masonry structures (M3.1—unreinforced masonry structures with flexible (wooden) floors, M3.4—unreinforced masonry structures with rigid (reinforced concrete) floors and M4—confined masonry structures) and one for adobe structures (M2) and wood structures (W1), respectively. Moreover, the residential buildings were divided into four categories as a function of the level of seismic design code: pre-code (for buildings built up to 1963, when the first official seismic design code was enforced in Romania), low-code (for buildings built in the period 1963–1977), moderate code (for buildings built in the period 1978–1992) and high-code for all the buildings built subsequently. The limits employed for defining the height regime are given below:

For RC structures:

- low rise structures (L)—1–3 stories;
- mid-rise structures (M)—4–7 stories;
- high-rise structures (H)— ≥ 8 stories.

For masonry, adobe and wood structures:

- low rise structures (L)—1–2 stories;
- mid-rise structures (M)—3–5 stories.

The well-known N2 method proposed by Fajfar (2000) was employed for the evaluation of the seismic performance of each structural typology in the study of Pavel and Vacareanu (2016) and Pavel et al. (2017a). Subsequently, the HAZUS (2012) model was used for the evaluation of seismic damage and losses (both economic and human). In the study of Pavel et al. (2017b), the macroseismic method proposed by Lagomarsino and Giovinazzi (2006) was applied. The vulnerability characteristics of some structural typologies were assessed based on damage data collected in the aftermath of the March 4, 1977 Vrancea intermediate-depth earthquake. The seismic demand was computed based on the procedure described in the previous chapter.

6 Results

The differences in seismic risk metrics shown in the previous section can be attributed first of all to the applied methodology. The use of macroseismic intensity as intensity measure is subjected to increased uncertainty levels as compared to the use of various ground motion parameters. However, the use of the design response spectrum anchored at a specific level of peak ground acceleration (computed so as to take into account the ground motion variability at city level) has also a major drawback in the sense that it induces significant displacements even in the case of lower magnitude seismic events. As shown in Fig. 2, only the largest Vrancea intermediate-depth seismic events generate significant long-period spectral ordinates and as such the seismic losses for lower magnitude earthquake are overestimated.

Figure 3 shows the average relative contributions of each building material (RC, masonry and other materials) to the total overall losses computed for an earthquake scenario with $M_W = 7.5$ and assessed using both the HAZUS (2012) approach and the macroseismic method of Lagomarsino and Giovinazzi (2006). The differences in the results obtained by the two approaches are limited, the most obvious one being that the contribution from RC structures is larger when using the HAZUS (2012) approach, while in the case of masonry structures the trend is exactly the opposite.

Additional differences obtained between the two methods are visible in Figs. 4 and 5. The contribution of low-code (LC) structures is larger, while the contributions from medium-code (MC) and high-code (HC) structures is smaller in the macroseismic method. In the case of the height regime, one can easily observe that the relative contribution of low-rise structures (LR) is larger in the case of the macroseismic method, while in the case of high-rise (HR) the relative contribution from HAZUS (2012) is larger.

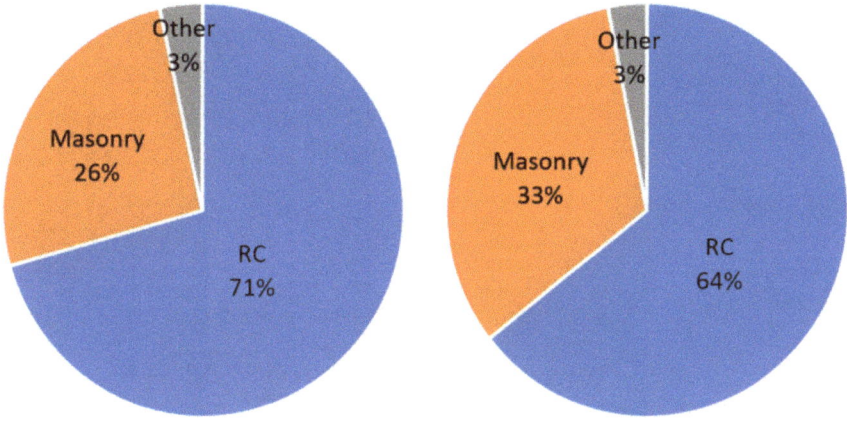

Fig. 3 Average relative contributions to total overall losses for a M_W 7.5 earthquake scenario as a function of building material using HAZUS approach (left) and macroseismic method (right)

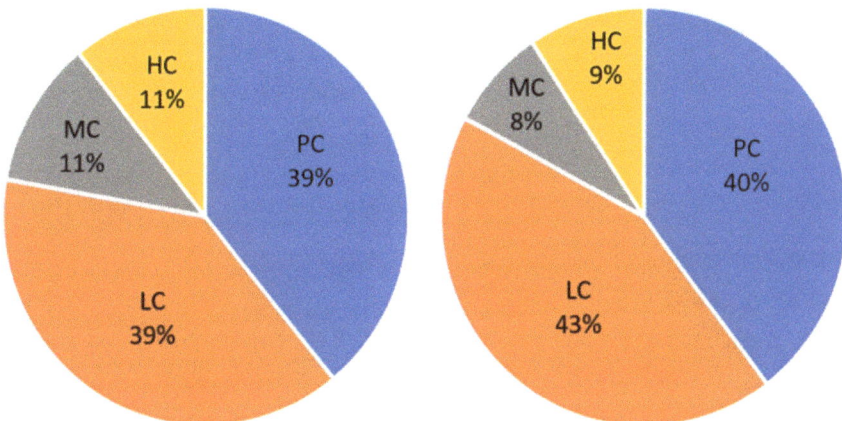

Fig. 4 Average relative contributions to total overall losses for a M_W 7.5 earthquake scenario as a function of level of seismic design code using HAZUS approach (left) and macroseismic method (right)

The economic losses and the number of affected people were evaluated for the same six deterministic earthquake scenarios in both the study of Pavel and Vacareanu (2016) and Pavel et al. (2017b). The mean values and the corresponding coefficients of variation are given in Tables 1 and 2. In general, one can observe that the mean values for the losses are larger when using the HAZUS (2012) approach. However, in the case of the number of affected people, the results from the macroseismic method of Lagomarsino and Giovinazzi (2006) are superior. In both cases, the variability from the macroseismic method (Lagomarsino and Giovinazzi 2006) is superior to the one from HAZUS (2012). The differences

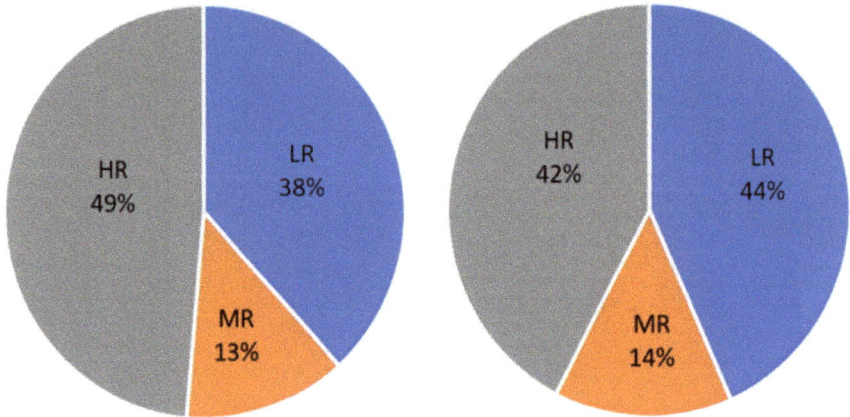

Fig. 5 Average relative contributions to total overall losses for a M_W 7.5 earthquake scenario as a function of height regime using HAZUS approach (left) and macroseismic method (right)

Table 1 Comparison of the statistics (mean value and coefficient of variation) for the economic losses obtained in the study of Pavel and Vacareanu (2016) and Pavel et al. (2017b)

Scenario no.	M_W, h, d	Economic losses from Pavel and Vacareanu (2016)		Economic losses from Pavel et al. (2017b)	
		Mean (bill. Euro)	COV	Mean	COV
1	$M_W = 7.0$, h = 90 km, d = 120 km	7.20	0.34	3.60	0.74
2	$M_W = 7.0$, h = 150 km, d = 180 km	5.04	0.41	1.78	0.82
3	$M_W = 7.5$, h = 90 km, d = 120 km	10.31	0.26	7.34	0.54
4	$M_W = 7.5$, h = 150 km, d = 180 km	7.68	0.32	3.93	0.72
5	$M_W = 8.0$, h = 90 km, d = 120 km	12.72	0.20	11.49	0.36
6	$M_W = 8.0$, h = 150 km, d = 180 km	10.09	0.26	7.18	0.55

between the two methods become smaller as the magnitude of the scenario earthquake increases. Nevertheless, the very large economic losses and number of affected people observed in the case of the largest magnitude earthquake scenarios obtained in both approaches are noteworthy.

The seismic risk results obtained in the three mentioned studies of Pavel and Vacareanu (2016) and Pavel et al. (2017a, b) should be also compared with some other recent researches with the same topic. For instance, the studies of Armas et al. (2016) or Toma-Danila and Armas (2017) provide a series of seismic risk metrics which might be used for comparison purposes. Based on the study of Armas et al. (2016) the number of deaths in the case of a M_W 7.8 earthquake scenario is 81,000,

Table 2 Comparison of the statistics (mean value and coefficient of variation) for the number of affected people obtained in the study of Pavel and Vacareanu (2016) and Pavel et al. (2017b)

Scenario no.	M_W, h, d	Affected no. of people from Pavel and Vacareanu (2016)		Affected no. of people from Pavel et al. (2017b)	
		Mean	COV	Mean	COV
1	$M_W = 7.0$, $h = 90$ km, $d = 120$ km	7086	0.94	9779	1.13
2	$M_W = 7.0$, $h = 150$ km, $d = 180$ km	3398	0.90	1888	0.97
3	$M_W = 7.5$, $h = 90$ km, $d = 120$ km	17,780	1.06	14,148	1.29
4	$M_W = 7.5$, $h = 150$ km, $d = 180$ km	8074	0.96	11,942	1.09
5	$M_W = 8.0$, $h = 90$ km, $d = 120$ km	38,282	1.02	46,161	1.29
6	$M_W = 8.0$, $h = 150$ km, $d = 180$ km	16,822	1.08	13,480	1.32

while an earthquake similar with the one from 1977 would produce 25,000 deaths. Toma-Danila and Armas (2017) provide results in terms of completely damaged buildings. Their analysis shows that a M_W 6.9 earthquake (similar with the May 30, 1990 event) would produce over 800 completely damaged buildings, while an earthquake scenario similar with the March 4, 1977 event would produce around 1800 completely damaged buildings. One can easily notice that the number of deaths in the case of a M_W 7.8 exceeds the number of affected people shown in Table 2. It is our opinion that the above-mentioned results are grossly over-estimating both the number of deaths and the number of collapsed buildings (the experience of the Vrancea earthquake of 1977, 1986 and 1990 has shown a totally different situation in Bucharest). Moreover, it is our opinion, that these results might induce an unwanted level of panic among the population. As such, we advocate against publishing seismic risk-related results which are neither verified by past experience or by other researches with the similar topic.

7 Feasibility of Pre-earthquake Strengthening of High-Rise RC Structures

The issue of feasibility of pre-earthquake strengthening of high-rise RC buildings was discussed in the paper of Pavel and Vacareanu (2017c). The life-cycle analysis was considered in the following manner: all the buildings having either PC or LC typology would be retrofitted so as to have a similar behavior with the HC typology. The benefit/cost ratio which basically show whether the strengthening is economically efficient or not is computed using the relation proposed by Kappos and Dimitrakopoulos (2008):

Fig. 6 B/C ratios for
strengthening of high-rise RC
buildings for various planning
horizons and for a discount
rate of 2% (Pavel and
Vacareanu 2017c)

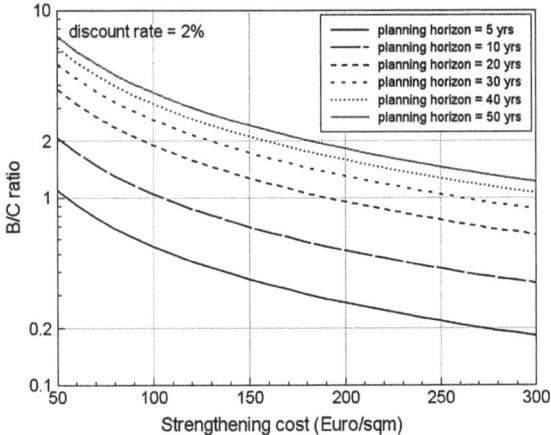

$$B/C = \frac{B_t + V_{DA}}{R_C - V_S} \qquad (1)$$

where B_t represent the benefits over the planning horizon, V_{DA} is the cost of deaths avoided, R_C is the retrofit cost and V_S represents the salvaged value of the building. The salvaged value is the increase in the building value due to seismic strengthening, viewed within a benefit/cost analysis as a future benefit (Kappos and Dimitrakopoulos 2008). The upper bound cost of human life for the case study of Thessaloniki (Greece) was taken by Kappos and Dimitrakopoulos (2008) as 500,000 €. Figures 6 and 7 show the results in terms of B/C ratios as a function of the strengthening cost (in €/m²) and one can see that the longer the planning horizon, the larger the cost of the strengthening of high-rise RC buildings is. Another observation is that as the discount rate increases, the cost of economically efficient strengthening decreases.

Fig. 7 B/C ratios for
strengthening of high-rise RC
for various planning horizons
and for a discount rate of 4%
(Pavel and Vacareanu 2017c)

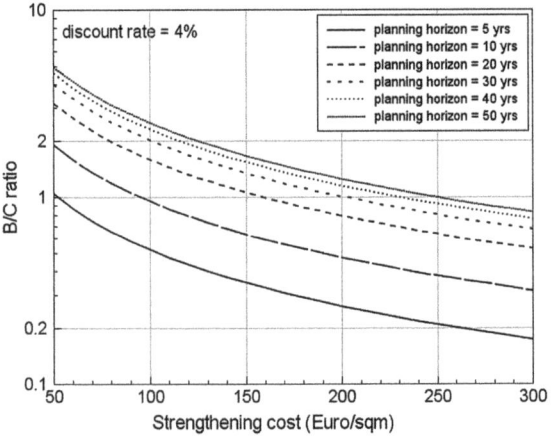

Fig. 8 Comparison of B/C
ratios for strengthening of
high-rise RC for a planning
horizon of 50 years and for a
discount rate of 4% (Pavel
and Vacareanu 2017c)

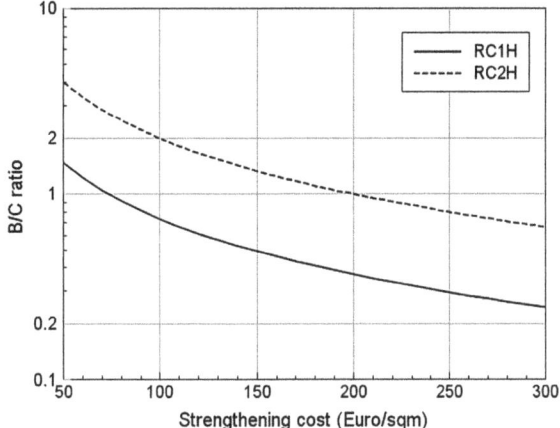

Figure 8 shows the benefit/cost ratio (B/C) in terms of strengthening costs for
high-rise frame structures (RC1H) and high-rise shear walls structures (RC2H). It is
noticeable the fact that in the case of RC walls, the strengthening is economically
feasible for cost up to 200 €/m², while in the case of RC structural walls it is
feasible only up to 80 €/m².

8 Conclusions

In this paper, some of the most recent results regarding the seismic risk and resilience analysis for Bucharest, obtained within the framework of the COBPEE
research project are highlighted. The studies of Pavel and Vacareanu (2016) and
Pavel et al. (2017a, b) have evaluated the seismic risk of Bucharest using two
approaches: a scenario-based approach and an approach based on a simulated
earthquake catalogue for the Vrancea intermediate-depth seismic source. In the first
two cases the seismic damage and losses was evaluated based on the well-known
HAZUS (2012) methodology. In the former study, the macroseismic method proposed by Lagomarsino and Giovinazzi (2006) was employed for seismic risk
evaluation. The results of the seismic risk analyses show that the mean economic
losses and are larger when using the HAZUS (2012) approach, while the number of
affected people from the macroseismic method (Lagomarsino and Giovinazzi 2006)
is superior. The variability of the results from the latter approach is larger than in
the former case. An ongoing research related to the seismic risk assessment for
residential buildings and whose results are expected to be shown at the 16th
European Conference on Earthquake Engineering in 2018 is related to the use of a
ground motion model specifically derived for Bucharest and which is able to
capture long-period spectral amplifications instead of the previous procedure of
anchoring the design response spectrum to a peak ground acceleration value

resulted from the ground motion model of Vacareanu et al. (2015a). In this way, the major drawback related to the increased spectral displacement demands induced by the use of a design response spectrum in the evaluation of the seismic performance is removed. The results of the analysis of the economic feasibility of pre-earthquake strengthening of high-rise RC structures show that the cost of the economically feasible strengthening ranges between 170 and 300 €/m^2 for a planning horizon of 50 years. In addition, it was observed that as the planning horizon increases, so does the cost of the economically-feasible strengthening. As the discount rate increases, the cost of the economically-feasible strengthening decreases. An additional result shows that in the case of high-rise RC structural wall typologies, the strengthening is economically feasible for costs more than two times larger than in the case of high-rise RC frames typologies.

Acknowledgements This work was supported by a grant of the Romanian National Authority for Scientific Research and Innovation, CNCS—UEFISCDI, project number PN-II-RU-TE-2014-4-0697. The authors gratefully acknowledge the financial support.

References

Armas I, Toma-Dănila D, Ionescu R, Gavris A (2016) Quantitative population loss assessment: seismic scenarios for Bucharest using 2002 census data. GI_Forum 1:30–40

Calotescu I, Pavel F, Sandulescu AM, Sibişteanu H, Vacareanu R (2016) Preliminary investigation on community resilience of Bucharest, Romania. In: Proceedings of the international conference on urban risks ICUR 2016, Lisbon, Portugal, paper no. 142

Calotescu I, Pavel F (2017) Seismic preparedness of the population of Bucharest, Romania: questionnaire results. In: Pavel F et al (eds) Proceedings of the 6th national conference on earthquake engineering & 2nd national conference on earthquake engineering and seismology, Bucharest, Romania, pp 225–232

Fajfar P (2000) A nonlinear analysis method for performance-based seismic design. Eq Spectra 16:573–592

Federal Emergency Management Agency (2012) Multi-hazard loss estimation methodology. Earthquake model—HAZUS MH 2.1. Technical manual, Washington, USA

Kappos AJ, Dimitrakopoulos EG (2008) Feasibility of pre-earthquake strengthening of buildings based on cost-benefit and life-cycle cost analysis, with the aid of fragility curves. Nat Hazards 45:33–54

Lagomarsino S, Giovinazzi S (2006) Macroseismic and mechanical models for the vulnerability and damage assessment of current buildings. Bull Earthq Eng 4(4):415–443

P100-1/2013 (2013) Code for seismic design—part I—design prescriptions for buildings. Ministry of Regional Development and Public Administration, Bucharest, Romania

Pavel F, Vacareanu R (2016) Scenario-based earthquake risk assessment for Bucharest, Romania. Int J Disast Risk Re 20:138–144

Pavel F, Vacareanu R (2017a) Spatial correlation of ground motions from Vrancea (Romania) intermediate-depth earthquakes. Bull Seismol Soc Am 107(1):489–494

Pavel F, Vacareanu R (2017c) Feasibility of pre-earthquake strengthening of RC buildings in Bucharest, Romania. In: Pavel F et al (eds) Proceedings of the 6th national conference on earthquake engineering & 2nd national conference on earthquake engineering and seismology, Bucharest, Romania, pp 375–382

Pavel F, Vacareanu R, Douglas J, Radulian M, Cioflan CO, Barbat A (2016) An updated probabilistic seismic hazard assessment for Romania and comparison with the approach and outcomes of the SHARE project. Pure Appl Geophys 173(6):1881–1905

Pavel F, Vacareanu R, Calotescu I, Sandulescu AM, Arion C, Neagu C (2017a) Impact of spatial correlation of ground motions on seismic damage for residential buildings in Bucharest, Romania. Nat Hazards 87:1167–1187

Pavel F, Calotescu I, Vacareanu R, Sandulescu AM (2017b) Assessment of seismic risk scenarios for Bucharest, Romania. Nat Hazards. https://doi.org/10.1007/s11069-017-2991-3

Pavel F, Vacareanu R, Arion C, Neagu C, Calotescu I (2017d) Recent developments regarding seismic risk and resilience assessment for Bucharest, Romania. In: Pavel F et al (eds.) Proceedings of the 6th national conference on earthquake engineering & 2nd national conference on earthquake engineering and seismology, Bucharest, Romania, pp 383–390

Toma-Danila D, Armas I (2017) Insights into the possible seismic damage of residentialbuildings in Bucharest, Romania, at neighborhood resolution. Bull Earthq Eng 15:1161–1184

Vacareanu R, Radulian M, Iancovici M, Pavel F, Neagu C (2015a) Fore-arc and back-arc ground motion prediction model for Vrancea intermediate depth seismic source. J Earthq Eng 19:535–562

Vacareanu R, Iancovici M, Neagu C, Pavel F (2015b) Macroseismic intensity prediction equations for Vrancea intermediate-depth seismic source. Nat Hazards 79:2005–2031, http://cobpee.utcb.ro/

New Archival Evidence on the 1977 Vrancea, Romania Earthquake and Its Impact on Disaster Management and Seismic Risk

Emil-Sever Georgescu and Antonios Pomonis

Abstract The March 4, 1977, Vrancea earthquake was Romania's greatest natural disaster in the 20th century. Today, forty years later, its impact, but also post-disaster actions and policies, continue to influence seismic risk. Based on lessons learnt, an improved seismic design code was introduced in 1978 for new construction, but in July 1977 for damaged buildings all strengthening projects were abruptly stopped by a high level political order that allowed only local—often cosmetic- repairs. After 1989, many state archives have been opened, helping us deepen our understanding of how disaster management was conducted in that time. This paper reveals new findings and data on building damage and human casualties. This new evidence suggests that nearly a quarter of the lives lost in Bucharest were not in collapsed multi-story buildings, but in other, yet to be investigated, locations. Also new evidence about the unidentified victims lifts popular doubts on this issue. Lastly, the authors analyze the July 4, 1977, government decision to end the seismic reinforcement of damaged buildings and the debate at the time between specialists and government officials.

Keywords Building damage and collapse · Casualties · Strengthening
Repair · Bucharest · Civic center

E.-S. Georgescu (✉)
National Institute for Research and Development—URBAN-INCERC,
Bucharest, Romania
e-mail: emilsevergeorgescu@gmail.com

A. Pomonis
Cambridge Architectural Research Ltd, Cambridge, UK
e-mail: antpomon@ath.forthnet.gr

© Springer International Publishing AG, part of Springer Nature 2018   281
R. Vacareanu and C. Ionescu (eds.), *Seismic Hazard and Risk Assessment*,
Springer Natural Hazards, https://doi.org/10.1007/978-3-319-74724-8_19

1   Introduction

During the first few months after the March 4, 1977, Vrancea earthquake (hereafter cited as "the 1977 earthquake") the local and international engineering community was well informed about its consequences and lessons learnt as nearly full transparency and openness prevailed over secrecy. Coverage of its impact was in the state-run and foreign press, in books in Romanian such as Buhoiu (1977), and in international reports and papers as Fattal et al. (1977), Jones (1977, 1986), Berg et al. (1980) and the World Bank (1978, 1983). Appreciation of prompt countermeasures, quick restoration of critical facilities and infrastructure, good behavior of the majority of structures designed according to the 1963 and 1970 seismic codes, were all emphasized. After the emergency period however, when international attention was diminished, government decisions about the recovery efforts turned out to be more problematic with long-term adverse effects on earthquake risk. The local engineering community knew only a small part of the debate behind key decisions as ambitious, often unrealistic, targets were set by the regime for the completion of building repairs, first by May 1 (International Labor Day) and later by August 23 (Romania's National Day). Eventually though, the technical, financial, moral, and political implications of the regime's post-earthquake recovery management crossed Romania's borders.

In 1979, Radio Free Europe broadcast how the regime imposed local repairs instead of overall strengthening of damaged buildings, outlined during the July 4, 1977, working meeting with the Political Executive Committee, leading architects and engineers, using text from engineer Gheorghe Ursu's letter (Povestea Vorbei nr. 195, 1979). The tragic case of his imprisonment and death in 1985 was better known only after the 1990s (Fundaţia "Gheorghe Ursu" 2017).

Archival sources collected by Dr. Karin Steinbrueck during her doctoral thesis research (Steinbrueck 2017), that were kindly shared with the authors, allowed us to deepen our understanding about how key disaster management decisions of the time were made and how they continue to influence seismic risk in Romania today. The press coverage and communiqués after meetings of the Political Executive Committee of the Romanian Communist Party (RCP) in 1977 (hereafter cited as CPEx), including Scinteia newspaper, the RCP Gazette, were evaluated with today's wisdom, to help us understand how the Ceauşescu regime's decisions evolved. CPEx meetings on March 5 (ANR CCRCP, Chancellery 18/1977), March 6 (ANR CCRCP, Chancellery 21/1977), March 7–11, as well as March 14 and 17; and large working conferences on March 22, 28 and 29 were analyzed. In addition, some declassified cables of the US Embassy in Bucharest of 1977 were also consulted (US Diplomatic or State Department cables 1977).

2 Earthquake Engineering and Engineering Seismology Lessons from the 1977 Earthquake

2.1 Earthquake Engineering Lessons

Despite the regime's tight control of information, the technical aspects on the causes of damage to buildings were discussed quite openly among specialists. In 1978, based on lessons learnt, a highly improved seismic design code was introduced. An important lesson, with significant international implications, was the realization that the seismic ground motion recording in the grounds of the Romanian Buildings Research Institute (INCERC) in NE Bucharest registered larger values than the 1970 and 1963 seismic design codes, and was one of the first recordings worldwide with significant long-period spectral content (Fattal et al. 1977; Berg et al. 1980). This helped engineers understand why damage and destruction in Bucharest were concentrated on multi-story buildings (up to 12 stories). As a result a highly improved seismic design code was introduced in 1978 (Balan et al. 1982; Georgescu 2003).

Beyond the extensive human casualties (90% of the earthquake's 1570 fatalities were in Bucharest), a long-term adverse effect of poor decisions and management of the time is that today in the city many multi-story buildings built before World War II, without seismic design consideration and damaged by the 1940 and 1977 earthquakes, remain occupied, even though deemed to be at significant risk and in urgent need of strengthening or replacement. Technical data on disaster patterns, crisis and recovery management were only gradually fully appreciated through research and publications including crucial data from internal reports (e.g., ICCPDC 1978), data on the losses per economic sector, and the human casualties' distribution. Significant Romanian studies and data addressing such important issues were presented in: Balan et al. (1982), Georgescu and Kuribayashi (1992), Georgescu (2003), Sandi et al. (2008) and more recently, in Georgescu and Pomonis (2007, 2008, 2010, 2011, 2012).

2.2 Engineering Seismology Lessons

Local situation of obsolete or damaged seismological equipment led the government to ask and/or accept international technical assistance and cooperation. Teams of seismologists from USA, Germany, Yugoslavia, China, USSR, Greece, Turkey, Iran and Venezuela arrived, while mobile seismic stations from University of Karlsruhe, Germany were deployed. A Working meeting of Ceaușescu with seismologists—researchers of the Center of Earth Physics and Seismology (CFPS—*Centrul de Fizica Pamântului și Seismologie*) was held on March 8, 1977, while a CPEx Meeting of March 10, reiterated seismological concerns of RCP leaders (ANR, CCRCP, Chancellery 28/1977).

New archival disclosures from early March 1977 show us poor or conflicting knowledge on key seismological issues within the RCP leadership and some earth-sciences researchers, primarily related to the difficulty of understanding the nature of long-distance ground motion patterns originating from the Vrancea seismic source zone. Some seismologists believed that the epicenter was, possibly, near Bucharest, idea embraced in the first days by the national leaders, in addition to anxiety about the shaking being related to a possible nuclear explosion alongside expectations on earthquake prediction capability, using animals.

We consider that the technical significance of the INCERC accelerogram was not fully appreciated and that efforts to try to explain to the regime's leadership its significance and relation with damage to high-rise structures could have changed attitudes and possibly may have lead to relaxed orders on strengthening. These issues will be discussed more at length in a future paper.

3 Earthquake Disaster Patterns—New Archival Findings

The first summary of consequences and related loss estimate was released by the authorities on March 9, estimating 6 billion Romanian Lei loss from circa 20,000 destroyed or damaged dwellings, 195 destroyed or damaged enterprises, 7000 families homeless, 1387 deceased, 10,396 injured. Search and rescue operations lasted more than ten days at 23 sites of major collapsed buildings in Bucharest. The Ministry of Interior (MI), issued situation reports once or twice per day, reporting the count of damaged and destroyed buildings, and the human casualties at each of the Bucharest major building collapse sites and in the country as a whole.

On March 14, 1977, the regime mandated a penal inquiry on the collapsed and badly damaged buildings. What is presently known, from a technical point of view, was included as engineering assessments of collapse causes for specific buildings in internal and published reports (Balan et al. 1982). At present we have no additional information about court procedures and decisions. Table 1 presents the destroyed and damaged buildings' situation as of March 18, 1977.

The lack of buildings at risk in Bucharest is of interest. This may be due to the tight political control and/or due to building evacuation orders. We do have evidence that in the immediate emergency phase (e.g. March 5, 1977, ANR CCRCP,

Table 1 Situation of damaged buildings as of March 18, 1977 (CNSAS Archive, D 011737/vol 105, p. 139)

Extent of damage	Romania	Out of which	
		Bucharest	Counties
Buildings & dwellings destroyed	7270	157	7113
Buildings & dwellings at risk	12,931	–	12,931
Buildings & dwellings damaged	45,746	2101	43,645

Chancellery 18/1977, p. 4) it was instructed that a commission of engineers for damage assessment be established, orders of damaged buildings evacuation enforced, and pre-1940 apartment blocks with cracks to be carefully assessed. Fear of aftershocks prompted authorities to recommend evacuation of damaged buildings.

Large search and rescue operations took place at the 23 major collapsed buildings in Bucharest. Bodies were stored for identification by authorities, relatives, or friends and the MI kept daily tallies of the situation for each site. The Attorney General and Interior Minister finalized data on March 31, presented in Table 2.

We learn that, 201 out of 1110 bodies recovered from the rubble of the 23 collapsed-occupied buildings remained unidentified at the end of March. Many of the unidentified victims were found at the collapse sites where fire, due to gas pipe ruptures, had occurred, with 195 of the 201 being in 10 building collapse sites. These victims were buried afterwards, according to proper procedures in Domneşti Cemetery with indication of the collapse location at the grave site. The legal procedure to identify the dead used individual files, photographs, and personal objects. The recovery and identification of the earthquake casualties in Bucharest was made according to international procedures (the handling was very similar to what happened in New Zealand after the February 22, 2011 Christchurch earthquake, when identification procedures continued for nearly three months related to just 2 major building collapse sites). In addition, 310 deaths in Bucharest were not linked to the 23 major building collapse sites.

Ceauşescu mentioned the total of 1570 deaths in his public speeches. A final tally of 1578 deaths was reported in BSSA 1978 with 1424 deaths in Bucharest (for territorial distribution of casualties, their causes and other details see Georgescu and Pomonis 2010, 2011, 2012). According to the Ministry of Interior archives, of the 1420 dead in Bucharest, 1110 (78%) died in the 23 collapsed buildings of which 97 were not registered residents. The deceased non-registered were only reported as a total, not by collapse site.

Of the remaining 310 victims in Bucharest, it was possible that they were in other collapsed non-residential buildings (e.g., office blocks Nestor and Carpaţi; Hotel Victoria; Faculty of Chemistry), damaged industrial workshops, smaller residential buildings, and in the streets killed by falling objects and debris. This issue has not been discussed until now and deserves further investigation. Related to this we report that US Embassy cable of March 31 indicated the death of three employees due to the collapse of the MTTC Computing Center building (US Diplomatic or State Department cables 1977).

Table 2 Number of deceased and situation of identification on March 31, 1977, (ANR CCRCP, Section of Administrative and Politics 12/1977, p. 43)

Fatalities situation	Place		
	Bucharest	Other localities	Total
Deceased	1420	150	1570
Of which identified	1219	150	1369
Not yet identified	201	–	201

Overall, according to the MI archive data, the lethality ratio (ratio of deceased over the number of resident-occupants at the time of the earthquake) was 50% in the 21 pre-1940 collapsed buildings. The lethality ratio in the two newer collapsed buildings (Bloc 30 built in 1962 and Bloc OD16 built in 1974) was much lower, at 16%. Inspection of the data against photos pointed to the fact that lethality was highest in 13 buildings that collapsed entirely leaving very few survival voids as well as in few buildings where fire followed the collapse. Overall lethality was 62% in these thirteen buildings. Overall lethality in the 10 buildings that partially collapsed was 20%.

The individual data for each of the 23 major collapse sites (registered residents, people inside at the time of collapse, people saved, people hospitalized, identified and unidentified victims, etc.) will be the subject of a future publication.

4 The Emergency and Recovery Period Versus Political Decisions with Long-Term Impact

4.1 The Post-disaster Response and the Succession of Actions at Superior Leadership Level

The post-disaster response of the Romanian State must be understood in its contemporaneous national and international context, as it was related to:

- The highly centralised nature of the time's communist regime, with little room for initiative and actions taken only upon specific orders from the leadership;
- the extensive damages and losses and the need of resources in construction sector, where damage was 69.4% of the total 2 billion USD losses (Word Bank 1978). A ratio of property loss of 5% of GNP or 1.63% of National Wealth was evaluated by Georgescu and Kuribayashi (1992) based on World Bank data. A reevaluation by Georgescu and Pomonis (2007, 2008) showed a new range of loss ratios with direct loss to GDP or GNP ratio in 1977 from 4.9 to 7.8%, while the range of total loss to GDP or GNP ratio in 1977 could be from 6.0 to 9.5%; using the total possible loss (US$4.54 billion), the range of total possible loss ratio to GDP or GNP in 1977 could be from 13.3 to 21.1%;
- the differentiation of Romania related to the crashing of the 1968 Prague Spring;
- the positions of RCP leadership regarding relations with the USSR and China;
- Romania's Most Favoured Nation status granted by the USA and good trade with UK and West Germany;
- the emerging labour unions and human rights activists, perceived as a threat by the regime within its wider Cold War context;
- Western media speculation about the risk of regime collapse as a result of the disaster.

The succession of regime leadership actions and opinions during the first three weeks is revealed by the Minutes of CPEx and official media, as follows:

- On March 5, 1977 CPEx Meeting—important buildings such as the building of the Central Committee of the RCP, Government (Victoria Palace), Council of State (Former Royal Palace) and Bucharest City Hall were damaged, there was concern, as leadership was aware of China's experiences just 7 months earlier—Tangshan earthquake, July 26, 1976—when a strong aftershock occurred in the second day. (Minutes of CPEx, March 6, 1977, ANR CCRCP, Chancellery 21/1977, p. 7).
- On March 5, 1977: President Carter informed US Embassy that "the United States stands ready to extend emergency assistance..."
- On March 6, 1977 CPEx Meeting (ANR CCRCP, Chancellery 21/1977)—Army and technical staff did not work properly on collapsed buildings sites. The lack of tools and skills for search and rescue was obvious, as the dispatched Army personnel was not trained or equipped for such operations (Romania's recent disaster response was after the 1970, 1975 floods, a totally different type of emergency).
- On March 6, 1977, US Embassy cables inform that Romania's Foreign Affairs Minister Macovescu requested long-term financial assistance that would help Romania maintain its independence vis à vis the Soviet Union;
- On March 7, 1977 CPEx Meeting—It is strongly advised that building damage-safety assessments should not be made in haste, while on March 8, 1977 CPEx Meeting the good organization of repair activity is stressed out.
- On March 9, 1977 CPEx Meeting fast repairs were ordered, so as all repair works would be completed by May 1-st—The International Labor Day!
- On March 9, 1977, US Embassy cables inform on the need and request of seismological equipment, since one third of CFPS apparatus was lost and many other damaged. Prof. Bruce Bolt is on his way to Romania. CFPS was in need of two seismic observation stations;
- On March 10, 1977, CPEx Meeting State of Emergency is lifted, except Bucharest, telecom services and Radio-Television; Ceaușescu criticized the "chaos" at airport where foreign aid is arriving; orders to start work on debris clearance, to strengthen even the buildings that may later be deemed as unsafe to use!; for the first time he expresses his opinion, thoughts about a new Bucharest Civic Center, somewhere on Victoriei Square or Calea Plevnei—Bd. Stirbei Voda because of damage to many buildings of RCP and State;
- Between 9 and 12 March, 1977, a USGS warning on a possible new Vrancea earthquake (broadcasted also by Radio Free Europe) triggered a diplomatic dispute with the USA. However, Ceaușescu did ask the US Embassy about possible visit of Charles Francis Richter. A large press campaign to combat fear of aftershocks started in the regime's newspaper (Scinteia).
- On March 11, 1977 CPEx Meeting it was discussed how the insurance compensations money could be used, while on March 14, 1977 CPEx Meeting urged to establish a team of highly qualified specialists and appreciated that the

activity of building damage classification and repair-strengthening is progress-ing well, the extrication of the last bodies from the ruins had ended, although debris were not all cleared.

- On March 22, 1977, US President Carter sent an official letter to Ceaușescu offering post-disaster assistance (further on, in April, the US Congress voted to appropriate 20 million USD to Romania) (US Congress Records 1977).
- The Decree No. 66/March 22, 1977 of State Council enforced a new seismic zoning map, but it was not published in the Government's Official Gazette! For Bucharest and many towns, the design MSK macroseismic intensity degrees were reduced in the 1978 earthquake code zoning map (STAS 11100/1-77), enforced by Decree No. 163/May 11, 1978.

For the goals of the present paper, we selected and detailed only the issues with technical and legal consequences on seismic risk from the most important minutes of March 22 working meeting (ANR CCRCP, Chancellery 42/1977) and CPEx meetings of March 30 (ANR, CCRCP, Chancellery 42/1977) and July 4, 1977 (ANR, CCRCP, Chancellery 21/1977; ANR, CCRCP, Economics 78/1977).

4.2 Minutes of March 22, 1977 Meeting with Architects and Constructors

In this meeting, Ceaușescu's introductory words made a clear connection between earthquake damage and the need to improve landmark buildings' site location and traffic patterns in Bucharest's future development. Since the earthquake propelled the proposal to build a new Bucharest Civic Center, on March 22 the idea was presented again at the working conference and its possible location. At first Ceaușescu was concerned about Victoriei Square and Victoriei Avenue, the need for a new Opera House and new monuments, but he shifted his remarks to the need for a Civic Center, stated Arsenal/Uranus Hill as the place for its construction, and gave 1984 as its target completion date (ANR, Chancellery 28/1977, 6–7, 12 reverse—13 in archives; Țiu 2014, p. 13).

The operation was proudly declared as "the first intervention of a large scale in the history of Bucharest." He mentioned some 1000–2000 apartments to be "re-placed" with new ones. He stressed, "we do not stumble about some demolitions, we shall do a new planning…and we will work as if on an empty place" (ANR, CCRCP, Economics 41/1977, p. 4). About the buildings on Magheru Blvd. (the city's most central boulevard), he was more accommodating, and accepted fewer demolitions and allowed reconstruction, but he also allowed the demolition of Dunarea Bloc and the adjacent historic Enei Church built in 1611 (ANR, CCRCP, Economics 41/1977, p. 8).

4.3   Minutes of March 30, 1977 CPEx and the IGSIC Instructions

Overall strengthening solutions were applied in the buildings that were deemed repairable during the first three months, with extensive and speedy works. Scinteia of March 18 published a technical paper of a professor of earthquake engineering of Institute of Constructions/presently Technical University of Civil Engineering of Bucharest, openly advocating the need of high qualification and enhanced safety by strengthening works. The drawings indicate consistent jacketing solutions. The advise is to not make rush decisions, to identify all damages, i.e. the cracks in depth and this should be done only by specialists (Scinteia, March 18, 1977).

On March 30, after visiting some of the 23 collapsed building sites (that by that time had been cleared of debris), Ceauşescu made statements considered to be a prelude for the extensive Bucharest demolitions that were to follow few years later (Boia 2016, Addenda p. 212). He is quoted as saying: "if we will demolish all of Bucharest it will be nice" (ANR, CCRCP, Chancellery 42/1977). Through our archival investigation we now understand that the issue of demolitions was stated openly on March 22 meeting of Ceauşescu with architects and constructors in relation with the perceived need for a new Civic Center.

Some discussions after collapse site visits on March 30 prompted IGSIC (the General State Inspectorate for Investments and Constructions) to send an official letter to all authorities, building design and research institutes, universities and construction companies, which stated the following: "According to the instructions given by the superior party and state leadership on March 30, 1977, we inform you that strengthening of old buildings shall provide at least the strength as before the 1940 earthquake, while for new buildings the strength and stability for which they were [initially] calculated" (IGSIC instructions letter 1977).

The term "strengthening" was used, although strictly speaking the intention was to bring a damaged building back to its pre-earthquake state even if a building had been constructed prior to the implementation of the first earthquake-resistant design code of Romania, established in 1963. But for engineers the words "at least" was a solution to design rather extensive interventions. Scinteia, May 26, 1977 reports that in Bucharest, "The work of strengthening and recovery of damaged buildings continues in sustained rhythm", while the photos show overall jacketing of columns:

- the report on Romarta Block (Calea Victoriei No. 60) is poignant—94 of 102 ground-floor columns strengthened with full jacketing in reinforced concrete, since existing members have reduced strength;
- pre-1940 structures safety assessment proved difficult and some 2–3 strengthening design variants were proposed; in Bucharest, from a total of 2100 buildings, around 1500 need extensive intervention and 600 to be repaired (none was undamaged).

4.4   Secret Police Surveillance of Structural Engineers Related to Official Guidelines on Strengthening

The surveillance of structural engineers was ordered because of expressed opinions about assessment, repair and strengthening that challenged the regime's decisions. The July 4, 1977 versus March 30 decision, discussed in the next section is quite relevant.

A Ministry of Interior (MI) report on March 17, 1977 mentions prestigious professors from "Ion Mincu" Institute/University of Architecture, as being against cosmetic repairs that hide the damage. Another MI Report on May 13, 1977 mentions engineers and researchers concerned about lack of unified viewpoint on strengthening solutions. Some USA specialists also mentioned the difficulties and time consuming nature of strengthening works. USSR specialists are quoted as remarking superficial strengthening works in many cases. It is relevant that USA and USSR specialists have the same opinions with Romanian specialists. Thus, we now clearly understand that the important issue of damage to structural members (apparent or hidden) that leads to the need to strengthen all columns in large occupancy buildings, was discussed extensively by various local and international specialists and that the regime's leadership was fully aware of these opinions.

4.5   Minutes of the July 4, 1977 CPEx Meeting and the July 6 IGSIC Telex, #11264 About Damaged Buildings' Strengthening and Repair

Many Bucharest RCP officials and specialists in charge of repair and strengthening of buildings damaged by the earthquake attended the July 4th meeting. Bucharest's Mayor Dincă presented a report on the impressive number of assessments, repairs, and strengthening completed. He accepted the blame that the work took too long (ANR, CCRCP Economics 78/1977, pp. 1–10). A number of 14,063 buildings were in need of repairs, of which 4510 with completed repairs and 3616 structures with advanced repair works, mostly houses; in addition 351 buildings over five stories had to be repaired and strengthened, of which 114 in process of execution.

Ceaușescu was irritated and complained to the meeting members that they and the city's leaders relinquished decisions about the repair efforts to "professors" and that the responsibility for what happened in Bucharest was left in "the hands of specialists" (ANR, CCRCP Economics 78/1977, p. 28). He expressed anger with the engineering assessment commissions and the degree of strengthening work, saying that they produced more damage than the earthquake itself. It was clear that Ceaușescu was the ultimate authority to decide if, where, and when full strengthening or demolitions would take place.

Almost all participants defended their actions by referring to the March 30 decisions i.e., the aim of bringing back the repaired structure's resistance as it would have been prior to the earthquake. Some of the participants referred to the new seismic zoning map enforced on March 22 as being a source for their understanding to increase the resistance of old buildings.

The crucial outcome of the July 4 meeting was that only local repairs were to be done from then onwards for the remaining circa 10,500 damaged buildings in Bucharest. Ceaușescu ordered that no apartment building be demolished or have a column repaired without special approval (ANR, CCRCP Economics 78/1977, p. 25). The new deadline for ending most of the repair works was set on August 23, 1977, the National Day (Ibid., p. 38).

On July 6 IGSIC telex No. 11264 was sent to the relevant authorities stating: "Working Order: According to the received instructions, we let you know that for the strengthening work for buildings damaged by the March 4, 1977, earthquake one will take into account strictly the local strengthening of damaged members, while for the remaining one shall make only repairs that are strictly necessary...the strengthening project is not permitted to introduce supplementary measures for earthquake safety of buildings...all designed and ongoing works will be in conformity with the present working order...any working orders that are contrary to the present one will be cancelled" (IGSIC Telex No. 1264/1977).

This order is the root of current seismic risk problems in Bucharest and other cities, since it removed entirely reference to a damaged building's seismic resistance and ordered only localized repairs. This crucial decision was not endorsed by any CC RCP Decision, Government Act or Decree. This decision strongly conflicted with Law No. 8 on Constructions Safety, passed barely a week before, on July 1, which put all responsibility and legal liability for building safety on designers and engineers (Georgescu 2003; Law no. 8/1977).

There was no press coverage of the July 4 meeting, but on July 8, a CPEx meeting reiterated these orders. A new deadline for all repair works was set for the end of 1977. On July 15 and August 17, new visits by Ceaușescu to the construction sites were intended to maintain the pressure for quick completion of the work. There is a striking difference between the articles of Scinteia from March to June 1977 presenting extensive number of strengthened buildings and the sudden changes in orders at July 4, 1977. Soon after July 4, 1977 there was a noticeable shift in Scinteia's writing and coverage, from recovery and strengthening towards the new style in architecture, leading also to the future Civic Center.

After July 4, some pressure is presumed to have been exerted on ICCPDC—INCERC to make compatible the new rules with technical constraints. Therefore, in the revised INCERC Instructions C 183/1977 on epoxy resin injections (12.07.1977) the width of repairable cracks was extended from 3 to 5 mm, excluding the provision of 1975 edition that cracks over 3 mm may be repaired only under INCERC technical assistance (INCERC Instructions 1975, 1977).

Some contemporary sources understood the regime's negative reactions as fear of the economic implications and "chaos" caused by extensive strengthening works. In available minutes, however, Ceaușescu and the CPEx members did not discuss the financial cost of structural repairs as an argument for or against one solution. On the contrary, the regime's July 4, 1977, decision to put an end to large-scale strengthening projects was consistent with the grand plans for a new Bucharest stated in March 22 and eventually to the razing of some 450 ha to make way for Ceauşescu's Civic Center in the 1980s (ANR, CCRCP Economics 78/1977; Steinbrueck 2017).

On December 7, Ceauşescu reported to the National Conference of Romanian Communist Party that March 4, 1977 earthquake losses exceeded two billion US Dollars, but the rapid recovery is a proof of superiority of Romania's socialist society. The comment of a US analyst (US Diplomatic or State Department cables 1977) was: "President's tone implied entire earthquake episode is now history and is no longer a consideration in Romanian forward planning".

5   Conclusions

New archival evidence and data allowed the authors to recover yet unpublished official data on building damage and human casualties in the 1977 Vrancea earthquake. The latter are highly relevant to preparedness and risk management in the present day. It was also possible to track the contradictory path of politically driven decisions, from March 10 to July 4, 1977, which concluded with the cessation of strengthening of the damaged buildings.

Sound technical arguments were replaced by Ceauşescu's allegations as they seemed to be a threat to his systematization of Romania and the new Civic Center vision that he was to implement in the 1980s. We may conclude the following regarding the contradictory leadership decisions during these four months:

- extensive damage to large-public buildings in the old Bucharest city centre prompted the first-initial plans for a new Civic Center. This task and related urbanism plans, unquestionably involving extensive demolitions, was accepted with aim to be achieved by the 1990s;
- the restoration of damaged multi-story residential blocks (mostly built prior to earthquake design principles introduced in 1963 or even prior to the big November 1940 Vrancea earthquake) was initially addressed in a framework of overall structural strengthening but after the July 4 drastic decision, only speedy repair works were allowed;
- Misguided and conflicting orders from a highly centralized regime that left little or no room for private initiative, pressure to provide housing to those left homeless as soon as possible, but also historical processes (as pre-1940 Bucharest went through a construction boom) left the city with the unenviable legacy of a very large number of high-occupancy residential buildings that have been damaged by the 1977 (and possibly the 1940) Vrancea earthquake that have not been strengthened. The majority of these buildings was built prior to 1940 without seismic design considerations. Most of these buildings, still occupied today, are at significant risk from future Vrancea earthquakes.

Acknowledgements The authors express their gratefulness to the NIRD URBAN-INCERC, Bucharest, Romania for providing a framework of research activity and related documentary data, as well as for the financing of Ministry of Research and Innovation within the Nucleu Programme CRESC, Project PN 16 10.01.01, "Integrated concept of seismic monitoring and instrumentation to characterize seismic safety…", Phase 3/2017.

A first-shorter version of this paper, published in Proceedings of The 6-th National Conference on Earthquake Engineering and 2-nd National Conference on Earthquake Engineering and Seismology, Bucharest, Romania, June 14–17, 2017, was co-authored with Dr. Karin Steinbrueck, Independent Scholar, Evanston, IL, USA. Without her archive research efforts this publication would not have been possible.

References

ANR, CCRCP (1977) Arhivele Naţionale ale României (ANR), Central Committee of the Romanian Communist Party (CCRCP): Section of Economics, 78/1977; Section of Economics, 41/1977 (Minutes of the meeting of March 22, 1977); Section of Administrative and Politics, 12/1977, 2/1977; Chancellery 18/1977 (Minutes of the meeting of March 5, 1977); Chancellery 21/1977 (Minutes of the meeting of March 6, 1977); Chancellery 28/1977 (Minutes of the meeting of March 10, 1977); Chancellery 42/1977 (Minutes of the meeting of March 30, 1977)

Balan S, Cristescu V, Cornea I (coordinators) (1982) The Romania earthquake of March 1977 (in Romanian, with English abstract). Editura Academiei, Bucharest, Romania

Berg GV, Bolt BA, Sozen MA, Rojahn C (1980) Earthquake in Romania March 4, 1977. An engineering report. National Academy Press, Washington, DC

Boia L (2016) The strange history of Romanian communism (and its unhappy consequences), in Romanian—Strania istorie a comunismului românesc (şi nefericitele ei consecinţe). Humanitas Publishing House, Bucharest

BSSA (1978) Seismological notes—March-April 1977. Bull Seismol Soc Am 68(6):1781–1783

Buhoiu A (Coordonator) (1977) Secunde tragice, zile eroice. Din cronica unui cutremur. Editura Junimea, Iaşi

CNSAS Archives (1977) Consiliul Naţional pentru Studierea Arhivelor Securităţii. Direcţia Cercetare, Dezvoltare, Publicaţii. Chancellery 9665 (D 011737/vol. 105)

Fattal G, Simiu E, Culver C (1977) Observation on the behaviour of buildings in the Romanian earthquake of March 4, 1977. US Department of Commerce, Sept 1977, NBS Special Publication 490

Fundaţia "Gheorghe Ursu" (2017). Accessed 1 May 2017

Georgescu ES (2003) Earthquake engineering development before and after the March 4, 1977, Vrancea, Romania earthquake. Symposium "25 years of Research in Earth Physics", National Institute for Earth Physics, 25–27 Sept 2002. Bucharest. St. Cerc. GEOFIZICA, Tomul 1, pp 93–107

Georgescu ES, Kuribayashi E (1992) Study on seismic losses distribution in Romania and Japan. In: Proceedings of 10-th WCEE, Madrid, Balkema, Rotterdam

Georgescu ES, Pomonis A (2007) The Romanian Earthquake of March 4, 1977: new insights in terms of territorial, economic and social impacts. In: International symposium on strong Vrancea earthquakes and risk mitigation, 4–6 Oct 2007, International Conference Center of the Parliament of Romania, Bucharest. Symposium jointly organized by the UTCB Bucharest and the CRC 461 University of Karlsruhe, Germany

Georgescu ES, Pomonis A (2008) The Romanian earthquake of March 4, 1977 revisited: new insights into its territorial, economic and social impacts and their bearing on the preparedness for the future. In: Proceedings of 14th World Conference on Earthquake Engineering, Beijing, China, 12–17 Oct 2008

Georgescu ES, Pomonis A (2010) Human casualties due to the Vrancea, Romania earthquakes of 1940 and 1977: learning from past to prepare for future events. In: Proceedings of Mizunami international symposium on earthquake casualties and health consequences, Mizunami, Gifu, Japan, 15–16 Nov 2010

Georgescu ES, Pomonis A (2011) Emergency management in Vrancea (Romania) earthquakes of 1940 and 1977: casualty patterns vs. search and rescue needs. In: Proceedings of TIEMS 2011—The International Emergency Management Society, the 18th annual conference, Bucharest, Romania

Georgescu ES, Pomonis A (2012) Building damage vs. territorial casualty patterns during the Vrancea (Romania) earthquakes of 1940 and 1977. In: Proceedings of 15th WCEE, Lisbon, Portugal

ICCPDC (1978) Internal Report on the Romania Earthquake of 4 March 1977 "Cutremurul din Romania din 4 martie 1977 si efectele sale asupra constructiilor". Editia interna ICCPDC, Bucuresti

IGSIC instructions letter (1977) The official letter of Council of Ministers/IGSIC—General State Inspectorate for Investments and Constructions to the Executive Committee of the People's Council of Bucharest Municipality. Registration number not readable. Written after March 30, 1977. Personal photocopy of E.S. Georgescu

IGSIC Telex No. 1264 (1977) IGSIC Telex number 11264/75/15817 from July 8, 1977. Registered by hand under No. 7387/11.VII 1977 at ICB—Institutul de Constructii Bucuresti (presently UTCB-Technical University of Civil Engineering of Bucharest). Personal photocopy of E.S. Georgescu

INCERC Instructions (1975) Instrucţiuni tehnice privind folosirea pastei de ciment şi a amestecurilor pe bază de polimeri la remedieri de defecte ale lucrărilor de constructii şi la lucrări speciale, indicativ C 149-75, aprobate cu ordinul IGSIC nr.190 din 8 decembrie 1975, p 107. In: Buletinul Construcţiilor, vol 1, 1976. Editat de M. C. Ind.—INCERC

INCERC Instructions (1977) Instrucţiuni tehnice privind injectarea fisurilor din elemente de beton armat cu răşini epoxidice,indicativ C 183-77, aprobate prin decizia ICCPDC nr. 88 din 12 iulie 1977, pp 118–130. In: Buletinul Construcţiilor, vol 6, 1977. Editat de: ICCPDC-INCERC

Jones BG (1986) Planning and management for the prevention and mitigation of natural disasters in a metropolitan context, vol 3. Planning for Crisis Relief. United Nations Centre for Regional Development, pp 31–42, 24–30 Sept 1986, Nagoya, Shizuoka and Tokyo Sessions. UNCRD, Japan

Jones BG, Avgar A (1977) A Protocol on the effects on urban systems of the earthquake in Romania of March 4 (unpublished paper)

Law No. 8/1977 (1977) Legea nr. 8/1977 privind siguranta constructiilor. Monitorul Oficial, nr. 64, July 9, 1977. Accessed 15 May 2017. http://www.legex.ro/Legea-8-1977-630.aspx

Povestea Vorbei, nr. 195 (1979) The Radio Free Europe program, Povestea Vorbei transcript based on a text from a letter sent by the engineer Gheorghe Ursu. Povestea Vorbei nr. 195, Aniversarea cutremurului by Virgil Ierunca, Sunday March 4, 1979, Romanian Broadcasting Department, box 3811, folder 2, Radio Free Europe/Radio Liberty, Romania Broadcast Department, 1960–1995, Hoover Institution Archives

Sandi H, Pomonis A, Francis S, Georgescu ES, Mohindra R, Borcia IS (2008) Development of a nationwide seismic vulnerability estimation system. In: Proceedings of the symposium thirty years from the Romania earthquake of March 4, 1977, Bucharest, Romania, 1–3 March 2007, CONSTRUCTII, No. 1/2008, pp 38–47. Retrieved from http://constructii.incerc2004.ro/Archive/2008-1/2008-1-5.pdf

Scinteia (1977) The Newspaper of Central Committee of Romanian Communist Party

Steinbrueck K (2017) Aftershocks: Nicolae Ceauşescu and the Romanian Communist Regime's Responses to the 1977 Earthquake. Unpublished Ph.D. dissertation, Northwestern University, Evanston

Ţiu I (2014) Sfera Politicii, Nr 178/martie-aprilie 2014

US Congress Records (1977). Foreign Assistance and Related Agencies Appropriations for 1978. 8 Sept 1977

US Diplomatic or State Department cables (1977) Wikileaks "dump"—"US library of public diplomacy," including telexes between the US Embassy Bucharest and the US State Department in DC and Belgrade. https://wikileaks.org/plusd/about/

World Bank (1978) Report No. P-2240-RO, Report and recommendation of the president of the international bank for reconstruction and development to the executive directors on a proposed loan to the investment bank with the guarantee of the socialist Republic of Romania for a Post Earthquake Construction Assistance Project, 17 May 1978

World Bank (1983) Report No. 4791, Project completion Report. Romania—Post Earthquake Construction Assistance Project, Loan 1581-RO, 18 Nov 1983

Earthquake Risk Awareness in Bucharest, Romania: Public Survey

Ileana Calotescu, Florin Pavel and Radu Vacareanu

Abstract During a period of eight months, an extensive survey was conducted on the population of Bucharest as part of the CoBPEE research project, financially supported by the Romanian National Authority for Scientific Research and Innovation. The focus of the survey was to evaluate the level of awareness and preparedness of the population for a hypothetical earthquake generated by the Vrancea intermediate-depth source, the level of expected damage of residential buildings as well the level of implication of the population in the aftermath of the seismic event. In total, 1000 responses were collected. This paper describes the main findings of the survey. Results show that although more than 60% of the respondents are aware of the possibility of occurrence of a major earthquake in Romania, a very limited number is prepared to deal with such a situation. Also, most respondents would feel the safest in a building built in the period 1978–1992 rather than in a building built in the past ten years, showing a lack of confidence in the construction industry. As far as post-earthquake attitude is concerned, the vast majority of residents agree to provide humanitarian assistance in various ways, financial aid being the last preference.

Keywords Questionnaire · Earthquake awareness · Seismic damage
Vrancea seismic source

I. Calotescu (✉)
Steel Structures Department, Technical University of Civil Engineering of Bucharest, Bucharest, Romania
e-mail: ileana.calotescu@utcb.ro

F. Pavel · R. Vacareanu
Seismic Risk Assessment Research Center, Technical University of Civil Engineering of Bucharest, Bucharest, Romania
e-mail: florin.pavel@utcb.ro

R. Vacareanu
e-mail: radu.vacareanu@utcb.ro

© Springer International Publishing AG, part of Springer Nature 2018   297
R. Vacareanu and C. Ionescu (eds.), *Seismic Hazard and Risk Assessment*,
Springer Natural Hazards, https://doi.org/10.1007/978-3-319-74724-8_20

1 Introduction

Within the framework of CoBPEE research project financially supported by the Romanian National Authority for Scientific Research and Innovation (CNCS UEFISCDI), a survey was proposed in order to investigate the risk awareness, preparedness and expectations of the population living in Bucharest in case of a hypothetical major seismic event originating in the Vrancea intermediate-depth seismic source.

Research on population awareness and behaviour in case of an earthquake (Armaş 2006; Joffe et al. 2013) has shown that important factors in determining seismic risk are not only social, economic, institutional and environmental, but also cultural and psychological. Education, earthquake recurrence or religion play an important role in people's behavior during an earthquake. Studies on people's perceptions about earthquakes (Ainuddin et al. 2014) conducted by means of questionnaires have shown that respondents are aware of the seismic risk, but not of the principles on which design codes rely. These are developed with the aim of ensuring the highest possible degree of safety, as well as profitability, but studies show that people living in seismic areas are increasingly interested in safer and more expensive buildings (Jaramillo et al. 2016; Calvi et al. 2014).

The CoBPEE survey was conducted over a period of eight months (February to September 2016) and was taken by 1000 respondents. The questions were grouped into four parts quantifying: the level of education and awareness of the population regarding the occurrence of a potential major earthquake in Romania; the importance of structural safety; the level of damage/losses expected by the population after a major earthquake; and the level of public involvement with respect to humanitarian help following a major seismic event. The questions were developed based on the hypothesis that Bucharest would be affected by a major earthquake generated by the Vrancea source which was described to respondents as an earthquake similar to the one that occurred on March 4th, 1977. Although only approximately 30% of the respondents have the recollection of this earthquake, it was extensively publicized in Romania and, as such, even younger people have knowledge on the extent of damage that it has produced at the time.

2 The Questionnaire

The questionnaire used within the framework of the CoBPEE research project was designed based on the four basic steps required in the design process (De Leeuw et al. 2008): (i) define objectives, (ii) select target population, (iii) establish the time interval required for designing the survey and (iv) establish the survey procedure.

The main objective of the survey was to identify the population opinion concerning several earthquake related topics such as seismic risk awareness, earthquake preparedness, vulnerability of existing buildings and population involvement.

The target population is represented by the people currently living in Bucharest, either as owners or tenants. An ample description of the target population is given in Sect. 3. The survey format includes demographic questions, multiple choice questions, scale questions as well as open-ended questions (Scaffer and Perser 2003; De Leeuw et al. 2008). The design time interval, including implementation and interpretation of data was approximately 8 months. Both a paper as well as an online version of the questionnaire were made available and data was collected using such methods as interviews, emails and dissemination through social media.

The questionnaire is structured into five parts, each dealing with a particular seismic-related issue such as earthquake preparedness, safety concerns, vulnerability of buildings and expected damage and, finally, population involvement in the aftermath of a possible major seismic event affecting Bucharest. The final part of the survey contains questions related to the demographics of the respondents. Figure 1 shows the cover and a page of sample questions of the CoBPEE questionnaire.

3   The Target Population

The target population selected for study is the population of Bucharest which is located at an epicentral distance of approximately 180 km from the Vrancea intermediate-depth seismic source (Fig. 2). This choice was based on the fact that

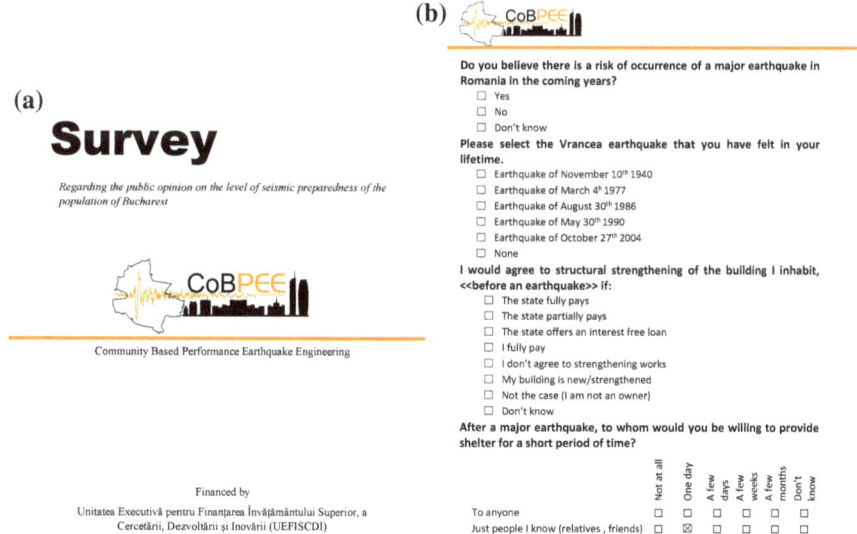

Fig. 1 The CoBPEE questionnaire: **a** cover; **b** sample questions

Fig. 2 Map of Romania showing Bucharest location with respect to Vrancea seismic source

Fig. 3 Sex distribution of the
respondents

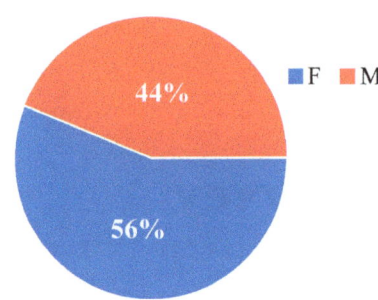

the capital is the most vulnerable city in Romania, having sustained the largest
number of casualties as well as building damage after the March 4th, 1977
earthquake.

The most recent Population and Housing Census which was conducted in 2011,
showed that Bucharest was inhabited by 1,883,425 residents at the time, out of
which approximately 54% were women and 46% were men. The margin of error of
the CoBPEE questionnaire, when compared to the 2011 Census, resulted ±3.1%.

The gender and age distributions of the respondents are presented in Figs. 3
and 4, respectively. The survey was addressed solely to adults aged over 18 years
old. Consequently, the age group 0–19 years old has minimal representation,
achieved only through those aged 18–19 years old. Also, the low participation of

Fig. 4 Age distribution of the respondents

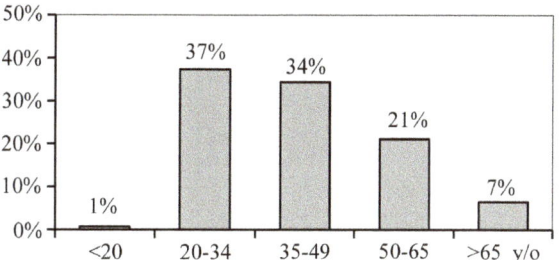

Fig. 5 Income level of the respondents

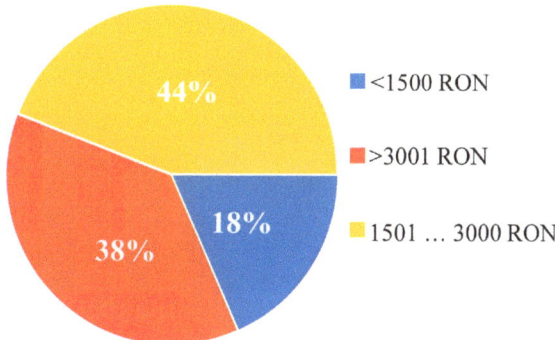

Fig. 6 Last level of education completed

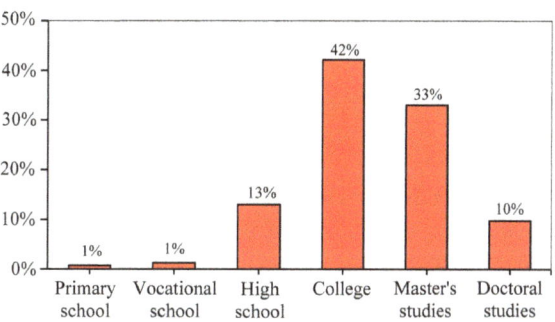

the group aged over 65 years old may be attributed to the low accessibility to the survey of this age group, as it was carried out mainly online.

Results show that, out of the 984 valid answers, 49% of the respondents have children whereas 51% don't. Figure 5 shows the level of income of the respondents, with over 40% having an average income of 1501–3000 RON (330–660 €) whereas 18% have an income smaller than 1500 RON (330 RON) per month. The majority of those who completed the CoBPEE questionnaire are people with higher education level, approximately 85% having completed at least college as their last level of education (Fig. 6). This may be due to the fact that the survey was conducted mainly online and people having a lower education level may not have proper access.

Fig. 7 Distribution of individual houses and blocks of flats with year of construction

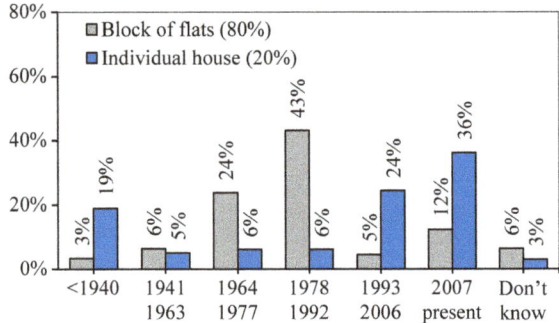

Fig. 8 Census 2011 and Survey 2016: comparison of distribution of the population by district

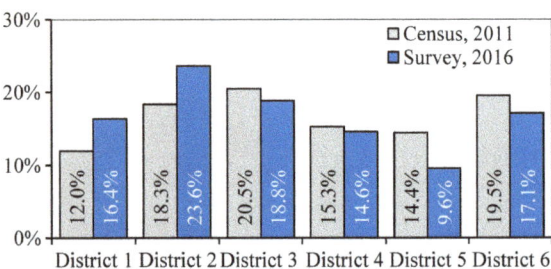

Out of the total number of respondents, roughly 80% live in blocks of flats whereas 20% live in individual houses. In Fig. 7 the distribution with the year of construction of the building for each category is presented. Most of the individual houses that the respondents inhabit were recently built (2007–present) whereas, in the block of flats category, 43% live in buildings built in the year interval 1978–1992 followed by 24% living in blocks of flats built between 1964 and 1977.

The capital city is divided into six districts, all being well represented in the CoBPEE survey. The comparative analysis of demographic distribution by districts (Fig. 8) between the 2011 Census and the CoBPEE survey shows that for three districts (3, 4 and 6) the percentage differences are of −1.7, −0.7 and −2.4%, respectively, while for the remaining three (1, 2 and 5) the differences are of +4.4, +5.3 and −4.8%, respectively.

A characterisation of the districts based on the year of construction of the building stock is given in Table 1, showing that Districts 3–6 are mainly

Table 1 Year of construction of buildings distributed by district

Year of construction	District 1 (%)	District 2 (%)	District 3 (%)	District 4 (%)	District 5 (%)	District 6 (%)
Prior to 1977	35	40	34	33	25	29
After 1977	62	54	59	61	71	65
Don't know	3	6	7	7	4	7
	100	100	100	100	100	100

characterized by buildings built after 1977 whereas most of the older buildings (built prior to 1977) are located in Districts 1 and 2 (Table 1).

4 Seismic Risk Awareness

The Vrancea intermediate-depth seismic source has generated two destructive earthquakes in the past century in November 10th 1940 (M_w = 7.7; focal depth h = 150 km) and March 4th 1977 (M_w = 7.4; focal depth h = 94 km). In recent years, several earthquakes having moment magnitude $M_w \geq 6.0$ have occurred such as the earthquakes of August 30th 1986 (M_w = 7.1, h = 131 km), May 30th 1990 (M_w = 6.9, h = 91 km) and October 27th 2004 (M_w = 6.0, h = 105 km). Seismic risk studies for Bucharest (Lang et al. 2012; Toma-Danilă et al. 2015; Pavel and Vacareanu 2016) have shown that for Vrancea earthquakes with $M_w \geq 7.5$, the direct losses for residential buildings in Bucharest can be of the order of several billion euros and the number of heavily damaged or collapsed buildings of the order of several thousand euros.

In order to quantify the seismic risk awareness of the respondents, they were asked to identify the earthquakes that they have felt during their life-time (Fig. 9). The purpose of this question is to evaluate their real life-experience of earthquakes and, whether having felt and earthquake makes them more aware of the risk. Results show that approximately 46% of the respondents have felt at least one major earthquake, this being, of course, correlated to the respondent's age. When inquired about the possibility of a major earthquake occurring in Romania in the coming years, 63% answered that they are aware of this possibility, as shown in Fig. 10. However, it is still worrying that the remaining 37% either don't know or don't believe that a major earthquake could occur.

When questioned about awareness related to prevention measures and behaviour in case of earthquakes, more than half of the respondents answered that they are informed to a small (35%) or a very small extent (24%) on what they need to do, according to specifications of the Emergency situation guidelines issued by the General Inspectorate for Emergency Situations. Only approximately 41% of the respondents are well informed. Also, a limited number of respondents (roughly 5%) have an emergency backpack prepared, containing items that would ensure their

Fig. 9 Responses to: *Please select the Vrancea earthquakes you have felt during your lifetime*

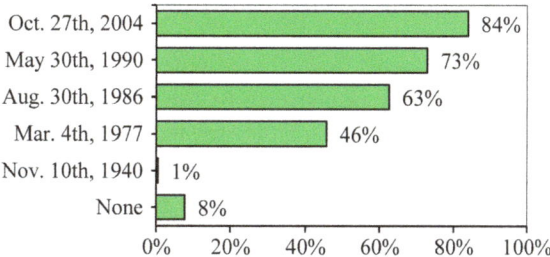

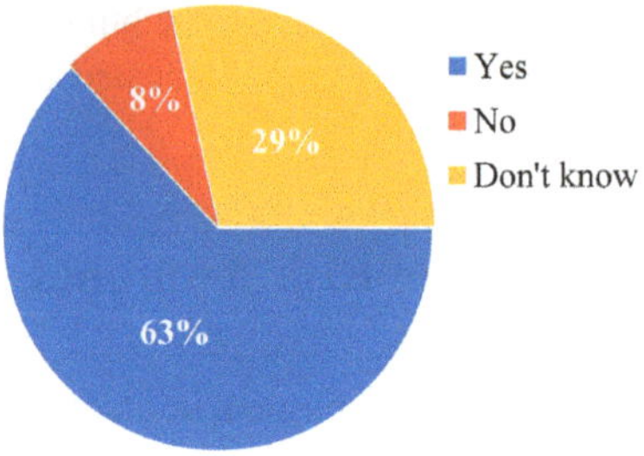

Fig. 10 Responses to
question: *Do you believe there
is a risk of occurrence of a
major earthquake in Romania
in the coming years?*

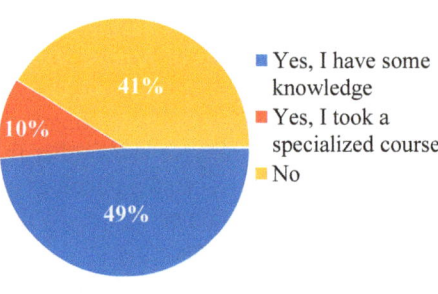

Fig. 11 Responses to: *Do
you know how to perform first
aid?*

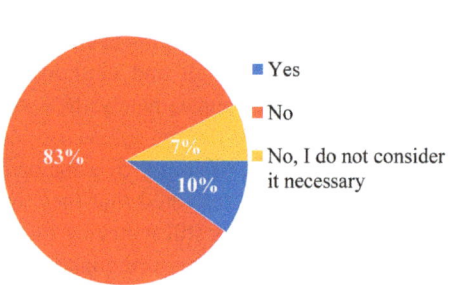

Fig. 12 Responses to
question: *Do you have a
meeting place with your
family in case of a major
earthquake?*

survival for approximately three days after a possible destructive seismic event,
whereas 95% of the respondents don't have such a survival kit.

In the event they need to perform first aid, only 10% of the respondents admitted
to having taken a specialized course whereas 41% have no notions related to this
issue (Fig. 11). So, a very limited number of people are well prepared to help a
wounded person in case of immediate need. Figure 12 shows answers to question:
Do you have a meeting place with your family in case of a major earthquake?. As it
might be noticed, the vast majority (83%) do not have such a meeting place in case
a major earthquake affects Bucharest. Considering the vulnerability of the building
stock of the capital, many older buildings are in danger of major damage or even
collapse so many people may be forced to find alternative housing. In the immediate aftermath of a natural disaster, it is not unusual for phone lines to stop working
so, getting in contact with relatives or friends might be impossible. Not having a
meeting place with the family shows that respondents are not well prepared to deal
with the emotional issues that might arise in such a situation.

Table 2 Cross tabulation: knowledge on first aid performance versus age of respondents

		Do you know how to perform first aid?			Total (%)
		Yes, I have some notions (%)	Yes, I took a specialized course (%)	No (%)	
Age group	<20	33.3		66.7	100
	20–34	45.7	13.6	40.7	100
	35–49	49.8	9.1	41.1	100
	50–65	48.4	7.4	44.2	100
	>65	55.2		44.8	100
Total		48.2	9.8	42.0	100

Table 2 shows the cross tabulation between answers obtained for question shown in Fig. 12 and the age distribution of respondents. As it might be noticed, the respondents aged 20–34 years old are the most prepared followed by those aged 35–49 years old. Results also show that the majority of respondents having taken a specialized course on performing first aid (10% of the total number of respondents) are those having higher education as last level of education completed (approximately 86%). Finally, in terms of income, as expected, people having higher income (more than 1500 RON = 330 € per month) represent the majority of those who have taken a specialized course on first aid.

5 Vulnerability of Buildings

Due to the old building stock of Bucharest associated to the seismic hazard of the Vrancea seismic intermediate-depth source, the Romanian capital is considered to be one of the riskiest cities in Europe. The residential building stock of Bucharest comprises of more than 130,000 buildings of which around 60% was built prior to the seismic event of 1977. In addition, roughly 20% of the existing buildings (mostly masonry and reinforced concrete buildings) are approximately 70 years old or even older (Calotescu et al. 2017). In this context, the CoBPEE survey aimed at identifying the opinion of the population of Bucharest with respect to the aspect of the seismic vulnerability of the building stock of the capital.

Concerning awareness related to possible damage produced to the buildings they inhabit, respondents were asked to identify the type of damage that they consider possible to occur to their buildings in case of a major earthquake. Figure 13 shows the type of damage respondents living in blocks of flats and individual houses expect to occur. It may be noticed that, in both categories, mainly minor to moderate damage is expected to occur.

When only blocks of flats are considered, the distribution of the damage type with the year of construction (Fig. 14) shows that, in each category, respondents expect minor to moderate damage to occur to the building they inhabit. Total

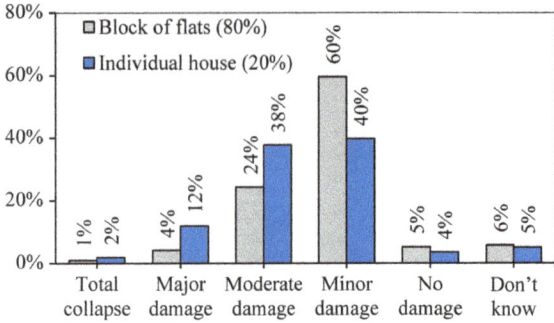

Fig. 13 Type of building versus expected damage

collapse and major damage are expected in a higher percentage by respondents living in buildings built prior to 1978. Based on Fig. 14 it may also be deduced that buildings built in the year interval 1993–2006 are considered to be the safest by the respondents, as, in this category, they don't expect any major damage or total collapse to occur. For the major damage category, a decreasing pattern with the year of construction may be observed whereas for the minor damage, the pattern decreases. However, for the recently built buildings (after 2007), the pattern is not consistent, showing that people generally don't trust the level of safety of new buildings.

In another question, the respondents were asked to specify the interval year of construction for a building they would feel the safest to live in. The intervals were chosen to correspond to the periods of enforcement of the various versions of the Romanian seismic design codes: before 1940, 1941–1963, 1964–1977, 1978–1992, 1993–2006 and after 2007. Results obtained for the category block of flats showed that 36% of the respondents selected the 1978–1992 interval, followed by 1964–1977 (20%) and 2007–present (17%). The other intervals, each obtained less than

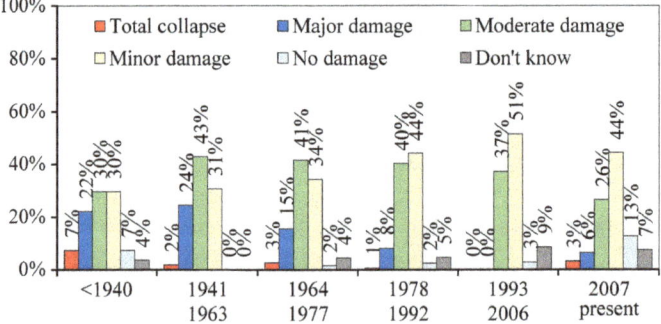

Fig. 14 Expected type of damage versus year of construction for blocks of flats

7%. This shows a lack of trust of the population on the safety level of the recently constructed buildings.

The structural strengthening of privately owned buildings in Bucharest is a controversial issue in Romania. There are countless cases where strengthening works had to be suspended or cancelled as a small percentage of owners refused to leave their homes for the duration of the works. The survey results confirm this fact. To the following question:

> Should the building you inhabit be proposed for strengthening, would you agree to a temporary relocation to allow for strengthening works?

5% of the respondents answered "No", 58% answered "Yes" whereas the remaining 37% answered "Not the case". The reasons given by those who would refuse a temporary relocation include the lack of financial capacity to rent alternative housing during the time of the consolidation works, mistrust in the quality of the works and in respecting the deadlines for their completion and belief that if their home was not damaged during the March 4th 1977 earthquake then the building is safe. Other reasons mentioned were: lack of time, the availability of additional housing (the owner does not live in the building that requires consolidation), and lack of personal comfort.

When inquired under what circumstances they would agree to structural strengthening of the building they inhabit, the majority of the respondents (29%) would agree only if the state would provide partial financial support whereas only 3% would assume the financial responsibility by fully paying for the works done as shown in Fig. 15.

An analysis of the financial responsibility for damage to privately owned buildings following a major earthquake shows that the majority of the respondents (52%) assigned this responsibility to the insurance company, 36% of respondents believe that the state should contribute financially either by assuming full financial responsibility or by granting a preferential loan, whereas only 8% of respondents believe strengthening works should be financially supported by the owner.

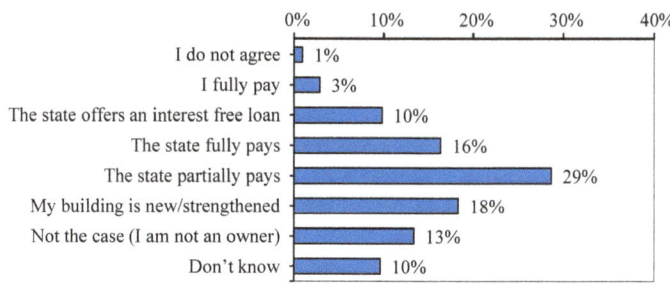

Fig. 15 Responses to question: *I would agree to structural strengthening of the building I inhabit ≪before an earthquake≫, if*

6 Population Involvement and Expectations

One of the issues targeted by the CoBPEE questionnaire was to evaluate the opinion of the population related to community involvement in the aftermath of a seismic event affecting Bucharest. When asked if they would be willing to provide shelter for a short period of time to people affected by the earthquake, 89% answered affirmatively to accommodating people they know for a period of one day to several months while 54% would agree to provide shelter to anyone, mostly for a few days (Fig. 16).

The type of damage that would determine the respondents to leave their homes in case of a major earthquake is presented in Fig. 17 and Table 3 respectively. As it might be noticed, damage like cracks in some or all structural elements, spalled concrete or partial collapse of the building are the main type that would determine

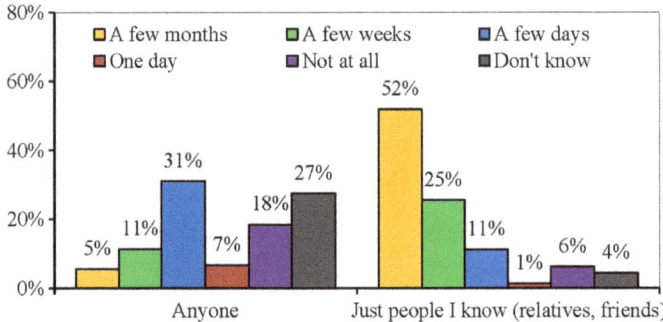

Fig. 16 Responses to question: *After a major earthquake, to whom would you be willing to provide shelter for a short period of time?*

Fig. 17 Responses to question: *What type of damage would determine you to leave your home after a major earthquake?*

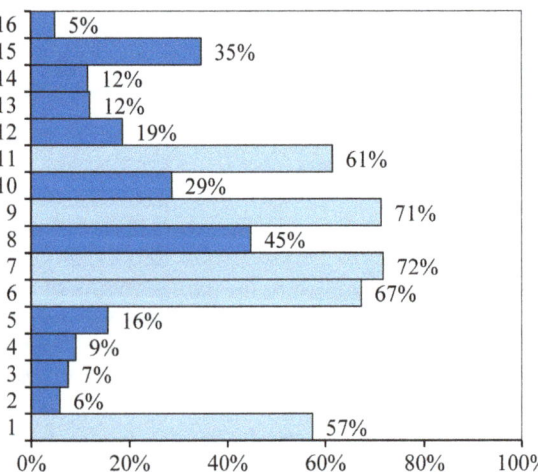

Table 3 Responses to question: *What type of damage would determine you to leave your home after a major earthquake?*

	Type of damage	Percent (%)
1	Hairline cracks in some structural elements	57.3
2	Fallen plaster	5.8
3	Hairline cracks around door openings and at the intersections of non-structural walls	7.5
4	Slight damage to non-structural elements	9.0
5	Broken windows	15.6
6	Cracks in all structural elements	67.3
7	Spalled concrete and visible reinforcement in most structural elements	71.7
8	Destroyed masonry walls	44.8
9	Partial collapse of the building	71.3
10	Destroyed partition walls	28.6
11	Hairline cracks in most structural elements	61.4
12	Cracks in masonry walls	18.6
13	Fallen tiles of suspended ceiling	12.0
14	Damaged non-structural walls	11.6
15	Damaged roof	34.7
16	Overturned furniture	4.9

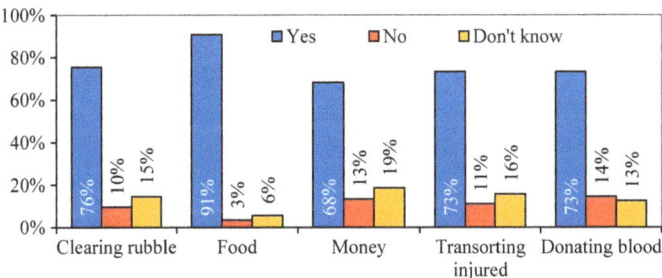

Fig. 18 Responses to question: *After a major earthquake, what type of humanitarian assistance would you offer?*

the respondents to reach for a temporary shelter. The type of shelter that the respondents would mainly choose is the home of a relative or friend.

Related to the type of humanitarian assistance, the results show that a large part of the population would agree to provide aid through various methods: from removing rubble to donating blood. However, of all methods of assistance, offering financial aid obtained the lowest percentage (Fig. 18).

Responses to questions on post-earthquake expectations in the event of a major earthquake, showed that most people would find it acceptable to wait only for one

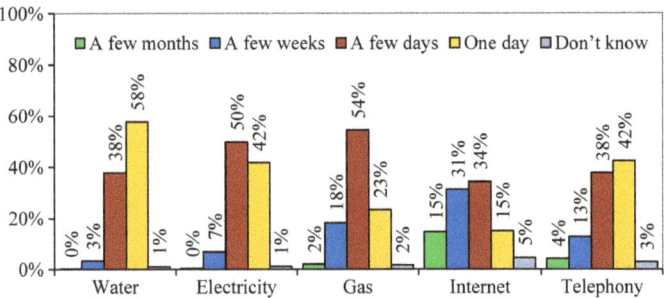

Fig. 19 Responses to question: *After a major earthquake, what period of time would you find it acceptable for water, gas, internet or telephone provision/services to be restored?*

day for utilities such as water or telephone services to return to functionality, while for electricity and gas, a period of a few days is acceptable for almost half of the respondents (Fig. 19).

7 Conclusions

In this paper, the results of a survey performed within the framework of the CoBPEE research project are presented and analysed. The main objective of the survey was to investigate the opinion of the population of Bucharest concerning their awareness, preparedness and expectations in case of a major seismic event originating in the Vrancea intermediate-depth seismic source. The survey targeted the population of Bucharest and was conducted over a period of eight months, reaching the intended sample size of 1000 respondents. The main findings in this research may be summarized as follows:

- most respondents would feel safest in a building built in the period 1978–1992 (40%) compared to only 27% who would feel safest in one built more recently (2007–present); this shows on the one hand the lack of public knowledge regarding the safety level currently adopted in the design codes and, on the other hand, the lack of trust in how these have been respected in the new construction projects.
- only 5% of the respondents have a survival kit prepared to use in case of a major earthquake;
- 10% of the respondents have specialized training for performing first aid.
- only 6% of the respondents would offer shelter to a stranger for several months and over 50% would offer shelter for the same period of time to a friend or relative;
- the most important public service as resulted from the questionnaire is the water, followed, in importance, by services such as the electricity and gas.

- the issue of structural steepening for the existing vulnerable housing was addressed through two questions: if those living in such housing would be willing to leave home temporarily to allow for strengthening works; and to which extent they would be willing to participate financially, as owners, to structurally strengthen the building. Most respondents believe that the state should contribute financially through partial (29%) or full (16%) contributions, or by granting interest-free loans (10%). This shows on one hand a lack of financial means and, on the other hand, a lack of accountability. Regarding those who refuse to leave their home, unexpectedly, most of them are living in the most vulnerable types of buildings (apartment buildings built during 1941–1963). Various reasons were given for this refusal, most cited being the inability to pay rent or the absence of temporary housing alternatives.
- over 80% of the respondents expect no damage, minor or moderate damage to their residence in the case of a major earthquake;
- as far as post-earthquake attitudes are concerned, the vast majority of Bucharest's residents agree to provide humanitarian assistance in various ways, of note being however that providing financial aid is their last preference.

Acknowledgements This work was supported by a grant of the Romanian National Authority for Scientific Research and Innovation, CNCS—UEFISCDI, project number PN-II-RU-TE-2014-4-0697. The financial support is gratefully acknowledged.

References

Ainuddin S, Mukhtar U, Ainuddin S (2014) Public perception about enforcement of building codes risk reduction strategy for seismic safety in Quetta, Baluchistan. Int J Risk Reduction 9:99–106

Armas I (2006) Earthquake risk perception in Bucharest, Romania. Risk Anal 26(5):1223–1234

Calotescu I, Pavel F, Săndulescu AM, Sibișteanu H, Vacareanu R (2017) Population perspective on the social impact of a strong earthquake affecting Bucharest. Math Model Civil Eng 13(3):1–9

Calvi GM, Sullivan TJ, Welch DP (2014) A seismic performance classification framework to provide increased seismic resilience. Perspect Eur Earthq Eng Seismol (Geotechnical, geological and earthquake engineering), vol 34, pp 361–400

De Leeuw DE, Hox JJ, Dillman D (2008) International handbook of survey methodology. European Association of Methodology, New York

Jaramillo N, Careno ML, Lantada N (2016) Evaluation of social context integrate into the study of seismic risk for urban areas. Int J Disaster Disk Reduction 17:185–198

Joffe H, Rossetto T, Solberg C, O'Connor C (2013) Social representation of earthquakes: a study of people living in three higly seismic areas. Earthq Spectra 29(2):367–397

Lang D, Molina-Palacios S, Lindholm C, Balan S (2012) Deterministic earthquake damage and loss assessment for the city of Bucharest, Romania. J Seismolog 16:67–88

Pavel F, Vacareanu R (2016) Scenario-based earthquake risk assessment for Bucharest, Romania. Int J Disaster Risk Reduction 20:138–144

Scaffer NC, Perser S (2003) The science of asking questions. Ann Rev Sociol 29:65–88

Toma-Danilă D, Zulfikar C, Manea EF, Cioflan CO (2015) Improved seismic risk estimation for Bucharest, based on multiple hazard scenarios and analytical methods. Soil Dyn Earthq Eng 73:1–16

A Plea for People Centered Perspectives on Seismic Risk Evaluation

Mihai Şercăianu, Ioana Nenciu, Bogdan Suditu, Marina Neagu, Roxana Popescu and Diana Murzea

Abstract The issue of seismic risk in Bucharest, a 1.7 mil people Capital, is very much ignored, considering the scale and the urgency. Seismic risk assessment needs to focus also on human losses, rather than on physical aspects like location, construction materials, etc. "How many people live in seismic risk housing?" is a question not even the municipality can give a recent evidence based answer to. The paper presents the results of ALERT—seismic risk mitigation project, started in Bucharest in February 2016. ALERT is based on an online data crowdsourcing platform regarding seismic risk housing (with particular focus on the number, needs and problems of their inhabitants). The Alert poll results, as of September 2017 indicate the severity of the problem. More than 8000 dwellers were counted in 80% of the highest seismic risk buildings (Seismic Risk Class I) and it is expected that by the end of the research the 10,000 people threshold will be long exceeded. The paper is intended to highlight insights on ownership, occupancy and management

M. Şercăianu (✉)
Urban Engineering and Regional Development Department,
Technical University of Civil Engineering of Bucharest, Bucureşti, Romania
e-mail: mihai.sercaianu@utcb.ro

I. Nenciu · M. Neagu
Bucureşti, Romania
e-mail: ioana.nenciu@mkbt.ro

M. Neagu
e-mail: marina.neagu@mkbt.ro

B. Suditu
Faculty of Geography, University of Bucharest, Bucureşti, Romania
e-mail: bogdan.suditu@mkbt.ro

R. Popescu
Risk and Vulnerability Expert, General Inspectorate for Emergency Situations,
Bucureşti, Romania
e-mail: popescu.roxana.mihaela@gmail.com

D. Murzea
Center for Innovation in Local Development, Bucureşti, Romania
e-mail: diana.murzea05@gmail.com

© Springer International Publishing AG, part of Springer Nature 2018
R. Vacareanu and C. Ionescu (eds.), *Seismic Hazard and Risk Assessment*,
Springer Natural Hazards, https://doi.org/10.1007/978-3-319-74724-8_21

of these buildings. The conclusion argues the need for a more human centric approach of seismic risk assessment to complement the engineering and geological perspectives.

Keywords Social vulnerability · Awareness · Housing · Ownership
Social vulnerability

1 Introduction

Bucharest is located in the most active seismogenic zone of Romania. Seismogenic areas are defined as areas with grouped seismicity, within which seismic activity and the field of stress are considered to be relatively uniform. According to Romanian National Institute for Earth Physics (NIEP) the identification of the long-term characteristics of the earthquake generation process in each seismic zone is of major importance for the estimation of seismic hazard. The concentration of population, old buildings or high heights in Bucharest and in the cities of the south and south-east of the country makes living under seismic risk one of the most important economic and socio-demographic vulnerabilities in Romania.

Statistics provide a numerical representation of a phenomena. In Bucharest, the seismic risk seen in terms of numerical data consists of a list of buildings divided in 4 risk classes, each defined by the magnitude of the potential damage caused by an earthquake. But these buildings are not just addresses or numbers on a list, but spaces that give shape to the city, spots we come across in our daily journey and, unfortunately, without involvement and concrete actions, a daily potential danger to our lives.

The issue of seismic risk in Bucharest is largely ignored, considering its magnitude and the degree of urgency. With nearly 1.7 million inhabitants and located only 170 km away from the Vrancea seismic source, Bucharest is considered to be the capital with the highest seismic risk in Europe. In Bucharest, there are 346 buildings classified as seismic risk class I—so-called "Red dot buildings"—in danger of collapse during a high intensity earthquake (>7 Richter scale), as of Bucharest City Hall official seismic risk inventory list published on the 22nd of June 2017. 175 of these buildings pose public danger, according to the official list of the City Hall of Bucharest. Up to 1997, seismic evaluations were prioritized for buildings with a minimum floor height of 5 floors and built before 1940, in a step-by-step process at the request of the municipality. After 1997, the expertise process was left in the hands of the owners, no longer requested by the municipality, and as a result, the number of buildings surveyed after 1997 and classified according to the seismic risk scale declined considerably (to a few dozen in the last 20 years).

The World Bank's report on housing shows (2015), based on data from the Ministry of Regional Development and Public Administration, that there are over 10 thousand housing units listed as seismic risk class I, more than half of these being located in Bucharest. In addition to these, there are also commercial and cultural spaces whose operating license has already been withdrawn, based on Law no. 282/2015 (law for modifying and completing the Government Ordinance no. 20/1994 on measures to reduce the seismic risk of existing constructions).

The paper is based on the results of Alert, a seismic risk awareness project, launched in Bucharest in February 2016. The project builds primarily on an online data collection platform regarding the situation of seismic risk housing in Bucharest. Data collected focus on the number, needs and challenges faced by those who live in such buildings. The project included discussions and debates on seismic risk with those affected, either through field research, but also by organizing 1-to-1 meetings and group debates with landlords, owners and block administrators.

The data collection for Alert project was done by university volunteers (teachers, students) as well as non-academic contributors (residents, administrators of seismic risk buildings, etc.). During the field research activity started in May 2016, 160 owners' associations were informed through informative letters about the initiative and data was gathered through the online questionnaire on www.seismic-alert.ro platform. The source of the information thus collected consists exclusively of declarations of the tenants/administrators of the buildings concerned, together with the information included in the Bucharest City Hall official list. The present article is based on the data collected in the Alert project until September 2017 concerning 295 buildings, approximately 80% of the 346 buildings classified as Seismic Risk Class I.

The aim of the paper is to switch the attention from the building to the residents, to argue that in the background of a seismic risk evaluation for a building, there is an owner/occupant a facing danger on a day by day basis, there are responsibilities and choices for certain types of dwellings.

To be able to support seismic risk mitigation it is also needed an understanding of the behavioral types and issues faced by the residents.

2 Legal Practice

Cost of housing expertise, according to Government Decision no. 709/1991, was financed from the state central administration budget at a rate of 75% for private property, respectively 50% for public property, provided that the owners of the buildings would be obliged to carry out the consolidation works recommended through technical expertise (Ministry of Regional Development and Public Administration 2016). This measure explains the number of technical evaluations carried out in 1992–1994.

In the context of public policies promotion aimed to public housing stock privatization, increasing demands of nationalized houses restitution and the sale of the nationalized houses administered by former ICRAL (Enterprise for Construction, Repair and Housing Management State Company), seismic risk reduction on existing constructions and establishment of responsibilities for structural intervention of residential buildings becomes for a temporary period of time a legislative priority for the Government.

In this context, was elaborated the Government Ordinance no. 20/1994 on measures for seismic risk reduction on existing constructions, normative act in force and currently, modified by Laws no. 84/2015 and no. 282/2015, being republished in 2011. As in the previous years, in the period following the approval of Government Ordinance no. 20/1994, on the initiative of ICRAL were made numerous technical expertise for the dwelling buildings to be sold to the tenants, in accordance with current legislation. It is necessary to be mentioned that through Government Ordinance no. 20/1994, was implemented the Annual Action Program for the design and execution of consolidation works on multi-storey dwelling buildings, classified by technical expertise report in seismic risk class I and presenting public danger, funding program currently managed by the Ministry of Regional Development, Public Administration and European Funds.

Socio-economic transformations in Romania in the 1990s and the impoverishment of a significant part of the urban population, modification of the system of financing the technical expertise and realization of measures for structural consolidation of buildings with seismic risk, diminishing the public funds for the housing sector and delaying the elaboration and promotion of the norms Methodological Methods of Application of Government Ordinance no. 20/1994 (established within 30 days from the date of approval of the ordinance, the norms were promoted 7 years later by Government Decision No. 1364/2001) determined the decrease of the economic capacity and the interest of citizens and authorities in strengthening existing buildings and increasing the safety of the residents of seismic-risk buildings.

23 years after the Government Ordinance no. 20/1994 became effective, the reduced number of buildings consolidated through the Annual Action Program for the design and execution of consolidation works on multi-storey dwelling buildings (Ministry of Regional Development and Public Administration 2015–2016), classified according to the technical expertise in the first class of seismic risk and posing public danger by the Romanian Government, as well as the non-implementation of the sanctions established by the mentioned normative act, was a negative signal for the targeted population, having as a result the lack of interest in the safety of the occupants and buildings with seismic risk, in addition to avoidance of specific legislation by the owners/associations of owners.

Of the evaluated seismic risk buildings classified as seismic risk class I with public danger, the most numerous (81 buildings) are high buildings with more than 6 floors (46.28%), followed by being of 4 and 5 storey buildings (50 buildings and 28.57% respectively). The 44 low buildings (GF, GF + 1) and 2 and 3 floors represent only 25.14% of the real estate. Regarding the 171 seismic risk buildings classified as seismic risk class I, there is a reversal of the representation ratios, a single high building (GF + 6 and above) being identified and the 162 buildings up to GF + 3 floors representing 94.72% of the identified real estate and included in this class of seismic risk (Fig. 1).

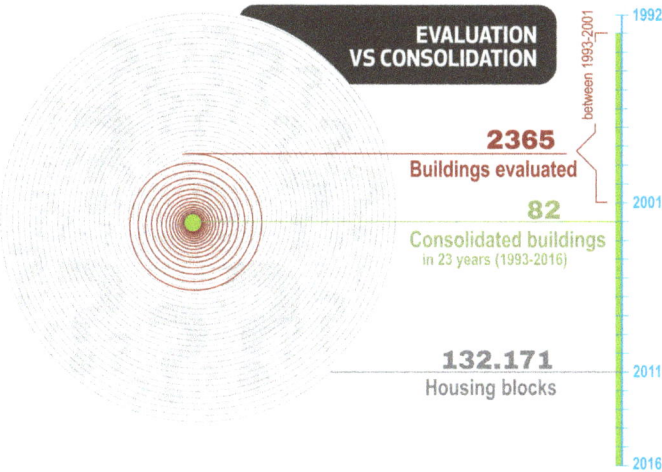

Fig. 1 Total evaluated buildings (includes risk categories classes according to P100-3/2008, emergency categories according to P100/92 but also those that have been evaluated but not classified)

Also the lists of seismically assessed buildings in Bucharest contains more than 400 buildings classified as Seismic Risk Classes II and III, and their deterioration may have worsened in the meantime. In addition to this list, more than 1500 buildings were assessed under the P100/92 in the early 90's. P100/92 norms frames seismic risk constructions in emergency categories. Emergency category U recommends a maximum start time for structural consolidation work: U1—maximum 2 years, U2—maximum 5 years, U3—maximum 10 years, U4—maximum 15–20 years. These buildings are ignored in the public discourse, but they can be a real danger in case of an earthquake. P100/92 is replaced by Romanian Code for the seismic assessment of existing buildings P100-3/2008. Without a new re-evaluation of the buildings, there cannot be made a link between Seismic Risk Classes and old Emergency Categories, we can speak at most about an approximation.

3   Translating a Built Environment Issue into a Social Issue (Alert Survey Results)

3.1   Occupation

Who and how many people are affected by the risk? Considering that we have now reached about 80% of the red dot buildings, unfortunately we expect to exceed the threshold of 10,000 people after we manage to reach all of these buildings in field research.

In addition:

- 23 buildings are demolished
- 13 buildings are completely unoccupied
- In the remaining inhabited buildings, there are 3607 apartments
- 81% is average occupancy
- 878 of the occupied apartments are inhabited by tenants.

The occupancy rate of the buildings, representing the percentage of flats inhabited by total apartments, ranges from 4 to 100%. Of the 199 buildings for which we have data entries, only 29 buildings were reported to be still occupied 100%. This reveals a considerable stock of vacant housing (about 1585 apartments), which suggests that many landlords have put themselves in safe shelters by moving out of these buildings. This can also lead to problems in the consolidation process, by the difficulty of locating the owners and arguing the necessity of investing in a property that has no use or capitalization for them (Figs. 2 and 3).

24% of the inhabited apartments are occupied by tenants, again indicating a category of landlords who are not affected by the seismic risk, but they exploit their properties economically. Vulnerability to risk is passed from the owner to the tenant, which could be explained—from a tenant's perspective—either by the total lack of knowledge of the risk or ignoring it, for the benefit of a potentially lower rent. From the owner's view, the fact that such apartments can still be economically valued by renting could explain the low interest and limited involvement of the owners in taking steps to strengthen the building.

On the other hand, 11% of the occupied apartments (representing 405 units) have utilities debts of over 6 months. From here, we deduce that many of those who still live in these buildings have low material capacities and probably little capacity to support the costs of seismic rehabilitation work.

3.2 Ownership

Who are the owners of these buildings? 68 of the buildings for which information was gathered include, amongst the owners, various public authorities such as the

Fig. 2 Number of residents in seismic risk class I public danger buildings according to alert survey

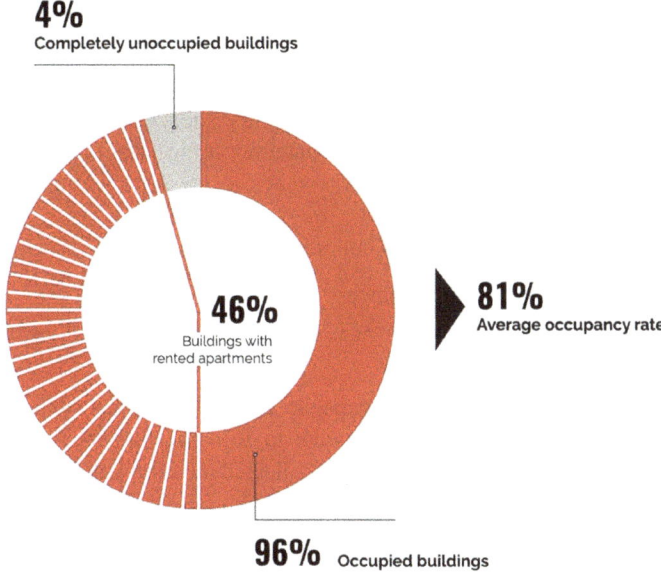

Fig. 3 Occupancy of evaluated seismic risk class I buildings

Administration of the Bucharest Real Estate Fund (AFI) and the Autonomous State Administration of the State Public Patrimony (RAPPS). As a matter of fact, it was found that 2 buildings are entirely in public property, containing 56 apartments.

Consequently, it is not wrong to state that the owner with the most seismic risk properties is the public administration. The survey has highlighted so far 429 apartments in publicly owned red dot buildings.

3.3 Condominium Management

How prepared are the owners to work together to retrofit the building? The effort to consolidate a red dot building requires co-operation and co-ordination among several owners, this being the most significant challenge of seismic retrofitting. For this reason, the ability to manage a condominium - as jointly owned property—is essential.

Unfortunately, for 20% of buildings (55) we did not identify the existence of a homeowners' association, which probably reveals the largest barrier in the implementation of a seismic consolidation program.

3.4 Functions

What (other) activities are taking place in these buildings? Functions, meaning the activities taking place in these buildings, amplify the danger and increase the number of potential victims. In buildings containing office and commercial spaces or cultural and social functions, the risk extends to employees and customers of those services.

So far the 110 analyzed buildings counted:

– 180 work spaces as a business office
– 120 commercial and service spaces
– 2300 dwelling units.

Some service spaces include high-capacity cultural functions such as cinemas and theaters. Areas, from University Square to Romană Square, brings together several such red dot buildings for which we collected data through the Alert survey, based on cultural facilities capacity (Bica 2013):

– Patria Cinema (Magheru 12–14) has a capacity of 1014 seats. It operates in a 10-storey building with, according to the data gathered, 60 permanently occupied dwellings.
– Cinema Pro (Ion Ghica 3) has a capacity of 541 seats. It operates in a 9-storey building with 62 homes, 49 permanently occupied, where we counted 66 people.
– Cinema Studio (Magheru 29) has a capacity of 380 seats and operates in a 10-storey building.
– The Notarra Theater (Magheru 20—a building newly listed in the lower class of Seismic Risk Class II) has a capacity of 584 seats. It operates in a 10-storey building with 60 permanently occupied homes where we counted 74 people.

3.5 Neighborhood

What is the risk for those in the immediate proximity of red dot buildings? As part of the Alert survey, we also collected information on the building vicinity. Such buildings can be wall-to-wall with commercial or cultural spaces with high traffic of people such as restaurants, terraces, but also public functions such as schools, kindergartens and so on. The damage registered to buildings with seismic risk also endangers the lives of the people in transit in their vicinity.

An example in this regard is the Academy Street, where Seismic Risk Class I Buildings with public danger (No. 15, 19) are across a 2 lanes street from the "Ion Mincu" University of Architecture and Urbanism, with intense pedestrian traffic (Fig. 4).

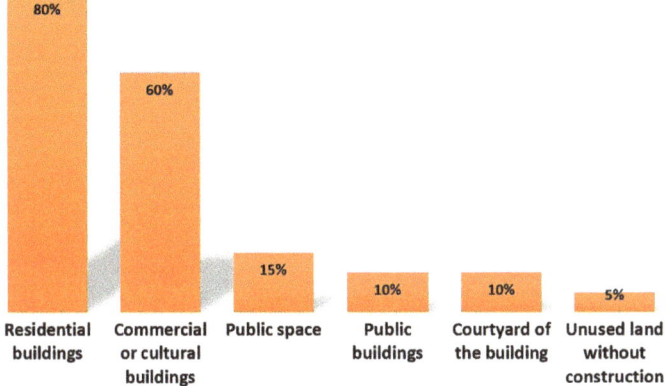

Fig. 4 Land use distribution on surveyed areas

60% of 110 red dot buildings we have surveyed regarding neighborhoods functions so far, have commercial or cultural functions, and 10% are next to public buildings. The vast majority (80%) are in the close vicinity of other residential buildings.

3.6 Heritage. Value Versus Danger

90% of the 346 buildings have an important architectural heritage value, being located in protected areas (with specific urban regulation). This aspect emphasizes both the value of the built environment and the importance of controlling interventions done on it. This includes consolidation interventions, so the heritage value of the building becomes a hindrance/obstacle to the consolidation process in the absence of a viable financial and technical mechanism. In such situations, real estate interests are in tension with restrictions on the construction work that can be done in these areas. The situation is even more critical for the 60 buildings listed as a historical monument classified as Seismic Risk Class I.

Some of these protected areas with a significant concentration of buildings with high seismic risk are: the Historical Center (Calea Victoriei—Lipscani area), Magheru Boulevard, Dacia Boulevard, Cotroceni, Aviatorilor etc. These are also highly circulated and densely populated areas, with public and commercial functions, adding to the vulnerability of the built environment, the insecurity and vulnerability of residents and users of these spaces.

As an illustration of the scale of this vulnerability, let's take an example of the Magheru Boulevard, located across the street. On this boulevard are 4 major cultural interest spots—Patria Cinema, Cinema Scala, Notarra Theater and Cinema Studio—3 of them closed due to the aforementioned law. Taken together, these cultural spaces have a capacity of more than 2800 seats, being located in 8–10-level buildings, with hundreds of housing units.

4   Social Vulnerability Perspective

4.1   Bucharest Context

In Romania, vulnerability is scarcely treated in all aspects of risk management, while efforts still focus on the issue of hazard. At Bucharest level, several studies addressing the issue of vulnerability to seismic risk have been developed within the academic community, proposing different methodologies to estimate and map out vulnerability, often taking into account the social, economic, structural (built-up) and functional dimensions (Armaş et al. 2017).

At the public sector level, the most recent and important initiative is the "Risk Assessment Methodology and Integration of Sectoral Risk Assessments" developed within the RO-RISK—Disaster Risk Assessment Project at national level, where vulnerability is taken in consideration in calculating the impact of each type of risk analyzed. The analysis is, however, carried out at the level of a territorial administrative unit and is, in essence, a statistical analysis, the results of which cannot be used in the Alert project. However, RO—RISK remains a reference project not only for the important step taken to introduce the concept of vulnerability in risk studies but also because it is a unique project in Romania, a project of collaboration between a public authority in the field, the General Inspectorate for Emergency Situations (project leader) and 13 research institutions representative for the types of risk considered.

Efforts to understand and evaluate this component of risk are essential to identify effective measures to reduce seismic risk. They not only involve structural intervention on vulnerable buildings, but also understanding how the city works and what are the main factors contributing to its vulnerability. Some examples could be: increasing the level of access to necessary data for vulnerability analysis, standardizing as much as possible the produced data, creating a framework for interdisciplinary dialogue, encouraging research into this direction. The results produced could help to more accurately identify the most vulnerable areas, taking into account the socio-economic level of the population, the access to resources and the infrastructure necessary for everyday life, but especially during a crisis, the development of awareness and information campaigns for the population specific to the vulnerability profile identified, etc.

In Alert project, the vulnerability study aims in identifying the most vulnerable areas at the level of Bucharest in terms of the socio-demographic dimension of vulnerability and territorial vulnerability. The methodology used is innovative, combining several elements from other research in Latin America, the Caribbean,

China, Iran and Turkey. Novelty for Bucharest is also a reading of a vulnerability reading on a smoother scale (1 km^2) compared to other studies done so far.

It is important to note that the interpretation of the obtained results must be viewed and understood in terms of the concept of vulnerability used. Thus, vulnerability is defined by the susceptibility of the exposed elements (population, infrastructure, etc.) to being affected by an event. Therefore, vulnerability is an already existing state of these elements in the face of a hazard, providing a measure of possible loss before the event occurs. The idea behind is that two exposed items that may be affected by an event of the same intensity will suffer a higher or lower level of loss depending on the level of vulnerability. The higher the vulnerability, the greater the losses. In this study, only the social dimension is considered having the population as the exposed element. Therefore, this does not mean that all the population with the highest vulnerability will be affected, but that these people may be the most affected. The impact level (e.g. number of injured and dead) could not be calculated in the absence of data necessary to assess seismic hazard.

4.2 Social Vulnerability Methodology

The methodology of analyzing socio-demographic vulnerability is a common approach in the international literature. Starting from a series of relevant indicators that capture the characteristics of the population considered vulnerable, they are brought to a common denominator and integrated into a vulnerability index at the level of each unit with an area of 1 km^2. The selection of indicators and their aggregation was based on the experience of the SIRV—TAB research project implemented in Haiti after the earthquake in 2010 where a similar vulnerability analysis was carried out using a grid as a unit of analysis (D'Ercole et al. 2013).

Thus, the unit of analysis used for Bucharest is 1 km^2, represented as a grid consisting of 300 units totaling an area of 240 km^2. The grid used is the one published by the National Institute for Statistics at national level and populated with data on the population of Romania by age group and gender according to the 2011 census. Also from the NIS was used other data regarding the building materials of the buildings. Another source of data is the Open Street Map (for the administrative infrastructure, transport, points of interest, etc.) and the City Hall of Bucharest regarding the buildings classified in seismic risk classes.

After preparing the data, about 10 vulnerability indicators were selected and an analysis of the main components was performed using the Analyze-it software. Very correlated or very little correlated indicators have been eliminated, retaining 5 indicators relevant to socio-demographic vulnerability. This analysis aimed at avoiding over-representation of a vulnerability factor through a series of indicators that express the same trend. For example, the gender factor, according to which the feminine population is more vulnerable than male, was removed due to the high degree of correlation with the young and old population. Finally, 5 indicators were retained: the total population aged 0–14; the total population aged over 65, the total of residential buildings built of wood, the total of residential buildings built from the parcel, the total of buildings classified in a risk class.

Young and elderly people are some of the most commonly used indicators to assess the vulnerability of the population, starting from the premise that they are more fragile during a crisis, and the risk of suffering serious injuries is greater. Also, their mobility is reduced, and they are dependent on those around them in case of evacuation, access to information is more difficult, requiring adapted materials and means of communication, especially for children.

The rest of the indicators were chosen to approximate the value of social vulnerability in the absence of other economic data. Thus, it was considered that the material from which a building is built, in this case wood and fanatic, and the fitting of a building into a risk class, reflects the socio-economic situation of the resident population. The wooden or fanatic houses are very fragile with low economic value, and they are inhabited by low income people. Buildings with seismic risk were chosen from the same perspective, although this reasoning does not apply to all buildings surveyed by the City Hall and is not an exhaustive list. There may be cases where people with a medium or medium to high income may be living in these buildings and are unaware of the risk they are exposed to. In addition to the fact that this may in itself constitute a vulnerability, we considered these cases to be less representative and we assumed that, as a rule, a person with a good economic situation will choose a home in a new building as soon as possible. This indicator was also retained to reflect the well-known vulnerability in the old center of Bucharest and which, statistically, was not visible in relation to the other indicators.

Next, in order to build the index and compare the chosen indicators, they were standardized using the min–max equation. Total values were converted to a scale from 0 to 1. Subsequently, the standardized indicators were aggregated using a simple average, resulting in the socio-demographic vulnerability index. This aggregation method assigns to all indicators the same importance and is one of the simplest ways to build a vulnerability index. It has been favored at the expense of others to reduce as much as possible the subjectivity in ranking the importance of indicators in relation to vulnerability and time and resource considerations. The results were classified in ArcGIS, using the natural method of Jenks, on a scale of 5 vulnerability classes. To improve the result, there are other classification methods that could better represent the spatial variability of vulnerability, but for the same reasons mentioned above, the simplest method was chosen.

4.3   Results

The pattern of spatial distribution of socio-demographic vulnerability at the level of Bucharest indicates a concentration on two axes NV-SE as shown in map no. X (Table 1).

In total, 8% of the population of Bucharest is ranked at the highest level of socio-demographic vulnerability and focuses on 11% of its area, totalizing 146,864

Table 1 The Bucharest area according to the level of socio-demographic vulnerability

Vulnerability level	Very high	High	Medium	Low	Very low
Total sq. km	26	38	51	66	59
Percent of total area of Bucharest (%)	11	16	21	28	25

Table 2 Total young and old population depending on the level of socio-demographic vulnerability in Bucharest

Vulnerability level	Very high	High	Medium	Low	Very low
Total vulnerable persons (elders and children)	146.864	180.120	131.169	37.129	9.266
Percent of the total population of Bucharest (%)	8	10	7	2	0.5

Table 3 Total population according to socio-demographic vulnerability in Bucharest

Vulnerability level	Very high	High	Medium	Low	Very low
Total resident persons	541.611	661.950	480.328	152.049	39.318
Percent of the total population of Bucharest (%)	29	35	26	8	2

inhabitants, according to Tables 2 and 3. It is only about children and the elderly, the selected indicators as the most representative. By calculating the total resident population in these areas characterized by a very important socio-demographic vulnerability, we find that 29% of the inhabitants of Bucharest live in a very high risk area (see Table 4). This situation is aggravated by the high level of density, considering the fact that they live on only 11% of the total area of Bucharest. The density factor was not taken into account in the analysis but it plays a major role in the quality of housing, further aggravating the specific vulnerability.

At the opposite end, the areas occupied by a population with the lowest degree of socio-demographic vulnerability are concentrated mainly in the northern area of Bucharest.

By grouping the two extremes, it is easy to see that more than half of the total resident population of Bucharest is characterized by a high socio-demographic vulnerability (64%) and is concentrated on about a quarter of its area (27%). On the other hand, only 10% of the total population is at a low vulnerability and occupies half of Bucharest (53%). These conclusions signal an alarming trend of bipolarization of the vulnerability profile of Bucharest, considering that only a quarter of the population (26%) is classified as a medium vulnerability and occupies 21% of its area (Fig. 5).

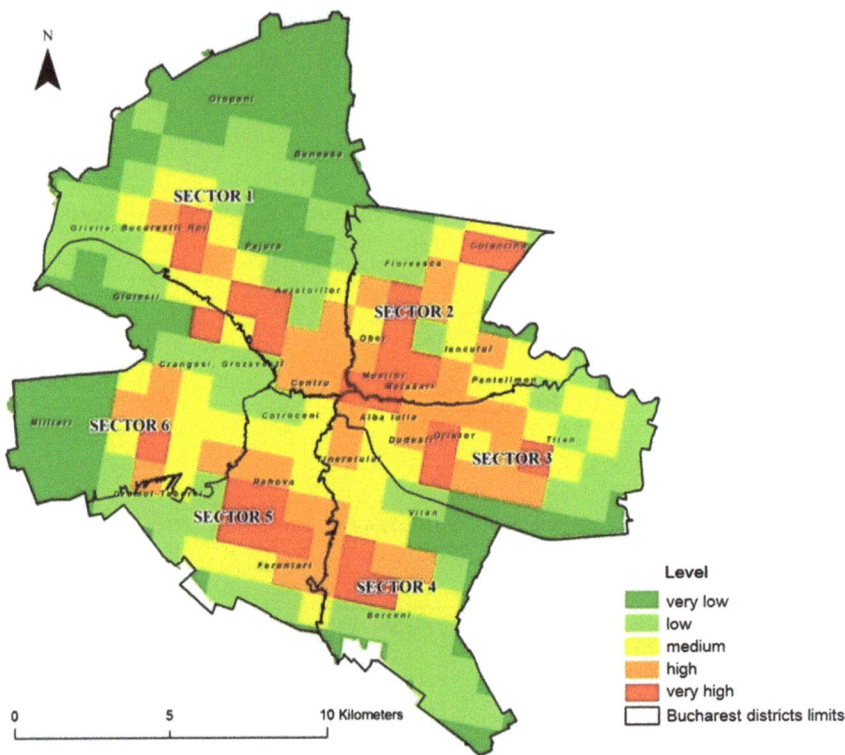

Fig. 5 The socio-demographic vulnerability of Bucharest (per sq. km)

5 Conclusions

The paper illustrates the results of Alert project survey, having analyzed a set of seismic risk class I buildings (approx. 80% of the listed buildings).

The data gathered on the Alert crowdsourcing platform highlights complex ownership issues and poor condominium management. We trust that a better understanding of the way such a building is currently used by its occupants, its functioning and vicinity provides essential information for better calibrated and more efficient risk mitigation policies. While engineering solutions and geological evaluations are a prerequisite for proper consolidation works, the social dimension of seismic housing is what, in the end, could impede or allow their implementation.

Insights provided by Alert showcase that the on field situation deems the current seismic consolidation program unfitted for the challenge it targets. For instance, as long as owners can economically exploit their properties, transferring the risk to renters, there is no stimulant for consolidation. This, together with the required consensus for collective action in order for such works to be undertaken, make the retrofitting investment non appealing and non-viable.

From the social vulnerability points of view, the results show a tendency to increase the proportion of the vulnerable population in relation to the total population of Bucharest and to concentrate it in small areas as a surface, usually located at the periphery. More than half of the total resident population of Bucharest is characterized by a high socio-demographic vulnerability (64%) and is concentrated on about a quarter of its area (27%), while only 10% are at a vulnerability level and occupies half of the city's surface (53%). This process is heavily influenced by the high density factor specific to the large concentrations of collective dwellings. Here, too, we find an aging majority population, which often exceeds the total number of children. Thus, corroborating these factors with a low socioeconomic level, it appears that Rahova, Ferentari and some areas in Colentina, Berceni, Militari, Bucharesti Noi-Grivita present all these characteristics and are the most vulnerable areas in Bucharest from a socio demographic point of view.

Taking into account the phenomenon of urban expansion, in which most of the cities in the world are confronted, we observe a process of marginalization and concentration, densification of the vulnerable population at the outskirts of Bucharest. Also, the center of the city has some characteristics, showing a high level of vulnerability, due to the presence of a large number of elderly people with a low socio-economic level, but the predominant factor is that of structural vulnerability, which was not taken into account in this study. This trend can be emphasized in the future, taking into account the rural–urban migration phenomenon we are witnessing. The new inhabitants, regardless of the socio-economic level, are considered vulnerable because they lack the knowledge of the risks and problems and the existing resources. Moreover, it is likely that these people will choose as a prime home a more affordable apartment located in already crowded areas, aggravating the already existing vulnerability. So, it is likely that the current trend will continue and intensify, with the possibility of emerging new areas of very high vulnerability or the expansion of the existing ones.

References

Armaş I, Toma-Dănilă D, Ionescu R, Gavriş A (2017) Vulnerability to earthquake hazard: Bucharest case study, Romania. Int J Disaster Risk Sci 8(2):182–195

Bică AG (2013) The seismic vulnerability of the Romanian built-up environment to earthquake action. Ph.D. thesis, University Construction Engineering Bucharest

City Hall of Bucharest, list of evaluated buildings. http://www.pmb.ro/servicii/alte_informatii/lista_imobilelor_exp/docs/Lista_imobilelor_expertizate.pdf. Accessed on 09 Sept 2016

D'Ercole R, Robert J, Hardy S, Ponte E, Vernier P, Popescu R, Etaby S, Salome G, Vicario T (2013) Utilisateurs des ressources de gestion de crise et problèmes. In: La population de Tabarre et ses vulnérabilités, Raportul proiectului SIRV-TAB, vol 2. COOPI, IRD, ECHO, Mairie de Tabarre, CNIGS, CIAT, DPC, Port-au-Prince, p 623

Ministry of Regional Development and Public Administration (2015–2016) Annual program actions for the design and execution of consolidation work to multi-storeyed dwelling buildings, classified by technical expertise report in the first class of seismic risk and posing public danger (red dot buildings)

Ministry of Regional Development and Public Administration (2016) Efficient and transparent co-ordination and selection of infrastructure financing from structural instruments and from the State budget for 2014–2020, SMIS code 48659—Component 4—Living in Romania. Towards a National Housing Strategy, http://www.mdrap.ro/lucrari-publice/-3175

Part III
Seismic Design and Structural Performance Assessment of Buildings

A New Structural Health Monitoring System for Real-Time Evaluation of Building Damage

Koichi Kusunoki

Abstract The author has developed a new method for evaluating the seismic performance of existing structures from measured accelerations based on the capacity spectrum method. This involves comparing the performance curve, which is the equivalent nonlinear behavior of a simplified single-degree-of-freedom system, and the demand curve, which is the relationship between the response acceleration and displacement spectra. Two telecommunication towers in Japan were instrumented in 2016, and their responses during several earthquakes have been recorded. This paper discusses the evaluation of damage during two earthquakes. Moreover, parameters such as the predominant period and the required performance are discussed. The proposed system evaluated both towers as being "elastic". The damping ratios of the towers are very low, which caused the oscillations to continue for more than 5 min after the mainshock of each earthquake because of long-period components of the seismic motion.

Keywords Structural health monitoring · Capacity spectrum method
Steel tower

1 Introduction

In 2000, a new structural calculation method known as the "response and limit-capacity calculation" was added to the Building Standard Law of Japan as a new option for evaluating the seismic performance of buildings. The method involves comparing the performance and demand curves of a building. The performance curve represents the building's nonlinear behavior as a simplified single-degree-of-freedom (SDoF) system, whereas the demand curve represents the force–deformation design criteria defined in the building code.

K. Kusunoki (✉)
Earthquake Research Institute, the University of Tokyo,
1-1-1 Yayoi, Bunkyo-ku, Tokyo 113-0032, Japan
e-mail: kusunoki@eri.u-tokyo.ac.jp

© Springer International Publishing AG, part of Springer Nature 2018
R. Vacareanu and C. Ionescu (eds.), *Seismic Hazard and Risk Assessment*,
Springer Natural Hazards, https://doi.org/10.1007/978-3-319-74724-8_22

331

Meanwhile, when a strong earthquake occurs, a quick inspection of the damage to buildings in the affected area is very important for reducing the damage caused by aftershocks. However, the current inspection method this involves structural engineers making visual assessments, which is both time consuming and subjective. To overcome these drawbacks, a new real-time method has been proposed (Kusunoki and Teshigawara 2004, 2003; Kusunoki et al., 2008, 2012) for evaluating building damage with inexpensive accelerometers based on the aforementioned response and limit-capacity calculation.

In 2014, two steel towers used for microwave telecommunication in Tokyo and Yokohama in Japan were instrumented and have been monitored since then. One tower was instrumented because it was seen to continue oscillating after the ground had ceased to shake during the 2011 Tohoku Earthquake. Since the towers were instrumented, two major earthquakes have occurred in the region. One was the Fukushima Offshore Earthquake of magnitude 7.4 at 0559 JST on November 22, 2016. At both tower sites, the maximum seismic intensity was recorded as 3 on the Japan Meteorological Agency (JMA) seismic intensity scale. The other earthquake was the northern part of Ibaraki Prefecture Earthquake of magnitude 6.3 at 2138 JST on December 28, 2016. Again, the maximum seismic intensity at each tower site was 3. By using the proposed system, both towers were evaluated as being elastic during those 2016 earthquakes. This paper discusses the damage evaluation results and design parameters such as the predominant period and the required performance.

2 Outline of Towers and Measurements

Hazawa Tower (Fig. 1a) is a cylindrical steel tower in Yokohama, Japan, that is 58 m tall and is instrumented with accelerometers (ITK-002, shown in Fig. 2a) at its base and top and at 1/3 and 2/3 of its total height, as indicated in Fig. 1a. The accelerometers are sampled at a rate of 100 Hz with a 24-bit A/D converter. The accelerometers are three-dimensional ones with a maximum measurable acceleration of 2450 cm/s^2 and a noise level of no more than 0.1 gal/s. The acceleration data are stored for 24 h on a server that is located in the tower and that is connected to the outside world through the cellphone network. The data can be accessed through the Internet.

Higashi Oshima Tower (Fig. 1b) is another cylindrical steel tower and is located in Tokyo, Japan. It is 63 m tall and is instrumented with accelerometers (IoLAM-01, shown in Fig. 2b) at its base and top and at half its total height. The accelerometers are sampled at a rate of 100 Hz with a 24-bit A/D converter. The accelerometers are three-dimensional ones with a maximum measurable acceleration of 3430 cm/s^2 and a noise level of no more than 0.1 gal/s. As with Hazawa Tower, the acceleration data are stored for 24 h on a server that is located in the tower, the server is connected to the outside world through the cellphone network, and the data can be accessed through the Internet.

(a) Hazawa Tower   (b) Higashi Oshima Tower

Fig. 1 Target towers

(a) ITK-002 for Hazawa Tower   (b) IoLAM for Higashi Oshima tower

Fig. 2 Instrumented sensors

3 Measured Accelerations

3.1 Fukushima Ken Oki Earthquake (November 22, 2016)

At 0559 JST on November 22, 2016, a strong earthquake of magnitude 7.4 occurred 30 km beneath the seafloor off the shore of Fukushima Prefecture in Japan. The earthquake epicenter and the tower locations are shown in Fig. 3. A JMA seismic intensity of 3 was measured at each site, and the responses of the towers during the earthquake were stored successfully.

Figure 4 shows the calculated transfer functions of the NS direction from the top, 2/3, and 1/3 heights to the bottom of Hazawa Tower. The predominant frequency was 0.6897 Hz, which corresponds to a period of 1.45 s. The amplification

Fig. 3 Earthquake epicenter (triangle) and tower sites (circles)

Fig. 4 Transfer functions
(NS direction)

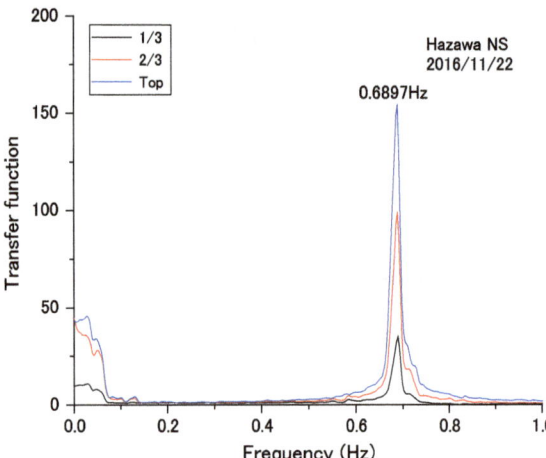

factor associated with the top of the tower was roughly 150, which means that the damping coefficient of the tower is very low.

Figure 5 shows the acceleration time histories of the predominant NS direction measured at the bottom and top of Hazawa Tower. As shown in Fig. 5a, the ground

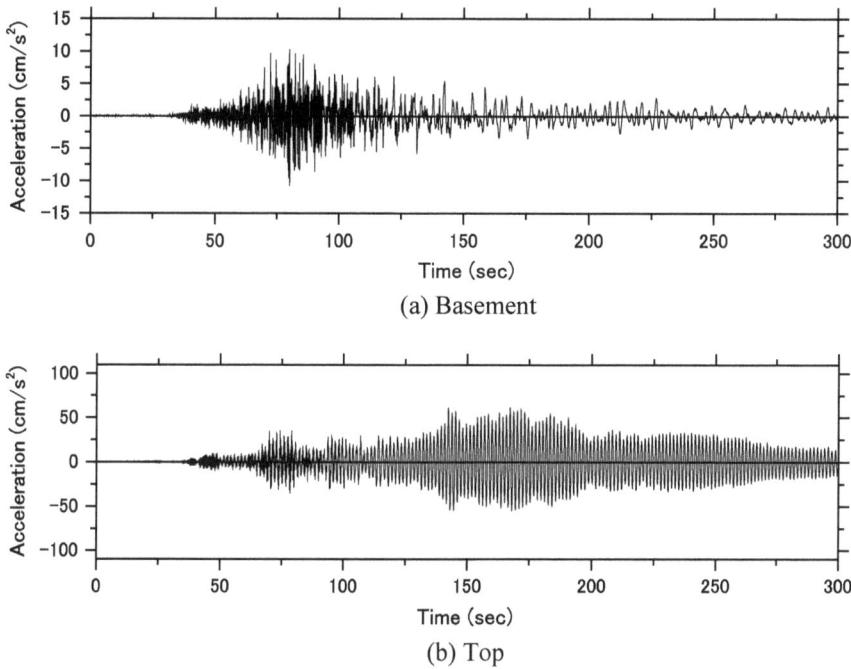

Fig. 5 Measured accelerations of Hazawa tower (NS direction)

acceleration reached its peak at roughly 80 s, but a small long-period component can be observed after 150 s. Because of this long-period component, the acceleration at the top reached its peak at roughly 170 s, at which time the acceleration at the bottom was very small. The long-period component may be why the tower was seen to continue oscillating after the ground had ceased to shake during the 2011 Tohoku Earthquake.

3.2 Ibaraki Ken North Earthquake (December 28, 2016)

At 2138 JST on December 28, 2016, a strong earthquake of magnitude 6.3 occurred 10 km beneath the ground surface in the north of Ibaraki Prefecture in Japan. The earthquake epicenter and the tower locations are shown in Fig. 6. A JMA seismic intensity of 3 was measured at each site, and the responses of the towers during the earthquake were stored successfully.

Figure 7 shows the calculated transfer functions of the NS direction from the top and half-way up Higashi Oshima Tower to the bottom. The predominant frequency was 0.415 Hz, which corresponds to a period of 2.41 s. The amplification factor associated with the top was roughly 175, which means that the damping coefficient of this tower is also very low.

Fig. 6 Earthquake epicenter (triangle) and tower sites (circles)

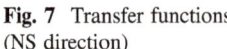
Fig. 7 Transfer functions
(NS direction)

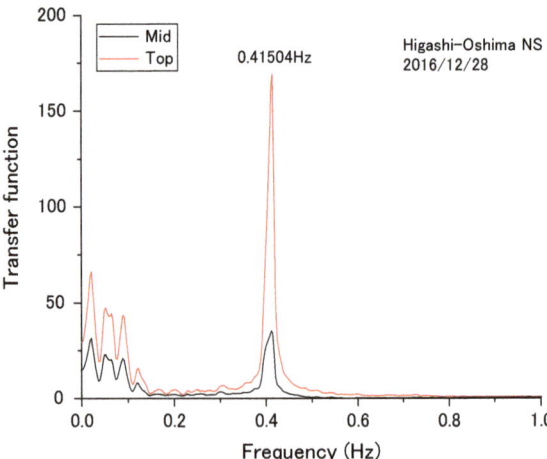

Figure 8 shows the acceleration time histories of the predominant NS direction measured at the bottom and top of Higashi Oshima Tower. As shown in Fig. 8a, the ground acceleration reached its peak at roughly 30 s, but a very small long-period component can be observed until roughly 200 s. Because of this long-period

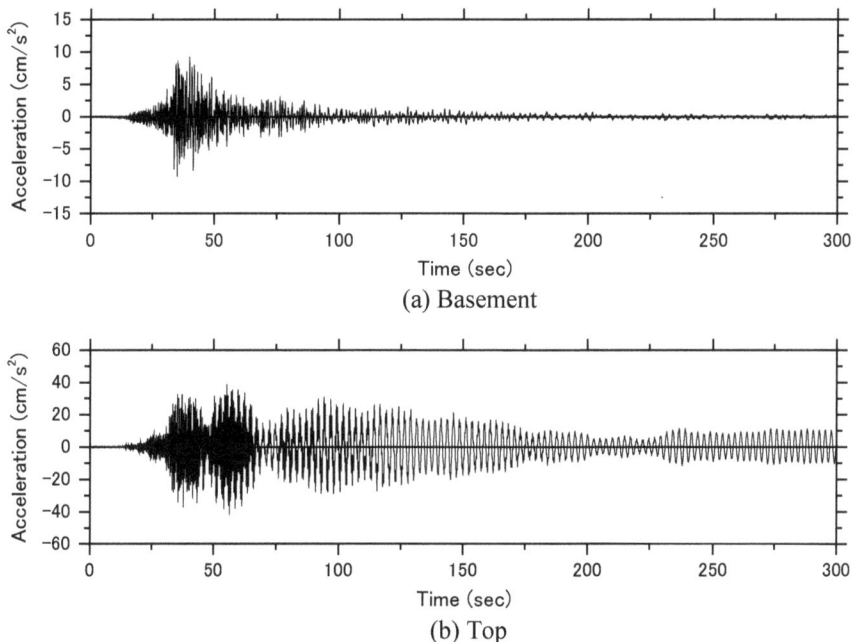

Fig. 8 Measured accelerations of Higashi Oshima Tower (NS direction)

component and the relatively light damping, the top reached its peak acceleration at roughly 55 s and was still shaking at 300 s, by which time the acceleration at the bottom was effectively zero.

4 Damage Evaluated from Measured Accelerations

The performance curve, which is the simplified base-shear–deformation relationship of an SDoF system, was derived based on the method given by Kusunoki (2016). The representative displacement $_1\Delta$ and the representative acceleration (force) $\left(_1\ddot{\Delta} + _1\ddot{x}_0\right)$ are calculated using Eqs. (1) and (2), respectively, where m_i is the mass of story i, $_1x_i$ and $_1\ddot{x}_i$ are the displacement and acceleration of the first mode at floor i relative to the basement, and $_1\ddot{x}_0$ is the ground acceleration of the first mode component.

$$_1\Delta = \frac{\sum m_i \cdot _1 x_i^2}{\sum m_i \cdot _1 x_i} \tag{1}$$

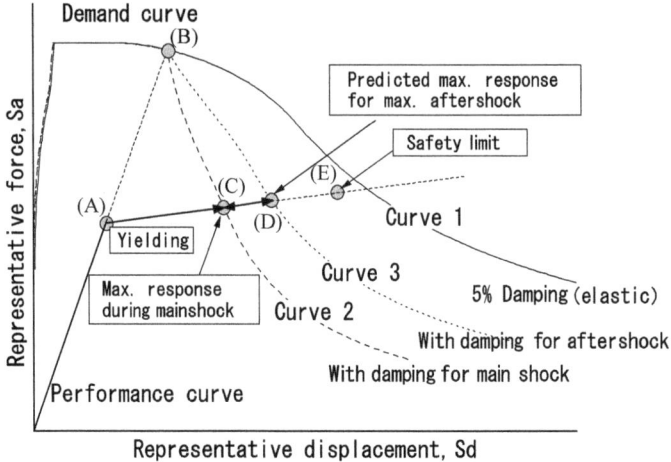

Fig. 9 Performance curve and demand curve

$$\left({}_1\ddot{\Delta} + {}_1\ddot{x}_0\right) = \frac{\sum m_i \cdot {}_1 x_i^2}{\left(\sum m_i \cdot {}_1 x_i\right)^2} \sum_{i=1}^{N} m_i \cdot {}_1 \ddot{x}_i + {}_1\ddot{x}_0 \qquad (2)$$

In the demand curve, the vertical axis represents the response acceleration spectrum and the horizontal axis represents the response displacement spectrum, as shown in Fig. 9. A viscous damping ratio of 5% is usually considered for the elastic range. When a structure shows a nonlinear response, which corresponds to the performance curve after point (A), an additional damping effect due to nonlinearity can be considered, and the demand curve is reduced accordingly as shown in Fig. 9. The intersection of the demand and performance curves at point (C) is the predicted maximum response point. With the assumption that the maximum aftershock is the same as the mainshock, the demand curve with a 5% damped maximum aftershock is the same as that for the mainshock. However, the effect of the additional damping due to nonlinearity is different because the input energy for both the mainshock and aftershock is almost double that for the mainshock only. Point (D) in Fig. 9 is then the predicted maximum response point during the aftershock. If the predicted point (D) does not go beyond the safety limit (E), the structure is evaluated as being safe for the maximum aftershock.

5 Demand Curves Derived from Measured Accelerations

There are K-Net seismograph stations close to the towers, roughly 5 km southeast of Hazawa Tower (KNG002) and roughly 2 km south of Higashi Oshima Tower (TKY013). Figure 10 shows the calculated demand curves in the NS direction from the accelerations measured at the bottom of each tower and at its nearest K-Net

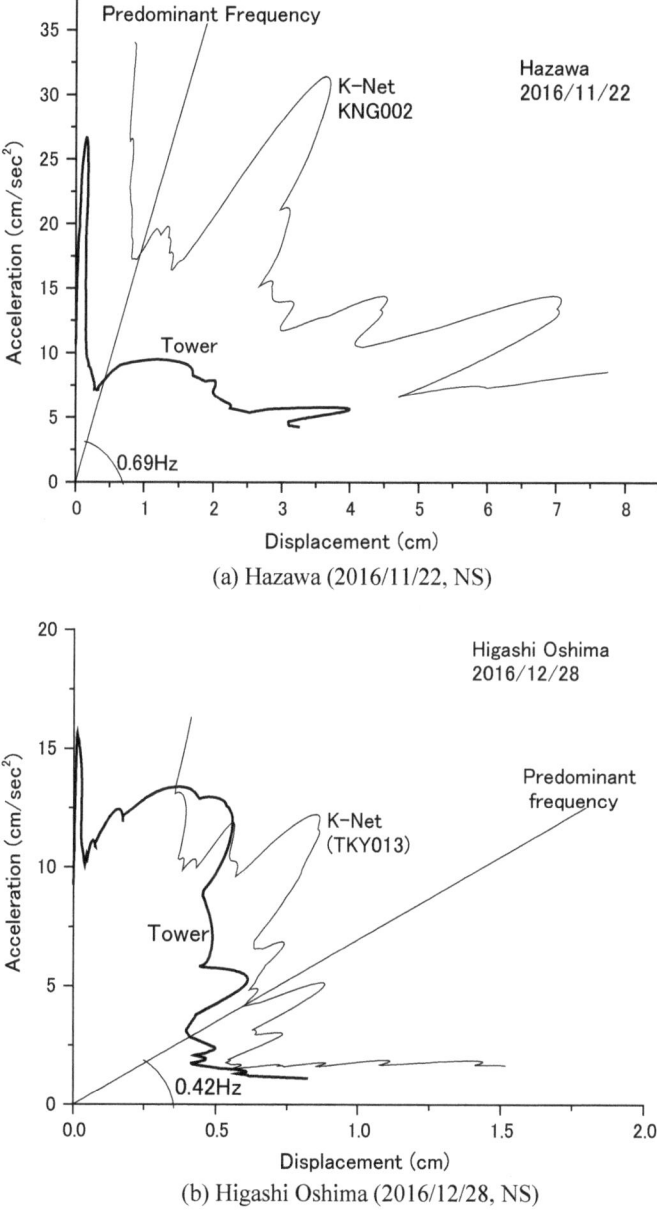

Fig. 10 Demand curves (at tower base and at nearest K-Net station)

station; the lines for the predominant frequencies are superimposed. In both cases, the demand curve for the acceleration measured at the bottom of the tower is smoother and smaller at the predominant frequency than that for the K-Net station,

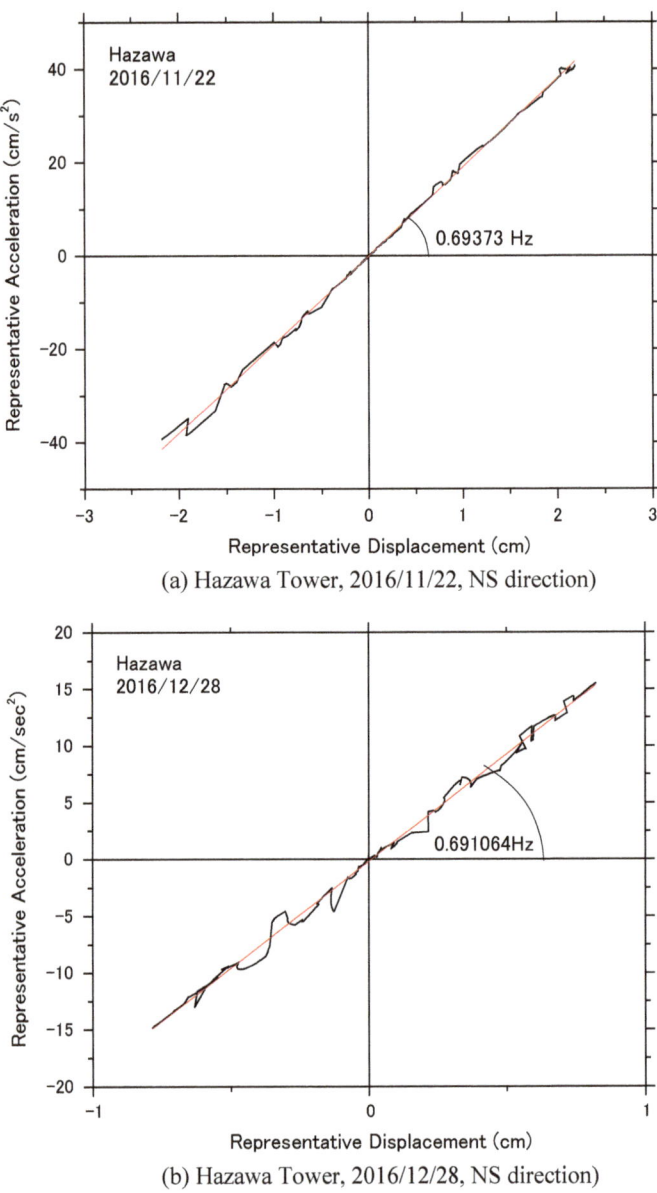

(a) Hazawa Tower, 2016/11/22, NS direction)

(b) Hazawa Tower, 2016/12/28, NS direction)

Fig. 11 Performance curves of Hazawa Tower

which was measured on the free field. From this, said it can be concluded that the effective earthquake force on a structure should not be defined according to the free-field record but rather according to the record from the base of the structure, which takes the soil–structure interaction into consideration.

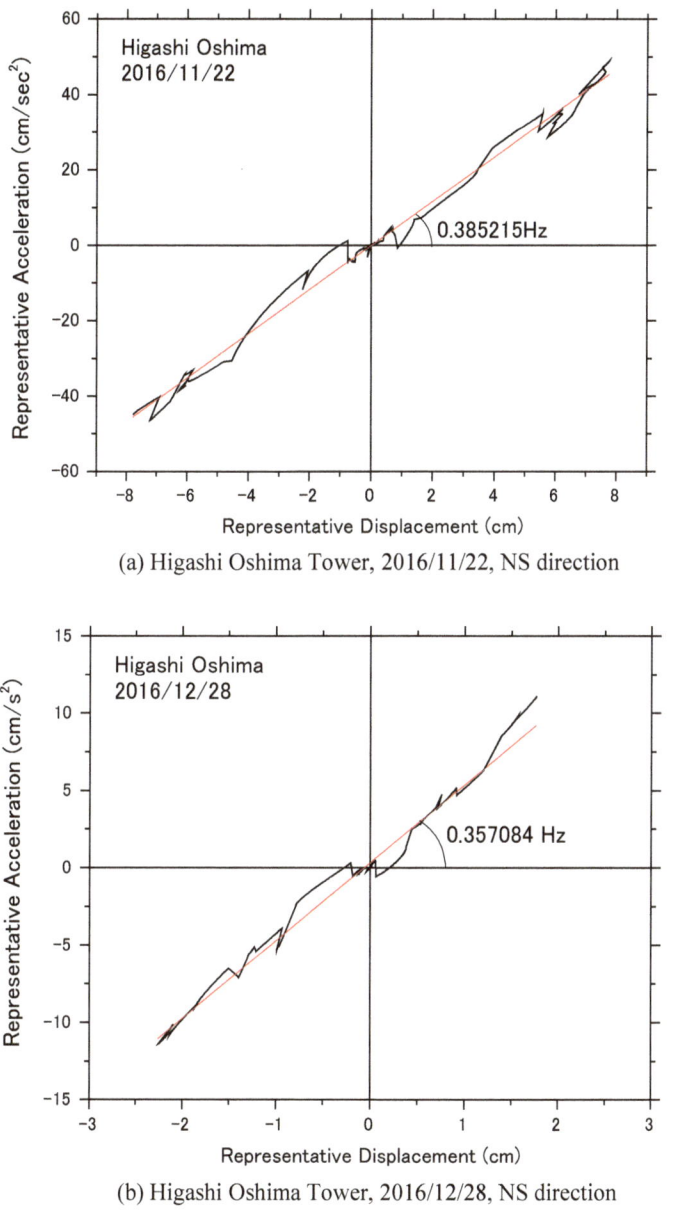

(a) Higashi Oshima Tower, 2016/11/22, NS direction

(b) Higashi Oshima Tower, 2016/12/28, NS direction

Fig. 12 Performance curves of Higashi Oshima Tower

6   Performance Curves Derived from Measured Accelerations

The calculated performance curves for Hazawa Tower and Higashi Oshima Tower during the Fukushima Ken Oki Earthquake the Ibaraki Ken North Earthquake are shown in Figs. 11 and 12, respectively. Figure 11a shows clearly that Hazawa Tower remained elastic with a predominant frequency of 0.69373 Hz for the Fukushima Ken Oki Earthquake and 0.691064 Hz for the Ibaraki Ken North Earthquake, which are very close to the value from the transfer function shown in Fig. 5. Having three sensors in Hazawa Tower ensured good accuracy.

In contrast, the performance curve shown in Fig. 12 for Higashi Oshima Tower fluctuates. The slope of the straight-line fit corresponds to a frequency of roughly 0.385 Hz for the Fukushima Ken Oki Earthquake and 0.357 Hz for the Ibaraki Ken North Earthquake, which are less than the value of 0.415 Hz obtained from the transfer function shown in Fig. 7. A possible reason for the observed fluctuations is that too few sensors were installed in the tower. It is recommended to increase the number of sensors to improve the accuracy of the evaluation.

7   Concluding Remarks

The dynamic response of two instrumented steel telecommunication towers that were in place during two major earthquakes in 2016 was investigated. The main results from the study are as follows.

- The measurement systems worked successfully.
- Because of a long-period component in each earthquake, both towers continued to oscillate even after the ground shaking had diminished or even ceased.
- The demand curves associated with the base of each tower were smaller than those of the free field (K-Net).
- The performance curves of the towers were derived successfully from the measured accelerations, which showed that the towers remained elastic.
- It is recommended that more accelerometers be placed in Higashi Oshima Tower.

Acknowledgements The author would like to thank the National Research Institute for Earth Science and Disaster Resilience for providing the data obtained by the Strong-motion Seismograph Networks (K-Net).

References

Kusunoki K (2016) Damage evaluation of a base-isolated building with measured accelerations during tohoku earthquake. In: The 16th world conference on earthquake engineering, digital

Kusunoki K, Teshigawara M (2003) A new acceleration integration method to develop a real-time residual seismic capacity evaluation system. J Struct Constr Eng 569:119–126 (in Japanese)

Kusunoki K, Teshigawara M (2004) Development of real-time residual seismic capacity evaluation system—integral method and shaking table test with plain steel frame. In: The 13th world conference on earthquake engineering, CD-Rom

Kusunoki K, Elgamal A, Teshigawara M, Conte JP (2008) Evaluation of structural condition using Wavelet transforms. In: The 14th world conference on earthquake engineering, CD-Rom

Kusunoki K, Tasai A, Teshigawara M (2012) Development of building monitoring system to evaluate residual seismic capacity after an earthquake. In: The 15th world conference on earthquake engineering, digital

Toward the Seismic Evaluation of "Carol I" Royal Mosque in Constanţa

Alexandru Aldea, Cristian Neagu, Eugen Lozinca, Sorin Demetriu, Sidi Mohammed El-Amine Bourdim and Federico Turano

Abstract Seismic protection of cultural heritage constructions is a priority in earthquake prone countries, due to their cultural, historical and touristic importance. The paper presents a first step toward the seismic evaluation of the "Carol I" Royal Mosque in Constanţa, Romania: ambient vibration measurements and the assessment of the existing damage state induced by previous earthquakes and/or other actions (index R2 according to the Romanian code for seismic evaluation of existing buildings P100-3/2008). The mosque was built in 1910–1913 and has a masonry structure with a 26 m height reinforced concrete dome. It has an approximately 40 m height RC minaret and thus it is between the first civil constructions using reinforced concrete in Romania. The construction experienced the major earthquakes of 1940 ($M_W = 7.7$) and 1977 ($M_W = 7.5$) and several other medium size events originating from Vrancea subcrustal seismic source without significant damage. However, the long-term climatic aggression had a negative impact on the structure. Ambient vibration measurements were performed in the minaret. The results will be used for the proper calibration of the computational model for linear analysis.

Keywords Cultural heritage · Minaret · Ambient vibration · Seismic evaluation

A. Aldea (✉) · C. Neagu · E. Lozinca · S. Demetriu
Technical University of Civil Engineering of Bucharest, Bucharest, Romania
e-mail: alexandru.aldea@utcb.ro

C. Neagu
e-mail: cristi.neagu@utcb.ro

E. Lozinca
e-mail: elozinca@utcb.ro

S. Demetriu
e-mail: demetriu@utcb.ro

S. M. El-Amine Bourdim
University Abdelhamid Ibn Badis, Mostaganem, Algeria
e-mail: sidimohammed.bourdim@univ-mosta.dz

F. Turano
Sapienza University, Rome, Italy
e-mail: federicoturano@libero.it

© Springer International Publishing AG, part of Springer Nature 2018   345
R. Vacareanu and C. Ionescu (eds.), *Seismic Hazard and Risk Assessment*,
Springer Natural Hazards, https://doi.org/10.1007/978-3-319-74724-8_23

1   Introduction

"Carol I" Royal Mosque is the largest mosque in Romania, built between 1910 and 1913 by the Romanian authorities at the request of King Carol I. Officially listed as a historical monument, architecture category, of national or universal interest/value, it is an important cultural asset, and a historic proof of the multicultural community in Constanța city. Located in the old city centre, the mosque is at close distance from the Archiepiscopate of Tomis, the Orthodox Cathedral (1883–1885), Sfantul Nicolae orthodox church (1889), Sfantul Anton de Padova Catholic Church (1938) and the Great Synagogue (1911). It's architect Victor Stefanescu used a mixture of Arabic and Egyptian styles, Fig. 1.

Seismic protection of cultural heritage constructions is a priority in earthquake prone countries, due to their cultural, historical and touristic value endangered by earthquake hazard. As a first step toward the seismic evaluation of the "Carol I" Royal Mosque ambient vibration measurements were performed.

The use of ambient vibration measurement for identification of structural modal parameters is a frequently used method in the last decades (Wenzel and Pichler 2005; Ivanovici et al. 2000; Quek et al. 1999; Giraldo et al. 2009, etc.). The method is applied on a wide typology of constructions: buildings (Guillier et al. 2016; Iiba et al. 2004; Aldea et al. 2007), steel towers (Castellaro et al. 2016), bridges (Bedon et al. 2016), nuclear power plants (Nour et al. 2016), etc. In case of minarets (Doğangün et al. 2006; Cakti et al. 2013; Sezen et al. 2008; Turkeli et al. 2014, etc.) ambient vibration measurements are considered for extraction of the modal parameters (natural frequencies, mode shapes and damping ratios) and for calibration of the numerical models. The results characterize the minarets' linear dynamic behaviour in case of low-amplitude vibrations.

Fig. 1 General view of "Carol I" Royal Mosque in Constanța, Romania (https://www.ghiduri-turistice.info/)

2 "Carol I" Royal Mosque

Engineer Ion Neculcea was the Mosque construction contractor who built in-between 1910 and 1913 one of the first masonry and reinforced concrete buildings in Romania. He carefully documented with photos the mosque construction. The reinforced concrete dome and minaret are often credited to engineer Gogu Constantinescu (a graduate of the École des Ponts et Chaussées in Bucharest). He had a wide international popularity being considered as one of the great inventors of that period (about 400 patents in Europe and US) with important contributions on reinforced concrete and steel structures for buildings, bridges and towers.

Documents reviewed by Pauleanu and Coman (2010) indicate a quite rapid degradation of the construction, already in 1925, some partial reparations being done. In 1940 the necessity of reparations is highlighted. This was also strongly underlined by the Great Mufti's Office of Muslim Community of Romania after an aviation bombing in 1942.

In 1957, UTCB's professor Aurel Beles (one of the founders of earthquake engineering in Romania) was called for evaluating the state of the construction and for proposing a reparation solution. In his report (Beles 1957), he stated that in the early age of reinforced concrete in Romania there was no sufficient experience related to the exposure of concrete to sea winds. Consequently, the concrete material suffered degradations questioning the mosque resistance and especially the resistance of the minaret. He found a significant concrete degradation on the Black Sea direction (N-E), in some regions the concrete cover falls down exposing the heavily rusted reinforcement. In many locations, the rusting of reinforcement bars induced concrete cracks and the separation of concrete from the reinforcement. Beles also found a poor concrete quality, with segregation, with non-homogenous composition allowing the infiltration of rain and snow water. The consequence of such water infiltrations combined with the effect of sea air with a high magnesium salt content had a destructive result on both the concrete and the reinforcement. The windows openings initially had metallic frames that rusted, cracked the concrete and allowed water infiltrations directly to the reinforcement from the window level to the base of the minaret. Consequently, a general reinforcement was recommended by Beles for preventing the collapse of the minaret. He proposed the casting of an outer reinforced concrete layer on all over the minaret height. This RC jacketing was intended to act as the main resistance element of the minaret, allowing to withstand permanent and accidental loads, especially wind action considered to be the most important load (a.n. the first earthquake resistant design regulation of Romania was introduced in 1963). Beles gave a precise description of the new RC composition and casting procedure with a peculiar attention to the adhesion with the old structure. He also recommended reparations at the balcony structure and at the staircase. Metallic windows were replaced with wooden ones, and plaster was applied all over the minaret, inside and outside. The reinforcement works and reparations were executed in 1957–1958.

According to Pauleanu and Coman (2010): in 1958 Beles inspected the site and carefully investigated the adhesion of the new casted concrete; in 1963 construction experts evaluated that the minaret design allows the simultaneous visit of 100 persons; in 1966 the Great Mufti's Office informs about serious water infiltrations in the masonry mosque, and water infiltrations in the reinforced concrete roof of the minaret balcony.

In 1993 rehabilitation was proposed and executed for the RC supporting structure of the minaret balcony roof.

In Figs. 2 and 3 are presented the layout of the mosque and a vertical cross-section.

The reinforced concrete dome (reaching a top height of 26.3 m) is supported by masonry walls and columns. The minaret has a circular shape, with a diameter of ~3.7 m at the base and ~3.0 m at the balcony level (at 24.8 m). The balcony is supported by 16 cantilever beams. The minaret has an interior spiral staircase up to the balcony level. Inside the minaret there is an internal circular reinforced concrete column having 1.75 m at the base and 0.63 m at the platform level. The reinforced concrete staircase is fixed at both ends in the outer minaret wall and in the inner tube. Up to ~9 m from the ground level the minaret is contained within the mosque itself. The outer wall has several small size windows regularly disposed from the base up to the balcony. The investigation performed by Beles in 1957 indicated an outer wall thickness of 20 cm at the base and 15 cm at the top.

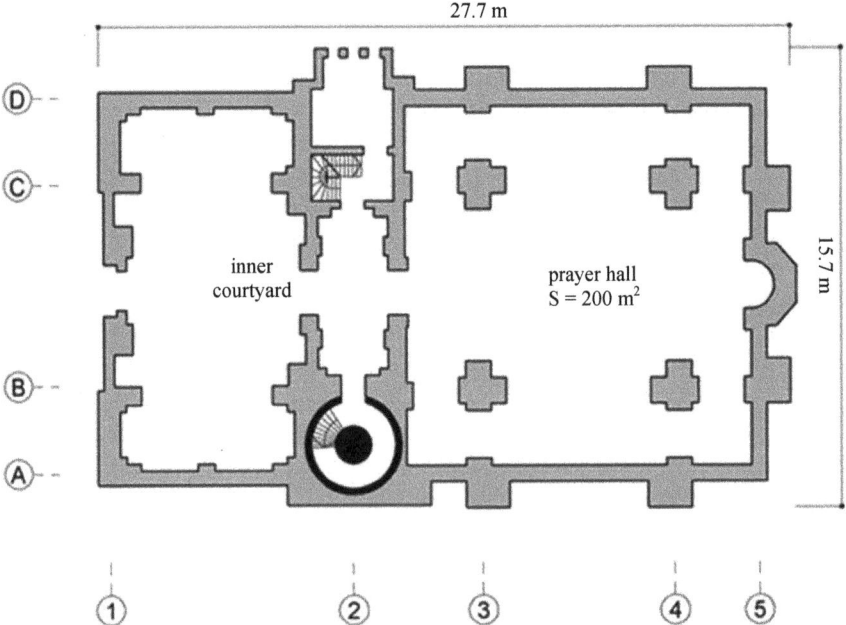

Fig. 2 The layout of "Carol I" Royal Mosque

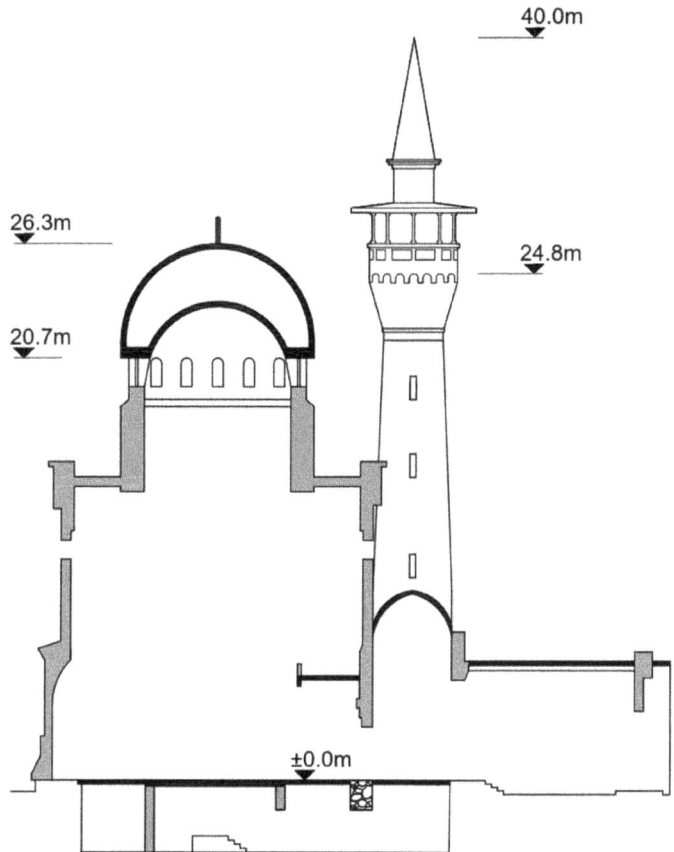

Fig. 3 Vertical cross-section of "Carol I" Royal Mosque

Beles also indicated that the reinforcement consisted of a system of 18 mm thick longitudinal bars located at the middle of the wall thickness, spaced at 20 cm at the base up to 15 cm at the top, with 8 mm thick circular stirrup at about 15–20 cm distance.

The construction experienced the major earthquakes of 1940 (M_W = 7.7) and 1977 (M_W = 7.5) and several other medium size events originating from Vrancea subcrustal seismic source (1986, M_W = 7.2, 1990, M_W = 6.9 and 6.4, 2004, M_W = 6.0) without significant damage.

The Modified Mercalli Intensity in Constanta was estimated as VII in 1940 and 1977, V in 1986, VI and V in 1990 and IV in 2004 (Konrod et al. 2013). The peak ground acceleration was estimated as $\sim$0.1 g in 1940 and 1977 (Vacareanu et al. 2015).

According to the Romanian code for seismic evaluation of existing buildings P100-3/2008, the Peak Ground Acceleration *PGA* level to be considered in Constanța is $PGA_{100year}$ = 0.16 g (value with 100 years mean return period). In the Romanian seismic design code P100-1/2013 the *PGA* level to be considered in

Constanţa is $PGA_{225year} = 0.20$ g. A recent probabilistic seismic hazard assessment study (Pavel et al. 2016) estimates a $PGA_{475year} = 0.22$ g at Constanţa.

3   Ambient Vibration Measurements

3.1   Equipment and Measurements Layouts

The ambient vibration measurement equipment used consisted of a 24 bits acquisition system and 1 s velocity sensors (frequency bandwidth 1–20 Hz) produced by Buttan Service-Tokyo & Tokyo Soil Research Co., Ltd, Fig. 4.

The ambient vibration measurements were performed in May 2017 using three sensor disposal layouts, Fig. 5, using simultaneously 6 to 8 velocity sensors. For each measurement layout, two samples of 3 min were recorded using a 100 Hz sampling frequency.

3.2   Data Analysis and Preliminary Results

Fourier spectral analysis was used to estimate modal frequencies of the minaret from ambient vibration time-history records. This approach in frequency domain is used for identifying modal characteristics (Iiba et al. 2004; Kohler et al. 2005; Demetriu et al. 2012).

In Fig. 6 is presented an example of time velocity records from layout 2.

The Fourier amplitude spectra for the records obtained in Layout 1 and Layout 2 sensors disposal schemes are presented in Figs. 7 and 8. In Fig. 9 are shown the spectra for the records in Layout 3.

Fig. 4 Equipment for ambient vibration measurements—velocity sensor and acquisition station (Buttan Service-Tokyo & Tokyo Soil Research Co., Ltd)

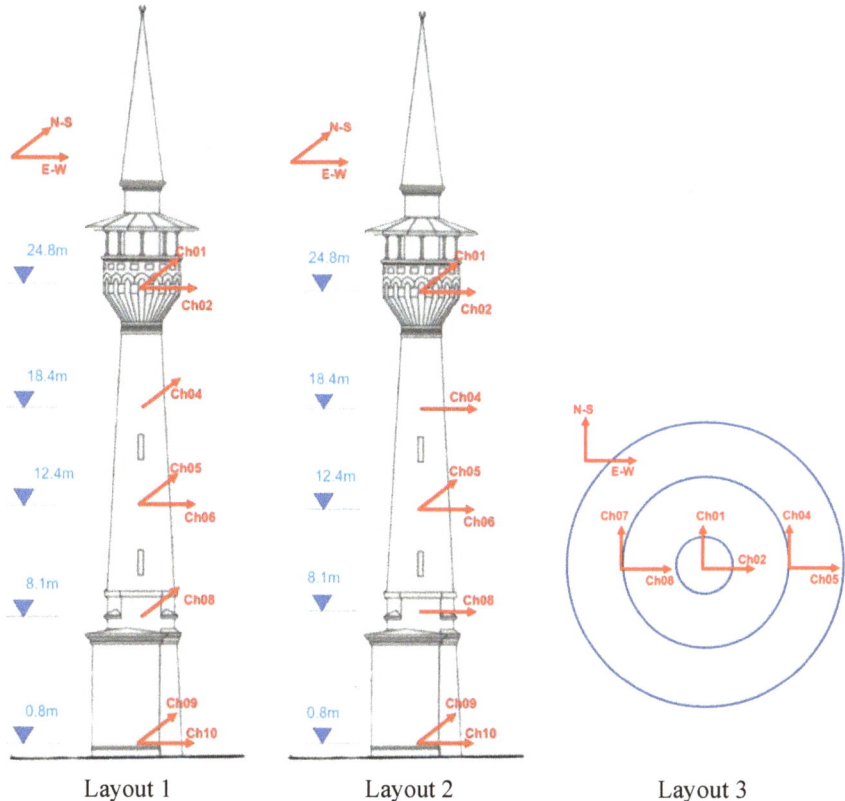

Layout 1   Layout 2   Layout 3

Fig. 5 Ambient vibration measurements—sensors layouts

The spectral peaks identified in Figs. 7, 8 and 9 indicate stable dynamic characteristics in all measurements and layouts.

As expected, the frequencies corresponding to peak spectral amplitudes on the both measurement directions show minor differences.

The first modal frequencies of the minaret identified from ambient vibration measurements are: $f_1 = 2.38$ Hz $(T_1 = 0.42$ s$)$, $f_2 = 3.80$ Hz $(T_2 = 0.26$ s$)$.

4   Damage Evaluation of the Structural System

In Romania, first provisions dealing with seismic evaluation of existing buildings were enforced by two chapters in the seismic code P100-1992 that was mainly used for the seismic design of new buildings. Based on these provisions a sustained activity of seismic evaluation was carried out for a large number of buildings, especially for those belonging to public institutions.

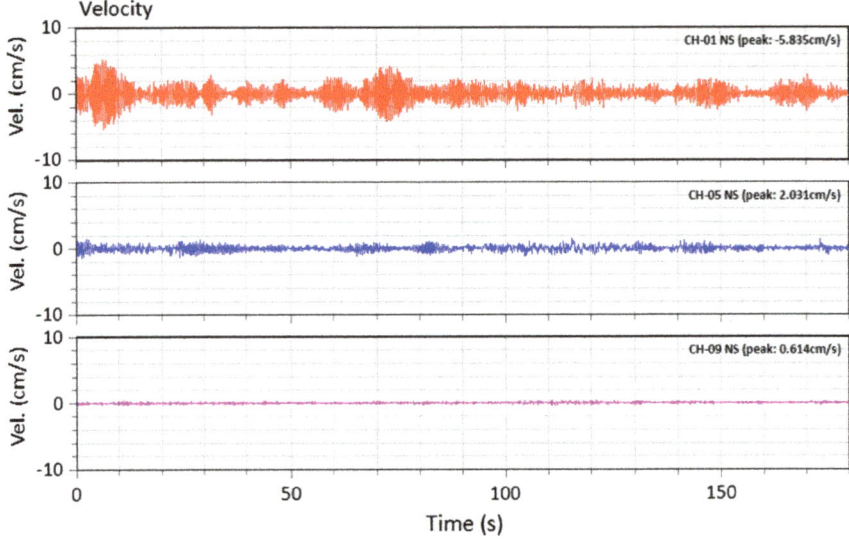

Fig. 6 Example of time-history velocity records—sensors Layout 2

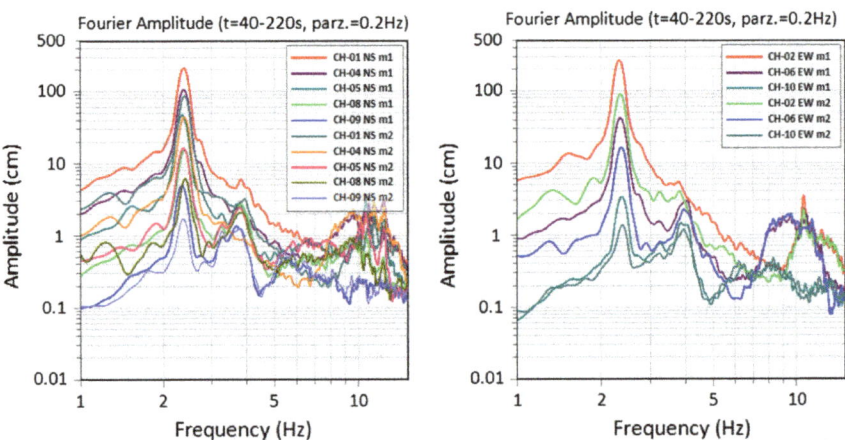

Fig. 7 Fourier amplitude spectra—sensors Layout 1

Based on the continuous progress in the field of earthquake engineering, these two chapters were upgraded in 2008 to an independent code that is presently used for seismic evaluation of existing buildings in Romania. The European Standard EN 1998-3:2005, as well as the concepts and provisions from FEMA documents were used to elaborate the Romanian code P100-3/2008, (Postelnicu et al. 2010).

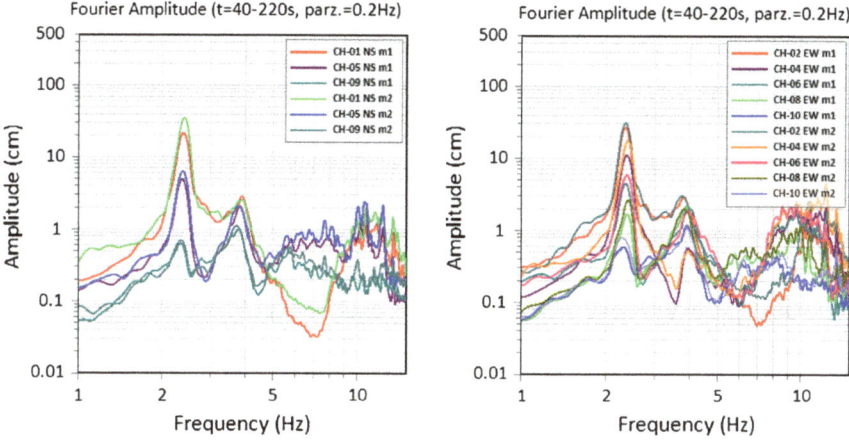

Fig. 8 Fourier amplitude spectra—sensors Layout 2

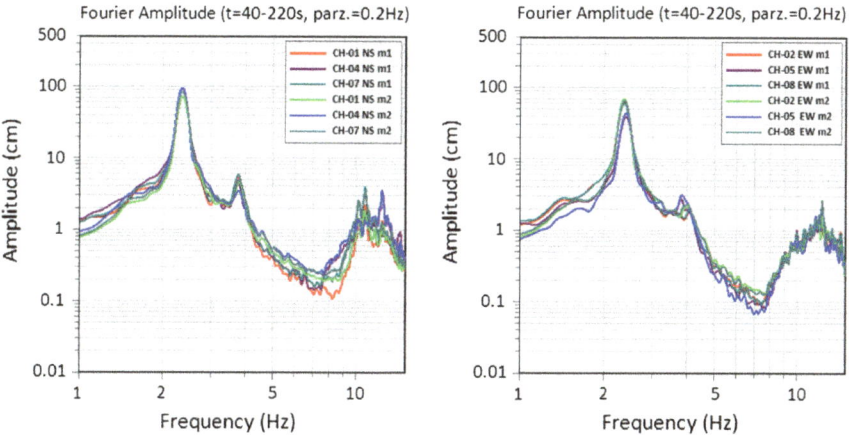

Fig. 9 Fourier amplitude spectra—sensors Layout 3

According to P100-3/2008, the seismic risk class of an existing building is established based on the values of three different factors associated to three evaluation criteria: (i) the qualitative evaluation that check the seismic conditions for the structural system and the details and connections of structural elements (index *R1*); (ii) the damage evaluation that aims to assess the existing damage state induced by previous earthquakes and/or other actions (index *R2*) and (iii) the quantitative evaluation, based on structural seismic analysis and explicit verifications of strength and deformability capacity of the structural elements in respect to the seismic requirements (index *R3*).

According to the provisions of the Romanian code P100-3/2008, in case of masonry buildings the in situ damage assessment aims to identify some types of degradations, while for RC structures there are other types of damages that should be discovered.

The in situ inspections of the Royal Mosque revealed significant damage in the structural members that reduces their strength to both gravity and lateral loads and also the overall lateral stiffness.

4.1 Damage Degree Index for Masonry Structure of the Mosque

Many vertical and inclined cracks were observed both on the interior and the exterior faces of the masonry walls of the mosque. While some of these cracks are related to the climatic factors, most of the inclined cracks were caused by horizontal tremors induced by the major earthquake events in the 20th century originating from Vrancea seismic source.

Two important wide-opened cracks can be easily identified on both the longitudinal exterior walls as they develop on the entire height of the mosque as can be seen in Figs. 10 and 11. Caused by differential settlement of the foundations, these cracks continue to develop in the bare ground flooring and in the interior masonry walls in the basement of the mosque.

Another visible phenomenon that causes wide-spread structural and non-structural damage is the rain intrusion through roof. During the site inspections, we were able to identify several zones where the ceramic tiles are broken and permits water infiltrations. This long-lasting phenomenon is responsible not only for damaging the wooden structure of the roof (Fig. 12), but also for significant degradations of the dome. The collapsed non-structural ceiling in Fig. 13 revealed a heavily damaged reinforced concrete beam where the entire concrete cover layer is expelled and the steel rebars are severely corroded.

Fig. 10 Wide-opened cracks in a masonry wall in the basement and in the bare ground flooring

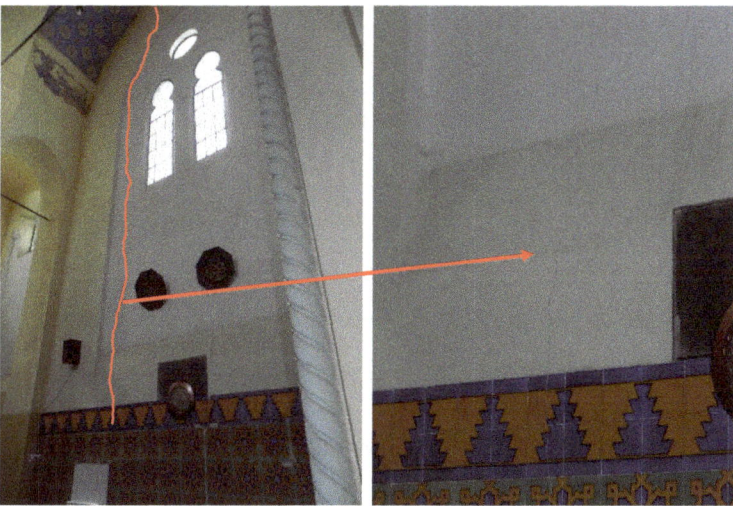

Fig. 11 Vertical cracks developed on the entire height of the masonry wall

Fig. 12 Broken tiles and damaged roof in the lower part of the dome

According to the provisions of the Romanian code P100-3/2008, in case of masonry structure of the mosque the in situ damage assessment aimed to identify the following types of degradations: inclined or X-shaped cracks in structural or parapet walls; vertical cracks in lintels and brick masonry arches; crushing of the bricks due to excessive local compression; vertical cracks at intersections that reveals the tendency for out-of-plane behaviour. Considering the extent and the severity of these degradations, different values for the damage degree index were determined according to the provisions from P100-3/2008 for horizontal and vertical masonry elements. Finally, the overall damage degree index for the masonry structure of the Royal Mosque resulted as $R_2^{mosque} = 50$ from a maximum value of 100 when the investigated building does not show any damage.

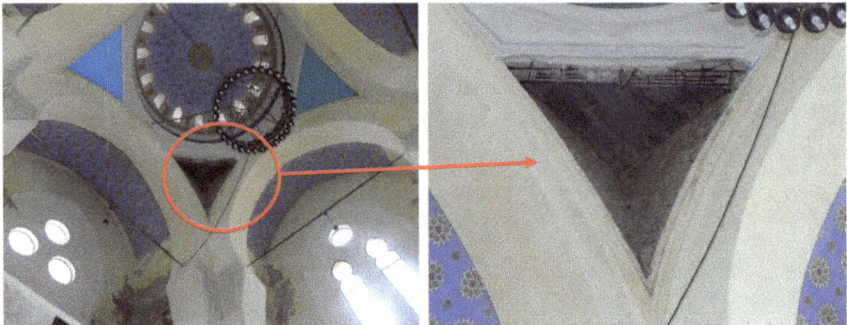

Fig. 13 Fallen ceiling that revealed a heavily damaged RC beam in the lower part of the dome

4.2 *Damage Degree Index for RC Minaret*

In its first decades, the RC minaret suffered progressive degradation due to the exposure to sea winds. Therefore, it was retrofitted twice in 1957–1958 and 1993. In consequence, nowadays, the RC minaret exhibits only some minor damages. Some superficial stress concentrations cracks were observed in the exterior RC jacket near the windows openings (Fig. 14) and few cracks associated with the freeze-thaw cycles were detected in the plastering layer of the rehabilitated RC supporting structure of the minaret balcony roof, as can be seen in Fig. 15.

In the case of the minaret, the site inspections focused on other types of damages, specific to the reinforced concrete structures normal or inclined cracks and

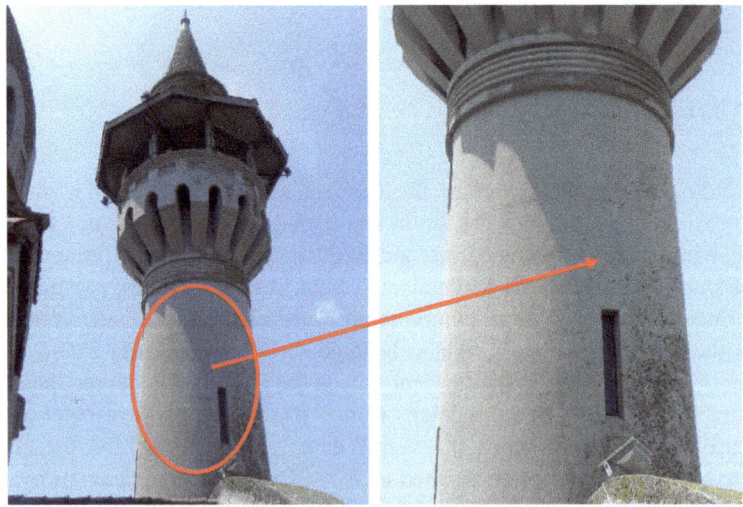

Fig. 14 Minor cracks in the exterior jacketing layer of the RC minaret of the mosque

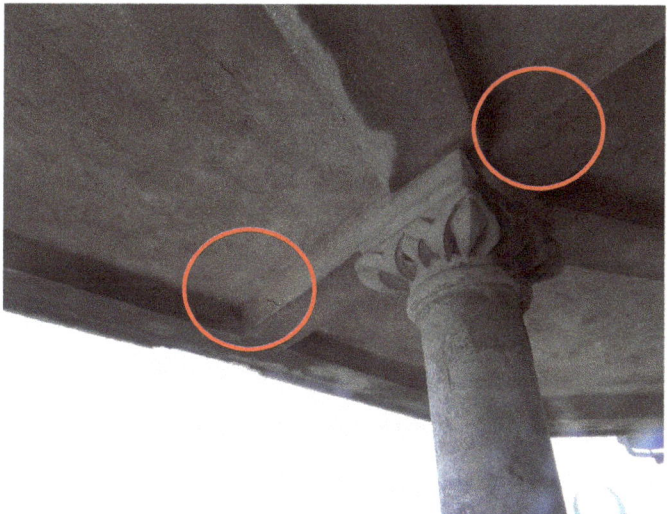

Fig. 15 Cracks in the plastering layer of the RC beams that supports the roof of the minaret balcony

residual deformations in the critical zones of the structural elements splitting cracks at lapped-splices bond failure of the rebars, spalling of concrete cover, etc. Penalizing the RC tower for the degradations caused by climatic factors and the major earthquakes in the 20th century, the damage degree index determined according to P100-3/2008 provisions was $R_2^{minaret} = 82$.

5 Conclusions

The "Carol I" Royal Mosque in Constanța was exposed to a long term climatic aggression, and several major and moderate Vrancea subcrustal earthquakes. It was subject of several reparations of different extent in 1925, 1957 and 1993. The visual inspection of the mosque revealed cracks in most of the masonry, some of them quite significant, and a continuous degradation due to water infiltrations.

The damage assessment revealed that nowadays the masonry structure of the mosque is much more affected than the already retrofitted RC minaret. Considering the minaret as an independent structure and taking into account only the damage degree index R_2, it should be ranked in the seismic class R_sIII that is associated to the buildings that after the design level earthquake might suffer some structural damage not affecting the overall structural safety. On the other hand, the damage degree index determined for the masonry structure of the mosque corresponds to seismic risk class R_sII that comprises those buildings that might suffer major structural and non-structural damages, but for which the structural collapse is slightly probable.

The numerous cracks observed during the site inspections reduce the overall lateral stiffness of the structure. Thus, the ambient vibration measurements become quite useful for calibration of the numerical models.

Future studies are necessary for establishing the extent of material degradation due to climatic actions and the countermeasures to be taken.

The ambient vibration measurements allowed the identification of modal frequencies of the 40 m height reinforced concrete minaret, characterizing its linear dynamic behaviour with small vibration amplitude range. No significant directional variations of the modal frequencies were observed. It is the first step toward the detailed and complete seismic evaluation of the Carol I Royal Mosque in Constanţa, Romania.

Acknowledgements The authors acknowledge the kind help of the Great Mufti's Office of Muslim Community of Romania, who accepted the study and provided all necessary support during the in situ investigation and ambient vibration measurements.

Japan International Cooperation Agency JICA is acknowledged for the donation of ambient vibration equipment. "Eugen Ionescu" and "Erasmus+" scholarship programs of Romanian Government and, respectively, of European Commission are acknowledged for supporting the foreign authors studies in Romania.

The present paper is an extended version (the damage evaluation was added) of the paper published in the Proceedings of the 6th National Conference on Earthquake Engineering & 2nd National Conference on Earthquake Engineering and Seismology, (Aldea et al. 2017).

References

Aldea A, Demetriu S, Albota E, Kashima T (2007) Instrumental response of buildings. Studies within JICA Project in Romania. In: Proceedings of ISSRR2007 International Symposium on seismic risk reduction, The JICA technical cooperation project in Romania, pp 157–170

Aldea A, Neagu C, Lozinca E, Demetriu S, Bourdim SM, Turano S (2017) Toward the seismic evaluation of "Carol I" Royal Mosque in Constanta—ambient vibration measurements. In: 6th national conference on earthquake engineering & 2nd national conference on earthquake engineering and seismology, Conspress Publishing House, pp 177–184

Bedon C, Dilena M, Morassi A, (2016) Ambient vibration testing and structural identification of a cable-stayed bridge. Meccanica 51

Beles A (1957) Technical report for the reinforcement of the minaret (in Romanian), p 3

Cakti E, Oliveira CS, Lemos JV, Saygili O, Gork S, Zengin E (2013) Earthquake behavior of historical minarets in Istanbul. In: Papadrakakis M, Papadopoulos V, Plevris V (eds) COMPDYN 2013, 4th ECCOMAS thematic conference on computational methods in structural dynamics and earthquake engineering, Kos Island, Greece

Castellaro S, Perricone L, Bartolomei M, Isani S (2016) Dynamic characterization of the Eiffel tower. Engr Struct 126:628–640

Demetriu S, Aldea A, Udrea A (2012) Modal parameters of a RC frame structure identified from ambient vibration measurements. 15th World Conference on Earthquake Engineering, Lisbon

Doğangün A, Acar R, Livaoğlu R, Tuluk OI (2006) Performance of masonry minarets against earthquakes and winds in Turkey. In: 1st international conference on restoration of heritage masonry structures, Cairo, Egypt

Giraldo D, Song W, Dyke S, Caicedo J (2009) Modal identification through ambient vibration: comparative study. J Eng Mech 135(8):759–770

Guillier B, Chatelain JL, Perfettini H (2016) Building frequency fluctuations from continuous monitoring of ambient vibrations and their relationship to temperature variations. Bull Earthquake Eng 14

Iiba M, Watakabe M, Fujii A, Koyama S, Sakai S, Morita K (2004) A study on dynamic soil structure interaction effect based on microtremor measurements of building and surrounding ground surface. Building Research Institute, Tsukuba, Japan, p 17

Ivanovic SS, Trifunac MD, Todorovska MI (2000) Ambient vibration tests of structures—a review. ISET J Earthq Techn 37(4)

Kohler MD, Davis PM, Safak E (2005) Earthquake and ambient vibration monitoring of the steel-frame UCLA factor building. Earthq Spectra 21(3):715–736

Kronrod T, Radulian M, Panza G, Popa M, Paskaleva I, Radovanovich S, Gribovszki K, Sandu I, Pekevski L (2013) Integrated transnational macroseismic data set for the strongest earthquakes of Vrancea (Romania). Tectonophysics 590:1–23

Nour A, Cherfaoui A, Gocevski V, Léger P (2016) Probabilistic seismic safety assessment of a CANDU 6 nuclear power plant including ambient vibration tests: case study. Nucl Eng Des 304:125–138

P100-1992, (Romanian) Normative for earthquake resistant design of housing, socio-cultural, agricultural and industrial constructions, MLPAT

P100-1/2013, (Romanian) Seismic design code—Part I provisions for building design, MDRAP

P100-3/2008, (Romanian) Code for the seismic evaluation of existing buildings, MDLPL

Pauleanu D, Coman V (2010) Royal mosque "Carol I" constanta 1910–2010. Ed. Ex Ponto Constanta, p 195

Pavel F, Vacareanu R, Douglas J, Radulian M, Cioflan CO, Barbat A (2016) An updated probabilistic seismic hazard assessment for Romania and comparison with the approach and outcomes of the SHARE project. Pure Appl Geophys 173:1881–1905

Postelnicu T, Lozincă E, Pascu R (2010) The New Romanian code for seismic evaluation of existing buildings. Sixth International Conference on Concrete under Severe Conditions, Environment and Loading, June 7–9, 2010, Mérida, Mexic

Quek ST, Wang WP, Koh CG (1999) System identification on linear MDOF structures under ambient excitation. Earthq Eng Struct Dynam 28:61–77

Sezen H, Acar R, Dogangun A, Livaoglu R (2008) Dynamic analysis and seismic performance of reinforced concrete minarets. Eng Struct 30:2253–2264

Türkeli E, Livaoğlu R, Doğangün A (2014) Dynamic response of traditional and buttressed reinforced concrete minarets. Engineering Failure Analysis

Vacareanu R, Radulian M, Iancovici M, Pavel F, Neagu C (2015) Fore-arc and back-arc ground motion prediction model for Vrancea intermediate depth seismic source. J Earthq Eng 19: 535–562

Wenzel H, Pichler D (2005) Ambient vibration monitoring. John Wiley & Sons Publishing company https://www.ghiduri-turistice.info/ghid-turistic-5-motive-sa-vizitezi-constanta

Damage Due to Earthquakes and Improvement of Seismic Performance of Reinforced Concrete Buildings in Japan

Masaki Maeda and Hamood Al-Washali

Abstract Japan, which is one of the most earthquake prone countries in the world, has suffered from damaging earthquakes repeatedly and learned lessons from damages. First, history of damages to existing reinforced concrete (RC) buildings due to previous earthquakes and seismic code revision are summarized. Secondly, basic concept and procedure of Japanese seismic evaluation method were outlined and seismic capacity index, I_s, of buildings suffered Kobe Earthquake. Strong correlation between damage level and seismic capacity index, I_s, was found. After the 1995 Kobe Earthquake, the law for promotion of seismic evaluation and retrofit was enforced based on the lessons learnt from the damage and investigation. Seismic evaluation and retrofit were widely applied to existing RC buildings in all over Japan and contributed to improvement of seismic capacities of existing RC buildings designed by old seismic code. The improvement was proved by recent major earthquakes such as 2011 Tohoku Earthquake and 2016 Kumamoto Earthquake. Typical damage pattern, failure modes and tendency in each earthquake were introduced and effectiveness of seismic evaluation and retrofit was discussed.

Keywords RC existing building · Seismic evaluation · Seismic retrofit Earthquake damage

1 Introduction

Japan, which is one of the most earthquake prone countries in the world, has suffered from damaging earthquakes repeatedly and learned lessons from damages. First, history of damages to existing reinforced concrete (RC) buildings due to

M. Maeda (✉) · H. Al-Washali
Faculty of Architecture and Building Science, Graduate School of Engineering,
Tohoku University, Aobayama 6-6-11, Aoba-Ku, Sendai, Japan
e-mail: maeda@archi.tohoku.ac.jp

H. Al-Washali
e-mail: Hamood@rcl.archi.tohoku.ac.jp

© Springer International Publishing AG, part of Springer Nature 2018 361
R. Vacareanu and C. Ionescu (eds.), *Seismic Hazard and Risk Assessment*,
Springer Natural Hazards, https://doi.org/10.1007/978-3-319-74724-8_24

previous earthquakes and seismic code revision are summarized. Secondly, basic concept and procedure of Japanese seismic evaluation method were outline and seismic capacity index, I_s, of buildings suffered Kobe Earthquake. Strong correlation between damage level and seismic capacity index, I_s, was found. After the 1995 Kobe Earthquake, the law for promotion of seismic evaluation and retrofit was enforced based on the lessons learnt from the damage and investigation. Seismic evaluation and retrofit were widely applied to existing RC buildings in all over Japan and contributed to improvement of seismic capacities of existing RC buildings designed by old seismic code. As a result of upgrade of seismic capacity for old buildings, damage to structural elements was remarkably decreased. Typical damage pattern, failure modes and tendency in each earthquake were introduced and effectiveness of seismic evaluation and retrofit was discussed.

The Japanese seismic design codes for buildings were revised in 1971 and 1981. Specifications such as maximum spacing of hoops of reinforced concrete columns were revised to increase structural ductility in 1971, whereas the verification on the ultimate lateral load carrying capacity of designed structure by limit state or pushover analysis considering deformation capacity of members was required in 1981.

2   History of Damaging Earthquake and Seismic Code Revision in Japan

Japan has suffered from earthquakes repeatedly and seismic code was revised based on the lessons. Table 1 shows history of damaging earthquakes and seismic code revision. The 1923 Great Kanto Earthquake induced catastrophic damage to buildings in Tokyo area and casualties of over 140,000. It resulted in introduction of seismic design to building design code in 1924. Japanese Building Standard Law was enacted in 1950, five years after the end of the World War II. Working stress design was the basic concept of seismic design with seismic force of 20% of building weight. Allowable stresses were nominal yield strength of rebar and two third of concrete compressive strength.

The 1968 Tokachi-oki earthquake induced heavy damage and collapse of many reinforced concrete (RC) buildings as shown Fig. 1. Brittle shear failure in columns were most typical damage and considered the main reason of collapse. Therefore, requirement of hoop spacing in columns were upgraded from 30 to 10 cm in order to prevent shear failure in 1971. At the same time, Japanese government established a national project for improvement of seismic capacity and seismic design including universities, national research institutes and construction companies. As a result of ten years project, ultimate state design was introduced to seismic code in 1981 revision. In addition, Standard for Seismic Evaluation of Existing Reinforced Concrete Buildings (Seismic Evaluation Standard) was published by Japan Building Disaster Prevention Association (JBDPA) in 1977.

Table 1 History of damaging earthquakes and seismic code revision in Japan

Year	Name earthquake and seismic code revision	Magnitude (M)	Casualty
1891	Nohbi EQ	8.0	7273
1923	Kanto EQ	7.9	140,000
1924	Building code revision (introduction of seismic design)		
1948	Fukui EQ	7.3	3895
1950	Building standard law (working stress design)		
1964	Niigata EQ	7.5	26
1968	Tokachi-oki EQ	7.9	52
1971	Seismic code revision		
1975	Ohita Chubu EQ	6.4	0
1977	Seismic Evaluation Standard	7.4	28
1978	Miyagi-ken-oki EQ		
1981	Revision of seismic code (ultimate strength design)		
1983	Nihon-kai Chubu EQ	7.7	104
1993	Kushiro-oki EQ	7.8	2
1994	Hokkaido-nansei-oki EQ	7.8	230
1995	Hokkaido-toho-oki EQ	8.1	0
	Sanriku Far-off EQ	7.5	3
	Kobe EQ	7.2	6434
1995	Law for promotion of seismic evaluation and retrofit		
2000	Revision of seismic code (Performance based design)		
2004	Niigata Chuetsu EQ	6.8	68
2008	Iwate Miyagi EQ	7.2	17
2011	Great East Japan EQ	9.0	18,446 (incl. missing)
2016	Kumamoto EQ	7.3	88

Fig. 1 Collapse of RC buildings in 1968 Tokachi-oki earthquake

Fig. 2 Damage statistics of
RC school buildings (1995
Kobe earthquake)

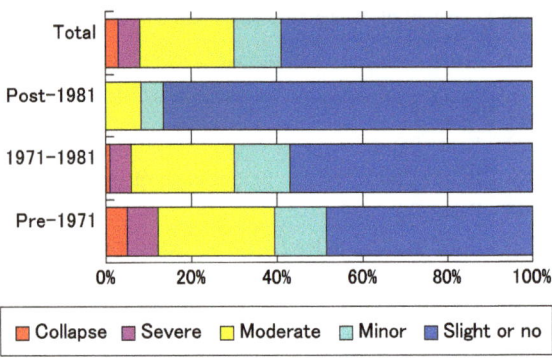

After code revision in 1981, number of casualty by earthquake disaster was
remarkably reduced, except by tsunami, as shown in Table 1. However, the 1995
Great Hansin Earthquake (Kobe Earthquake) revealed vulnerability of existing
buildings designed by old seismic code. Figure 2 shows damage statistics for RC
school buildings in affected area (Okada et al. 2000). Significant difference in
damage ratio was found between construction ages. Forty percent of the buildings
before 1971 suffered moderate or severer damage. On the other hand, damage to the
building after 1982 was limited and ninety percent remained slight or less damage.
It is obvious evidence of a success of improvement of seismic capacity of buildings
in japan by the seismic code revision in 1971 and 1981. Moreover, importance of
seismic evaluation and retrofit was widely recognized. Law for Promotion of
seismic Evaluation and Retrofit was enacted at the end of 1995, and it was a start
point of general application of seismic evaluation and retrofit for existing buildings.

3   Seismic Evaluation and Retrofit for Existing
RC Buildings

The Japanese Seismic Evaluation Standard (JBDPA 1977) consists of three procedures of different levels, i.e., first, second and third level procedures. The first
level procedure is the simplest but most conservative since only the sectional areas
of columns and walls and concrete strength are considered to calculate the strength,
and the inelastic deformability is neglected. In the second and third level procedures, the ultimate lateral load carrying capacity of vertical members or frames is
evaluated using material and sectional properties together with reinforcing details
based on field inspections and structural drawings.

In the Standard, the seismic performance index of a building is expressed by the
I_s index for each story and each direction, as shown in Eq. (1)

$$I_s = E_0 \times S_D \times T \qquad (1)$$

where, E_0: basic structural seismic capacity index calculated from the product of strength index (C), ductility index (F), and story index (I_s) at each story and each direction when a story or building reaches the ultimate limit state due to lateral force, i.e.,

Strength index C index of story lateral strength calculated from the ultimate story shear in terms of story shear coefficient.

Ductility index F index of ductility calculated from the ultimate deformation capacity normalized by the story drift of 1/250 when a standard size column is assumed to fail in shear. F depend on the failure mode of the structural members and their sectional properties such as bar arrangement, shear-span-to-depth ratio, shear-to-flexural-strength ratio, etc. In the standard, F is assumed to vary from 1.27 to 3.2 for ductile columns, 1.0 for brittle columns and 0.8 for extremely brittle short columns (shear-span-to-depth ratio less than 2).

ϕ index of story shear distribution during earthquake estimated by the inverse of design story shear coefficient distribution normalized by base shear coefficient. A simple formula of is basically employed for the i-th story level of an n-storied building by assuming inverted triangular shaped deformation distribution and uniform mass distribution.

S_D factor to modify E_0-Index due to stiffness discontinuity along stories eccentric distribution of stiffness in plan, irregularity and/or complexity of structural configuration, basically ranging from 0.4 to 1.0.

T reduction factor to allow for the deterioration due to age after construction fire and/or uneven settlement of foundation, ranging from 0.5 to 1.0

Figure 3 shows the relationship between the second level seismic performance indices I_{s2} of RC school buildings suffered from Kobe Earthquake and construction age. The Seismic Evaluation Standard recommends as the demand criterion of the second level procedure that I_{s2} Index higher than 0.6 should be provided with buildings to prevent major structural damage or collapse. This criterion is based on the correlation study from the past earthquake damage and the calculated indices for the damaged buildings. Past experiences of 1968 Tokachi-Oki, 1978 Miyagi-ken-Oki and other earthquakes reported that buildings with I_{s2} indices higher than 0.6 suffered from moderate or less damage. As can be found in Fig. 2, I_{s2} indices for most of the buildings constructed before 1971 were less than 0.6, whereas they were more than 0.6 for those constructed after 1981. As mentioned earlier, the Japanese seismic design codes for buildings were revised in 1971 and 1981.

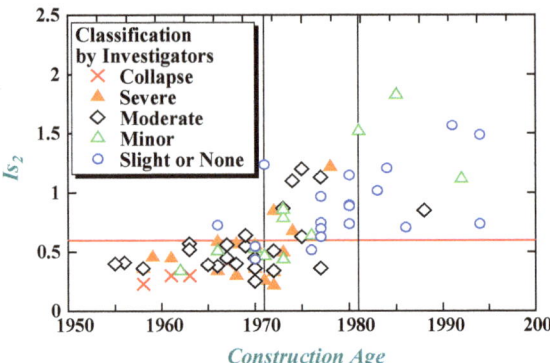

Fig. 3 I_s-index and
construction age of RC school
buildings (1995 Kobe
earthquake)

The results shown in Fig. 3 indicated that seismic capacities of reinforced concrete
school buildings in Japan were successfully improved due to the revisions of
seismic design codes.

4 Damage to RC Building Due to Recent Major Earthquakes

4.1 Damage Due to 2011 Great East Japan Earthquake

The Great East Japan Earthquake struck Tohoku region on March 11, 2011. Huge
tsunami attacked coastal area and more than 20,000 people were killed. On the
other hand, damage to building structure was relatively limited. That is attributed to
seismic evaluation and retrofit of vulnerable buildings that has been widely applied
to existing buildings, as mentioned in Chap. 2, after the 1995 Kobe Earthquake. As
a result, more than 90% of school buildings in Miyagi prefecture, which are located
in the centre of Tohoku region and nearest prefecture from epi-center, are reported
to satisfy the criteria of the seismic evaluation standard (Maeda et al. 2012a, b).

Figure 4 shows the damage ratio of 520 RC school buildings in Miyagi
Prefecture suffered from the 2011 East Japan Earthquake. In the figure, first generation (pre-1971) and second generation (1972–1981) buildings are classified into
two groups; (a) un-retrofitted and (b) retrofitted or evaluated safe. Tendency of
damage ratio of un-retrofitted is similar with those found in the 1995 (Fig. 1),
although damage ratio of severer damage is fewer. Most of the buildings, which
suffered from serious damage, were designed and constructed before 1981, and
especially those before 1971 had extensive damage. On the other hand, most new
buildings designed according to the current seismic codes enforced in 1981 showed
fairly good performance and prevented severe structural damage. Most of the
buildings before 1981, if retrofitted or evaluated safe, escaped damage as ca be

Fig. 4 Damage statistics of RC school buildings in Miyagi prefecture

Fig. 5 *Is*-index and construction age of RC school buildings (2011 East Japan earthquake)

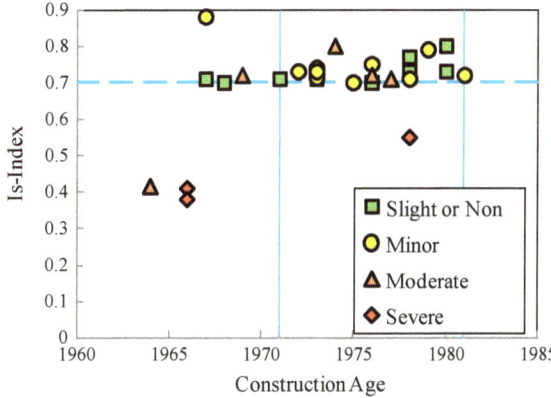

found in the Fig. 4. It is an evidence of effectiveness of seismic evaluation and retrofit.

The Seismic Evaluation Standard recommends as the demand criterion that I_s-Index higher than 0.6 should be provided to prevent major structural damage or collapse. I_s-Index of school buildings is demanded higher value (0.7) than normal buildings. It is because that school buildings require not only the security of safety but also the security of function to use buildings without repairing structural damage after big earthquake. As can be found in Fig. 5, I_s-Indices for most of the buildings were more than 0.7 and prevented severe structural damage even if they were old buildings. Figure 6 shows the relationship between I_s-Index and damage level indices R-Index proposed in "Standard for Post-earthquake Damage Level Classification of Reinforced Concrete Building" (JBDPA 1990). A good correlation was observed between calculated I_s-index and observed damage. Most buildings with I_s-values lower than 0.6 were vulnerable to moderate and severe damage. Most of the buildings with I_s-values higher than 0.7 avoided severe damage and had minor and slight damage (R > 80). I_s-Index of 0.7 is generally regarded as an effective demand criterion for screening seismically vulnerable buildings.

Fig. 6 *Is*-index and damage level of RC school buildings (2011 East Japan earthquake)

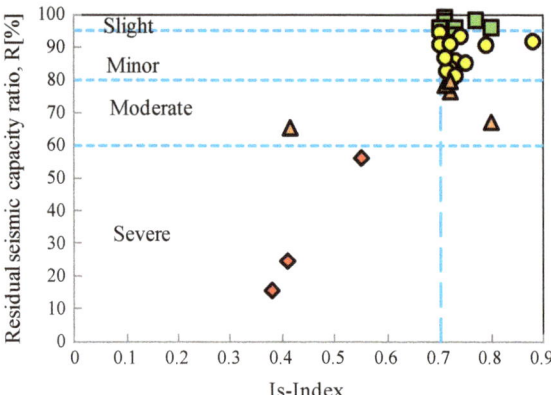

As mentioned above, most of seismically retrofitted RC buildings performed well against the 2011 East Japan earthquake. However, few retrofitted building suffered moderate damage. Figure 7 shows three storied RC building of an elementary school in Sendai city constructed in 1974. The building is divided by expansion joint into west side and east side. Seismic evaluation was carried out to both sides. According to the seismic evaluation, the East side building needed to be retrofitted and the West side was evaluated to have enough seismic capacity and no retrofitting was needed. The East side building was retrofitted by adding framed steel braces and shear walls. By this earthquake, the retrofitted building had only minor damage. On the other hand, the west side building had shear failure in its short columns as shown in Fig. 7b. Shear failure of those short columns was allowed in the seismic evaluation because axial loads could be redistributed to other columns and the building didn't collapse. However, the school couldn't continue using the east side of the building. This issue of functionality is one of important problems.

(a) General view of the building   (b) Shear failure of short columns

Fig. 7 Damage to RC school buildings in 2011 East Japan Earthquake

(a) General view of the building   (b) Damage to non-structural RC walls

Fig. 8 Damage to a residential building in 2011 East Japan Earthquake

Another issue was damage to non-structural elements. Figure 8a shows an overview of an eleven storied SRC residential building which was evaluated as I_s-index of larger than requirement (0.6). However, shear failure of non-structural concrete wall were observed in lower stories as shown in Fig. 8b. These damage induce problems in functions of the building. Although damage to structural elements were quite limited, inhabitants evacuated from this building because of interruption of service and the building was finally demolished. It suggest functionality is getting more important for performance of buildings structure from the resilience point of view.

4.2 Damage Due to 2016 Kumamoto Earthquake

The Kumamoto earthquake on 16th of April 2016 Magnitude of 7.3 is the most recent major earthquake in Japan. Human causalities were 88 death and 2137 injured. In general, damage to buildings that satisfied the seismic evaluation code where limited. Most of the damage observed was similar to those of observed in the past earthquakes.

Figure 9 shows the damage of 7-story residential RC building design by old seismic code and having soft first story (piloti) used as parking lot. Such damage was commonly observed in previous earthquake in Kobe earthquake 1995.

Figure 10 shows 5-story RC building used as city hall. The story collapse occurred at the 4th story. This building was also designed by the old seismic code (before 1981) and was evaluated to have low seismic capacity I_s index. Future plans of either retrofit or rebuild was under consideration before the earthquake.

(a) General view of the building   (b) Collapse of first story

Fig. 9 Damage to a residential building in 2016 Kumamoto Earthquake

Fig. 10 Damage to city hall
building in 2016 Kumamoto
Earthquake

Figure 11 shows the damage of another RC building where the damage to its
structure elements was slight. However, the non-structural secondary walls where
greatly damaged. This repeated damage, rise the importance of revaluating serviceability limits due to non-structural elements.

(a) General view of the building   (b) Collapse of non structural RC walls

Fig. 11 Damage to a residential building in 2016 Kumamoto Earthquake

5 Conclusions

In this paper, overview of damaging earthquakes, seismic code and evaluation standard in Japan was presented. Lessons learned and findings are summarized:

(1) There is great improvement of safety of both existing and new buildings and success in limiting earthquake damage.
(2) Heavy damage to public school buildings in recent earthquakes were limited because almost all buildings satisfied the current design code owing to wide application of seismic evaluation and retrofit.
(3) In recent earthquakes, damage was concentrated to un-retrofitted buildings designed based on old seismic code. This emphasize the necessity of speeding up the process of assessment and retrofit of buildings.
(4) Many buildings lost immediate occupancy because of damage to non-structural components even though they escape structural damage. The functionality is getting more important for performance of buildings structure from the resilience point of view.

References

Japan Building Disaster Prevention Association, JBDPA (1977, revised in 1990, 2001) Standard for seismic evaluation of existing reinforced concrete buildings (in Japanese)

Japan Building Disaster Prevention Association, JBDPA (1990, revised in 2001, 2015) Standard for post-earthquake damage level classification of reinforced concrete building

Maeda M, Al-Washali H, Suzuki K, Takahashi K (2012a) Damage of RC building structures due to 2011 East Japan earthquake. ASCE Structures Congress 2012 (CD-ROM)

Maeda M, Takahashi K, Al-Washali H, Tasai A, Kitayama K, Kono S, Nishida T (2012b) Damage survey on reinforced concrete school buildings in Miyagi after the 2011 East Japan earthquake. In: 15th world conference on earthquake engineering, (paper ID:2591)

Okada T, Kabeyasawa T, Nakano Y, Maeda M, Nakamura T (2000) Improvement of seismic performance of reinforced concrete school buildings in Japan-Part 1 damage survey and performance evaluation after 1995 Hyogo-ken Nambu earthquake. In: Proceedings of 12th world conference on earthquake engineering,(CD-ROM) (paper no. 2421)

Development of Two Types of Buckling Restrained Braces Using Finite Element Modelling

Ciprian-Ionut Zub, Aurel Stratan, Adrian Dogariu, Cristian Vulcu and Dan Dubina

Abstract Buckling Restrained Braces (BRBs) provide structures with large lateral deformation capacity, strength and stiffness. Throughout the years, BRBs proved to be a viable solution, based on their increasing usage as structural fuse elements. A BRB consist of a steel core introduced into a buckling restraining mechanism (BRM). The main BRB's feature is its quasi-symmetric and stable hysteretic behaviour when cyclically loaded. In addition to normal braces, BRBs have their compressive behaviour improved by confining core's transversal deformations, and thus restraining the core's global buckling. Due to the fact that most BRBs are proprietary and rather manufactured than designed, design codes (P100-1/2013) require their experimental qualification. This code's regulation represents an impediment in using the BRBs on a wider scale. To overcome these problems, a set of typical BRBs were developed with capacities corresponding to typical steel multi-storey buildings in Romania. The paper presents the pre-test numerical simulations performed. The numerical finite element model development using Abaqus 6.14 software and its calibration based on experimental data are presented in detail. The numerical study was used to develop "conventional" and "dry" solutions by analysing several different solutions. The main difference between the

C.-I. Zub (✉) · A. Stratan · A. Dogariu · C. Vulcu · D. Dubina
Faculty of Civil Engineering, Department of Steel Structures and Structural Mechanics, Politehnica University Timisoara, Timisoara, Romania
e-mail: ciprian.zub@student.upt.ro

A. Stratan
e-mail: aurel.stratan@upt.ro

A. Dogariu
e-mail: adrian.dogariu@upt.ro

C. Vulcu
e-mail: cristian.vulcu@upt.ro

D. Dubina
e-mail: dan.dubina@upt.ro

© Springer International Publishing AG, part of Springer Nature 2018
R. Vacareanu and C. Ionescu (eds.), *Seismic Hazard and Risk Assessment*,
Springer Natural Hazards, https://doi.org/10.1007/978-3-319-74724-8_25

two is the absence of concrete in the case of "dry" solution. The cyclic performance of the BRBs was assessed in terms of compression adjustment factor and global performance.

Keywords Buckling restrained braces · Experimental testing · Numerical simulations

1 Introduction

Pioneered in the 1970s by Yoshino et al. (Xie 2005), who conducted the firsts tests on "shear wall with braces", and then extensively developed and tested in Japan as steel braces encased in a steel [Kimura et al. (1976) in Xie (2005)] or reinforced concrete [Mochizuki et al. (1980) in Black et al. (2002)] tube during the eighties and nineties, Buckling Restrained Braces (BRBs) are widely used due to their excellent energy dissipation capacity and quasi-symmetric cyclic behaviour. Conceptually, a BRBs consists of a dissipative steel core encased in a buckling restraining mechanism (BRM) that confines the core, preventing global buckling. The core is decoupled from the BRM by using an unbonding interface (material or gap, depending on the type of BRM), allowing it to achieve higher compression modes. Depending on the type of the BRM, the BRBs can be "conventional" (BRM is a concrete filled steel tube) or "dry" (BRM is a steel assembly).

Many BRB designs were developed, tested and patented in Japan, USA, Canada, Taiwan (Xie 2005). In Europe, just few tests were performed up to date (D'Aniello et al. 2014; Bordea 2010; Dunai et al. 2011). The use of BRBs in Europe is not regulated by Eurocode 8 (2004) yet. However, general design rules, as well as manufacturing and testing requirements of BRBs are available in EN 15129 (2010). The national anti-seismic design code of Romania, P100-1 (2013), introduced provisions for design of steel structures with BRBs since January 2014. The code requires experimental qualification of BRBs used in practical applications, either project-specific, or based on existing experimental evidence.

This paper describes the numerical pre-test program on a set of BRBs suitable for typical low-rise and mid-rise buildings in Romania. Several low and mid-rise steel buckling restrained braced frames (BRBFs) were designed according to P100-1 (2013), resulting a range of BRB resistances from 150 to 840 kN. ANSI/AISC 341-10 requires that the yield strength of tested BRBs be in the range 0.5–1.2 times the nominal capacity of the prototype BRB. Therefore, BRBs with strength of 300 and 700 kN were selected, which would cover the target strength range of 150–840 kN. The aim of the numeric investigation was to observe and evaluate the plastic deformation capacity of the BRBs. A set of main parameters were analysed in order to identify the optimal configuration that will lead to the desired behaviour of the BRB. The following parameters were analysed: nominal resistance (N_{pl}), critical elastic axial load of the buckling-restraining mechanism (N_{cr}), core aspect ratio, gap size configurations. The performance of the BRBs was assessed in terms of compression adjustment factor (β) and global performance.

2 Conceptual Development of BRBs

2.1 Core Geometry

The main component of a BRB is the steel core. It consists of several distinct segments: connection zone, elastic zone, transition zone, plastic zone (Fig. 1). The plastic zone is the segment were plastic deformations are intended to take place and it represents the dissipative part of the core. The transition zone is an intermediate segment between the plastic and elastic zones, and needs a careful design and fabrication, to prevent stress concentrations and brittle failure modes. A numerical study was carried out using ABAQUS software (ABAQUS 2014) to investigate the optimal geometry of the transition zone, by analysing different geometries of the core.

Two different solutions for the core's geometry were analysed: (1) by machine cutting from a steel plate or (2) by using a compact cross-section (rectangular, square or round) with stiffeners welded at the ends. The residual stresses caused by the technological processes were not modelled in Abaqus, and therefore only the influence of the geometrical configuration of the core with respect to the stress concentration phenomenon was analysed. Several concepts (Fig. 2a–d) were

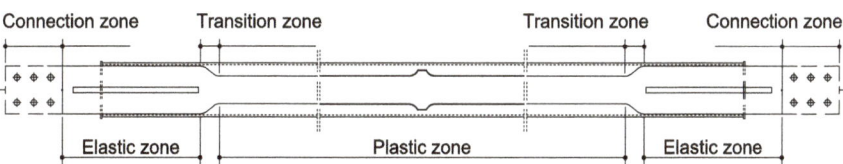

Fig. 1 The zones of the BRB's core

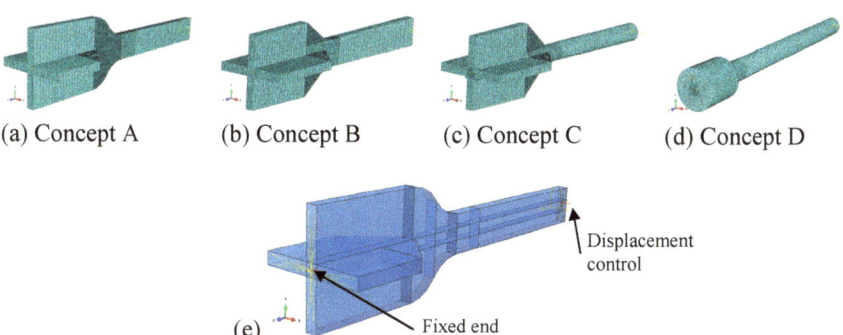

(a) Concept A   (b) Concept B   (c) Concept C   (d) Concept D

Fig. 2 Discretization (**a, b, c, d**) and boundary conditions (**e**)

investigated to evaluate the susceptibility to brittle fracture of the transition zone and of the plastic zone close to it. In a simplified way, the assessment was done in terms of maximum stresses and equivalent plastic strains. The following cases were analysed:

Concept A: steel core obtained by machine cutting of a steel plate, using fillet radius to reduce the stress concentrations. *Concept B*: steel core obtained by welding four stiffeners at the ends of a rectangular steel profile. *Concept C*: steel core obtained by welding four stiffeners at the ends of a round steel profile. *Concept D*: similar to Concept C, except that the four stiffeners were substituted by a circular "stiffener".

All the parts of the concepts were considered as being made of a S355 steel. In Abaqus, an isotropic hardening material model was used to define the plastic behaviour: the yield strength was considered $f_y = 1.25 * 355 = 444$ MPa, the ultimate strength was $f_u = 1.25 * 510 = 638$ MPa corresponding to a strain $\varepsilon_u = 15\%$.

The numerical analyses were performed using the "Dynamic Implicit" procedure available in Abaqus. A quasi-static force control analysis was performed for all models. The force, F, was applied at the free-end, increasing monotonically up to the value of 95% of the tensile resistance of the cross-section of the core $(0.95 * f_u * A_c)$. A fine mesh was adopted for the transition zone (mesh size 2 mm), and C3D8I and C3D6 finite element type were used for the discretization. Two boundary conditions were used: the enlarged end section was considered fixed and the reduced one was a roller support in order to allow the free movement of the core in the longitudinal direction (Fig. 2e).

Figure 3 shows the distribution of the von Misses stresses (σ_M) and the equivalent plastic strains (PEEQ) in the four models, while Table 1 shows the peak values of the PEEQ in the transition zone and in the core. Sensitivity of the model to brittle fracture was associated with large values of PEEQ.

It can be observed that in the case of *concept A* there is a gradual transition of stresses from the reduced to the enlarged cross-section, PEEQ in the core being the smallest among the four models (9.24%). Due to the fact that the stiffeners are positioned in the elastic zone and the transition zone is smooth, there is no susceptibility to brittle failure.

For the *concept B*, stress concentrations can be observed in the core next to the stiffeners, which correspond to PEEQ values of 20.5% more than twice the one in concept A. On the other hand, in the transition zone the PEEQ values are slightly smaller than the ones in the case of concept A (7.82%).

In the case of *concept C*, stress concentrations in the core occur in a similar manner to concept B, but the PEEQ values are significantly smaller (13.49%). Smaller demands are also noticed in the transition zone of this model (4.73%).

In the case of *concept D*, stress concentrations in the core are slightly reduced with respect to case C (PEEQ = 11.06%), while in the transition zone PEEQ values are almost the same (4.28%). This solution is however rather theoretical, as it has difficulties in being adopted as a transition zone for a BRB.

Concept A clearly represents the best solution in terms of minimising the potential of brittle fracture in the core close to the transition zone. Concept C provides reasonable alternative to concept A when it is desired to optimize the costs by reducing

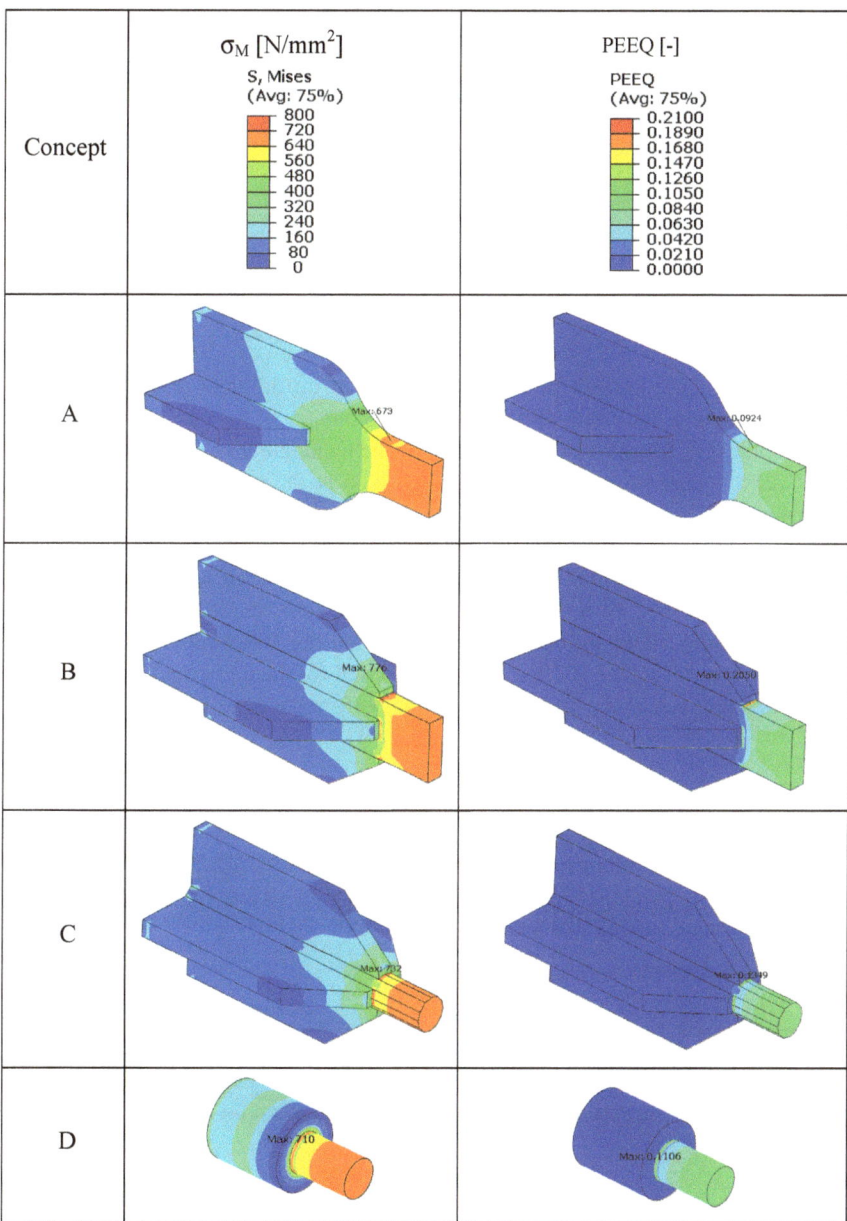

Fig. 3 FEM results for the transition zone for different concepts of the core geometry

the machining and using instead stiffeners welded to a compact as-rolled core. It is assumed that similar results would be obtained in the case of a core made of a square profile to the ones obtained for the concept C. Concepts B (stiffeners welded to slender plates) is to be avoided as it is susceptible to brittle failure.

Table 1 Evaluation of the performance of the transition zone

Concept	Concept description	PEEQ in the transition zone (%)	PEEQ in the core (%)
A	Machined plate + 2 stiffeners	9.24	9.24
B	Rectangular profile + stiffeners	7.82	20.50
C	Round profile + 4 stiffeners	4.73	13.49
D	Round profile + 1 CHS stiffener	4.28	11.06

2.2 BRB Solutions

The outcome of the numerical analyses performed on different geometries of the transition zone was the basis for choosing the geometry of the core. Two solutions were adopted: (1) a machined plate with variable rectangular cross-section (*concept A*) and (2) a square profile with stiffeners welded at the ends to improve stability and assure an elastic response for the core's end zones (similar to *concept C*). As regarding the buckling restraining mechanism (BRM) two solutions were proposed: a "conventional" BRM consisting of a steel CHS profile filled with concrete; and a "dry" BRM consisting of a steel assembly made of four steel SHS profiles welded together by using fillers and steel plates. Based on the above-mentioned solutions for the core and BRM, the BRB solutions proposed for the numerical investigations for optimization and evaluation of the cyclic performance are presented in Fig. 4.

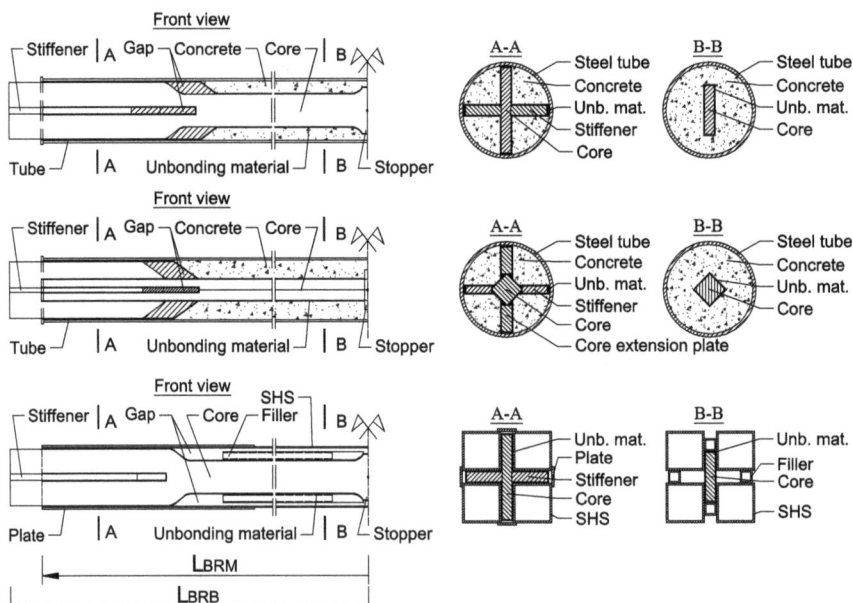

Fig. 4 BRB solutions: front view, cross-sections

To provide a reference for the development of different core geometries, the length of the core (L_{BRB}) was kept fixed for all BRB solutions. The first BRB solution is a "conventional" BRB, having the core made of a machined plate with variable rectangular cross-section and two stiffeners welded at each end to improve the stability of the elastic zones. Due to machine cutting of the core, the solution has the great advantage of adjusting the capacity to the desired one. The second BRB solution is also a "conventional" BRB, having the core made of a steel square profile with four stiffeners welded at each end to assure an elastic response and to improve the stability of the core's end zones. This solution has the advantage of simplifying the technology needed to obtain the core, but has the disadvantage of not being able to have the capacity adjusted to the required one due to the limited range of sizes for the steel sections. The third BRB solution is a "dry" BRB, having the core similar to the first solution but the BRM is a steel assembly in order to have a steel-only BRB.

3   Calibration of the FEM Model

Numerical investigations were performed using Abaqus/CAE 6.14. The numerical model adopted was calibrated against experimental data available in Dunai et al. (2011). The calibration had the purpose of providing information on specific input parameters (vertical and horizontal core to concrete gap, contact law definition, value of the friction coefficient) that are necessary for modelling the BRBs. Using the BRB's geometry from Dunai et al. (2011), a finite element model was constructed (Fig. 5a) and good predictions were obtained. Further details about the calibration process can be found in Zub et al. (2017). The numerical model could capture all the main characteristics of the tested BRB: initial elastic stiffness, yield force and displacement, strain hardening, dissipated energy (Fig. 5b).

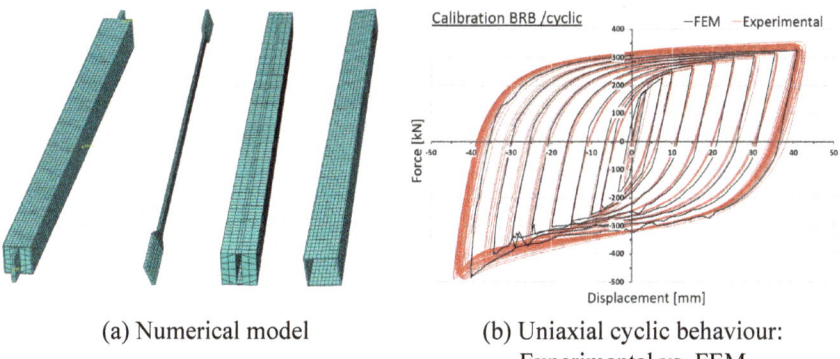

(a) Numerical model

(b) Uniaxial cyclic behaviour:
Experimental vs. FEM

Fig. 5 Calibration of FEM BRB model

4   Analysis of BRB Solutions

4.1   Calibration of the Material Model

In Abaqus, the combined isotropic/kinematic hardening plasticity model was used to model the steel components. The model consists of a nonlinear kinematic (Fig. 6a) and a nonlinear (exponential law) or multilinear (tabular data) isotropic hardening component (Fig. 6b). A very good prediction can be obtained with this model for both monotonic and cyclic analyses (Fig. 7). The report of Kaufmann et al. (2001) was used for the calibration of steel material's cyclic properties, and then scaled up to provide properties for a S355 mild carbon steel ($f_y = 355 \times 1.25 = 444$ MPa, $f_u = 510$ MPa). The calibration procedure described in Abaqus (2014) documentation was used. The isotropic hardening component of the model was defined using the cyclic hardening suboption available for plasticity models. Equivalent stress—equivalent plastic strain data pairs were obtained from the calibration process and used for the definition of a multilinear isotropic hardening model. Since mild carbon steel exhibits a yield plateau, the definition of the isotropic model had to include an initial softening segment up to the end of the plateau, followed by a hardening segment.

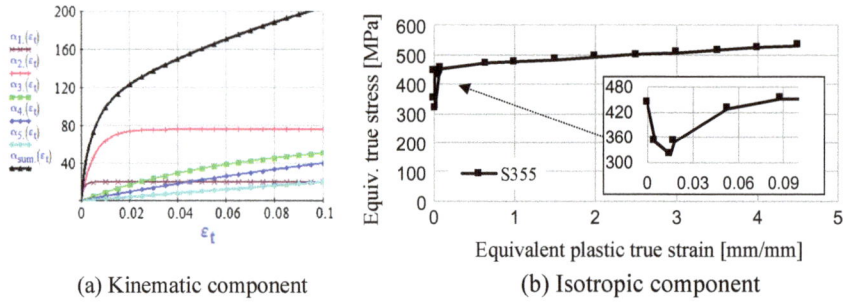

(a) Kinematic component   (b) Isotropic component

Fig. 6   Combined isotropic/kinematic hardening model

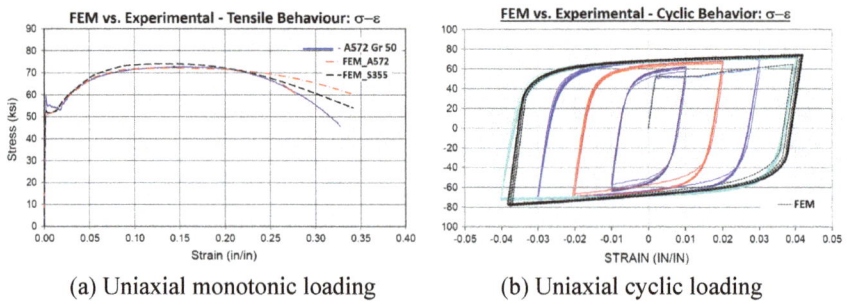

(a) Uniaxial monotonic loading   (b) Uniaxial cyclic loading

Fig. 7   Calibration of material model for monotonic and cyclic uniaxial loading

4.2   Modelling an Analysis of "Conventional" BRBs

4.2.1   Finite Element Model

Based on the approach used in the calibration process, similar models were constructed for all "conventional" BRBs (Fig. 8). The core assembly was modelled using C3D8I elements, which are appropriate to model bending with contact interactions. The concrete part was modelled using C3D8R. The steel tube was modelled using shell elements, S4R, due to its reduced thickness. Gaps were placed near the transition zones to allow for free movement of the core in compression. A general contact was defined, and the contact domain consisted of two selected surface pairs that were assigned different contact properties as follows: the core–to–concrete interaction was defined as having the tangential behaviour defined as "Penalty" with a friction coefficient of 0.1 and the normal behaviour set to "Hard" contact; the concrete-to-steel casing interaction had the same properties except the friction coefficient which was set to 0.4. Also, a coupling constraint was defined at each end of the core, by connecting a reference point to a surface using the "Continuum distributing" coupling type, allowing for free transversal deformation of the selected surface. The BRB was considered pinned at both ends, while the load was applied in displacement control. The BRBs were cyclically loaded by using the AISC 341-10 loading protocol (Fig. 9) where Δ_{by} represents the

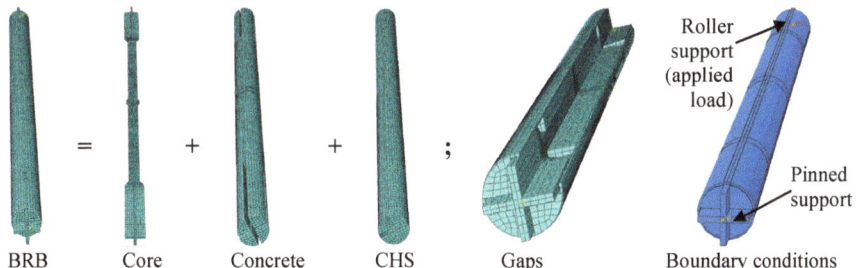

Fig. 8 "Conventional" BRB. Finite element model. Assembly, parts, gaps and boundary conditions

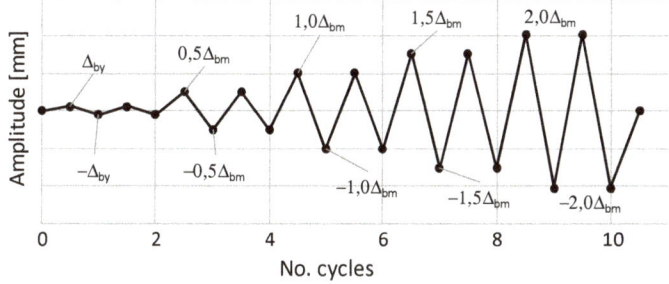

Fig. 9 Loading protocol

displacement corresponding to the yielding of the core (Δ_{by} obtained from a monotonic tensile analysis), while Δ_{bm} represents the displacement corresponding to the design story drift, in this case considered as a function of core's strain ($\Delta_{bm} = \varepsilon_c \times L_c$, where: ε_c was the strain level in the BRB's core and L_c was the length of the core's plastic zone).

Geometrical imperfections were included into the numerical model by using the first buckling mode as a deformed shape, scaled up to obtain a bow imperfection of $L_{BRM}/1000 = 3.7$ mm. The "Dynamic Explicit" procedure was used for the main model and mass scaling to reduce the computational time. The FEM model was validated by checking the artificial energy to be less than 1% of the internal energy.

At the basis of the development of "conventional" (C) BRBs several requirements were set, as follows. Two *nominal resistances* were requested, $N_{pl,3} = 300$ kN (3), $N_{pl,7} = 700$ kN (7), resulting two rectangular (R) cross-sections for the cores: 14×60 mm and 20×99 mm, respectively, made of S355 steel. To evaluate the influence of the *core shape*, with respect to the amount of friction forces developed at the interface of core with the BRM, an alternative square (S) cross section solution was proposed for both capacity types: 30×30 mm, 45×45 mm, respectively. As regarding the *BRM*, for the 700 kN with rectangular cross-section model, two capacities were investigated to validate the design methodology proposed by the literature. Therefore, two models were created: the first model had the ratio $N_{cr}/N_{pl} = 1.5$ (1) [value proposed by Watanabe et al. (1988)], and the second had the ratio $N_{cr}/N_{pl} = 3$ (3) [practice design, Black et al. (2002)]. It should be mentioned that all the other models had the BRM designed for $N_{cr}/N_{pl} = 3$.

4.2.2 Influence of Core Shape

Square core-cross section model (CS73) performed slightly better in terms of compression adjustment factor, with $\beta = 1.13$ compared to the rectangular model (CR73) with $\beta = 1.16$. Also, the square model proved to possess a slightly larger ductility capacity (Fig. 10a). The difference between the two models was more visible when analysing the maximum PEEQ values (Fig. 11): for the CR73 model the value of $PEEQ_{max} = 1.723$ was 30% higher than in the case of CS33 ($PEEQ_{max} = 1.3335$). This aspect would lead to a lower low cycle fatigue resistance for CR73 in comparison with CS73. For both models, the failure mode was a ductile one by rupture of the core in tension.

4.2.3 Strength of BRM

The critical elastic axial load of BRM, had a small influence from the β factor point of view. A 2.58% difference was noticed when comparing the results from the model CR73 ($\beta = 1.16$) that had the BRM designed for $N_{cr}/N_{pl} = 3$, with the results from the model CR71 ($\beta = 1.19$) that had the BRM design to $N_{cr}/N_{pl} = 1.5$ (Fig. 12). The main objective of investigating this parameter was to determine the

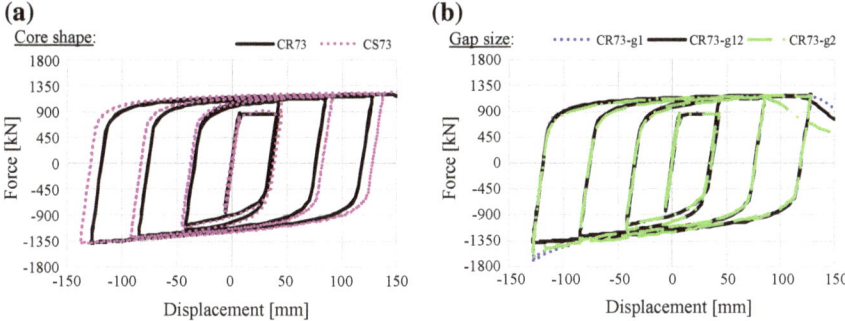

Fig. 10 "Conventional" BRB 700 kN. The influence of the core shape (**a**) and gap size (**b**)

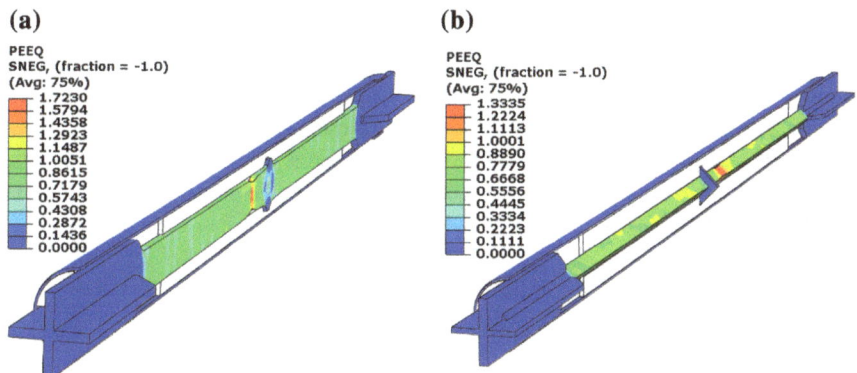

Fig. 11 PEEQ for "conventional" BRB 700 kN: **a** rectangular—CR73; **b** square—CS73

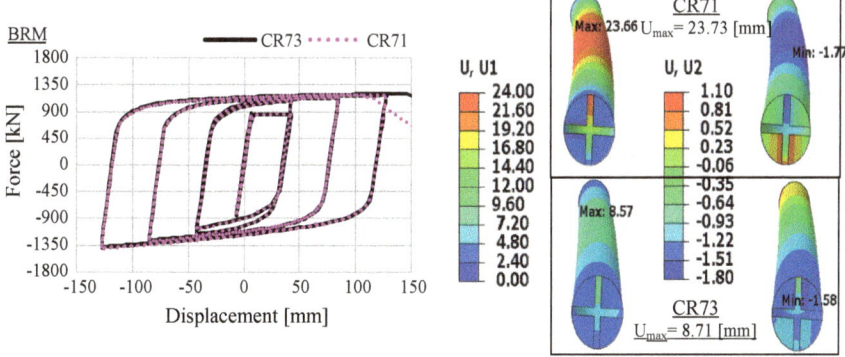

Fig. 12 "Conventional" BRB 700 kN: the influence of the critical elastic axial load of BRM

global behaviour of the BRB, with respect to maximum lateral deformation of the BRM under maximum compression force. Even though for CR71 a lateral deformation of 23.73 mm of the BRM was recorded (at the mid-length), compared to 8.71 mm of the CR73 model, both steel tubes remained in the elastic domain. Both models had a ductile failure mode by rupture of the core in tension.

4.2.4 Gap Size

The value of the core–to–BRM gap sizes for the through-thickness direction (g_t) and the through-width direction (g_w) was also studied. The influence of the gap layout was analysed with respect to the β factor. The lowest β value (1.16) was obtained for the $g_t = 1$ mm and $g_w = 2$ mm (model CR73-g12). For the models heaving a uniform gap of 1 mm (CR73-g1) and 2 mm (CR73-g2), values of $\beta = 1.43$ and $\beta = 1.38$, respectively were obtained. According to AISC 341-10 (2010) regulations, the value of β should be greater than 1 and less than 1.3. Therefore, only CR73-g12 fulfilled the code's requirements (Fig. 10b). It must be mentioned that to a FEM gap equal to 1 mm/2 mm corresponded a real gap of 2 mm/3 mm, respectively. Based on those observations, the BRBs equipped with rectangular cores were designed with different values for the gaps (greater value for g_w). BRBs with square cores were provided with a uniform gap of 2 mm.

4.3 Modelling an Analysis of "Dry" BRBs

4.3.1 Finite Element Model

"Dry" BRBs (D) were modelled similarly to "conventional" ones. Parts with thin walls (tubes) were modelled using S4R shell elements, and C3D8I elements were used to model parts made of thick plates (core, stiffeners). "Tie" constraints between edge-pairs (Fig. 13) were used to simulate the welds. Two *nominal resistances* were used for design, $N_{pl,3} = 300$ kN (D33-a), $N_{pl,7} = 700$ kN (D73-a), with similar cross-sections as in the case of "conventional" ones. The demand was to provide a lighter BRM, made of steel parts only, robust enough to withstand global buckling. Even though the initial solution of BRM made of four SHS welded together globally fulfilled the requirements, locally, instead, it did not perform well up to the maximum compression capacity of the BRB due to local damage of the thin walls of the SHS. The same phenomenon was notices for both capacity cases, D33-a and D73-a (Figs. 14 and 15). Therefore, the thin SHS were strengthened.

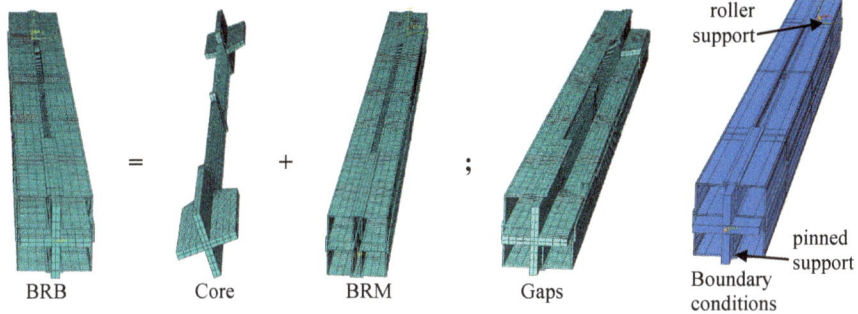

Fig. 13 "Dry" BRB. Finite element model: assembly, parts, gaps and boundary conditions

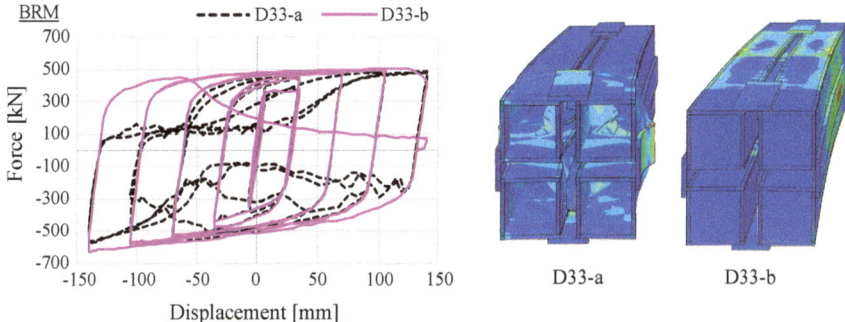

Fig. 14 "Dry" BRB 300 kN: cyclic performance of different BRM solutions

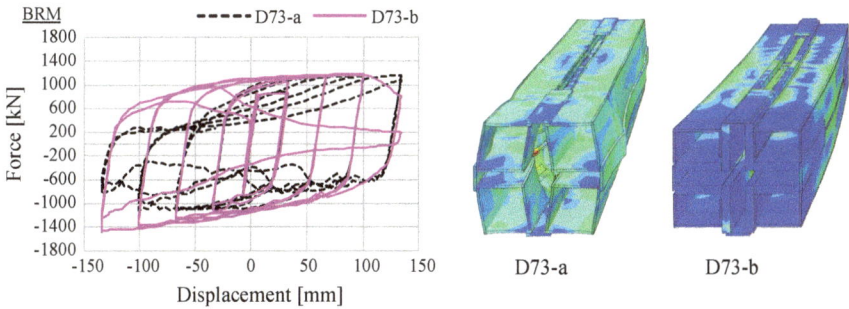

Fig. 15 "Dry" BRB 700 kN: cyclic performance of different BRM solutions

4.3.2 Influence of BRM

In the case of the 300 kN BRB, the tubes were filled with concrete (D33-b). The new BRM configuration proved to be stable and a $\beta = 1.25$ was obtained (Fig. 14). In the case of the 700 kN BRB, additional steel plates were placed on the sides of the core, with stiffened slotted ends (D73-b). The new BRM configuration proved to be stable and a $\beta = 1.26$ was obtained. However, compared to the "conventional" BRB, it could be observed that the former dissipated less energy, was less stable and had a smaller ductility (Figs. 10 and 15).

5 Conclusions

The main objective of the research presented in this paper was the development of two BRBs solutions, "conventional" and "dry", based on finite element numerical analyses. Two nominal capacities were chosen for the design, 300 and 700 kN. Based on the numerical investigations performed on the geometry of the transition zone of the core, rectangular and square core cross-sections were used to investigate the core's shape influence on the behaviour of the BRB. The resistance of the BRM was also analysed by investigating two different N_{cr}/N_{pl} ratios (1.5 and 3). In the case of "conventional" BRBs, the one with square core cross-section had a better cyclic performance than the one with rectangular cross-section, while the model with stronger BRM ($N_{cr}/N_{pl} = 3$) had a better performance than the one with $N_{cr}/N_{pl} = 1.5$. In the case of "dry" BRBs, the initial solution proved to have a poor cyclic behaviour, local damage of the SHS's thin walls leading to BRB's failure. Therefore, the initial solution was modified as follows: in the case of the 300 kN BRB, the tubes were filled with concrete, and in the case of the 700 kN BRB, additional steel plates were placed on the sides of the core, with stiffened slotted ends. The behaviour of the final "dry" solutions was significantly improved, but when compared to the "conventional" solutions, the former ones dissipated less energy, were less stable and had a smaller ductility. The outcomes of the numerical analyses were used to develop and design the engineering solutions that are going to be tested in view of prequalification.

Acknowledgements The research leading to these results has received founding from the MEN-UEFISCDI grant Partnerships in priority areas PN II, contract no. 99/2004 IMSER: "Implementation into Romanian seismic resistant design practice of buckling restrained braces." This support is gratefully acknowledged.

References

ABAQUS (2014) ABAQUS documentation. Dassault Systèmes, Providence, RI, USA
ANSI/AISC 341-10 (2010) Seismic provisions for structural steel buildings. American Institute of Steel Construction, Inc. Chicago, Illinois, USA

Black C, Makris N, Aiken I (2002) Component testing, stability analysis and characterization of buckling-restrained unbonded bracesTM. PEER Report 2002/08. Pacific Earthquake Engineering Research Center, College of Engineering, University of California, Berkeley

Bordea S (2010) Dual frame systems of buckling restrained brace. Ph.D. thesis

CEN—European Committee for Standardization (2004) Eurocode 8: design of structures for earthquake resistance. Part 1: general rules, seismic actions and rules for buildings. Eurocode 8-1/2008

CEN—European Committee for Standardization (2010) Anti-seismic devices. EN 15129:2010

D'Aniello M, Della Corte G, Landolfo R (2014) Finite element modelling and analysis of "all-steel" dismountable buckling-restrained braces. Open Constr Build Technol J 8:216–226

Dunai L, Zsarnóczay A, Kaltenbach L, Kálló M, Kachichian M, Halász A (2011) Type testing of buckling restrained braces according to EN 15129—EWC800—Final report

Kaufmann EJ, Metrovich B, Pense AW (2001) Characterization of cyclic inelastic strain behavior on properties of A572 Gr. 50 and A913 Gr. 50 rolled sections. ATLSS report no. 01-13

P100-1/2013 (2013) National seismic design code—part I—design provisions for buildings

Watanabe A, Hitomi Y, Saeki E, Wada A, Fujimoto M (1988) Properties of brace encased in buckling restraining concrete and steel tube. Proc 9th World Conf Earthquake Eng 4:719–724

Xie Q (2005) State of the art of buckling-restrained braces in Asia. J Constr Steel Res 61:727–748

Zub CI, Dogariu A, Stratan A, Dubina D (2017) Pre-test numerical simulations for development of prequalified buckling restrained braces. CE/Papers, Ernst & Sohn/Wiley 1(2–3):3404–3413. https://doi.org/10.1002/cepa.395

Static and Dynamic Approaches on the Low-Rise RC Frames Capacity Evaluation and Damage Quantification

Victor-Adrian Păunescu and Mihail Iancovici

Abstract The seismic capacity evaluation is a key-tool for practitioners in the standard post-design stage of building structures. This also serves as a major component for performing higher level analysis modules as seismic vulnerability analysis, risk analysis, loss estimation and resilience analysis. The seismic capacity of a structure is dependent of the applied load pattern as well as of nonlinear modelling of members. The paper examines the dependency of capacity curve parameters of typical low-rise regular reinforced concrete frames as a large proportion from actual building stock in Bucharest, using both nonlinear static and dynamic analysis approaches as well as discrete and distributed available plasticity models for reinforced concrete members. The sensitivity analysis of global seismic damage index based on the probabilistic approach using HAZUS methodology (2007) is then performed. A more accurate prediction of structural capacity in conjunction with integrated tools for the assessment of the damage states up to progressive collapse associated to various earthquake scenarios and time-domain analysis approach would truly lead implementing a real performance-based seismic design into practice.

Keywords Seismic capacity · Pushover curve · Incremental dynamic analysis Damage index

1 Introduction

The seismic capacity evaluation of a structure is an essential analysis module in the standard post-design stage, in the performance evaluation of an existing building. The nonlinear static approach using first non-adaptive and then adaptive lateral

V.-A. Păunescu
Ideea Proiect, Bucharest, Romania

M. Iancovici (✉)
Department of Structural Mechanics, Technical University
of Civil Engineering of Bucharest, Bucharest, Romania
e-mail: mihail.iancovici@utcb.ro

© Springer International Publishing AG, part of Springer Nature 2018
R. Vacareanu and C. Ionescu (eds.), *Seismic Hazard and Risk Assessment*,
Springer Natural Hazards, https://doi.org/10.1007/978-3-319-74724-8_26

loading schemes serves to perform the seismic capacity evaluation (Elnashai 2001; Chopra and Goel 2002).

However, in the last decades the raw information on the structural capacity provided by static nonlinear analysis is not only a basic evaluation tool of structural design for a practitioner. It serves for performing higher level structural analyses as vulnerability and risk analysis (HAZUS 2007), loss estimation mapping and resilience analysis. A true integrated performance-based seismic design framework is not possible without accurate structural capacity evaluation based on a large number of time-domain analyses. Moreover, the current analysis and design practice must be augmented not in a too distant future by a comprehensive progressive collapse analysis module in a so-called "what if" analysis type.

The seismic capacity of a structural model is represented by the so-called "pushover" curve that synthetically emphasizes both- linear and nonlinear behaviour stages of a structure under lateral monotonic increasing load compatible with the actual seismic forces. However the performance of a structural system essentially depends of the input ground motion at a particular site and of the nonlinear modelling of members (CEB 1996). Due to its inherent limitations, important features of the actual input motion as the frequency content, amplitude, phase as well as the effect of cyclic loading on the hysteretic-type behaviour of members are not properly taken into account in a static nonlinear approach.

While the nonlinear dynamic approach is primarily used for design checking purposes rather than a method to evaluate the structural capacity, the Incremental Dynamic Analysis approach proposed by Vamvatsikos and Cornell (2002) has the advantage to catch the input ground motions features at their full information. The effect of deterioration under cyclic loads is accounted for while the computational effort is much more higher.

The objective of the paper is to study the effect of static and dynamic loading on the seismic capacity evaluation of low-rise RC frame structures using two types of nonlinear models i.e. discrete and distributed. The global damage index is then computed using the HAZUS (2007) methodology framework.

2 Pushover Analysis Methods

The seismic capacity of a structural system is represented through the base shear force-tip displacement/drift/interstory drift relationship. In the non-adaptive loading scheme, the lateral load increases monotonically under the fundamental mode shape pattern, either in a force-controlled or displacement-controlled approach (Chopra and Goel 2002).

The full adaptive pushover procedure (Elnashai 2001) reconsiders at each loading step the change in the lateral load pattern that takes into account the modification of pseudo-vibration modes due to extent of deterioration. This however requires a larger amount of computational work.

Later on a uniform lateral loading pattern was proposed in order to reproduce the evolution of pseudo-mode shape or the soft or weak story during loading (Antoniou and Pinho 2004). The multi-modal approach considers the contribution of the higher vibration modes (Chopra and Goel 2002). Pietra (2008) studied the influence of lateral load distribution during the loading process over the seismic capacity for few cases of building structures.

As the highest level analysis tool, the Incremental Dynamic Analysis (IDA) consists of selecting a set of appropriate input ground motions, incrementally changing the input acceleration level in a multi-step nonlinear time-history analysis and finally plotting the peak response parameters in a so-called "dynamic pushover" curve of the nonlinear model.

3　The Input Ground Motions for the Incremental Dynamic Analysis (IDA)

The selection of an appropriate set of acceleration input ground motions able to drive significantly the structure into the nonlinear range and realistically reproduce its evolution from linear-elastic up to ultimate limit state is an essential issue for the IDA framework.

In this paper a number of four significant seismic events ($M_w > 6$) are considered to select a total number of 15 appropriate acceleration records (Păunescu 2013; Table 1; source Pacific Earthquake Engineering Research Center—PEER, http://peer.berkeley.edu/smcat/).

The input ground motion parameters are given in Table 2 altogether with the statistical analysis results. In Table 2, ε is the Cartwright-Longuet & Higgins frequency bandwidth indicator, δ is the frequency content shape factor, t_d is the total duration of motion and E is the maximum cumulative energy of ground acceleration (Kramer 1996).

The selection of input ground motions from the database follows the frequency content criteria, so that the predominant period of ground motion T_p to be in the range of the sway fundamental vibration period of model. Other approaches like scaling or generating input ground motions (Katsanos et al. 2010) can be considered for IDA purpose.

The frequency contents of acceleration records are represented in the Fig. 1.

Table 1 Seismic events

Event	Date	Magnitude, M_w	Records
Kern, USA	7/21/1952	7.3	1
Loma Prieta, USA	10/18/1989	6.9	3
Northridge, USA	01/17/1994	6.7	8
Kobe, Japan	01/16/1995	6.9	3

Table 2 Input ground motion parameters

Record	Event	PGA (m/s^2)	PGV (m/s)	PGD (m)	T_p (s)	ε	δ	t_d (s)	E (m^2/s^3)
GAR270	Northridge, USA	1.01	0.10	0.02	0.68	0.91	3.19	34.98	101.30
BLF296	Northridge, USA	1.26	0.06	0.02	0.68	0.86	4.71	21.82	64.61
BVA195	Northridge, USA	1.84	0.24	0.06	0.68	0.95	3.66	28.73	230.70
MAN000	Northridge, USA	1.97	0.14	0.02	0.68	0.94	3.38	34.98	94.42
NWH360	Northridge, USA	5.79	0.97	0.37	0.68	0.91	2.60	39.98	245.88
KJM000	Kobe, Japan	8.06	0.81	0.18	0.69	0.88	2.42	47.98	291.88
DMH000	Loma Prieta, USA	0.96	0.10	0.02	0.69	0.91	3.64	39.94	170.76
JAB220	Northridge, USA	0.96	0.07	0.03	0.69	0.87	4.70	34.98	85.98
PRS000	Loma Prieta, USA	0.98	0.13	0.04	0.69	0.96	2.86	39.95	163.52
LBC360	Northridge, USA	0.50	0.04	0.04	0.70	0.86	3.87	49.98	27.73
TOT090	Kobe, Japan	0.74	0.08	0.05	0.70	0.85	2.05	77.98	91.01
KJM090	Kobe, Japan	5.87	0.74	0.20	0.70	0.89	2.47	47.98	234.34
YBI000	Loma Prieta, USA	0.28	0.04	0.02	0.71	0.94	3.88	39.95	56.22
TAF021	Kern, USA	1.53	0.15	0.16	0.71	0.84	3.52	54.15	142.94
LOS270	Northridge, USA	4.73	0.45	0.13	0.71	0.93	3.31	19.98	276.41
	Mean	2.43	0.28	0.09	0.69	0.90	3.19	40.89	151.85
	Standard deviation	2.43	0.31	0.10	0.01	0.04	4.71	14.12	86.03
	COV	1.00	1.14	1.11	0.01	0.04	0.23	0.35	0.57

One can be observed a significant variability of peak ground motion parameters while the variability of frequency bandwidth indicator is lower.

The scaled input motions to be used for IDA analyses are generated based on the code linear-elastic absolute acceleration response spectrum (P100-1 2013; Păunescu 2013) and used in the analyses. The Peak Ground Acceleration (PGA) scaling factors are provided in the Table 3 for each loading step S_k (k = 1,..., 10).

Later, the variability of response parameters is represented function of input motion load step.

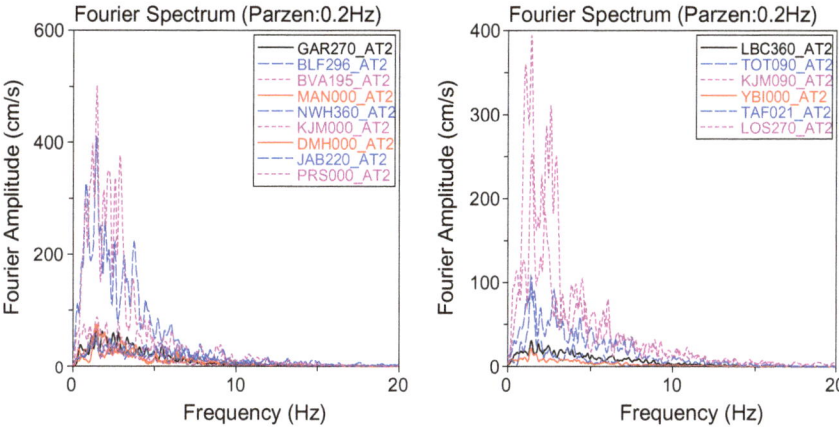

Fig. 1 Fourier amplitude spectra for 15 selected record components (ViewWave 2.21 software)

Table 3 PGA scaling factors for IDA

Record	S1	S2	S3	S4	S5	S6	S7	S8	S9	S10
BLF296	1.15	1.45	1.76	2.06	2.37	2.67	2.98	3.28	3.59	3.89
BVA195	0.55	0.79	1.02	1.26	1.49	1.73	1.96	2.20	2.43	2.67
DMH000	0.60	1.10	1.60	2.11	2.61	3.11	3.61	4.12	4.62	5.12
GAR270	0.75	1.20	1.66	2.11	2.57	3.02	3.47	3.93	4.38	4.83
JAB220	1.20	1.63	2.07	2.50	2.94	3.37	3.81	4.24	4.68	5.11
KJM000	0.10	0.16	0.21	0.27	0.33	0.38	0.44	0.50	0.55	0.61
KJM090	0.10	0.18	0.26	0.35	0.43	0.51	0.59	0.67	0.75	0.84
LBC360	2.00	2.86	3.73	4.59	5.45	6.32	7.18	8.04	8.91	9.77
LOS270	0.15	0.25	0.35	0.45	0.54	0.64	0.74	0.84	0.94	1.04
MAN000	0.65	0.85	1.06	1.26	1.47	1.67	1.87	2.08	2.28	2.49
NWH360	0.10	0.18	0.27	0.35	0.43	0.52	0.60	0.68	0.76	0.85
PRS000	0.60	1.09	1.58	2.08	2.57	3.06	3.55	4.04	4.54	5.03
TAF021	0.50	0.80	1.10	1.40	1.70	2.00	2.30	2.60	2.91	3.21
TOT090	0.50	1.18	1.87	2.55	3.24	3.92	4.61	5.29	5.97	6.66
YBI000	2.25	3.93	5.60	7.28	8.95	10.63	12.30	13.98	15.66	17.33

4   Nonlinear Models for Reinforced Concrete Members

A large amount of literature describes the results of various research conducted to accurately reproduce the nonlinear behaviour of reinforced concrete members and the associated damage states, and to accommodate the available computational resource (CEB 1996).

The most powerful analysis tool is recognized to be the nonlinear 3D FEM approach. This however has the major disadvantage of large computational effort that may exceed easily, not for very large models, the available software and hardware resources.

Thus two basic approaches are considered in the current practice: (i) the discrete plasticity model and (ii) the distributed plasticity model.

The discrete plasticity model is the simplest one. In the link model, the inelastic deformations are concentrated pointy at the element's ends only, in the potential plastic zones. A number of six uncoupled nonlinear springs with zero length, are willing to catch through predefined hysteretic rules, the structural response. The hysteretic rules can be either piecewise- or smooth-like formulations, able to represent primarily the stiffness and strength deterioration under cyclic loading. The advantages of this type of model are the simplicity, low computational effort and easy control. However, the lack of correlation between axial and flexural stiffness, some major simplifications in the modelling (e.g. the N-M interaction) when compared to the real observed behaviour requires the need to investigate alternate higher complexity models. The link model is currently used in the Applied Element Method (AEM) proposed by Tagel-Din and Meguro (2000, 2001) to reproduce the seismic progressive collapse and a variety of structural engineering applications that use to reproduce the ultimate limit state (impact, blast etc.). The alternate approach for discrete plasticity is the well-known hinge model.

The most promising distributed plasticity model from the point-of-view of simplicity of modelling, accuracy and computational effort is the fibre model (Mark 1976). It consists first of a section discretization in fibres, and then the fibre slices are distributed along the element's length. The constitutive law is not explicitly requested but derived from the fibres responses based on the Bernoulli's assumption of plane section. The response is obtained by first integrating over the fibres and then the integration follows the slices distribution along the element. The bi-axial flexure is explicitly considered in a coupled manner.

A computational compromise in between these two models is the finite length plastic hinge model (Scott et al. 2006), a counterpart of the fibre model.

The three models briefly reviewed above are used to study the modelling effect on the static and dynamic pushover curves. This is accompanied by a statistical analysis on the pushover parameters.

4.1 Study Case

A six story regular RC frame structure serves for the purpose of this study (Fig. 2). A 18 m height frame structure was designed according to the P100-1 (2013) Seismic Design Code provisions for Bucharest site conditions ($a_g = 0.30$ g, $T_c = 1.6$ s) for high ductility demand class (q = 6.75).

The modeling and the analysis were performed using the features of SeismoStruct 2016 software (www.sesimosoft.com) which accommodates all the nonlinear models previously discussed.

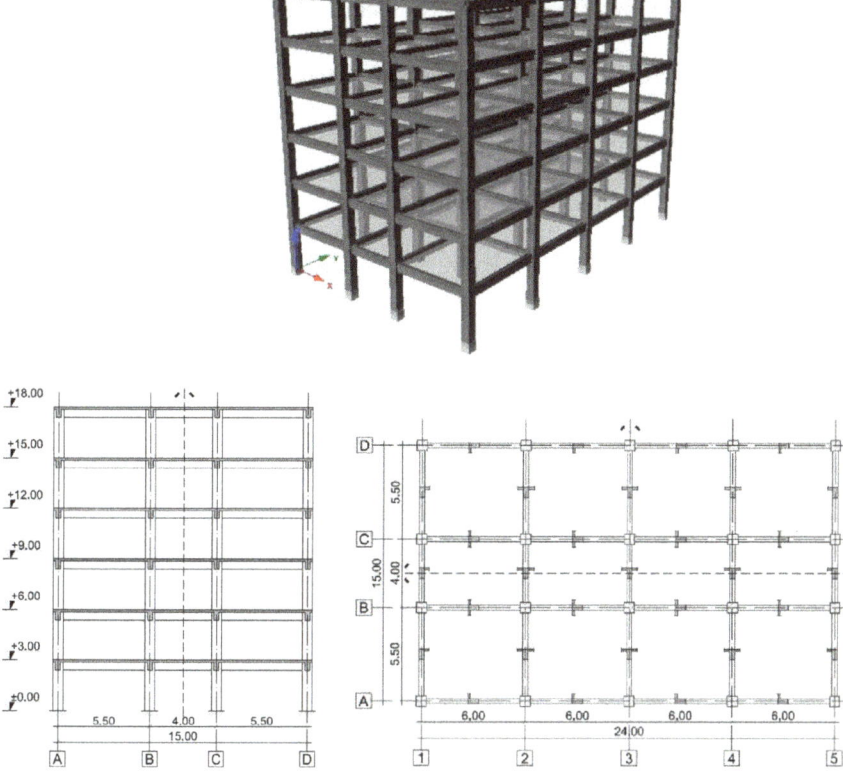

Fig. 2 3D view of building (SeismoStruct 2016 software graphics, upper figure), transverse section and current floor plan view (lower figures)

The concrete class is C25/30 and the steel is of BST500S grade. The columns are of 60×60 (cm) cross section and the beams are of 30×60 (cm). The slab is 13 cm thick. The concrete stress-strain relationship uses the Mander et al. (1988) model as for the steel is used a bilinear stress-strain relationship.

The fundamental natural periods in the x-, y- and z- directions are $T_{1x} = 0.76$ s (sway), $T_{1y} = 0.73$ s (sway), $T_{1z} = 0.21$ s (torque), respectively. The fundamental sway vibration modes damping ratios are set to 5% and a Rayleigh proportional damping model was used to construct the damping matrix of the structure.

Two types of nonlinear model for RC are considered in this study: a discrete plasticity model, using the basic link model approach, and a distributed plasticity approach using (i) the finite length plastic hinge model (inelastic force-based plastic hinge frame element infrmFBPH) and (ii) the fibre model (inelastic force-based frame element type infrmFB), all incorporated in the SeismoStruct 2016 software platform.

For the distributed plasticity fibre model, a number of 200 fibers were selected to accurately represent the stress-strain distribution. Additionally, six integration sections along the member's length were chosen. For the finite length plastic hinge model, a 10% integration length at the member's ends was considered.

The discrete plasticity link model requires the definition of an independent force-deflection relationship for each of the six degrees of freedom associated. The axial, shear and torque components were modelled linearly, as for the biaxial flexural components the modified Takeda model was chosen. The parameters to be provided (XTRACT 3.0.5 software) are the yield force F_y, the initial stiffness, the post-yield stiffness ratio (chosen as 0.05), the inner stiffness degrading ratio (chosen as 0.9) and the outer stiffness degrading ratio (chosen as 0.4).

By performing the nonlinear static and dynamic incremental analyses for the y-direction only (Fig. 2), static and dynamic pushover curves were obtained. The pushover curves corresponding to the three plasticity models are two pushover methods are represented in the Fig. 3.

Additionally, the variability of base shear force and of tip displacement is represented for each of the incremental dynamic loading steps (Fig. 3).

One can be noted that for the same plasticity model, the uniform loading type tends to provide the upper bound of base shear force at almost same displacement capacity level while the adaptive load-type tends to provide the lower bound of it.

The average IDA curve is best approximated by the triangular incremental load distribution. Moreover the variability of IDA curves covers fairly well the nonlinear static results.

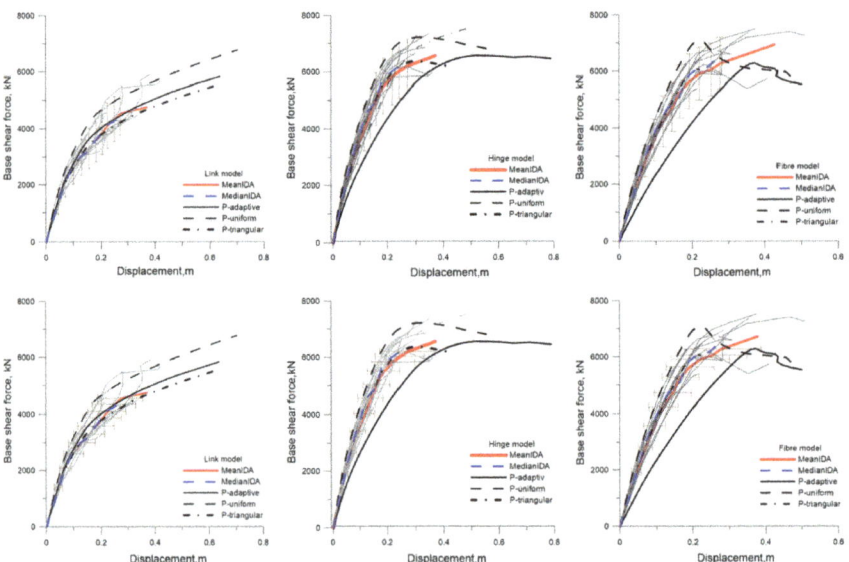

Fig. 3 Pushover curves for three plasticity models and two pushover methods; IDA variability of base shear force (upper figures) and tip displacement (lower figures)

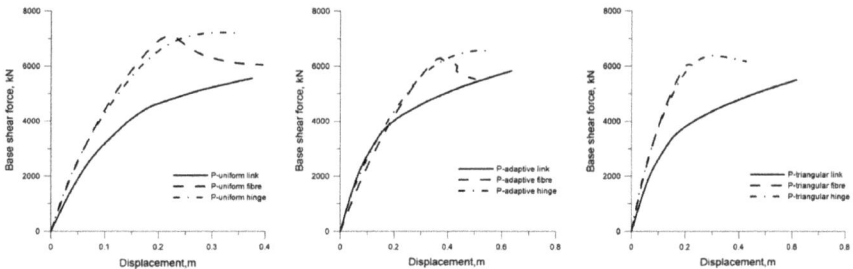

Fig. 4 Pushover curves for three plasticity models and three static lateral loading patterns

Table 4 Statistics of peak base shear force and tip displacement

	IDA link		IDA hinge		IDA fibre	
	F_b (kN)	D_{top} (m)	F_b (kN)	D_{top} (m)	F_b (kN)	D_{top} (m)
Mean	4546	0.27	6214	0.27	6282	0.28
Standard deviation	804	0.08	980	0.10	932	0.11
COV	0.18	0.31	0.15	0.40	0.16	0.37

For the three plasticity models and same static lateral loading patterns the static pushover curves are represented in the Fig. 4.

For the same static loading pattern, the link model provides the lower bound while the distributed plasticity models, the upper bound. This is true for the mean IDA curves as well (Fig. 3).

The dispersion of dynamic analyses results tends to increase with the input acceleration demand. The statistics of IDA curves for the three nonlinear type models are given in the Table 4.

The link model provides the lower bound of mean base shear force and mean tip displacement. The force variability is however the highest while the tip displacement is the lowest one.

5 Probabilistic Seismic Damage Assessment

The probabilistic seismic damage quantification and loss estimation for typical structures is currently performed based on the well-known HAZUS methodology (2007). HAZUS framework is dependent of knowledge on the structural capacity and seismic demand.

The first information required is the maximum response of the actual structure. Thus, the actual capacity curve is converted into an equivalent SA-SD bilinear curve for the nonlinear SDOF system associated to the fundamental vibration mode of the actual structure (CSM; Freeman 1998; ATC-40 1996; Fajfar 1999; Iancovici 2001; Iancovici et al. 2002).

The inelastic seismic demand given in SA-SD format response spectrum (ADRS) is obtained by reducing the linear-elastic spectral ordinates, according to the ductility level. The intersection point between these two representations–demand and capacity yields the performance point i.e. the actual nonlinear response spectral acceleration-spectral displacement pair (Freeman 1998; Fajfar 1999; ATC-40; FEMA-440; FEMA-356). The spectral displacement thus obtained is then further used in the loss estimation methodology.

The second information required consists of appropriate fragility functions—i.e. the probability that a structural system of being in- or being out of a certain damage state. The fragility functions are modelled as log-normal distributions given by median and log-standard deviation (β_{SD}) pairs associated to a certain structural typology (i.e. C1 M-type in this case), design code level (i.e. high code in this case) and height regime (i.e. low-rise in this case).

In the IDA approach the variability of spectral response parameters includes an intra-model variability component $\sigma_{intramodel}$ and an inter-model variability component $\sigma_{intermodel}$, respectively. Assuming that the results obtained using the three nonlinear models are statistically independent, the total standard deviation of spectral pairs of the equivalent SDOF median IDA curve at each acceleration loading step can be obtained using the SRSS approach i.e.

$$\sigma = \sqrt{\sum \sigma_{intramodel}^2} \qquad (1)$$

In the Fig. 5 the intra-model and the total standard deviation of spectral acceleration and spectral displacement of the equivalent SDOF system's capacity are represented corresponding to the each acceleration loading step in the IDA analysis framework.

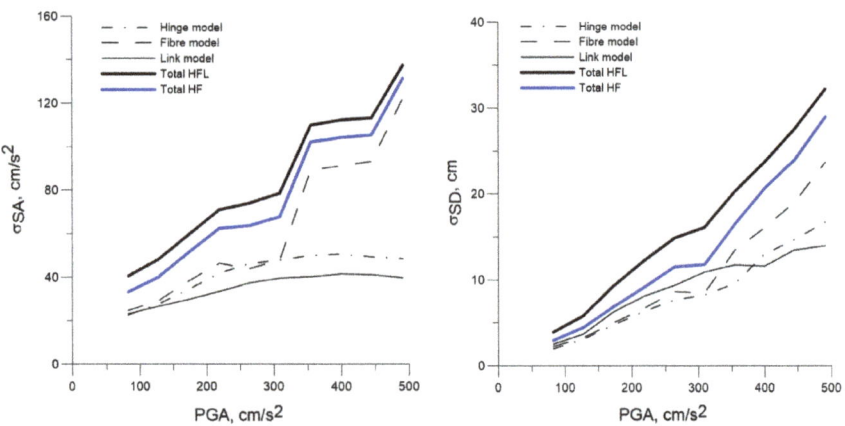

Fig. 5 Standard deviation of the spectral accelerations (left figure) and spectral displacements (right figure) from the equivalent SDOF IDA curves

The variability of spectral response parameters increases with the input acceleration level. The link model provides the lowest variability of spectral acceleration as well as the finite length hinge model provides the lowest variability of spectral displacement. The influence of the link's model contribution is also emphasized in the Fig. 5. The largest variability for higher input acceleration levels is brought in by the fibre nonlinear model.

In terms of log-standard deviation of spectral displacement as a measure of variability around de median values in the HAZUS methodology, these can be obtained similarly by using the SRSS approach. The variation of log-standard deviation of spectral displacement β_{SD} are represented corresponding to each acceleration loading step from the IDA approach (Fig. 6).

The results obtained in this study follow the values suggested by HAZUS methodology based on a larger amount of processed data (HAZUS 2007; Table 5).

The Capacity Spectrum Method (CSM) procedure for the IDA finite length hinge model only is represented in the Fig. 7. The SA-SD parameters of the equivalent SDOF bi-linear capacity curve are computed using the relationships provided by ATC 40 (1996) provisions. The demand spectrum is provided by the linear-elastic absolute acceleration response spectrum from P100-1 (2013) code for Bucharest site conditions.

The associated damage state probabilities corresponding to four pre-defined damage states defined by median and β-pair values of spectral displacements (HAZUS 2007) are then computed (©Matlab routine; Fig. 8).

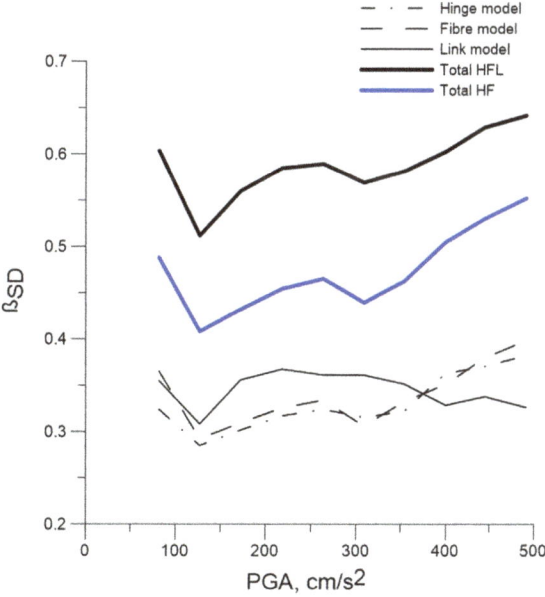

Fig. 6 Log-standard deviation of spectral displacements from the equivalent SDOF IDA curves

Table 5 Median and β pairs of spectral displacement SD (C1M low-rise building, high-code; HAZUS 2007)

	Slight damage	Moderate damage	Extensive damage	Collapse
Median SD, cm	3.81	7.62	22.86	60.96
β_{SD}	0.68	0.67	0.68	0.81

Fig. 7 Capacity spectrum method (CSM) for P100-1 (2013) code demand for Bucharest site conditions

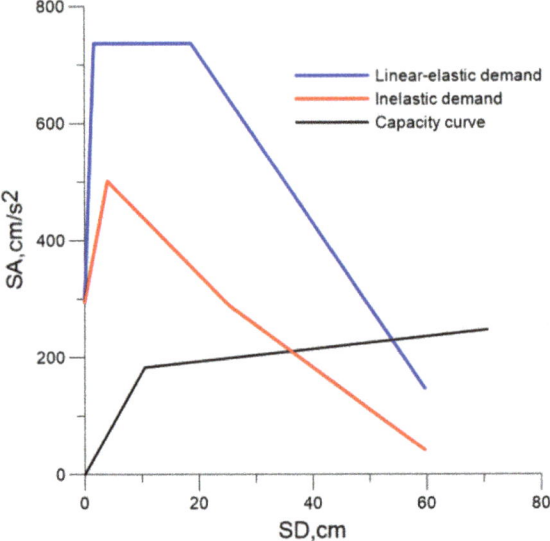

Fig. 8 Fragility functions for the IDA hinge model and associated damage state probabilities

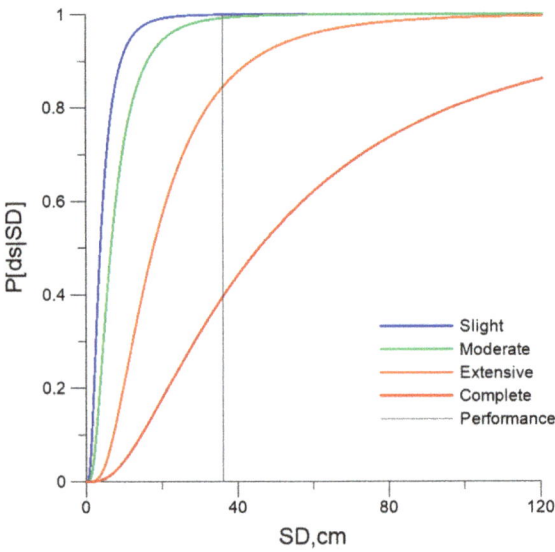

Fig. 9 Damage probabilities associated to P100-1 (2013) seismic code demand for Bucharest

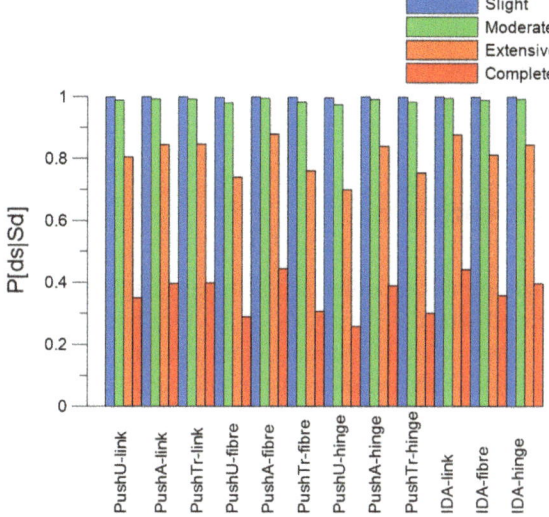

Fig. 10 Normalized damage indices for P100-1 (2013) seismic code demand for Bucharest

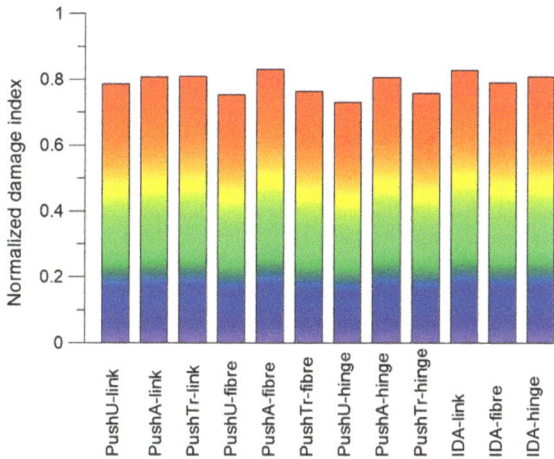

The distribution of damage probabilities (following the damage ranking 0—none, 1—slight, 2—moderate, 3—severe, 4—complete) for all the nonlinear models and loading patterns are given in the Fig. 9.

The normalized associated damage indices (1 means collapse) over the nonlinear models and lateral loading patterns used are represented in the Fig. 10.

The overall damage index distribution indicates an extensive damage state of the building for the specified seismic demand. The influence of the nonlinear model and lateral load distribution pattern is not negligible.

The uniform static lateral loading pattern and the finite length hinge model provides the lowest overall damage index (2.92) while the IDA link model, the highest one (3.32). These can be viewed as lower and upper margins for current seismic loss estimation analyses.

6 Conclusions

While significant difference between static nonlinear approach and dynamic one is pointed out, the variability of IDA response covers fairly well the static-based response variability coming from the lateral loading pattern used. The triangular static load distribution tends to reproduce best the dynamic analysis results in their mean and median sense.

The probabilistic seismic fragility analysis emphasizes that the choice of the nonlinear model type and the seismic loading pattern may influence the global damage index in a range of 15%. This might even more influenced by considering the effect of spatial correlation, input ground motion selection that must be carefully investigated especially for tall flexible building structures.

A more accurate prediction of structural capacity in conjunction with integrated tools for the assessment of the damage states up to progressive collapse associated to various earthquake scenarios and time-domain analysis approach would truly lead implementing a real performance-based seismic design into practice.

Acknowledgements Part of the research presented in the paper was supported through the Seismic Risk Evaluation Center (CCERS) of Technical University of Civil Engineering of Bucharest (UTCB) by the RO-RISK Project SIPOCA 30/2016 funded by the Ministry of Regional Development and Public Administration and co-funded by the European Social Fund. The authors highly acknowledge this support.

References

Antoniou S, Pinho R (2004) Advantages and limitations of force-based adaptive and non-adaptive pushover procedures. J Earthq Eng 8(4):497–522

ATC (1996) Seismic evaluation and retrofit of concrete buildings, ATC-40 report, vols 1 and 2. Applied Technology Council, Redwood City, California

CEB (1996) RC frames under earthquake loading. State of the Art report. Thomas Thelford. ISBN 072772085 6

Chopra AK, Goel RK (2002) A modal pushover analysis procedure for estimating seismic demands for buildings. Earthq Eng Struct Dyn 31:561–582

Elnashai AS (2001) Advanced inelastic static (pushover) analysis for earthquake applications. Struct Eng Mech 12(1):51–69

Fajfar P (1999) Capacity spectrum method based on inelastic demand spectra. Earthq Eng Struct Dyn 28:979–993

FEMA-356 (2000) Prestandard and commentary for the seismic rehabilitation of buildings, Federal Emergency Management Agency

FEMA-440 (2005) Improvement of nonlinear static seismic procedures, ATC-55 Draft, Washington

Freeman SA (1998) The capacity spectrum method as a tool for seismic design. In: Proceedings of the 11th european conference on earthquake engineering, Paris, France

HAZUS-MH 2.1 (2007) Multi-hazard loss estimation methodology. Earthquake model estimates earthquake damage and loss to buildings. Department of Homeland Security, Federal Emergency Management Agency, Mitigation Division, Washington DC

Iancovici M (2001) The assessment of the reinforced concrete building structures seismic performance based on the elements performance. Bull Int Inst Seismolog Earthq Eng 239–251

Iancovici M, Fukuyama H, Kusunoki K (2002) The assessment of the reinforced concrete building structures based on the seismic performance concept. Fifth international congress on advances in civil engineering, Turkey, vol 1, pp 555–564

Katsanos E, Sextos A, Manolis G (2010) Selection of earthquake ground motion records: a state-of-the-art review from a structural engineering perspective. Soil Dyn Earthq Eng 30: 157–169

Kramer S (1996) Geotechnical earthquake engineering. Prentice Hall Inc, NJ

Mander JB, Priestley MJN, Park R (1988) Theoretical stress-strain model for confined concrete, J Struct Engi 114(8):1804–1826

Mark K (1976) Nonlinear dynamic response of reinforced concrete frames. Department of Civil Engineering, Massachusetts Institute of Technology, Cambridge, MA (Res. Rep. R76-38)

P100-1 (2013) Code for seismic design—part I—design prescriptions for buildings. Bucharest

Păunescu VA (2013) Abordări statice şi dinamice în determinarea curbei de capacitate a structurilor. Lucrare de disertatie, Universitatea Tehnică de Constructii Bucuresti (in Romanian), p 60

Pietra D (2008) Evaluation of pushover procedures for the seismic design of buildings. M.Sc. Thesis, Rose School, Pavia, Italy

Scott MH, Fenves GL (2006) Plastic hinge integration methods for force-based beam-column elements. J Struct Eng ASCE 132(2):244–252

SeismoStruct (2016) A computer program for static and dynamic nonlinear analysis of framed structures. Available online at www.seismosoft.com

Tagel-Din H, Meguro K (2000) Applied element method for simulation of nonlinear materials: theory and application for RC structures. Jpn Soc Civil Eng (JSCE) 17(2):137–148

Tagel-Din H, Meguro K (2001) Applied element simulation of RC structures under cyclic loading. J Struct Eng ASCE 127(11):137–148

Vamvatsikos D, Cornell CA (2002) Incremental dynamic analysis. Earthq Eng Struct Dyn 31 (3):491–514

Nonlinear Design Optimization of Reinforced Concrete Structures Using Genetic Algorithms

Pricopie Andrei and Cimbru Iulian-Valentin

Abstract The aim of this study is to identify the optimal cross sections and reinforcement area of the beams and the columns in a frame structure in order to improve their seismic response. A six-storey reinforced concrete structure, modelled using fibre finite elements to represent the nonlinear behaviour, was studied. The structure was first designed according to existing codes (the equivalent static load method) and the total quantity of material computed (concrete and reinforcement). Secondly, a genetic algorithm was used in order to determine the optimized beam and column cross section dimensions along with the longitudinal reinforcement area for both types of elements. The optimization process aims to minimize the cost of the reinforced concrete structure while at the same time improving or at least maintaining the seismic response of the structure. A set of constraints was imposed for the optimization process: the minimum reinforcement area, the ratio between the beam height and width and the maximum drift of the structure at both the ultimate and service state, according to design codes. Dynamic time history analysis is used to determine the maximum drift of the structure. The analyses are carried out for three ground motion records which represent the seismic conditions given by the Vrancea source in Bucharest.

Keywords Structure optimization · Seismic design · Nonlinear analysis Multi objective

P. Andrei (✉) · C. Iulian-Valentin
Technical University of Civil Engineering of Bucharest, Bucharest, Romania
e-mail: andrei_pricopie@yahoo.com

C. Iulian-Valentin
e-mail: cimbruiulian@yahoo.com

© Springer International Publishing AG, part of Springer Nature 2018
R. Vacareanu and C. Ionescu (eds.), *Seismic Hazard and Risk Assessment*,
Springer Natural Hazards, https://doi.org/10.1007/978-3-319-74724-8_27

1 Introduction

Reinforced concrete structures have known a large usage since the last century and the FEM software are more capable, in the last years, to predict with high accuracy the structural response under static or dynamic actions.

At the same time, only few of these software programs are able to size structural elements and even less of them can realize the structural design optimization for nonlinear analysis.

In the area of reinforced concrete linear design optimization, Balling and Yao (1997) used genetic algorithms to optimize concrete-section dimensions and the number, diameter, and topology of the reinforcing bars, while others (Pezeshk et al. 2002) used them for the steel structures design optimization based on American design codes (AISC). Fadaee et al. (1996) used the Optimality Criteria (OC) method in order to minimize the cost of the concrete, steel and formwork for the structure.

Gharehbaghi and Khatibinia (2016) studied the seismic response of two structures (9 and 18 stories) under 10 accelerograms through a time-history analysis, using particle swarm optimization and an intelligent regression model.

Genetic algorithms are widely used in different areas, but one of the disadvantages of this method is that the variables to be optimized are chosen from a continuum interval, instead of a discrete one (Rajeev and Krishnamoorthy 1998). Hejazi et al. (2013), addresses using genetic algorithms to improve the nonlinear dynamic analysis of a structure. The genetic algorithm optimizes the properties of viscous dampers in order to improve the seismic response of the structure. The obtained results are positive, improving the seismic response of the studied reinforced concrete structure.

However, most studies do not address an optimization procedure which can be directly applied to the design of a structure. To that extent the current study aims to investigate the opportunity of using genetic algorithms in order to design a reinforced concrete frame.

In the first stage of the current study, the structure was dimensioned using static linear analysis (equivalent load method) according to design codes (P100-1/2013; EN 1998-1:2004) and this model was considered the reference model, whose results will be compared with the results of optimized models.

In the second stage, through a nonlinear time-history analysis, two optimizations functions were studied:

- Seismic response optimization through minimizing the maximum drift (structure cost was limited to the cost of the reference model);
- Structure cost optimization (maximum drift was limited to the maximum allowable drift from P100-1/2013, for both two limit states: service state, SLS and ultimate state, ULS).

2 Reference Model

2.1 Static Analysis

The analysed structure has 6 stories with the height of 3 m and the span of 6 m on both directions and will be modelled as a plane reinforced concrete frame. The building is located in Bucharest, where $a_g = 0.30$ g and $T_c = 1.6$ s for a MRI of 225 years (P100-1/2013).

The considered live load was 2.5 kN/m^2, while the permanent load was considered uniformly distributed on the slab and has a value of 3 kN/m^2. The analysis was performed according to Eurocodes (EN 1998-1:2004). The employed materials were BST500S for reinforcement and C30/37 as structural concrete.

Using the lateral equivalent force distribution method, the drift verification (P100-1/2013) was done and the structural elements were designed. The service limit state verification was the critical verification. The results of the drift of the structure are presented in Fig. 1 (Left). The resulting rebars are presented in Fig. 1 (right). Since the frame is symmetric only 2 spans of the four are represented with the rebar being symmetric, the position of the text of the rebar corresponds to the position on the cross section. Because the resulting efforts from the elastic analysis have only a marginal variation on each storey beams, the upper and lower rebars have been considered constant on each storey beam. The columns have a square section of 800 mm with 16 rebars of 25 mm diameter, while the beams dimensions are 600×300 mm.

In order to assess the cost of the structure the price of concrete, longitudinal reinforcement and formworks were accounted for, as materials. The cost of labour is not included, however for comparison purposes required by the genetic algorithm the labour cost would be proportional to the quantity of material and will not influence the outcome of the optimization process. The structure cost was estimated using the below unit prices (transversal reinforcement was ignored).

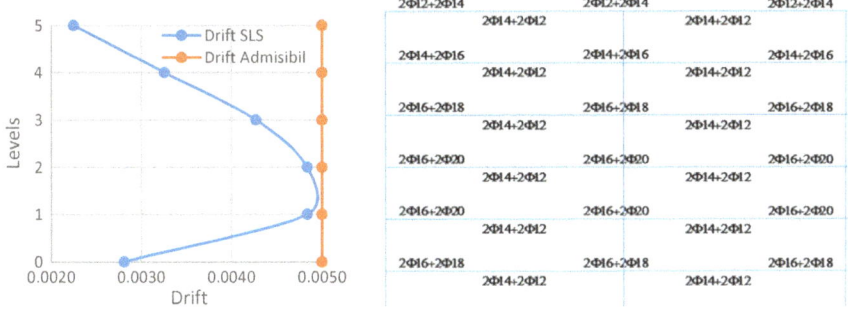

Fig. 1 Reference model. SLS drift verification (left) and beam longitudinal reinforcement (right)

$$Cost = (Concrete_{columns} + Concrete_{beams}) * 220\,\text{MU/m}^3 + (Reinforcement_{columns}$$
$$+ Reinforcement_{beams}) * 4.4\,\text{MU/kg} + (Formwork_{columns} + Formwork_{beams})$$
$$* 50\,\text{MU/m}^2$$

Total structure cost for the reference model: 52,131 MU (Monetary Units, 1 MU ≈ 0.222 €).

2.2 Dynamic Nonlinear Analysis

The dynamic nonlinear analysis was done using OpenSees (2015), which was launched from MATLAB, where the input files were generated by a genetic algorithm. The optimized models' results were validated using Seismostruct v7.0.3 (2017), which uses a similar nonlinear fibre model for the elements and the results can be displayed using the graphical interface. The Seismostruct model is presented in Fig. 2.

The seismic action was considered as the peak interval of 4 s, in terms of P.G.A., for 3 representative seismic records (I.N.C.E.R.C. 4th of March 1977, 30th of August 1986 and 31st of May 1990). The records were scaled linearly in terms of P. G.A. (in order to correspond to the P.G.A. of 225 years MRI used in P100-1/2013) and was applied at the frame base on the longitudinal direction. As only the maximum response in terms of displacements and rotation is required for the genetic algorithms, the accelerograms were truncated to the interval which yields the maximum displacement.

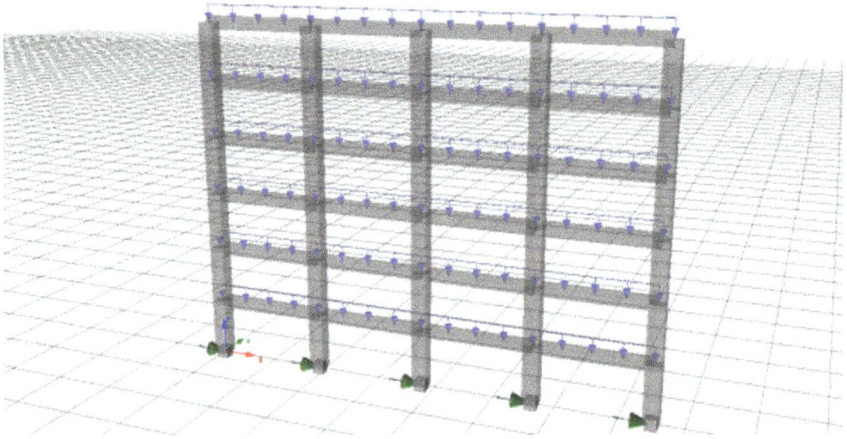

Fig. 2 Seismostruct structural model

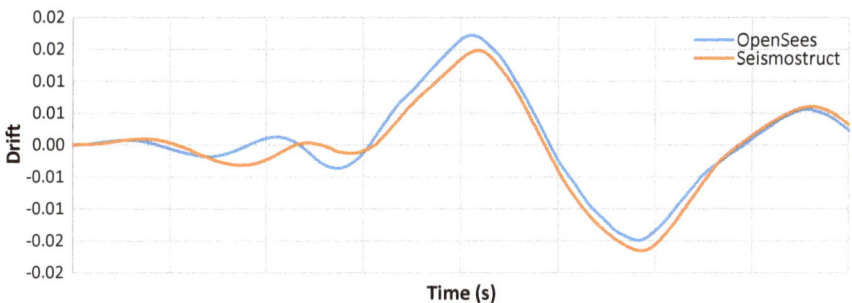

Fig. 3 Comparrison between Seismostruct and OpenSees drift verification at the fifth floor

The material constitutive laws considered for concrete and reinforcement were quite similar for the calibration of the finite element analysis model in OpenSees: OpenSees—Concrete02 (Fedeas Material) and Steel02, Seismostruct—Mander et al. and Bilinear steel model.

It was assumed that the elements failure will occur only as an effect of the bending moment, which is a general accepted hypothesis when capacity design is used.

The plastic mechanism which forms corresponds to design code requirements as plastic hinges are developed at the base of the columns and at the ends of the beams. One of the second storey columns yielded during the 4th March 1977 accelerogram however the plastic rotation is insignificant and this was considered acceptable.

The results obtained from both OpenSees and Seismostruct were similar. In terms of displacement the difference in results was of 14.8%, which validated the OpenSees model of the structure (Fig. 3).

3   Genetic Algorithm (GA)

GAs, which belongs to evolution algorithms category, is a search method based on the genetic natural selection process, using techniques like: inheritance, mutation, selection and crossover. Because of their generality, the utility of these algorithms is infinite (Fraser 1957).

The main concept of GAs is that the best environment-adapted individuals will survive. The adaptability capacity is quantified using a fitness function, or objective function. The optimum is the minimum value of the objective function.

In a GA, a population (of individuals) evolves to better solutions from a generation to another. Each populations' individual has a set of unique properties, which, in Genetics, is the equivalent of the DNA based on chromosomes (or genes). In this case, their equivalents are the variables which express the dimensions and reinforcements of the structural elements.

The technique called mutation refers to the modification of only one of the DNA's chromosomes, while the crossover refers to the concatenation of two complementary parts of two DNAs. In this way, each variable can be modified using these techniques.

Evolution starts from a random generated population, with variables, between imposed limits and it is an iterative process. Each iteration's population is called population and for each population, the objective function (seismic response or cost) is evaluated. The best individuals are selected from the current population and their DNA is modified in order to obtain the next population through mutation or crossover.

The genetic algorithm routine was coded in MATLAB. It uses a number of 16 variables: 12 for the beam reinforcement areas (superior and inferior, considered independent on each story), 2 for beams section (considered identical on all stories), one for columns section and one for the columns longitudinal reinforcement diameter (considering 16 bars placed on the section perimeter).

The following design restrictions were incorporated into the algorithm in order to generate only structures which are compliant with design codes and engineering practice: Constructive conditions (minimum reinforcement percentage, minimum reinforcement area), according P100-1/2013 were taken into account for both the beams and the columns. Namely for the columns the longitudinal reinforcement percentage is at least 1%. The ratio between the beam width and height is limited to at least 1/3. Lastly, the inferior reinforcement area in the beams will be at least 1/2 of the superior reinforcement area. These restrictions have been imposed in the genetic algorithm using a series of inequalities which are checked and maintained by the algorithm.

Typical design conditions were accounted for by imposing an interval of values for the following variables: beams reinforcement area: 534–1068 mm^2, beams height h_b : 550−900 mm, beams width (b_b) 250–300 mm, columns section height (h_{col}) 600–900 mm, columns longitudinal bar diameter ($\varnothing_{col}$) : 15−28 mm.

This routine is able to write an input OpenSees file for each set of values generated by the GA (each individual of the population), to calculate the cost, to launch the dynamic analysis for each accelerogram and to calculate the maximum ULS drift (objective function) based on the OpenSees output files. An outline of the algorithm is shown in Fig. 4.

The GA in the first phase generates randomly (between the imposed limits taking into account all the conditions imposed) 30 series of numerical values which will be the first population of 30 individuals.

Then an input file for OpenSees is written, the nonlinear analysis is done and then the cost is calculated for each individual. The objective function is evaluated and through the specific methods a new population is generated and this loop continues until the convergence conditions are reached and the optimized model is found.

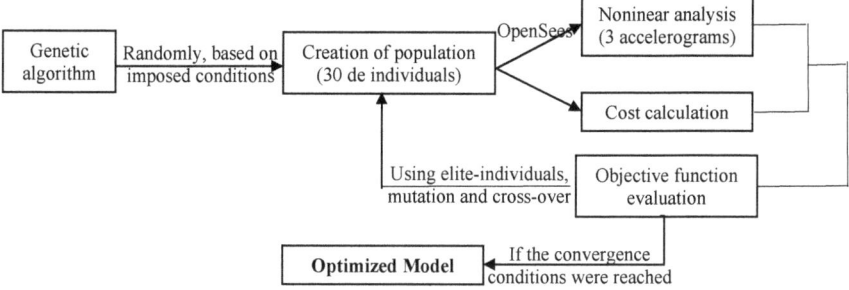

Fig. 4 Genetic algorithm scheme

4 Structural Seismic Response (ULS Drift) Optimization

In this phase, the structural seismic response was optimized compared with the reference model and the ULS drift value was considered as GA objective function. Because it was expected that the drift will decrease as the quantity of material increases, which would result in a higher cost, the cost was limited to the value of the reference model (52,131 MU).

$$\text{Objective function: Drift}_{\text{ULS}} = \begin{cases} \text{Drift}_{\text{ULS}}, & \text{if Cost} \leq \text{Cost}_{Reference\ model} = 52,131\ RON \\ \text{Drift}_{\text{ULS}} * 1.5, & \text{if Cost} > \text{Cost}_{Reference\ model} = 52,131\ RON \end{cases}$$

For the optimization process, some limits were imposed on each variable in order to narrow down the search space and account for design restrictions. As such the minimum percentage of rebar was considered for both beams and columns. Also, the area of rebar in the lower part of the beam is at least half of the area of rebar on the upper part of the beam.

For this optimization, 356 models were analysed and the results are presented below with the allowable ULS drift and the cost of the reference model. It can be easily observed that the beam height and reinforcements were increased and in this way for the same cost, a drift optimization was achieved (38%/63%), even if the values of the columns chord rotations and first level relative accelerations increased.

This optimization type can be efficiently used when one critical parameter of the seismic response (in this case, the ULS drift) has to be optimized, by considering it as objective function. So, the ULS drift was optimized by 38%. In Fig. 6 left, the results of the model are presented along with the elements which have yielded. Negligible plastic deformations appear in the second level columns (blue) compared with the ones in the first level columns (red, Fig. 6, right). Similar behaviour was registered for the reference model. The value of the rotation can be neglected.

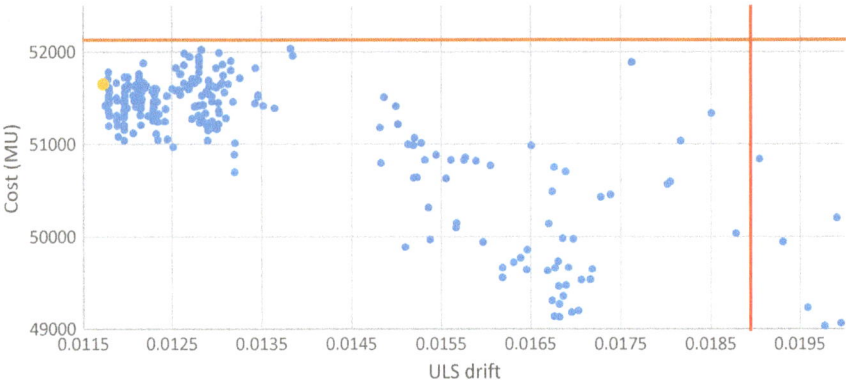

Fig. 5 ULS drift optimization

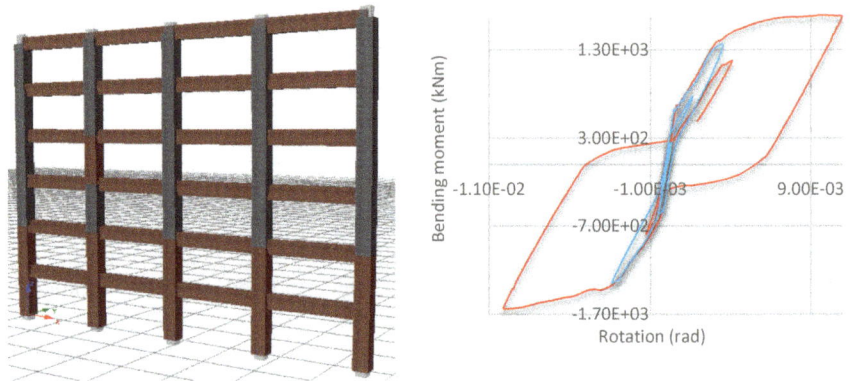

Fig. 6 Optimized model—elements in post—elastic stage (left); hysteretic curves for the first and second level column (right)

The ultimate plastic rotation has been checked in all of the plastic hinges and is compliant for all the structural elements. In Fig. 5, the results of the genetic algorithm in terms of cost and drift are presented. The horizontal and vertical lines represent the values of cost and drift for the reference model. In the below table (Table 1) $\sum A_{rb}$, represents the total area of reinforcement in the beams, h_b, b_b are the dimensions of the beam, $h_{col}, \varnothing_{col}$, are the dimension and reinforcement in the column, θ_{col}, θ_b represent the maximum column and beam rotations, acc_{rel} represents the maximum relative acceleration. The results of the optimization process indicate that improved behaviour in terms of drift response can be achieved by substantially increasing the height of the beam and the dimensions of the columns while reducing the amount of reinforcement in the columns. The results indicate a reduction in drift while the acceleration is increased.

5   Cost Optimization

In this phase of the study, the structure cost (objective function) will be optimized, simultaneously with the SLS and ULS drift verification according P100-1/2013. For this optimization, two nonlinear analyses (one for each limit state) are done for each model and each considered accelerogram.

$$\text{Objective function : Cost} = \begin{cases} \text{Cost, if } \text{Drift}_{\text{ULS}} \leq \text{Drift}_{\text{all,ULS}} = 0.025 \text{ } and \text{ } \text{Drift}_{\text{SLS}} \leq \text{Drift}_{\text{all,SLS}} = 0.005 \\ \text{Cost} * 1.5, \text{ if } \text{Drift}_{\text{ULS}} > \text{Drift}_{\text{all,ULS}} = 0.025 \text{ } or \text{ } \text{Drift}_{\text{SLS}} > \text{Drift}_{\text{all,SLS}} = 0.005 \end{cases}$$

952 models were analysed and the results are presented in Fig. 7 with the allowable SLS drift and the cost of the reference model.

In Fig. 8, the decrease of the objective function, represented by the cost of the structure, is presented during the generations:

The results are presented in Table 2. The optimized model has a cost of 47890 MU, (8.1% less than the reference model), while the SLS drift was 0.00497 (7.4% smaller than the reference model). Again, it can be easily observed that the beam height increased compared with the reference model. Looking at the failure mechanism, we can see that the reinforcement yields in the columns at the first two levels with negligible plastic deformations, similar with the previous optimization (Fig. 6, right) and the chord rotation verification is checked in all the structural elements. So, for the optimized model, by increasing the beam height, the cost was reduced with 8.1%. In Fig. 7 the vertical and horizontal line represent the cost and drift of the reference model.

The columns chord rotation increases simultaneously with the decreasing of the column section height and reinforcement. Again, the variable which changes the most is the height of the beam. Using the cost as an objective function produces an economy of 8.1%, while also marginally improving both drifts and relative acceleration. The optimized model managed to decrease the amount of rebar used by almost 20%.

Table 1 Results of the U.L.S. optimization process

	Input parameters					Results					
	$\sum A_{rb}$ (mm^2)	h_b (mm)	b_b (mm)	h_{col} (mm)	$\varnothing_{col}$ (mm)	Cost (MU)	ULS drift	SLS Drift	θ_{col} (rad)	θ_b (rad)	acc$_{rel}$ (m/s^2)
Reference model	8330	600	300	800	25	52,131	0.019	0.005	0.016	0.017	4.5
Optimized model	9642	860	250	894	16.3	51,655	0.011	0.002	0.019	0.015	6
Change (%)	15.8	43.3	−16.7	11.8	−34.8	−0.9	−38.0	−63.0	9.4	−18.4	33.3

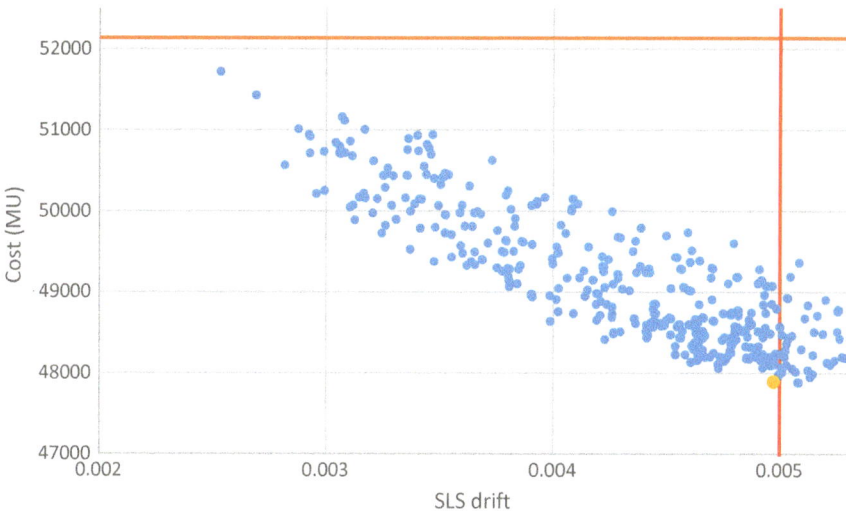

Fig. 7 Cost optimization

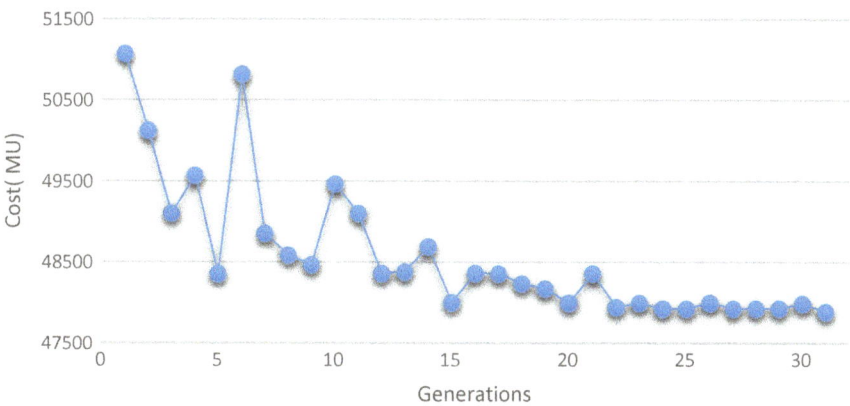

Fig. 8 Decrease of objective function for each generation

Table 2 Results of the Cost optimization process

Input parameters					Results					
$\sum A_{rb}$ (mm²)	h_b (mm)	b_b (mm)	h_{col} (mm)	$\varnothing_{col}$ (mm)	Cost (MU)	ULS drift	SLS Drift	θ_{col} (rad)	θ_b (rad)	acc_{rel} (m/s²)
Reference model										
8330	600	300	800	25	52,131	0.0189	0.0054	0.016	0.019	4.5
Optimized model										
6693	815	250	763	20.2	47,890	0.018256	0.005	0.017	0.017	4.3
Change (%)										
−19.7	35.8	−16.7	−4.6	−19.2	−8.1	−3.4	−7.4	6.3	−10.5	−4.4

6 Conclusions

The possibility of successfully using GAs in order to optimize the dimensions of a reinforced concrete structure for a nonlinear seismic response was demonstrated. A straight forward implementation of the usual process which occurs during the design process of a structure has been optimized using G.A. The algorithm presented accounts for the most important practical design restrictions given by the sizes of the elements of the structure, the minimum percentage of rebar and the ratio of top to bottom rebar in a beam, by including in the algorithm the imposed conditions. The genetic algorithm is also adapted in order to solve the problem by manipulating the fitness function.

To that extent the design process does not involve optimizing only one function like cost or seismic response but actually tries to find a balance between the two functions. Using genetic algorithms, this can be achieved by using multi-objective optimization, which involves adding more complexity to the problem. The current procedure bypasses this problem by imposing conditions on both cost and response of the structure, while actually still using a single function for optimization.

The study investigated two fitness functions, one aiming to optimize the cost of the structure, while keeping the displacement response within code limits, and another which aims at optimizing the response of the structure while keeping the cost of the structure close to a conventional design. The results of the optimization process are presented in Table 3.

The seismic response optimization improved the ULS drift by 38% (the SLS drift was improved with 63%) for the same cost, while for the cost optimization, the cost was reduced with 8.1% taking into account the all the seismic code verifications.

For each optimization realized, the main purpose was achieved and the reference model shows differences compared with the optimized models. This proves that the seismic response of the static linear calculated structures can be significantly improved, while the cost can also be optimized with a percentage of 8.1%.

For both optimizations objective function, an increase in beam height improves the seismic behaviour in terms of drift while decreasing or at least maintaining the cost. For the drift optimization, an increase in the element dimension both for the beam and for the column is beneficial accompanied by an increase in beam reinforcement and a decrease in column reinforcement. In this case, the cost of the structure is maintained. In the case of cost optimization, the rebar quantity has been

Table 3 Results of the optimization process

	Cost (MU)	ULS drift	SLS drift
Reference	52,131	0.0189	0.0054
Drift response optimization	51,655	0.0117	0.002
	−0.9%	**−38.0%**	**−63.0%**
Structure cost optimization	47,890	0.018256	0.005
	−8.10%	**−3.40%**	**−7.40%**

Bold values represents the change in cost ULS drift and SLS drift

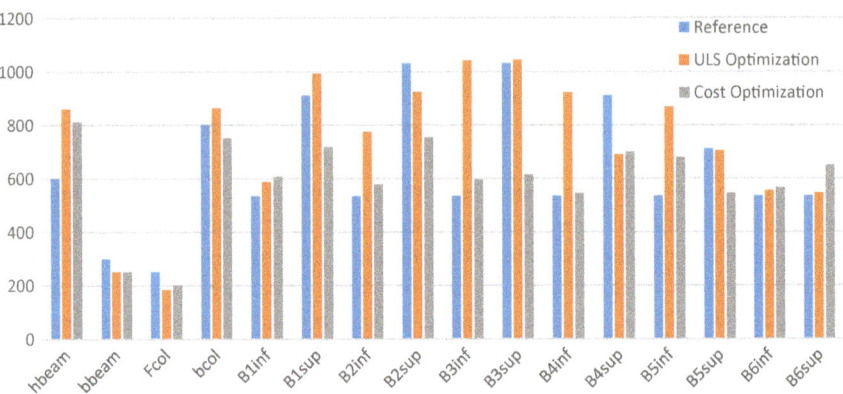

Fig. 9 Input and Output values for the reference model and the optimized models

reduced by almost 20% while registering marginal improvement 3, 7% for S.L.S. drift and U.L.S. drift.

Beam dimensions (height) for the optimized models are large, compared with the values usually used in design. Usually in the design process the dimensions of the beam are smaller, compared to the results in order to have weak beams and develop plastic hinges only at the base of the columns. However, for the analysed model the failure mechanism is still the one recommended by the code. In the same time for the other input values a general trend line cannot be observed during the optimization. Because the beam dimension increase at the same time with the decreasing of the column and/or reinforcement, the column reinforcements from the second level yield, however the plastic rotation is insignificant.

In Fig. 9, there are presented the input parameters for the 3 analysed models (height and width of the beam, the diameter of longitudinal bars[1] and the height of the column section and 12 values for the inferior and superior beam longitudinal reinforcement area).

The limitations of the study are the fact that only flexural rebars in the beams and columns have been considered. The cost of the transverse reinforcement was not taken into account and should be subject to further investigation. Also, the nodes of the structure have been assumed to behave elastically. This hypothesis is generally true however increased beam dimensions given by the optimization process can lead to plastic behaviour in the nodes of the structure. Although it is not likely for the present structure, for other structures it might be significant. A multi-objective genetic algorithm can also be used in order to validate the results obtained by the present classic single objective genetic algorithms.

---

[1]The column longitudinal bar diameter was amplified by 10 in this graphical representation.

References

Balling R, Yao X (1997) Optimization of reinforced concrete frames. J Struct Eng-ASCE 2:193–202

CR0-2011–Cod de proiectare. Bazele proiectarii structurilor in constructii

EN 1998-1:2004—Design of structures for earthquake resistance. Part 1: General rules, seismic actions and rules for buildings

EN 1992-1-1:2006—Design of concrete structures. Part 1-1: General rules and rules for buildings

Fadaee M, Grierson D (1996) Design optimization of 3D reinforced concrete structures. Struct Optim 12(2):127–134

Fraser A (1957) Simulation of genetic systems by automatic digital computers. I. Introduction. Aust J Biol Sci 10:484–491

Gharehbaghi S, Khatibinia M (2016) Optimal seismic design of reinforced concrete structures under time history earthquake loads using an intelligent hybrid algorithm—earthquake engineering and engineering vibration. Earthquake Eng Eng Vibr 14:97–109

Hejazi F, Toloue I, Jaafar MS, Noorzaei J (2013) Optimization of earthquake energy dissipation system by genetic algorithm. Comput Aided Civ Infrastruct Eng 28:796–810

MATLAB and Statistics Toolbox 8.1 (2012) The MathWorks, Inc., Natick, Massachusetts, United States

OpenSees 2.5.0, Open System for Earthquake Engineering Simulation (2015). University of California, Berkeley, United States

P100-1/2013—Cod de Proiectare Seismică—Prevederi pentru cladiri

Pezeshk S et.al (2002) State of the art on the use of genetic algorithms in design of steel structures, recent advances in optimal structural design. Structural Engineering Institute, ASCE

Rajeev S, Krishnamoorthy C (1998) Genetic algorithm-based methodology for design optimization of reinforced concrete frames. Comput Aided Civ Infrastruct Eng 13:63–74

Seismostruct v7.0.3 (2017) Seismosoft, Earthquake Engineering Software Solutions

Presentation of Structural Systems and Characteristic Parameters of Romanian Buildings for Application of the RVS Method

Claudiu-Constantin Stere, Claudiu-Anton Ursu
and Vasile-Virgil Oprişoreanu

Abstract In current codes there is no rapid method of seismic evaluation of buildings, as the three utilised methods are intensively demanding in terms of time, human resource and materials. Due to this reason, it is necessary to adopt and develop a method based on international codes, applicable to the structural systems in our country. As a starting point, the rapid assessment procedure published by the Federal Emergency Management Agency in 1988 is presented. To calibrate the method, 335 buildings were analysed; the vulnerability risk of these buildings was already known, previously certified by approved technical experts that used one of the three methodologies indicated in current seismic evaluation codes. Based on the research, an equivalent point system is created and applied to the structural systems suitable for rapid visual screening (RVS) and also to the characteristic parameters which influence the structural behaviour under seismic actions. Based on the point system applied, the building vulnerability can be determined using the rapid method of seismic evaluation. Moreover, a database is created which allows assessing which buildings require priority for a more detailed analysis through calculations.

Keywords Earthquake · Risk · Vulnerability · Seismic evaluation
Structural system

1 Introduction

The seismic activity in Romania is dominated by subcrustal earthquakes in Vrancea region. This area represents an active source of earthquakes, known for circa 500 years, having specific characteristics which are unique on a global scale. Knowing the country's seismic activities and the effects these had over existing

C.-C. Stere (✉) · C.-A. Ursu · V.-V. Oprişoreanu
Technical University of Civil Engineering of Bucharest, Bucharest, Romania
e-mail: claudiu.stere@alliedengineers.ro

C.-A. Ursu
e-mail: claudiu.ursu@alliedengineers.ro

© Springer International Publishing AG, part of Springer Nature 2018
R. Vacareanu and C. Ionescu (eds.), *Seismic Hazard and Risk Assessment*,
Springer Natural Hazards, https://doi.org/10.1007/978-3-319-74724-8_28

421

buildings, one could say that most of the buildings are vulnerable to seismic actions. This leads to the requirement of evaluating existing buildings and to determine their vulnerability to a potential earthquake.

The first requirements regarding seismic performance of existing buildings were published in code P100-92, chapters seven and eight; MLPAT (1992). Over the period of applying the methodologies of P100-92, the requirements for seismic evaluation have been developed, being updated by the publication of the regulation "Code of seismic evaluation of existing buildings: P100-3/2008"; MDRAP (2008).

The process of seismic assessment in Romanian code involves splitting buildings in classes of seismic risk following some procedures of which complexity varies based on the importance class of the structure, the level of information, etc.; Pantea (1977). If the building evaluation through calculations is realized by engineers with an inappropriate level of knowledge, uncertainties could arise following the analysis; Pantea (1977). Moreover, the wide range of existing structures creates a complex process of determining the risk class of buildings, which requires a lot of time, human resource, materials, etc. Therefore, the currently analysed range of existing buildings is relatively small compared to the whole. These are the reasons which lead to the requirement of a rapid assessment method, allowing to create a database which characterizes all the buildings, with the purpose of prioritizing the requirement of an assessment based on calculations and the codes.

2 The Rapid Assessment Method Based on FEMA 154

The rapid assessment method of existing buildings was first published in the Federal Emergency Management Agency (FEMA) 154 in 1988, being constantly updated and developed. The current version is from 2002 and is described in detail in: "Rapid Visual Screening of Buildings for Potential Seismic Hazards: A Handbook". The method presented by FEMA 154 is based on a structural compliance of buildings, which is realised in a relatively short time, approximately 20–25 min, with the inspection being held from the exterior with no access required inside the building. To determine the seismic vulnerability, the buildings are characterised by a point based system (a score) from which the probability of collapse is determined. The score is initially defined by a baseline score, which results from the structural system of the building, which is then adjusted using influence factors representing parameters such as: building height, vertical irregularity, irregularity in plan, construction year and soil type.

In this way, the final score is calculated with the following equations:

$$S = BSH \pm SMs \qquad (2.1)$$

Table 1 Values of the final score S and the associated probability of collapse; FEMA (2007)

Final score (S)	Probability of exceedance (%)
4.0	0.01
3.5	0.03
3.0	0.10
2.5	0.32
2.0	1.00
1.5	3.16
1.0	10
0.5	32

In Eq. 2.1, BSH is determined by:

$$BSH = -\log 10 \, (Pcollapse) \tag{2.2}$$

where, *Pcollapse* represents the probability of collapse of a building under seismic action of the design earthquake defined by the american code FEMA.

In Table 1, the final score is presented in relation to the probability of collapse. For example, a final score of S = 2.0 means that the probability of collapse of a building under the design seismic action is 10–2 or 0.01 (1%). In practice, the procedure is to fill in standard forms based on the zone seismicity, the person completing the form being required a minimum level of technical knowledge.

Due to the simple nature of the procedure and the high level of accessibility, the result was impressive: 70,000 buildings from the USA were investigated using the procedure. Through this an important database was created which defines the current situation of existing buildings. Based on the obtained results, in 1997 was presented the first system regarding management of urgency situations based on using Geographical Information Systems (GIS) represented by the HAZUS method; HAZUS (2000).

3 Rapid Assessment Methodology of Buildings in Romania

The rapid assessment method of buildings in Romania has the basis of the FEMA. The first step is to determine the structural system of the building to be assessed. Due to the fact that the method is based on a visual inspection from outside the building, the structural system is determined based on engineering judgment, the date of construction, the architectural details and the design codes that were applied at the time of construction. Additionally, a series of parameters that affect the behavior of the building based on structural system are identified, which also help towards determining the grade of seismic vulnerability. Calibration of allocated points for structural system types and characteristic parameters was realised through research of 335 buildings from Bucharest, for which the seismic vulnerability was already known from analysis and calculations of authorised engineers, in compliance with the regulations and codes in use. The buildings used in the study were randomly selected from the list presented on the website of The Bucharest Council; PMB (2014).

3.1 Structural Systems

Based on the variety of types of existing buildings in Bucharest, real issues arise in determining a precise structural system; Lungu (2002) and Bostenaru (2009). To simplify the method, the entire variety of structural types is grouped in three main categories, which are:

- *Structural masonry*

The main feature of buildings with structural masonry is the high level of vertical and in plan irregularity; the system mainly consists in blind walls arranged together in groups of two and working in one direction. This category of structural systems was generally used in very old buildings, usually one storey and sometimes with partial basement or partial upper storey, having structural masonry walls and timber floors. These buildings are mostly bungalows limited in height (Ground+1 or Ground+2) used as family residential buildings, or limited height residential buildings (Ground+3 to Ground+5) (Fig. 1).

Fig. 1 Buildings with structural masonry systems

- *Reinforced concrete frames and masonry*

Based on general arrangements drawings, the structural system is defined by irregular arrangements of columns and beams, with different layouts on different stories which do not form a regular spatial frame. It is observed that the vertical load paths are often taken through cantilevers or transfer beams, highly eccentric columns and variable cross sections through the height. This category is mainly formed by buildings with Ground+3Stories to Ground+6S, using a split of structural masonry walls 38–42 cm thick and interior R.C. columns and beams, or R.C. frame structures for residential flat buildings of Ground+6stories to Ground+12stories. The behaviour of such structural systems was catastrophic in the 04 March 1977 earthquake, achieving a negative record for most R.C. frame buildings collapsed in a single earthquake; ICCPDC (1978) (Fig. 2).

- *Reinforced concrete*

In this category of structural systems fall those buildings made entirely from R. C. frames which are: tall and very tall buildings starting from Ground+10S, made generally as standard residential flats, up to Ground+14,18S for structural systems with in situ diaphragms or precast planks and medium tall buildings Ground + 6,8S having an R.C. frame structure. Reinforced concrete buildings behaved well in the 04 March 1977 earthquake, regardless if they used in situ or precast diaphragms or R.C. frames; ICCPDC (1978) (Fig. 3).

Fig. 2 Reinforced concrete frame buildings with masonry panels

Fig. 3 Reinforced concrete frames

3.2 Structural Parameters Characteristic to Buildings

Similar to the procedure described in FEMA 154, the proposed method is based on obtaining a final scoring for a building, with the scope of determining the grade of visual conformity to seismic actions "RVS" and the provisional assignment to one of the four classes of seismic risk. The final score will be determined based on parameters which affect the seismic structural behaviour favourably or unfavourably. To determine the structural parameters which affect the seismic structural behaviour, the study analysed the impact which the 04 March 1977 earthquake had on existing buildings; Fattal et al. (1977). These structural parameters are presented as follows:

- *In plan irregularity*

In order to determine if a building presents in plan irregularity, the following aspects were considered: the building has to be roughly symmetrical in relation to two orthogonal directions and the building has to have a compact shape and a regular contour. If these two requirements are not met, the building was considered to be irregular in plan. In order to determine the level of irregularity, a sketch of the form of the building in plan is made and added in the respective section of the evaluation form. As an example, a few forms with in plan irregularity are presented in Fig. 4:

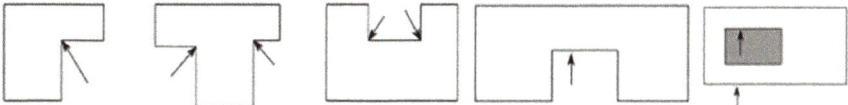

Fig. 4 Different types of in plan irregularities: L, T, U shapes, or the presence of large interior lightwell gardens; FEMA 154 (2007)

- *Vertical irregularity*

A few aspects of a building's conformity to vertical regularity are presented: the structural system has to develop vertically uniform, without large variations from the foundations to the top of the building and if there are steps this should be at the top floors, such as buildings with French roofs, dimensions and continuity of vertical structural elements to be constant to the top of the building, etc. Buildings with offsets only at the French roofs, or with partial level offsets without vertical elements discontinuity are not considered irregular.

Also in this category of structures with vertical irregularity, are buildings that have different functions of the ground floor relative to the upper levels. For example, commercial spaces at the ground floor or large openings created for access made either for carriages for old buildings, direct car access to the backdoor garden for newer buildings, or access to a back car park. For these types of buildings, it is common to observe stopping the vertical elements in the opening areas (Fig. 5).

- *Buildings with different function ground floor relative to the upper levels*

The parameter indicated by a different functionality of the ground floor relative to the upper levels represents another structural deficiency given by vertical irregularity, but is treated separately due to the big influence this has over the structural behavior to seismic actions (Fig. 6).

- *Position relative to neighbouring buildings*

The scoring for this parameter depends on the position the building has to neighbouring structures, which can be either: separate, corner and interior building.

Fig. 5 Buildings with different forms of vertical irregularity

Fig. 6 Buildings with different function ground floors relative to the upper levels

Fig. 7 The influence of neighbouring buildings when the joint is not appropriately sized from seismic requirements

A building is considered to be separated when it does not have any other buildings built against it. A building is considered in a corner position if it is part of a series of "terraced" buildings, without having an appropriately sized joint between them and is at one of the ends of the series. Buildings with an interior position are the ones which are part of such a series but it are not placed at the end.

Other parameters used for the application of rapid assessment are: the role of the slab as a rigid diaphragm, presence of bay-windows, observed defects to principal structural elements and eventual strengthening works to these (Fig. 7).

- *The role of the slab acting as a rigid diaphragm*

The role of a rigid diaphragm is another very important parameter influencing buildings' behaviour to seismic actions, as this represents the mechanism of distributing the horizontal loads to the vertical stability elements. Therefore, the presence of rigid diaphragms is a major advantage. However, there are an important number of buildings, especially the ones built before 1940, which due to the codes of practice corresponding to that period do not meet this requirement. Due to the nature of this parameter this aspect is hard to assess, so factors such as the construction year, the structural system and building height have to be considered. For example, small buildings Ground+2S, built before 1940 can be assessed as having timber or shallow masonry floors which do not form rigid diaphragms.

- *Structural degradations*

Structural defects have a significant influence towards the behaviour against an upcoming earthquake. Based on the study carried after the 04 March 1977

Fig. 8 Different types of structural degradations

earthquake it has been observed that most of the collapsed buildings have been previously affected by the 1940 earthquake with no repair interventions carried out afterwards; Fattal et al. (1977) (Fig. 8).

4 Scoring System

After calibrating the results obtained from studying the 335 buildings, by comparing the vulnerability determined through certified technical expertise MLPAT and vulnerability determined through the proposed rating system RVS, the following scoring results were obtained, Table 3.

Following the rapid visual inspection process, a building receives a final score based on factors that influence the seismic response. These factors, defined before as structural parameters can have a favorable effect (the purpose of rigid diaphragm or the presence of local strengthening) or unfavorable (all other parameters).

Establishing the score for each parameter is determined using the same algorithm (score for each parameter). This is done by technical experts using expertize techniques, applied for pre-inspected buildings which have been previously classified as grade I, II and III of seismic risk. After applying the proposed algorithm for each separate building, they are checked against the previous proposed seismic risk classification, which should match the new proposed classification (e.g. by applying the proposed algorithm, and based on assumptions 1 and 2, a class I risk building should still be classified as grade I).

Core method assumptions for final scoring of each separate parameter:

- Assumption 1: is the assumption in which a building can receive a maxim score of 100 points. This is the case of a building which meets all the unfavorable parameters and none favorable ones. Therefore, based on the visual inspection, such a building is classified as the weakest in terms of seismic performance.
- Assumption 2: relies on establishing limits for classification of each inspected building based on their final score. The chosen limits are presented in Table 2, with the option of modifying these values for a better calibration of each parameter.

Table 2 Scoring system for RVS of existent buildings; Stere et al. (2015)

Class	I	II	III	IV
Limits	100...60	59...30	29...10	9...0

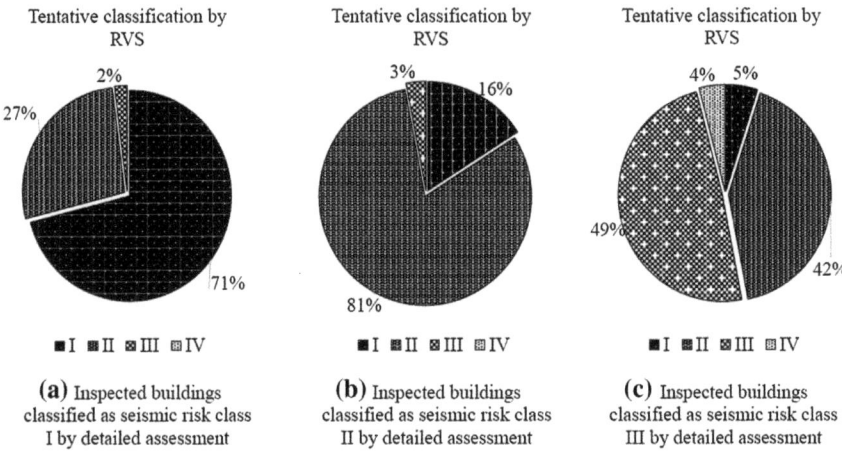

(a) Inspected buildings classified as seismic risk class I by detailed assessment

(b) Inspected buildings classified as seismic risk class II by detailed assessment

(c) Inspected buildings classified as seismic risk class III by detailed assessment

Fig. 9 Seismic risk classes, (based on proposed algorithm), for visual inspected buildings class I, II, III given by the expert engineer; Stere et al. (2015)

The proposed algorithms are not random, they are based on the effects produced by the two major earthquakes on 04 March 1977 and 10 November 1940.

The analysis results are presented in Fig. 9.

The pie charts in Fig. 9, lead to the following conclusions regarding rapid visual screening:

- From all class I seismic risk buildings, inspected by certified experts, the proposed procedure classed 71% in class I, 27% in class II, 2% in class III and none in class IV.
- From all class II seismic risk buildings, inspected by certified experts, the proposed procedure classed 16% in class I, 81% in class II, 3% in class III and none in class IV.
- From all class III seismic risk buildings, inspected by certified experts, the proposed procedure classed 5% in class I, 42% in class II, 49% in class III and 4% in class IV.

The results could be justified as acceptable, excepting class III seismic risk, for which the proposed procedure considered only 49% of the buildings as being the same class, 5% in class IV and all others in classes I and II. This indicates that the method is conservative. Moreover, the fact that the number of buildings in class III is relatively smaller than classes I and II also indicates conservatism.

Table 3 Rvs limits for seismic risk classes

Assignment of provisional seismic risk classes			
I	II	III	IV
RVS values			
< = 40	41–70	71–90	91–100

The seismic performance level of a building based on visual inspections, is defined by a list of parameters and conditions which have to be met by the existing building condition. The RVS method is applied according to Eq. 3.1, and based on the result, this is assigned to one of the four classes of seismic risk. Moreover, a low RVS value will dictate priority for that building to be inspected through detailed calculations, over another with a higher RVS value.

$$Rvs = 100 + \sum \Omega i Ri \tag{3.1}$$

where, i = S, N, P, Av, FB, Ac, D (see also Table 4).

- Rvs = Performance level of the building to seismic actions.
- Ωi—Weighted factor for the favorable and unfavorable parameters used for calculating Rvs. Ωi values are shown in Table 3 for each considered parameter.
- Ri—Each parameter score. Ri values are also presented in Table 3.
- S = Score based on the structural system category and building height.
- N = Score based on building irregularity.
- P = Score based on building position in relation to adjacent buildings.
- Av = Score based on defects of structural or non-structural elements, in relation to building height.
- FB = Scoring based on cantilever terraces and 'soft stories'.
- Ac = Scoring based on the building year.
- D = Scoring based on rigid diaphragm presence and local consolidation.

After determining the" Rvs" parameter, the building is provisionally assigned to a seismic risk class, as in Table 3.

For the simplicity of assessment of Romanian buildings, the method can be applied in accordance to the workflow indicated Table 5.

Table 4 Scoring system for RVS of existent buildings; Stere et al. (2015)

i	Structural characteristic	R_i			Ω_i
1	Structural typologies: concrete-masonry, masonry or concrete (S)	20	Masonry and concrete	$n < 4$	0.5
				$6 \leq n < 9$	0.75
				$9 \leq n$	1
			Masonry	$n < 4$	0.75
				$4 \leq n < 6$	1
			Concrete		0.5
2	Structural irregularities: vertical or horizontal (N)	20	Horizontal		0.75
			Vertical		0.25
			Horizontal and vertical		1
			None		0
3	Position in respect with other buildings (P)	5	Corner		1
			Interior		
			Independent		0
4	Visible damage in structural elements (Av)	25	Visible damage	$n < 4$	0.4
				$4 \leq n < 6$	0.6
				$6 \leq n < 9$	0.8
				$9 \leq n$	1
			None		0
5	Different architectural configuration of the first floor and/or presence of oriel or bow windows (FB)	10	Both		1
			Just bow or oriel windows		0.5
			Just different configuration		0.5
			None		0
6	Construction year, y (Ac)	20	$y \leq 1940$		1
			$1940 \leq y < 1950$		0.75
			$1950 \leq y < 1963$		0.5
			$1963 \leq y < 1977$		0.25
			$1977 \leq y$		0
7	Rigid horizontal diaphragm and/or local retrofitting of structural elements (D)	−15	Both		1
			Rigid diaphragm		0.67
			Local retrofitting		0.33
			None		0

n—the total number of floors

Table 5 RVS method workflow

1. Find building address	
↓	
2. Prepare worksheet	
↓	
3. Identify building on site	
↓	
4. Visual inspection for 10–15 min. The length of this inspection could vary based on the level of access and visibility the engineering surveyor has	
↓	
5. Fill in the worksheet	
↓	
6. Prediction of the building year based on indicative structural system and architectural style. When not possible, further information is sought from the owner	
↓	
7. Determination of the seismic performance level parameter "$\mathbf{R_{vs}}$"	
↓	
8. Assignment of the building in one of the risk classes	
↓	↓
Class I and II priority buildings for detailed calculation assessment according to "P100-3/2008" based on the value of 'RVS'	Class III and IV Lower priority buildings for detailed assessment according to "P100-3/2008", compared to Class I and II
↓	↓
Application of an evaluation methodology according to "P100-3/2008", for determination of a seismic risk class	Buildings not considered vulnerable to seismic risk

5 Conclusions

As a country with a high seismic risk, Romania as well as other countries developed towards seismic design, a rapid seismic assessment method is necessary to determine the vulnerability of existing buildings. This method is used to prioritize the process of analysing buildings through detailed calculations based on the current codes. The rapid visual screening method is easy to apply even for technical staff with minimum level of knowledge, the required time being approximately 20–25 min to analyse a building.

The already existent international rapid visual screening methods can not be applied in their current form to the existing buildings in Romania, as in each country there are different variations of structural systems, different use of materials and many other parameters which are specific to our country. Therefore it was necessary to adapt and develop this method.

Identifying the structural system type of a building from the exterior, together with identifying the characteristic structural parameters through a rapid inspection, make the presented method applicable to the existing buildings in Romania.

Regarding further research studies, there would be useful to further calibrate the presented method based on types of structural systems (such as precast structural walls) or based on buildings' importance and type (schools, hospitals, etc.).

References

Bostenaru-Dan M (2009) Architectural layout and the seismic vulnerability of housing building of the modern movement and contemporary currents I:57–66

Fattal G, Simiu E, Culver C (1977) Observations on the behavior of buildings in the Romania earthquake of March 4, 1977. NBS Special publication 490, US Department of Commerce/National Bureau of Standards, United States

FEMA 154 (2007) Rapid visual screening of buildings for potential seismic hazard: a handbook. Report No. FEMA 461, Washington DC, US

HAZUS (2000) Natural loss estimation methodology. http://www.fema.gov/hazus/hazus99.htm

ICCPDC (1978) Cutremurul din România din 4 martie 1977 şi efectele sale asupra construcţiilor, SINTEZA MONOGRAFIEI Bucureşti

Lungu D, Aldea A, Arion C, Cornea T, Vacareanu R (2002) Risk UE, WP1: European distinctive features, inventory, database and typology. In Proceedings of the international conference October 24–26, Bucharest, Romania

MDRAP—Cod de evaluare seismică a clădirilor existente – P100-3:2008 Bucureşti, 2008

MLPAT—Normativ pentru proiectarea antiseismică a construcţiilor de locuinţe, social-culturale, agrozootehnice şi industriale P-100-92 Bucureşti, 1992

Pantea A, Constantin PA (2013) Re-evaluation of the macroseismic effects produced by the March 4, 1977, strong Vrancea earthquake in Romanian territory. Ann Geophys 56(1):1–12

PMB (2014) Lista clădirilor expertizate tehnic, încadrate în claele de risc seismic I, II, III, IV şi categorii de urgenţă. http://www.pmb.ro/common/site_map.php?sbm_id=133

Stere C, Geangus R, Cotofana D, Albota E, Popa V (2015) Rapid visual seismic screening procedure for building stock in Bucharest. Carpatian J Earth Environ Sci 10(4):219–226

Structural Modeling and Elastic Calibration Based on Experimental Results for a Sensitive Torsion Structure

Claudiu-Anton Ursu, Claudiu-Constantin Stere
and Vasile-Virgil Oprişoreanu

Abstract Reinforced concrete structures exhibiting tri-dimensional effects such as torsion and nonlinear response are a main concern in the field of earthquake research. The purpose of the paper is to present an example of structural modeling of a reinforced concrete structure sensitive to torsion and how to calibrate the structural response to the results of the experimental scale test using Perform 3D software. The analyzed structure is part of the scientific research program "Seismic Design and best-estimate Methods Assessment for Reinforced Concrete Building subjected to Torsion and non-linear effects" (SMART 2013), financed by "Commissariat a l'Energie Atomique et aux Energies Alternatives" (CEA) and "Electricité de France" (EDF). The scope of the program SMART 2013 was to analyze, compare and validate the design and the modelling methods used in evaluating the response of reinforced concrete structures subjected to high intensity seismic loads. The structure, tested using a seismic shake table, was built on a smaller scale (1/4), and it represent a part of a building that is used to host nuclear facilities. The geometry and the dimensions of the specimen were chosen in order to highlight the torsional-effects. To assess the behavior of vibrating mass structures were used both synthetic accelerograms and some real accelerograms. Calibration elastic structure, and thus determine the stiffness and damping characteristics, was based on the elastic acceleration response spectrum. The experimental results revealed a high degree of accuracy of the model in terms of overall behavior simulation of the real structure.

Keywords Structural modelling · Torsion · Elastic calibration
Seismic evaluation

C.-A. Ursu (✉) · C.-C. Stere · V.-V. Oprişoreanu
Technical University of Civil Engineering of Bucharest, Bucharest, Romania
e-mail: claudiu.ursu@alliedengineers.ro

C.-C. Stere
e-mail: claudiu.stere@alliedengineers.ro

© Springer International Publishing AG, part of Springer Nature 2018   435
R. Vacareanu and C. Ionescu (eds.), *Seismic Hazard and Risk Assessment*,
Springer Natural Hazards, https://doi.org/10.1007/978-3-319-74724-8_29

1 Introduction

The development of the finite element method has widened the possibilities to determine the static or dynamic response for various structures by using this method in both, research programs and current design practice. This is possible due to an increase in the degree of knowledge of the behavior of structural elements, knowledge gained through tests on elements or on full scale structures, to which, is added the progress of the design methods and computing means. For this purpose, there have been carried a series of experiments developed by numerous Research Centers or Universities.

The aim of this paper is to present the modelling and the elastic calibration of a structure sensitive to torsion, tested previously on the shake table, during the research program SMART 2013. The main objectives of the research program were to compare and to validate the hypothesis used in the analytical assessment of the dynamic response for concrete structures subject to seismic action, with the consideration of the 3D configuration and the non-linear behavior.

The research program was divided in four stages which entailed the construction of the analytical model, the numeric calibration in correlation with the elastic behavior of the specimen SMART 2013 and the evaluation of the structural behavior in non-linear domain based on the previous elastic calibration.

The analytical model of the structure has been developed using the software Perform 3D. The column, the beams and the slab, have been modelled as bar elements. The non-linearity of these elements has been provided by the nodal plastic hinges placed at the edges. In case of the vertical concrete walls, a model with fibers has been used. The constituent fibers had a non-linear behavior for both, bending and shear force. The results obtained analytically, by using this type of modelling with macroelements has revealed a good approximation of the results obtained by experimental tests.

2 The Research Program SMART 2013

The specimen tested on the shake table was a scale model, ratio 1:4 and it had three storeys, a trapezoidal shape in plan and it was constructed entirely of reinforced concrete (see Fig. 1). Additional information in relation to the specimen tested and the research program SMART 2013 can be found at Richard and Chaudat (2013a, b).

The structure subject to the tests was designed in accordance with norms Autorité de Sûreté Nucléaire (ASN) and French nuclear regulation (RFS V.2 G) for a peak ground acceleration of PGA = 0.20 g and assuming a critical damping fraction of $\xi = 5\%$. The structural system was made of reinforced concrete individual walls, some of them with openings and a column of 20×20 cm. The concrete walls and the slabs had a thickness of 10 cm.

 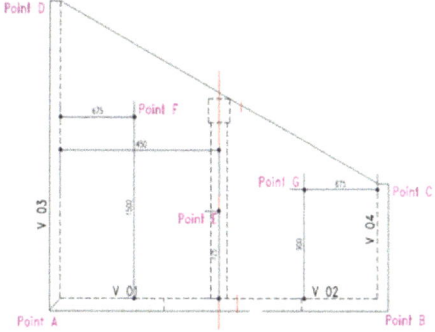

Fig. 1 The specimen on the shaking table and the plan view

The foundations for the concrete walls consisted in reinforced concrete strips, 65 cm wide and 25 cm deep. The foundation system ensured the fixing at the top level of the vibration table and this aspect has been considered in the analytical model.

The main features of the tested structure were chosen with the following purposes:

- to obtain an irregular structure to study the torsional behavior;
- the natural frequencies associated to the vibration modes to be in the range of 6–10 Hz;
- to develop cracks areas at design levels but have no significant crack at ½ design level;

The information in relation to the geometry of the structure, the construction materials, the features of the shake table and the dynamic loads used during the tests have been provided to Universities or Research Centers throughout the world, with the declared purpose to develop an analytical model able to reproduce the structural behavior of the real structure. The structural materials used in the construction of the specimen were defined in the analytical model in accordance with the provisions of Eurocode 2 (EC2). The concrete grade was C30/37 and the steel type FeE500-3 which has a characteristic yield strength of ReH = 500 MPa. The self-weight of the specimen was 11.90 tons. The total mass of the tested structure, including the additional mass at slabs levels, was about 45 tons.

The reinforced concrete specimen, tested in the SMART 2013 Experimental Program, was subjected to dynamic loads. The loading sequences are presented in Table 1. Three of this sequences represent seismic actions (one synthetic and two real accelerograms). The real accelerograms are recordings made during the Northridge (South California) earthquake from 17th of January 1994.

The testing sequences were divided into 9 runs, with the purpose to ensure the control of the vibrating table during the seismic tests (Table 2).

In order to assess the structural behavior of the specimen, the acceleration produced by the dynamic actions induced during the tests were measured and

Table 1 The nominal input ground motion—seismic sequences (Richard and Chaudat 2013a, b)

Sequences	PGA x (g)	PGA y (g)	Magnitude	Type
0	0.1	0.1	–	Synthetic—White noise
1	0.2	0.2	–	Synthetic—Design signal
2	1.78	0.99	6.7	Real—Northridge earthquake
3	0.37	0.31	5.2	Real—Northridge after shock

Table 2 The nominal input ground motion—seismic sequences (Richard and Chaudat 2013a, b)

RUN	PGA (g)	Type
6	0.1	Synthetic—White noise
7	0.1	Scaled—Design signal step 1
9	0.2	Real—Design signal nominal
11	0.2	Scaled—Northridge earthquake step 1
13	0.4	Scaled—Northridge earthquake step 2
17	0.8	Scaled—Northridge earthquake step 3
19	1.78	Real—Northridge earthquake nominal
21	0.12	Scaled—Northridge after shock step 1
23	0.37	Real—Northridge after shock nominal

recorded on three orthogonal directions (Ox, Oy, Oz). The instrumented points were located in various parts of the structure, on both: inside and perimeter (Fig. 1).

3 Numerical Model

The analytical model has been elaborated by using the finite element software, Perform 3D and it comprises only the structural elements above the top level of the foundation system. The strip foundations under the concrete walls and the shake table have not been included in the analytical model. The vertical elements of the structure have been assumed fixed above the foundations. Due to the contribution to the overall rigidity, the concrete slab has been modelled with bar elements which have been rigid connected with the walls in order to reproduce the coupling effect. Also, at each level it has been assumed a rigid diaphragm. The load on the slab has been uniformly distributed to the replacement beams (the bar elements).

The mass has been declared as a nodal property based on the associated loaded area. Within the analysis that have been performed it has been chosen the Rayleigh model for damping, in which, the damping matrix represents a linear combination between the mass matrix and the elastic rigidity matrix. The fractions of critical damping used for the structure have been obtained as a result of the elastic calibration of the analytical model.

3.1 Modelling of the Reinforced Concrete Walls

The modelling of the concrete walls, the main elements to resist to lateral loads, has been made by using a finite element with a non-linear behavior for both: bending and shear force. This type of finite element, "General Wall Element" is available within the software Perform 3D.

The finite element analysis models the behavior of the walls under the following situations: eccentric bending and shear force for vertical walls, bending and shear force for the coupling beams, as well as forces transferred through the strut and tie model.

The general wall element has 4 nods, which have 6 degrees of freedom each, resulting in 24 DOFs for the element. Out of these, there are 8 DOFs that represent in-plane displacements (Fig. 2). The wall undergoes constant bending deformations, based on the value of the bending moment at the middle of the element.

In order to model the eccentric bending and shear force behavior the wall element is composed out of five layers:

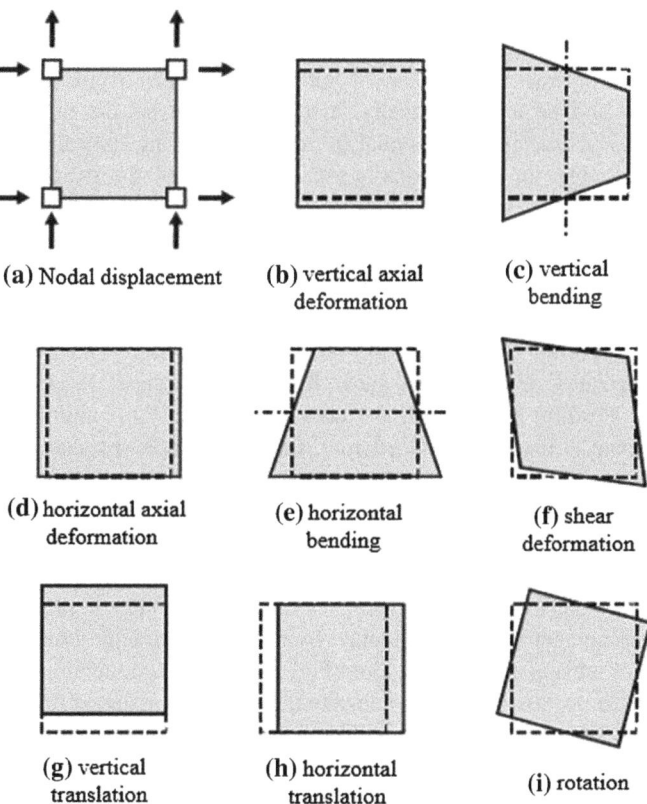

(a) Nodal displacement

(b) vertical axial deformation

(c) vertical bending

(d) horizontal axial deformation

(e) horizontal bending

(f) shear deformation

(g) vertical translation

(h) horizontal translation

(i) rotation

Fig. 2 General wall element in plane deformations (perform, components and elements)

- The axial-bending layer which models the eccentric bending in the vertical direction. This layer consists of steel fibers and concrete, for which the stress-deformation curves were pre-defined. This approach allows the neutral axis to form in the compression zone when the cracking of concrete occurs. The width of this layer may vary by inputting different widths for each fiber.
- The axial-bending layer which models the eccentric bending in the horizontal direction. This layer also consists of steel fibers and concrete. The two layers that model the axial-bending are perpendicular one to the other, therefor there is no direct interaction between them. The deformations from the eccentric bending in the vertical layer are associated only with the vertical displacements of the nodes, without influencing the horizontal displacements. In a similar manner, the deformations of the horizontal layer do not lead to vertical displacements of the nodes.
- The conventional shear layer which models the shear force behavior based on tangential shear stresses. This layer implies a constant shear stress and constant width along the element. The shear force behavior depends on the material characteristics that define this layer only with rigidity to tangential stresses. The layer does not increase the rigidity and resistance of the layers that model the bending behavior.
- Two layers that model de compressed diagonals. Each layer is coupled with constant compressive stresses and a constant width. The diagonal layer behaves as a strut and tie model. Usually, in the wall analysis the diagonal layer is ignored by setting its width equal to zero, considering only the conventional shear layer. One of the reasons is the complexity of the strut and tie model which may lead to overestimate the resistance to shear force. In the case of our analysis these layers were not considered.

The layers have different ways of behavior. The layers interact with each other, as they are connected at the nodes. The interaction between the layers defines the behavior of the wall element. In the case of a reinforced concrete wall which undergoes eccentric bending and shear forces the concrete develops multi-axis stresses. The Perform 3D wall element does not consider these multi-axial stresses. The multi axial stresses are divided into uniaxial stresses associated to different layers. This wall model considers an elastic behavior for the out of plane stresses.

The bending and shear nonlinearity have been defined through the constitutive laws of the materials, the stress–strain curves for steel and concrete (σ–ε). In case of the structural behavior to shear, it has been considered only the conventional shear, neglecting the contribution of the layer which replicates the compressed strut.

The reinforcement has been modelled by using a curve with three domains and symmetric for tension and compression (Fig. 3a). Also, the reduction of the overall capacity due to buckling has been neglected. It has been assumed that between the concrete and the reinforcing steel will be a perfect adherence. The parameters of the stress–strain curve have been determined after testing the steel reinforcement. The tests were carried out by the organizers of the SMART 2013 Research Program.

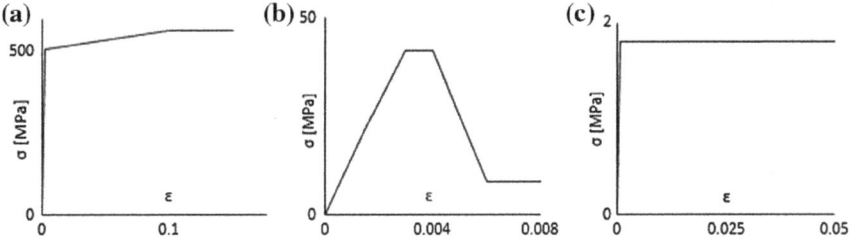

Fig. 3 Stress-strain curves for steel (**a**), concrete (**b**) the material which models the behavior for shear force (**c**)

In case of the concrete, the characteristic curve had also three domains, with reduction of strength. This curve reproduces the recommended curve within EC2 (Fig. 3b). The tension strength of the concrete has been considered, in order to include the cracking. The structural failure of a concrete element subject to compression forces will occur for a deformation higher than 0.004. The reduction of the compression capacity of a concrete element will vary linearly for deformations within the interval 0.004 and 0.006, reaching a minimum value of 20% from the total value.

In case of the material which defines the structural behavior to shear, it has been introduced a bilinear curve (Fig. 3). Due to the complex mechanism through which the concrete resist the shear, the real value of the elastic modulus is lower than the value of the shear modulus, especially for cyclic actions where this mechanism tends to reduce in terms of strength and rigidity.

The shear modulus has been obtained subsequently, through the elastic calibration of the analytical model. In case of the resistance to conventional shear, it has been considered the contribution of the reinforcement for shear resistance only. The value of the capacity obtained is 1.80 MPa. The contribution of the concrete has been neglected due to the variation of the axial forces during the seismic action.

For the concrete, there is also defined a cyclic degradation factor. This factor considers the energy dissipation through compression. In Fig. 4 there is presented the hysteretic behavior model for the compressed concrete fibers. The rigidity at unloading is always equal to the initial elastic rigidity. The model controls the energy dissipation by controlling the rigidity at reloading. In the case the cyclic

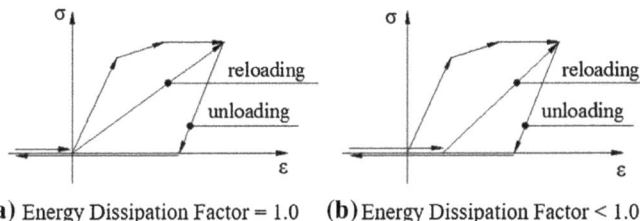

(**a**) Energy Dissipation Factor = 1.0   (**b**) Energy Dissipation Factor < 1.0

Fig. 4 Behavior of concrete material in compression (perform, components and elements)

degradation factor is equal to +1, the reloading starts from origin and ends at the previous maximum point. For a value of 0, the reloading follows the unloading route, without energy dissipation.

The value of the cyclic degradation factor resulted after the calibration of the element with a tested wall in an experiment at Technical University of Civil Engineering of Bucharest (UTCB) in 2006. In this research, there were tested 5 walls with different reinforcement ratios, especially for the horizontal reinforcement. Out of the 5 tested walls only one wall has a reinforcement ratio higher than 0.3%. This wall was used for the calibration process.

The wall dimensions are the following: 1.8 m height, 1.8 m length and 0.1 m width. The materials are designed according to the Romanian design codes from that time: Concrete Class Bc20 (with a maximum compressive strength of 12.6 MPa obtained after tests) and OB37 steel (fyd = 210 MPa). The characteristics of the materials have been obtained after testing multiple probes.

The cyclic behavior was modeled by defining multiple cyclic push-over analysis. The displacement increment corresponds to a 0.15% drift. The maximum drift value in the last push over analysis is 1.20%. The vertical loading was maintained constant, at a value of 310 KN, while the horizontal loading varies in time.

After the calibration process it was obtained a degradation factor with a value of 0.6. The graphic from Fig. 5b presents the force-displacement plots obtained from the experimental tests (red) and from the finite element analysis (black).

3.2 Modeling of the Bars Elements

The column, the beams and the strips of the slab were modelled as bar elements. The elastic rigidity of them has been considered as half of the rigidity provided by the gross section. The non-linear behavior was defined through the nodal plastic hinges defined at the edge of the elements. For the calculation of the bending

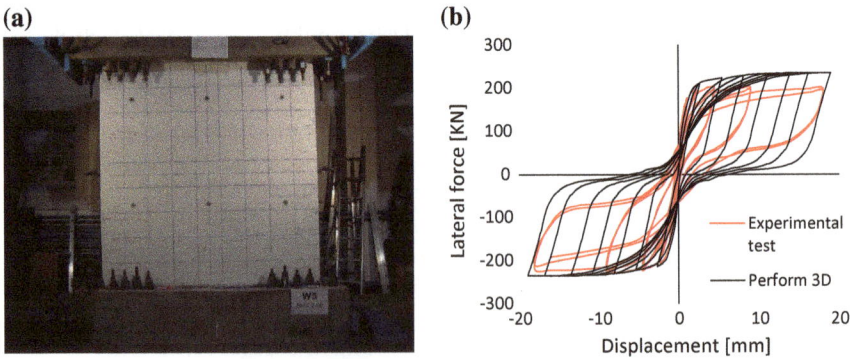

Fig. 5 Wall elevation (**a**) and F-d relationship (**b**) for wall specimen used for calibration

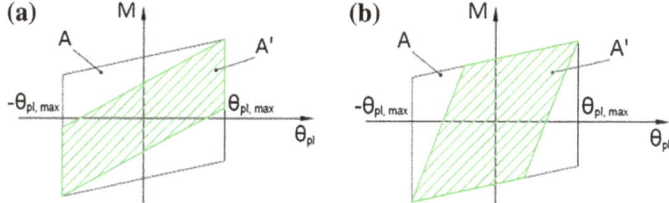

Fig. 6 The cyclic model for plastic hinges existing in Perform 3D: the significance of the energy factor (**a**) and the significance of the unloading stiffness factor (**b**)

capacities it has been used the strength of the materials resulted from tests and provided by the lead of the research program.

The cyclic behavior model implemented in the PERFORM 3D analysis for the plastic hinges is based on a hysteretic law that has at the basis the bilinear model with kinematic consolidation. In the definition of the model the following parameters are introduced:

- the energy factor β (with values between 0 and 1) for which the rigidity at loading and unloading remains the same, but the unloading elastic range is reduced to provide the required energy degradation (Fig. 6a).
- the unloading stiffness factor α (with values between +1 and −1), for which the unloading stiffness is reduced to provide the required energy degradation.

With M there was denoted the bending moment, with A and A' the areas under the curve and with θ the rotation.

The energy factor was considered 0.5, while the unloading stiffness factor was considered equal to 0. In any case the contribution of the columns and beams to the lateral forces is low.

4 Elastic Calibration. The Assessment of the Damping and Rigidity Properties

In the dynamic of the structures, the elastic analytical model has three defining components: the inertial component which is referring to the distribution of mass on the directions corresponding to the degrees of freedom, the dissipation component (damping component) represented by the damping matrix and the elastic component which characterize the deformability of the structure and implicitly, the return elastic forces. Taking into account the fact that for the analytical model the distribution of the mass is fixed, the calibration has to be performed only for the damping and the elastic component.

The elastic calibration has been carried out by using the elastic response spectrums in terms of absolute accelerations and considering the following statement: the peak values of the spectral accelerations measured in various points of a building, in the direction of the degrees of freedom, are located in the vicinity of the natural vibration periods of the building as long as the seismic motion amplifies the associated modal responses. The response spectrums had a fraction of the critical damping of $\xi = 5\%$. The accelerograms that have been used in the calibration process correspond to the run 007 of the SMART 2013 research program.

For each of the monitored locations, it has been calculated the response spectrums of the recorded accelerations which have been subsequently compared with the response spectrums of the accelerations resulted from the dynamic linear analysis carried by using the software Perform 3D. The amplitudes of the response spectrums have been calibrated by adjusting the value of the damping introduced in the analytical model depending on the period of vibration.

Following on from the elastic calibration, it has been established the parameters of the elastic component:

- the elastic modulus of the concrete (Young modulus): E = 10 GPa (a value equal with 0.3 Ecm, where Ecm represents the recommended value of the elastic modulus in EC2, for the concrete grade C30/37).
- the shear modulus: G = 3 GPa (a value of about 0.09 Ecm);

In order to determine the damping properties, it has been selected two periods of vibration Ti și Tj, where:

$$T_i = 0.3T_1 \quad T_j = T_1$$

for which it has been obtained different values for the fraction of the critical damping though the calibration: $\xi i = 0.055(5.5\%)$ and $\xi j = 0.095(9.5\%)$.

The modal shapes for the first three modes of vibrations of the structure are represented in Fig. 7. The first mode of vibration represents a predominant

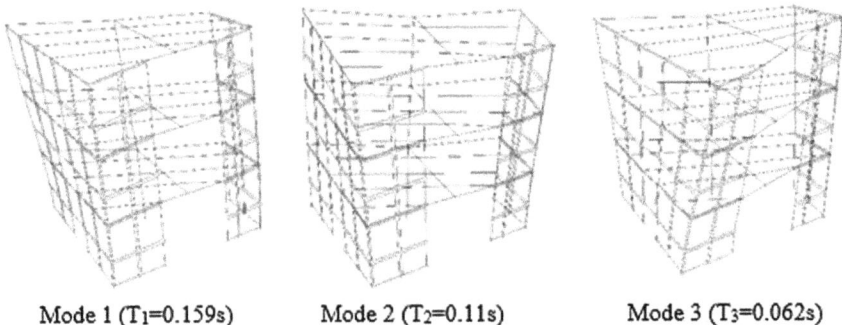

Mode 1 (T₁=0.159s)   Mode 2 (T₂=0.11s)   Mode 3 (T₃=0.062s)

Fig. 7 The modal shapes for the first three modes of vibrations determined by using Perform 3D

translation on X direction, the second mode represents a predominant translation on Y direction and the third mode represents an overall torsion.

The elastic spectrums of the accelerations obtained by tests and numerically subsequent to the elastic calibration of the analytical model in the points A, B, C and D at third level of the structure are presented by comparison in the Figs. 8, 9, 10, 11, 12, 13, 14 and 15.

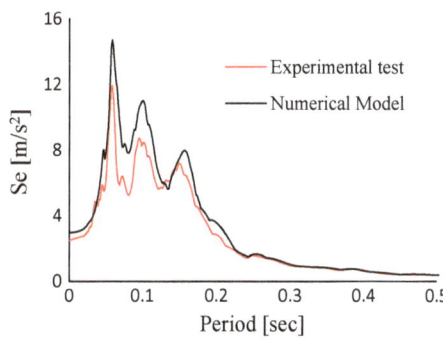

Fig. 8 Elastic response spectrums in terms of accelerations for point A on X direction

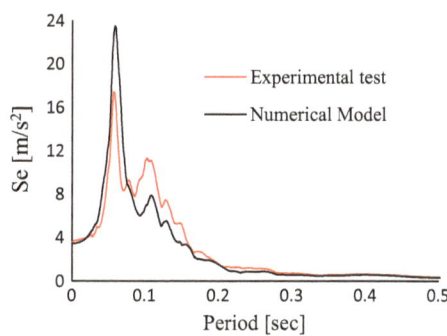

Fig. 9 Elastic response spectrums in terms of accelerations for point A on Y direction

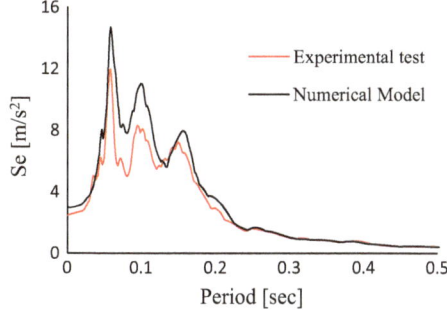

Fig. 10 Elastic response spectrums in terms of accelerations for point B on X direction

Fig. 11 Elastic response spectrums in terms of accelerations for point B on Y direction

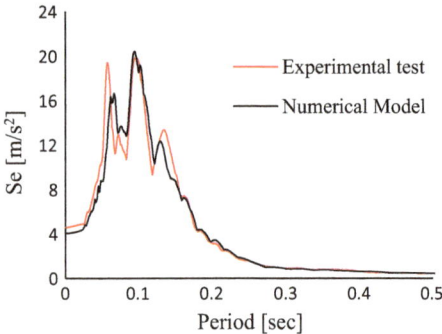

Fig. 12 Elastic response spectrums in terms of accelerations for point C on X direction

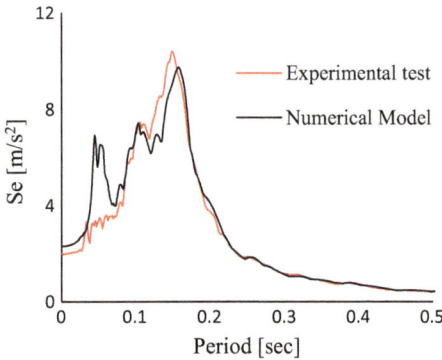

Fig. 13 Elastic response spectrums in terms of accelerations for point C on Y direction

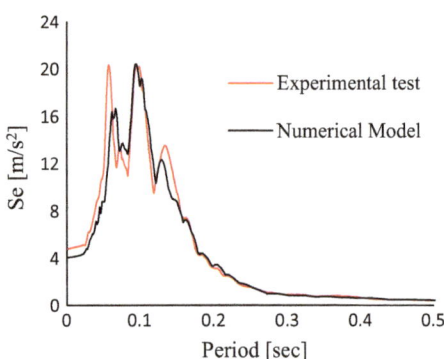

In Figs. 16, 17, 18 and 19 there are displayed the time variations for the displacements in the two main directions of the building for the extreme points of the building (points B and D). For the loading it was considered the accelerograms of the "007" run.

Fig. 14 Elastic response spectrums in terms of accelerations for point D on X direction

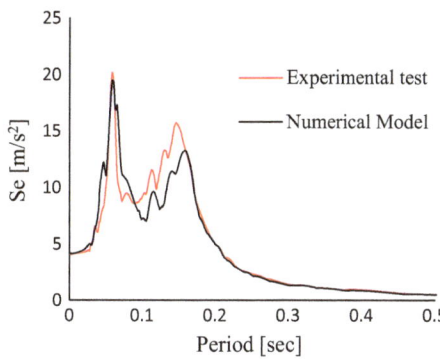

Fig. 15 Elastic response spectrums in terms of accelerations for point D on Y direction

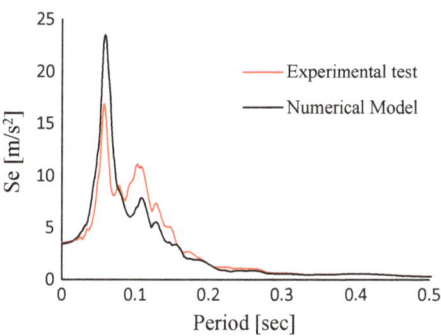

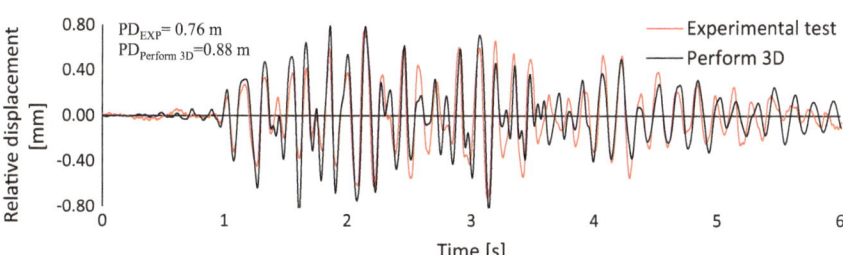

Fig. 16 The graphical representation of the displacements for point B in the "X" direction

PD$_{Exp}$ denotes the maximum absolute value of the displacement obtained during the tests;

PD$_{Perform 3D}$ denotes the maximum absolute value of the displacement obtained from the analytical model.

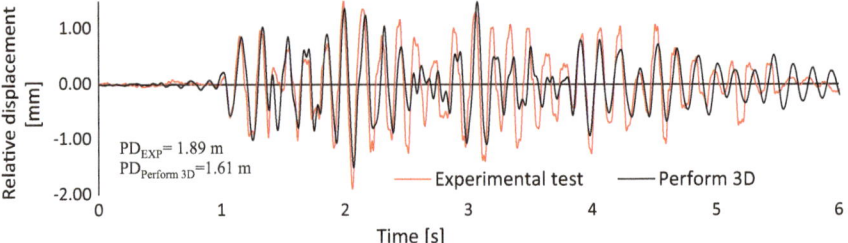

Fig. 17 The graphical representation of the displacements for point D in the "X" direction

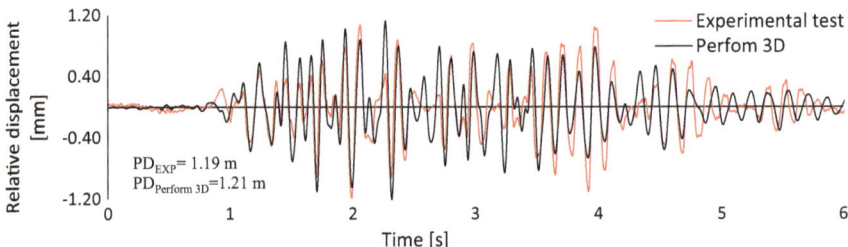

Fig. 18 The graphical representation of the displacements for point B in the "Y" direction

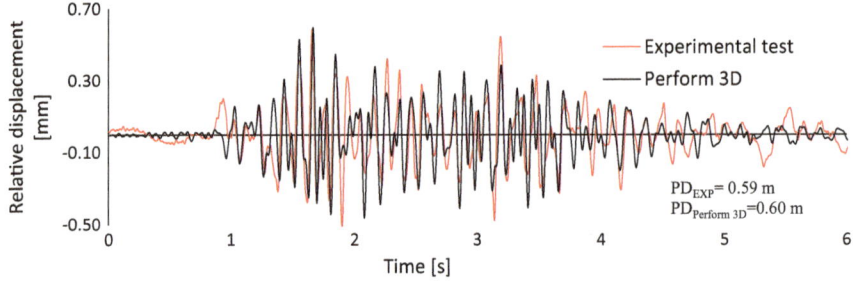

Fig. 19 The graphical representation of the displacements for point D in the "Y" direction

5 The Structural Behavior of the Structure Subject to Seismic Sections of High Intensity, Subsequent to the Elastic Calibration of the Analytical Model

The seismic actions of high intensity to which the specimen of SMART 2013 has been tested, correspond to the runs 13, 17 and 19 respectively. The sequences of testing were described in Table 2. The most important run is the run 19 because it represents the recorded accelerogram during the Northridge earthquake (1994, California). A parameter that allows the assessment of the overall structural

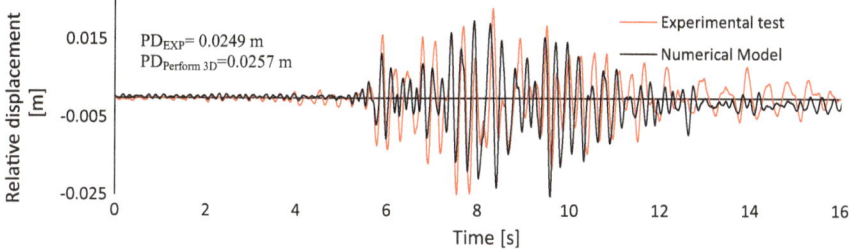

Fig. 20 The graphical representation of the displacements for point A in the "X" direction

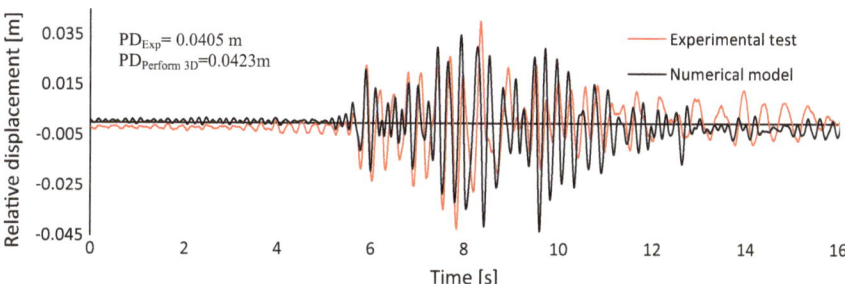

Fig. 21 The graphical representation of the displacements for point C in the "X" direction

behavior for structures subject to seismic actions is the maximum displacement recorded at the building top. The Figs. 20 and 21 illustrate the displacements versus time, recorded on the top of the structure, on the X direction for the points A, B, C and D (see Fig. 1 for the location of the points). The analysis of the SMART 2013 structure has indicated that it can be included easily in the category of buildings sensitive to torsion. The main reason for this is that in the X direction, the maximum displacement exceeds with 45% the average of the maximum and minimum displacements.

6 Conclusion

This paper has proved that the utilization of macroelements in the modelling of a concrete structure leads to satisfactory results and that the analytical model can be easily calibrated based on tests results. Following on from the elastic calibration of the analytical model based on the response values of the specimen to dynamic actions of low intensity, it has been determined the damping and rigidity properties. The values of the elasticity modulus for concrete (longitudinal and transversal modulus) have resulted smaller that the values recommended for the structural design. This aspect can be explained by the low level of axial forces throughout the elements of the structure. The fractions of the critical damping obtained during the tests were higher than the assumed value of $\xi = 5\%$, typically used for concrete structures in the current design practice. However, the damping is very important in the elastic domain but in case of large deformations in the plastic domain the influence of damping decreases substantially. The elastic calibration has highlighted the contribution brought by the concrete slab to the overall rigidity of the structural system and the fact that the omission of it can lead to an overestimation of the fundamental period of vibration.

References

EN 1992-1-1: Eurocode 2: design of concrete structures—Part 1: general rules and rules for buildings
Perform 3D. Computers and structures. http://www.csiberkley.com
Richard B, Chaudat T (2013a) Presentation of the SMART 2013 international benchmark. http://www.smart2013.eu
Richard B, Chaudat T (2013b) Data acquisition project. http://www.smart2013.eu

Influence of Infill Masonry Walls in the Seismic Response of Buildings: From Field Observations to Laboratory Research

**Humberto Varum, António Arêde, André Furtado
and Hugo Rodrigues**

Abstract The seismic performance of infill masonry walls is a topic of growing importance due to the significant number of collapses observed through the recent earthquakes. Nowadays is recognized by the scientific community the influence of these elements in the structural response of reinforced concrete structures subjected to seismic actions. The infills out-of-plane (OOP) behaviour depends on a series of variables and there is a lack of experimental data to understand and predict their expected seismic performance. There is a need of data to calibrate numerical models and to understand the effect of each variable such as type of masonry, boarder constrains, slenderness, previous in-plane damage and insufficient support width in the infills OOP capacity. The present chapter pretends to overview some considerations regarding the performance assessment of infills OOP performance such as based on experimental tests and numerical modelling results. Additionally, a brief literature and international codes recommendations review on this topic will be presented and discusses and will help to understand the importance of the infills seismic behaviour on the performance assessment of reinforced concrete structures.

Keywords Calibration · Experimental tests · Numerical modelling
Seismic behaviour

H. Varum (✉) · A. Arêde · A. Furtado
CONSTRUCT-LESE, Faculdade de Engenharia,
Universidade do Porto, Porto, Portugal
e-mail: hvarum@fe.up.pt

H. Rodrigues
RISCO, School of Technology and Management,
Polytechnic of Leiria, Leiria, Portugal

© Springer International Publishing AG, part of Springer Nature 2018
R. Vacareanu and C. Ionescu (eds.), *Seismic Hazard and Risk Assessment*,
Springer Natural Hazards, https://doi.org/10.1007/978-3-319-74724-8_30

1 Introduction

In recent years, increased interest is denoted in studying the infill masonry (IM) walls' influence in the seismic response of existing buildings, which can be favourable or not, depending on several phenomena, detailing aspects and mechanical properties, namely the relative stiffness and strength between frames and masonry walls, the type or lack of connection between masonry and surrounding structures, etc. (Hermanns et al. 2014).

From surveys on damaged and collapsed RC buildings in recent earthquakes many buildings having suffered severe damage or collapse exhibited poor performance due to IM panels. It is observed that in-plane (IP) behaviour of IM can prevent the development of out-of-plane (OOP) strength mechanisms by arching effect. By contrast, in most cases the major damages were found in non-structural elements, particularly in clay IM, including diagonal cracking, OOP collapse or detachment of surrounding RC frames (the latter taking place in early earthquake instants) due to absence of or deficient connection to that frames. These damage types often require high investments, either for the repair process or for demolition and reconstruction, resulting in economic problems related with interdiction of building use.

Different authors (Furtado et al. 2016a, b) reported that the OOP performance and capacity of IM walls can be strongly influenced by the following issues: connection between the panel and surrounding RC frames; connection between the internal and external leaves (in the case of two-leaf IM walls); insufficient support width due to constructive procedures adopted for thermal bridges' prevention and, last but not the least, the existence of previous in-plane damages. Moreover, IM walls OOP collapse can also introduce plan and/or height vertical stiffness irregularity which can induce formation of mechanisms such soft-storey or torsion, likely to originate building collapse. Considering the lack of experimental studies conducted during the last years regarding this behaviour and considering the common use of hollow clay bricks with horizontal perforation in Portugal, it is of utmost importance to validate some proposed retrofitting strategies in the literature and develop new ones to improve the OOP behaviour and prevent the collapse of these IM walls' types.

For assessing RC frame structures, the nonlinear behaviour induced by earthquakes demands and the influence of IM walls should be considered. Different modelling techniques can be found in the literature (Asteris et al. 2011a, b, 2013) for the response simulation of infilled frames, from refined micro-models to simplified macro-models, the former involving high discretization level of the IM panel and the later relying in simplifications aiming at representing the IM global behaviour with a few structural elements and mechanical parameters. In many cases, for non-linear analysis of complex structures under earthquake action, it is not suitable to adopt refined models. Thus, for simulating the response of frame structures with IM walls, taking into account the interaction between them, the adoption of simplified models is unavoidable (Smyrou et al. 2011; Rodrigues et al. 2010). This is further confirmed in FEMA356 (2000) which recommends assessing the structural response of buildings, considering the infill panels represented by an equivalent diagonal strut model.

The main objective of the present chapter is to highlight the seismic behaviour of IM walls, particularly when subjected to OOP loadings. Firstly, a brief review of code provisions regarding the IM walls seism perforce will be presented, discussing the most important issues on this field. Then, the major results of an experimental campaign of quasi-static OOP tests on full-scale IM walls that was carried out on the Laboratory of Earthquake and Structural Engineering (LESE) will be presented and discussed. Finally, a simplified numerical approach to simulate the combined IP and OOP behaviour of IM walls subjected to earthquakes will presented.

2 Code Provisions and Recommendations on IM Walls Seismic Performance

Some international codes recommend various formulations for the analysis the IM walls for both in-plane and out-of-plane directions. For instance, FEMA274 (1997) specifies that masonry infill panels shall be represented as equivalent diagonal struts and may be placed concentrically across the diagonals, or eccentrically to directly evaluate the infill effects on the columns. It specifies strength requirements for column members adjacent to infill panels. The shear force demand may be limited by the moment capacity of the column with reduced length. EC8 (2003) specifies that the period of the structure used to evaluate base-shear stress shall be the average between periods for the bare frame and for the elastic infilled frame. Frame member actions are then determined by modelling the frame without the struts. Irregular infill arrangement (in plan and elevation) is addressed with important recommendations to avoid the formation of soft-storeys and torsion-effects. Moreover, designing techniques are suggested to account for irregularities, such as increase of accidental eccentricities use of three-dimensional analysis. Regarding irregularities in elevation, if a refined model is not used, the code suggests the computation of a magnification factor to increase the seismic actions on columns (only).

Regarding the lateral load shared between infill walls and frame, EC8 (2003) does not make a reference to the infill walls, considering only that the frame system should resist totally the vertical loads, and to have a 65% base-shear capacity—50% as de minimum for other types of structure—of the total lateral loading on the building. For the serviceability limit state, it is recommended the control of lateral deformation between storeys (drifts, dr). For buildings with brittle non-structural elements, it should be limited to $0{:}005\ h = \mu$. For buildings with ductile non-structural elements the drift is limited to $0{:}0075\ h = \mu$, or $0{:}010\ h = \mu$ if the building does not have non-structural elements (h is the height of the storey, and μ

is the reduction coefficient ranging from 0.4 to 0.5, depending on the importance class). Due to the nature of the infill masonry walls, non-structural and brittle elements, the limit to use should be $0{:}005 \ h = \mu$.

There are some international codes that provides some recommendations on the OOP capacity of infills, as well as the indications given about the design demand acting on them. For example the Italian Building Code (NTC08 2008) gives some indications regarding OOP seismic action on infill considered as non-structural elements, but no provision aimed at determining their OOP capacity; FEMA306 (1998) provides some recommendations on infills OOP strength, but no indication on their maximum displacement capacity.

Eurocode 6 (2005), in Sect. 6.3.2, proposes an expression (Eq. 1) to calculate the lateral strength of masonry walls in which arching action can occur; this relationship can be extended, eventually, to infill panels:

$$q = f_d \left(\frac{t}{t_a}\right)^2 \tag{1}$$

In this relationship f_d is the design compressive strength of masonry in the direction of arching thrust while la is the panel dimension in the same direction. The maximum OOP load is the one that equilibrates the maximum thrust that can form in the masonry wall thickness determined from:

$$N_{ad} = 1.5 f_d \left(\frac{t}{10}\right) \tag{2}$$

FEMA 273 (1996) and FEMA 356 (2000) provide some indications concerning the ultimate OOP displacement of infills with reference to different Limit States. A 2% OOP drift is set as maximum displacement for Immediate Occupancy Limit State: this drift value corresponds with the opening of visible cracks on the panel surface; with reference to the Life Safety Limit State a 3% OOP drift is fixed as limit displacement: this drift value corresponds with high possibility of detachment and expulsion of at least part of the infill. FEMA356 (2010) sets a 5% OOP drift as maximum displacement at Collapse Prevention. These indications are effective for both new and existing buildings.

FEMA273 (1996) lists the conditions that allow considering arching action in the assessment of infills OOP strength, such as the effectiveness of the infill connection to the surrounding frame, its columns and beams stiffness and strength, and the panel slenderness. Among these statements, the one referring to the infills boundary conditions seems to be the most significant. In fact, the analysis of the experimental database presented in Sect. 4 shows that, even for panels with slenderness greater than the value proposed as upper limit for the arching action effectiveness, the best strength prediction was provided by relationships based on that resistant mechanism. Under the above-mentioned conditions, it is possible to express the lateral strength of the infill as Eq. 3:

$$Q = \frac{0.7 \times f_m \times \lambda_2}{(h/t)} \tag{3}$$

in which f_m is the lower bound of the compressive strength of masonry calculated by dividing by 1.6 (by 1.3 according to FEMA356 2000) its average compressive strength; λ_2 is a slenderness parameter. FEMA274 (1997) points out that the previous expression is the relationship by Angel simplified to evaluate a lower bound of the infills lateral strength. To compute it, FEMA306 (1998) provide Angel's relationship without modifications. This means that FEMA306 (1998) consider OOP strength reduction due to IP damage explicitly, even if it is not stated how to set the IP drift at which the OOP strength should be assessed.

3   Experimental Characterization of the Infills OOP Seismic Behaviour

3.1   Introduction

The experimental testing of infilled frames to OOP loadings started in 1988 with Moghaddam et al. (1988) with four steel infilled frames on a biaxial shake table test. Two small and two larger walls were tested with the aim of evaluate the effect of use reinforcement bars on the horizontal bed joints. The authors concluded that the infills' presence reduced the displacements of the frame even after the infills' cracking. It was observed that the reinforcement reduced the vulnerability of the panel by the improvement of the arching mechanism phenomena. Dawe and Seah (1989) tested 9 full-scale concrete infill walls subjected to uniform normal pressure applied through airbags. The authors concluded that the IM walls ultimate loads increased for larger panel thicknesses; however, for smaller panel length and height it was observed a strength reduction. No significant influence of the openings on the IM panels OOP strength. Finally, the authors remarked that the horizontal reinforcement bars provided higher OOP ductility. Angel et al. (1994) carried out an experimental campaign composed by combined IP and OOP tests of RC frames infilled with brick and concrete blocks. The strategy was to submit first the specimens to IP demands and to cause different levels of damage, and then the specimens were subjected to OOP monotonic distributed loadings applied by airbags. From these tests, the authors concluded that the panels' OOP strength depends highly of the slenderness ratio, masonry compressive strength but not from the tensile strength. The authors observed that cyclic loadings within the elastic region of the panel did not affect the stiffness of the panel. The IP shear demand combined with panel gravity load increase slightly the initial OOP stiffness but the OOP strength is not affected. Nevertheless, the previous IP cracking reduced the OOP strength for slender panels. Repairing techniques were tested and it was observed that the increase of the damaged panels OOP strength was achieved. Calvi and

Bolognini (2001) carried out OOP tests with and without previous IP damage, with and without reinforcement. Four OOP loading tests were carried out with the aim of assess the OOP vulnerability of traditional and slightly IM walls for different level of damages induced by IP action. It was observed that the panel' state of damage play an important role on the OOP response of the panel, and that higher levels of damage increase the OOP vulnerability of the panel.

From the analysis of the experimental efforts made throughout the different studies on the literature, the following main conclusions can be drawn:

- High panel slenderness can result on poor OOP performances, since it is observed that the arching mechanism developed after the maximum strength do not occur for panels with high slenderness;
- Previous IP damage reduce the OOP initial stiffness, strength and can lead to fragile OOP expulsions. This effect is due to the loss of the boarder constrains that were modified, since the detachment of the panel from the surrounding frame occurred and a rigid body behaviour occurs when subjected to OOP loadings;
- The masonry compression strength revealed to be more important to the formation of the arching mechanism than the tensile strength.

At the Laboratory of Earthquake and Structural Engineering (LESE) an experimental campaign composed by five full-scale IM walls was carried out with main purpose of analyze any effect of the panel support width, axial load on columns and previous IP damage (Furtado et al. 2016a, b). The test results will be presented and discussed in terms of damage observed throughout the tests and the cracking pattern.

3.2 Specimens' Detailing and Testing Campaign Description

The specimen dimensions are 0.15 × 4.80 × 3.30 m respectively thickness, length and height, with columns sections 0.30 × 0.30 m and the top and bottom beams 0.30 × 0.50 m. For the RC frame specimen construction, three different bar diameters were used, from the same lot, namely ø6 mm, ø10 mm and ø16 mm. Five IM walls (M1, M2, M3, M4 and M5) were built with hollow clay horizontal brick, as frequently adopted in the Southern Europe and particularly in Portugal. The mortar adopted was an industrial pre-dosed M5 class ("Ciarga" type) with the following composition. No plaster was adopted in both panels.

The main characteristics of each specimen are summarized below:

- M1: One-leaf panel (thickness: 150 mm), aligned with the external face of the support beam, monotonic test with 300 kN in the top of each RC column, totally width supported;

- M2: One-leaf panel (thickness: 150 mm), aligned with the external face of the support beam, cyclic test with no axial load in the top of each RC column, totally width supported;
- M3: Double-leaf panel, composed by one external leaf with thickness of 150 mm and an internal leaf of 110 mm. Subjected to a previous in-plane test until reach 0.5%. After the in-plane test the internal panel was removed and the external one was subjected to OOP cyclic test with no axial load in the top of each RC column;
- M4: One-leaf panel (thickness: 150 mm), aligned with the external face of the support beam, cyclic test with axial load in the top of each RC column of 270 kN (constant throughout all the test), panel width totally supported;
- M5: One-leaf panel (thickness: 150 mm), aligned with the external face of the support beam, cyclic test with no axial load in the top of each RC column, 2/3 width supported on the bottom beam;

The contact between both specimens and the surrounding columns and the bottom beam is provided by approximately 1 cm layer of mortar. Regarding the contact between the top beam, half-brick and mortar are used to fill the gap that resulted from the IM wall construction.

3.3 Test Setup

The OOP test consisted on the application of a uniform distributed pressure, throughout the entire panel under tested, through nylon airbags. With this procedure, it is pretended to mobilize the entire infill panel considering all the distributed inertia forces that results from a seismic excitation. The uniform load applied through all the infill panel is reacted against a self-equilibrated steel structure composed by five vertical and four horizontal alignments that are rigidly connected to the RC frame with steel re-bars in twelve previous drilled holes (Fig. 1). Between the self-equilibrated steel structure and the RC frame it was inserted twelve load cells that allow the monitoring of the forces transmitted along the experimental test. In front of the self-equilibrated steel structure it was placed a wooden platform to resist the airbags pressure and transfer it to the structure and to the tested panel. The self-equilibrated system uses the RC frame bending stiffness and strength to react to the OOP forces developed from the application of the pressure on the panel. This OOP test setup can be adaptable to specimens with different geometries, different types of masonry materials and existence of openings. As disadvantage is the impossibility of perform complete cyclic tests. With this test setup only charge-discharge loadings can be carried out. In the top of each column, the axial load was applied through hydraulic jacks inserted between a steel cap placed on the top of the columns and an upper HEB 200 steel shape, which, in turn, was connected to the foundation steel shape resorting to a pair of high-strength rods per column. Hinged connections were adopted between these rods and the top and foundation steel shapes.

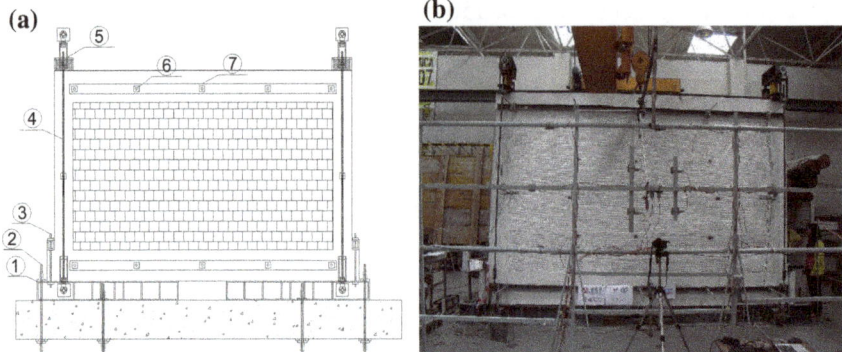

Fig. 1 OOP test setup: **a** front view schematic layout, **b** general front view. 0–strong floor, 1—foundation steel shape, 2—high-strength rods (ø30 mm) fixing the foundation steel shape to the reaction slab, 3—steel rod (ø20 mm) connecting the RC frame to the foundation steel shape, 4 —vertical high-strength rods (ø30 mm) to apply axial load, 5—steel cap, 6—steel rods (ø20 mm) connecting the RC frame and the reaction structure, 7—distributing load plate

Fig. 2 Comparative analysis of global results: global force displacement envelope

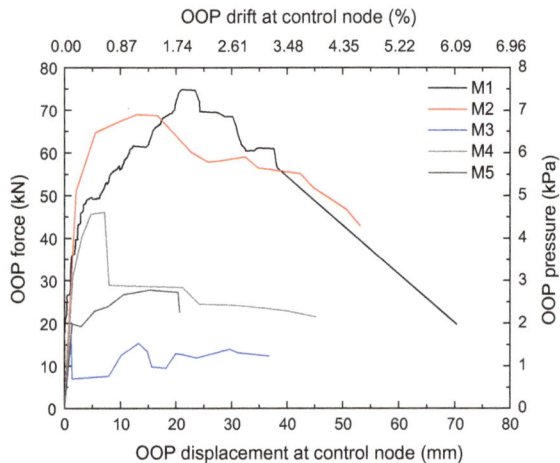

3.4   Test Results

From the force-displacement envelopes illustrated in Fig. 2 the following observations can be performed:

- Comparing the specimens M1 and M4 it can be observed that the first obtained 50% higher OOP capacity. Besides the same axial load in the top of the columns and being subjected to different loading solicitations (monotonic and cyclic respectively) the main differences that justify the OOP capacity can be associated to the mortar properties of each test;

- Regarding the tests M2 and M5, both tested cyclically and without axial load in the columns it can be observed that M5 reached 2 times and half lower OOP force than the M2 justified by the support conditions of the panel;
- From the tests it was observed that the M1 obtained less 7.4% initial stiffness than the M4 specimen, and the test M2 obtained 58.6% higher initial stiffness than M5.

In Table 1 is summarized the maximum strength (F_{max}), OOP displacement at the maximum strength ($D_{OOP,Fmax}$), initial stiffness (Ki) and the failure mode observed on each specimen.

The variables that affected more the OOP response of the specimens were the previous damage and the reduction of the width support of the panel. The previous damage which is more representative of the real behaviour of IM panels during an earthquake, revealed to be the most vulnerable condition. The previous detachment of the panel from the surrounding frame leads to fragile OOP ruptures and reduced about 60% the OOP strength capacity of the panel. From the test results it was observed that the reduction of the panel width support leads to OOP instability of the panel. The OOP strength was significantly reduced, however the arching mechanism provided the sufficient capacity to not occur the panel collapse. The application of the axial load on the top of the columns modified the cracking pattern observed and a vertical cracking was observed followed by the detachment of the panel from the top and bottom beams. On the tests with axial load on the top of the columns, a pronounced strength degradation was observed after reached the maximum ones. From the comparative study it was verified that the monotonic test with application of the axial load on the columns seems to define an envelope of the cyclic tests. These combined variables reached on higher OOP strength capacity, for larger OOP displacements.

Regarding the influence of the axial on the columns prior to carrying out the tests for OOP, revealed a decrease of the OOP stiffness of the panel although combined with an increase in load bearing capacity of the same panels. This increase of the confinement, resulted in a different failure mode when compared with the tests without axial load in the columns. The application the axial load on the top of the columns, without increasing the same during the test leads to the panel acquire a markedly brittle failure. The definition of the stiffness degradation curve, allows to visualize that such non-structural elements begins to lose rigidity to share OOP when the request starts to be transmitted to the panels. This loss of stiffness will be accentuated depending on the behavior exhibited by the panel, that is, the case presents a slow and gradual failure or an instantaneous rupture/collapse.

Table 1 Comparative study: initial stiffness, OOP displacement at maximum strength, OOP maximum strength and failure mode

Specimen	K_i (kN/m)	$D_{OOP, Fmax}$ (mm)	F_{max} (kN)	Failure mode	Cracking pattern
M1	13,481	21.1	75.9	Vertical cracking at the middle of the panel and detachment between the panel and the top and bottom beams	
M2	26,088	16.6	69.8	Trilinear cracking, extending to the bottom corners of the panel. Deflection pronounced at middle height of the panel	
M3	12,507	1.3	17.9	It was not observed any visible cracking. However, and due to the detachment of the right, left and top borders of the panel from the RC elements, the panel behaved as a rigid body	
M4	14,583	7.2	46.0	Vertical cracking slightly to the left of the central alignment of the panel	
M5	10,797	15.2	27.8	Trilinear cracking, slightly to the right of the central alignment of the panel until the mid-height and then extending to the bottom corners of the panel	

4   Numerical Simulation of the IM Walls Seismic Behaviour

4.1   Introduction

Recent advances have been achieved regarding modelling approaches to infill masonry walls to simulate their real contribution to the structural seismic behaviour. Different modelling approaches are available in the literature, such as detailed modelling strategies where the panel is discretized into numerous elements to consider the local effects in detail (Asteris et al. 2013) and simplified macro-models based on strut model concepts (Asteris et al. 2011a, b). The main advantage of the first group is the fact that the local effects related to cracking, crushing, and contact interaction can be captured and simulated; however, this approach requires several parameters and involves high computational effort. Simplified macro-models use the concept of simulating the IM wall through an equivalent strut, and these models have recently been extended to multi-strut models to consider different infill panel behaviours. In 1960, Polyakov proposed (1960) the concept of the equivalent strut, which was later modified by Holmes. Different proposals that simulate the cyclic infill panels' in-plane behaviour with good accuracy can be found in the literature (Rodrigues et al. 2010; Crisafulli and Carr 2007). Recently, some authors provided strut models with the capacity to represent the combined in-plane and out-of-plane behaviour, for example, Kadysiewsko and Modalam's proposal (Kadysiewsko and Modalam 2009), which modelled an IM wall through one diagonal beam element pinned at the edges and provided with a lumped mass in the centre that was active only in the out-of-plane direction. This section pretends to describe a simplified numerical model that was developed to represent the IM walls combined IP and OOP behaviour.

4.2   Simplified Macro-modelling Approach

The numerical modelling approach proposed here is a simplified strut macro-model that is an upgrade of the equivalent bi-diagonal compression strut model proposed by Rodrigues et al. (2010) and later upgraded and implemented in OpenSees (Furtado et al. 2016a, b). The model considers the interaction of the masonry panel behaviour in both directions; the occurrence of panel damage in one direction affects its behaviour in the other direction. Each panel is numerically simulated by four support strut elements, with rigid behaviour, and a central strut element, where the non-linear hysteretic behaviour is concentrated (Fig. 3a). This simplified macro-model can be applied in OpenSees (Mckenna et al. 2000) in association with the available OpenSees materials, sections, and element commands. The infill model was composed of four elastic beam columns for the diagonal elements and one nonlinear beam column element for the central element (Fig. 3) with six

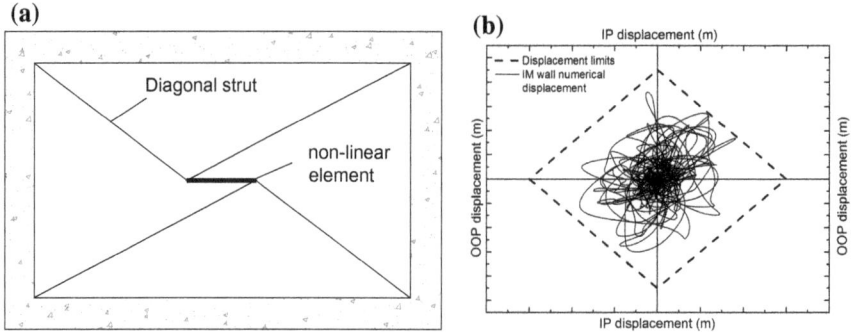

Fig. 3 IM walls simplified numerical modelling approach: **a** schematic layout, **b** IP and OOP interaction

degrees of freedom. The idealization of the central element's non-linear behaviour is characterized through a multi-linear curve, defined by eight parameters, representing: (a) cracking; (b) yielding; (c) maximum strength, corresponding to the beginning of crushing; and (d) residual strength. The Pinching4 uniaxial material model was used to represent the hysteretic behaviour of the infill panel and was attributed to the central element. This uniaxial material is used to construct a material that represents a "pinched" load-deformation response and exhibits degradation under cyclic loading. Cyclic degradation of strength and stiffness occurs in three ways: by unloading stiffness degradation, reloading stiffness degradation, and strength degradation. The model has the main advantage of considering the combined IP and OOP behaviour of the panel. The central element is joined to the diagonal struts through two nodes in which the OOP mass is lumped. The OOP behaviour is assumed to be elastic-plastic with strength and stiffness calculated according to Kadysiewski and Mosalam's approach. Also in this case, part of the model is an algorithm that removes the elements representative of the infill from the structural model if its IP and OOP displacement history exceeds an interaction domain in terms of ultimate displacement. The domain is linear and assumes that, for the undamaged panel, the maximum in-plane drift is equal to 1.5% while the maximum OOP drift is equal to 3% (Fig. 3b).

4.3 Seismic Assessment of a 8 Storey Infilled RC Structure Considering IM Walls OOP Behaviour

With the aim of evaluate the influence of the evaluate the seismic assessment of RC buildings with different considerations regarding the IM walls modelling subjected to seismic actions, one eight-storey building was studied. The building has plant dimensions of 20 m × 15 m, which consists of 4 × 5 m modules, with a storey height of 3 m. The building was designed by the Portuguese Laboratory of Civil Engineering (LNEC) as part of a study on the seismic design of buildings, in accordance with the existing code rules in Portugal. A 3D model was generated in the computer software OpenSees (Mckenna et al. 2000). A set of three building configurations was selected according to the IM modelling strategies adopted: (i) bare frame model (BF) which does not consider the presence of the IM walls; (ii) in-plane model (IP) which considers the presence of the IM walls in the external perimeter of the building, and only the IP behaviour is considered; (iii) OOP model (IP_OOP) which considers the presence of the IM walls in the external perimeter of the building and both the IP and OOP behaviour interaction. The 3D models were subjected to incremental dynamic analysis (IDA) to develop fragility curves according to the Monte Carlo proposal.

Limit state criteria based on the maximum inter-storey drift were selected for the present study. To determine the inter-storey drift limits, a set of 6 values proposed by Rosseto and Elnashai (2005) was fixed, as described in Table 2.

From the resulting vulnerability curves, it can be observed that the difference between the performance of the three numerical models namely for the moderate, extensive and collapse damage states (Fig. 4). As concerns moderate damage, it is observed that the BF model is the most vulnerable model and that the IM walls with only IP behaviour protect the building. Extensive damage occurs for lower peak ground acceleration values (<0.4 g) more quickly for the BF model and at >0.4 g for the IP_OOP model.

Finally, it is observed that the OOP behaviour of the IM walls is critical in terms of the collapse damage state. In fact, clear differences are observed between the IP_OOP model compared with the BF and IP models. This fact increases the need to consider the OOP behaviour of the IM walls in the seismic safety assessment of the existing buildings, and consequently in the numerical models.

Table 2 Inter-storey drift limits for infilled RC frames according to Rosseto and Elnashai's (2005) proposals

Damage state	Inter-storey drift (%)
Slight	0.05
Light	0.08
Moderate	0.30
Extensive	1.15
Partial collapse	2.80
Collapse	>4.40

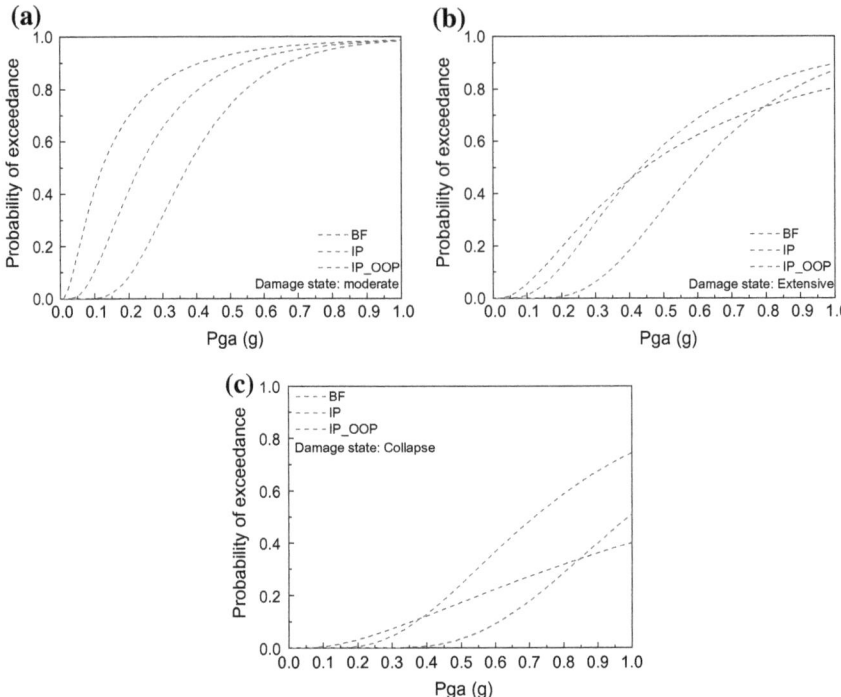

Fig. 4 Vulnerability curves for **a** moderate, **b** extensive and **c** collapse damage states

5 Conclusion

This chapter presents a research work regarding the IM walls seismic behaviour, such experimental and numerical, and their interaction with the RC structures. From this chapter the main conclusions that can be achieved is that in the assessment of existing buildings, and in the design of new buildings:

- Consideration of the masonry infill walls in the structural design (based on simple checking rules/procedures after the structural design) should be enforced;
- Attention should be given to the stiffness differences between the 1st storey and the upper storeys (storey height, dimensions and position of openings, distribution of masonry infill walls);

From the test results, and from numerical analysis of the RC building studied, with the simplified macro-model that simulates the IP and OOP behavior of IM walls, it can be concluded:

- The large IP shear demands that IM walls are subjected to are likely to increase their out-of-plane vulnerability;

- The OOP collapse of infills can result in serious human and material consequences, as observed in recent earthquakes.

So, there is a need to consider the OOP behavior of IM walls in the seismic safety assessment of existing RC structures.

Acknowledgements This work was financially supported by: Project POCI-01-0145-FEDER-007457—CONSTRUCT—Institute of R&D In Structures and Construction funded by FEDER funds through COMPETE2020—Programa Operacional Competitividade e Internacionalização (POCI)—and by national funds through FCT—Fundação para a Ciência e a Tecnologia on the research project POCI-01-0145-FEDER-016898—ASPASSI—Safety Evaluation and Retrofitting of Infill masonry enclosure Walls for Seismic demands.

References

Angel R, Abrams D, Shapiro D, Uzarski J, Webster M (1994) Behavior of reinforced concrete frames, with masonry infills, Civil Engineering Studies, Research Series No. 589, UILU-ENG, Department of Civil Engineering, University of Ilinois, USA, pp 94–2005

Asteris P, Antoniou S, Sophianopoulos D, Chrysostomou C (2011) Mathematical macromodelling of infilled frames: State of the art. J Struct Eng 137:1508–1517

Asteris PG, Antoniou ST, Spophianopoulos DS, ASCE M, Chrysostomou CZ (2011b) Mathematical macromodeling of infilled frames: State of the art. J Struct Eng 137:1508–1517

Asteris P, Cotsovos D, Chrysostomou C, Mohebkhah A, Al-Chaar G (2013) Mathematical micromodelling of infilled frames: State of the art. Eng Struct 56:1905–1921

Calvi G, Bolognini D (2001) Seismic response of reinforced concrete frames infilled with weakly reinforced masonry panels. J Earthq Eng 5:153–185

Crisafulli F, Carr A (2007) Proposed macro-model for the analysis of infilled frame structures. Bull New Zealand Soc Earthq Eng 40:69–77

Dawe J, Seah C (1989) Out-of-plane resistance of concrete masonry infilled panels. Can J Civ Eng 16:854–864

Eurocode 8 (2003) Design of structures for earthquake resistance - Part 1-1: General rules, seismic actions and rules for buildings, B. European Committee for Standardization, Belgium

Eurocode 6 (2005) Part 1-1—General Rules for buildings—Rules for reinforced and unreinforced masonry, European Committee for Standardisation, Brussels

FEMA273 (1996) NEHRP guidelines for the seismic rehabilitation of buildings; FEMA 274, Commentary. Ed: Federal Emergency Management Agency, Washington (DC)

FEMA274 (1997) NEHRP commentary on the guidelines for the seismic rehabilitation of buildings. FEMA-274, Applied Technology Council, Washington, USA. Ed: Federal Emergency Management Agency, Washington (DC)

FEMA306 (1998) Evaluation of earthquake damaged concrete and masonry wall buildings: basic procedures manual. FEMA-306—Applied Technology Council, Washington, USA

FEMA356 (2000) Prestandard and commentary for the seismic rehabilitation of buildings. Ed: Federal Emergency Management Agency, Washington (DC)

Furtado A, Rodrigues H, Arêde A, Varum H (2016a) Experimental evaluation of out-of-plane capacity of masonry infill walls. Eng Struct 111:48–63

Furtado A, Rodrigues H, Arêde A, Varum H (2016b) Simplified macro-model for infill masonry walls considering the out-of-plane behaviour. Earthq Eng Struct Dynam 45:507–524

Hermanns L, Fraile A, Alarcón E, Álvarez R (2014) Performance of buildings with masonry infill walls during the 2011 Lorca earthquake. Bull Earthq Eng 12:1977–1997

Kadysiewski S, Mosalam KM (2009) Modeling of unreinforced masonry infill walls considering in-plane and out-of-plane interaction. Pacific Earthq Eng Res Center, PEER 2008/102

Mckenna F, Fenves G, Scott M, Jeremic B (2000) Open system for earthquake engineering simulation (OpenSees). Ed. Berkeley, CA

Moghaddam H, Dowling P, Ambraseys N (1988) Shaking table study of brick masonry infilled frames subjected to seismic actions. In 9th World Conference on Earthquake Engineering Tokyo, Japan

NTC08 (2008) Decreto ministeriale 14 gennaio 2008—Norme Tecniche per le Costruzioni NTC2008. Supplemento ordinario n. 30 Gazzetta Ufficiale 4 febbraio 2008, n 29 (in Italian)

Polyakov S (1960) On the interaction between masonry filler walls and enclosing frame when loading in the plane of the wall. Transl Earthq Eng, 36–42

Rodrigues H, Varum H, Costa A (2010) Simplified macro-model for infill masonry panels. J Earthq Eng 14:390–416

Rosseto T, Elnashai A (2005) A new analytical procedure for the derivation of displacement-based vulnerability curves for populations of RC structures. Eng Struct 27:397–409

Smyrou E, Blandon C, Antoniou S, Pinho R, Crisafulli F (2011) Implementation and verification of a masonry panel model for nonlinear dynamic analysis of infilled RC frames. Bull Earthquake Eng 9:1519–1534

Code Based Performance Prediction for a Full-Scale FRP Retrofitted Building Test

Alper Ilki, Erkan Tore, Cem Demir and Mustafa Comert

Abstract Lack of adequate ductility in substandard reinforced concrete buildings is a major reason for significant amount of life and economic losses experienced during major earthquakes. Laboratory tests realized at member level (i.e. column and beam tests) show that external wrapping of potential plastic hinging regions of columns with Fiber Reinforced Polymer (FRP) sheets, can significantly enhance the ductility capacity. Thus, major seismic retrofitting documents available worldwide interpret this retrofit approach as a viable alternative and lay the rules for retrofit design with these materials. In this study, a very recent experimental activity that included the full-scale testing of a substandard building retrofitted with the above mentioned approach is briefly presented. Then, the performance prediction of the Turkish Seismic Design Code (2007) for this building is evaluated. Finally, a revision is proposed for the FRP effective rupture strain value defined by this code.

Keywords Ductility · Earthquakes · Laboratory tests · Full-scale testing

1 Introduction

Many existing buildings had suffered from past earthquakes due to inadequate ductility caused by the lack of confinement reinforcement at the potential plastic hinging regions of reinforced concrete (RC) members. Particularly columns under high axial loads tend to exhibit limited deformation capacity and brittle behaviour that may lead to total collapse of the building. Retrofitting with fiber reinforced polymers (FRPs) is an effective way to enhance the seismic performance of these

A. Ilki (✉) · C. Demir
Faculty of Civil Engineering, Istanbul Technical University, Istanbul, Turkey
e-mail: ailki@itu.edu.tr

E. Tore
Department of Civil Engineering, Balikesir University, Balikesir, Turkey

M. Comert
RISE Engineering and Consultancy, Istanbul, Turkey

© Springer International Publishing AG, part of Springer Nature 2018
R. Vacareanu and C. Ionescu (eds.), *Seismic Hazard and Risk Assessment*,
Springer Natural Hazards, https://doi.org/10.1007/978-3-319-74724-8_31

buildings in terms of ductility. External confinement with FRP sheets significantly improves deformation capacity of RC members and ensures ductility enhancement of the structural system. This retrofitting method for ductility enhancement is well investigated through member level laboratory research studies in the last two decades and design procedure of the method are covered by some of the design documents (i.e., ACI 440.2R-17 2017; fib Bulletin 14 2001; TSDC 2007). Only a limited number of building level research projects are available for evaluation of recommendations for modelling and analysis of retrofitted buildings (Balsamo et al. 2005; Di Ludovico et al. 2008; Garcia et al. 2014; Tore et al. 2017).

In this paper, recommendations of Turkish Seismic Design Code (TSDC 2007) on RC columns jacketed with FRP sheets are evaluated considering the results of a full-scale test performed very recently by the authors (Tore et al. 2017). For this purpose, the analytical predictions made by following the TSDC (2007) approach are compared with the experimental behaviour. During the numerical analyses, nonlinear fiber sections with FRP confined concrete material properties are used for modelling of FRP retrofitted structure and nonlinear static pushover analyses are performed. The comparison of analytical and experimental results shows that TSDC (2007) recommendations for modelling of FRP confined concrete give conservative results particularly in terms of the deformation capacity. To provide a more efficient use of the retrofit material, modification of the assumed FRP rupture strain is proposed.

2 TSDC (2007) Recommendations for FRP Confined Concrete

TSDC (2007) covers the design rules for shear capacity, concrete compressive strength and ductility enhancement of existing reinforced concrete building members with external FRP confinement. Enhancement in concrete axial behaviour is considered by using a bilinear stress-strain model for FRP confined concrete and this model can be used during the nonlinear analysis (Fig. 1). Accordingly, axial stress and strain values of the transition point of the model are unconfined concrete

Fig. 1 TSDC (2007) FRP confined concrete model

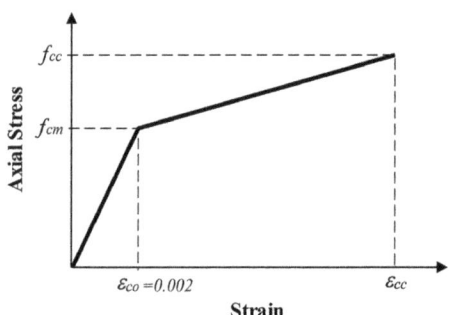

compressive strength (f_{cm}) and 0.002 strain. Confined concrete strength (f_{cc}) and strain of the ultimate point (ε_{cc}) are calculated by Eqs. 1 and 2. Effective lateral confinement pressure (f_l) is predicted by Eq. 3, where κ_a is the shape efficiency factor, ρ_f, ε_f and E_f are volumetric ratio, effective rupture strain and elastic modulus of FRP respectively. Shape efficiency factors for different types of cross sections can be calculated by Eq. 4. FRP material properties, curvature of surfaces, installation methodology, loading type and conditions are among the parameters that affect the ultimate rupture strain of FRP. In TSDC (2007), the FRP rupture strain is conservatively considered as the minimum of 0.004 and 0.5 ε_{fu} (50% of the ultimate strain given by the manufacturer), Eq. 5. Generally, the 0.004 strain value governs the value of ε_f. Additionally, other conditions, such as an upper limit of 2 for cross-sectional aspect ratio (h/b) and minimum corner rounding radius (r_c) of 30 mm are also recommended for ensuring an effective confinement. It should be noted that the above mentioned FRP confined concrete model in TSDC (2007) is based on the experimental works carried out by Ilki et al. (2002, 2003, 2004) on low and medium strength concrete prisms.

$$f_{cc} = f_{cm}(1 + 2.4(f_l/f_{cm})) \geq 1.2f_{cm} \tag{1}$$

$$\varepsilon_{cc} = 0.002\left(1 + 15(f_l/f_{cm})^{0.75}\right) \tag{2}$$

$$f_l = \frac{1}{2}\kappa_a\rho_f\varepsilon_f E_f \tag{3}$$

$$\kappa_a = \begin{cases} 1 & \text{circular sections} \\ \left(\frac{b}{h}\right) & \text{square sections} \\ 1 - \frac{(b-2r_c)^2 + (h-2r_c)^2}{3bh} & \text{rectangular sections} \end{cases} \tag{4}$$

$$\varepsilon_f = \min\left\{\begin{array}{c} 0.004 \\ 0.50\,\varepsilon_{fu} \end{array}\right\} \tag{5}$$

3 Outline of the Test Building

A large scale testing programme, which included the testing of two full-scale three-story buildings, have been conducted by the authors to demonstrate the efficiency of FRP retrofitting of substandard RC buildings. The test buildings had two bays in the loading direction and one in the perpendicular direction (Fig. 2). As also generally observed in many substandard buildings, the columns were designed to be weaker than beams and had a poor confinement reinforcement of 8 mm diameter stirrups with a spacing of 320 mm. Infill walls, 25 cm thick slab, and self-weight of the structural system were the dead loads. Additional concrete weighing blocks were used for increasing axial loads of columns to a critical level for member

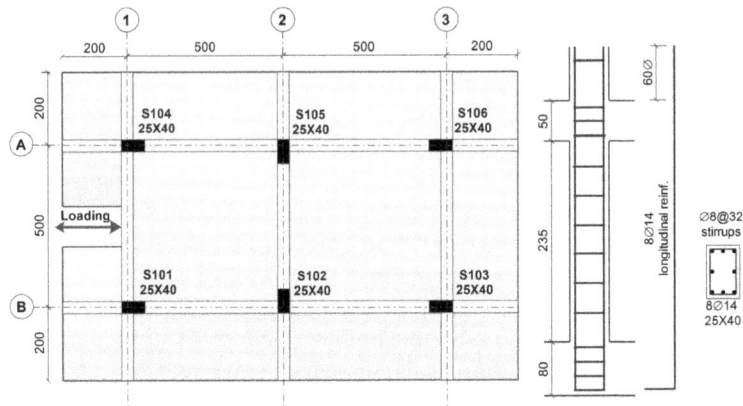

Fig. 2 Plan view of the building and column reinforcement details (dimensions are in cm)

Table 1 CFRP mechanical properties

Tensile strength (MPa)	Elastic modulus E_f (GPa)	Rupture strain ε_{fu} (%)	Effective thickness (mm)	Unit weight (g/cm^3)
4900	240	2.00	0.165	1.8

ductility. At the lowest story, column axial load ratios without reinforcement contribution to axial capacity of columns were approximately 25% for S101-S104, 45% for S102-S105 and S103-S106 columns. Gaps between the infill wall panels and the reinforced concrete members were intentionally left for eliminating the effect of walls on structural behaviour. Concrete compressive strength, obtained from compression testing of 100 × 200 mm cylinders, for the first story columns was approximately 18 MPa at the day of testing. Deformed bars with 447 and 444 MPa yield strength were used for longitudinal and transverse bars, respectively. Geometric details of the buildings and reinforcement details of the columns are presented in Fig. 2.

External confinement with CFRP sheets were applied to the first and second story columns for ductility enhancement. The first story column upper and lower ends were wrapped with 5 plies and the second story column end regions were wrapped with 3 plies of CFRP with 60 cm height to enhance deformation capacities of potential plastic hinging regions. Additionally, two plies of CFRP sheets were wrapped to remaining height of columns for increasing shear capacity. Corners of the columns were rounded to 30 mm radius and 200 mm overlapping length was applied at the last wrap of the sheets. Mechanical properties of CFRP, as provided by the manufacturer, are given in Table 1.

Lateral loading was applied to the first and second stories by using three servo-controlled hydraulic actuators with 300 kN load and 800 mm displacement capacities. One actuator was attached to the first story and the other two were attached to the second story slab levels. Throughout the test duration, ratio of the lateral loads applied to the first and second stories was kept constant as 0.5. Firstly, static cyclic lateral loading was applied incrementally from 0.125 to 0.9% first story drift ratio levels. Then, the last step of the lateral loading was applied as a pushover loading after actuators were demounted from the building. For the retrofitted building, the pushover loading was applied up to 15% first story drift ratio. Even at 15% drift, no partial or total collapse occurred though a 70% loss in lateral strength (mainly due to second order effects) took place. Test was terminated at that displacement level due to actuator stroke limitations. The general appearance and deformed shape of the retrofitted building at the end of the test are given in Fig. 3. Behaviour of the building in terms of base shear versus the first story drift ratio relationship is presented in Fig. 4. The building exhibited significantly ductile behaviour up to 6% drift ratio of the first story where the lateral strength loss of was approximately 20%.

4 Analytical Study

4.1 Modelling of the Test Building

Nonlinear analyses of the test building were conducted by using the SAP 2000 v.18.2 structural analysis software. Beams and columns of the building were modelled with frame elements and shell elements were used for modelling of stiff floor slabs. Rigid floor diaphragm and beam column joints were defined for each story level. Additional loading blocks were input as area loads.

The nonlinear behaviour of the structural members was modelled according to TSDC (2007). Strong beams were modelled with plastic hinges at the end of the frame elements. Columns were modelled with nonlinear fiber type hinges located at the top and bottom ends of the elements. The remaining parts of the columns and beams, in between the assigned plastic hinges, were modelled by using elastic frame elements with cracked section properties. Modelling of a column with nonlinear fiber hinges and elastic frame element are schematically presented in Fig. 5. Material stress-strain relationships were assigned to each steel or confined concrete fiber in the hinge section. The plastic hinge length of FRP confined columns and beams was considered as half of the cross-section depth as also recommended by TSDC (2007). It should be noted that, as also indicated by Jiang et al. (2014), parameters such as FRP jacket thickness may affect the plastic hinge length of FRP confined columns.

In this study, in order to evaluate the effective FRP rupture strain recommended by TSDC (2007), different values of ε_f are considered and predictions under each assumption are compared with the experimental results. For this purpose, FRP confined concrete strength and corresponding strain values of the bilinear confined concrete model were calculated by Eqs. 1 and 2 for three different FRP rupture

Fig. 3 General views of the retrofitted building **a** before and **b** after the test

strain assumptions as summarized in Table 2. Trilinear stress-strain relationship based on steel tension tests and calculated bilinear stress strain relationships of confined concrete for different FRP rupture strains are presented in Fig. 6. Behaviour of reinforcement steel in tension and compression were assumed to be same due to prevention of bar buckling with CFRP confinement.

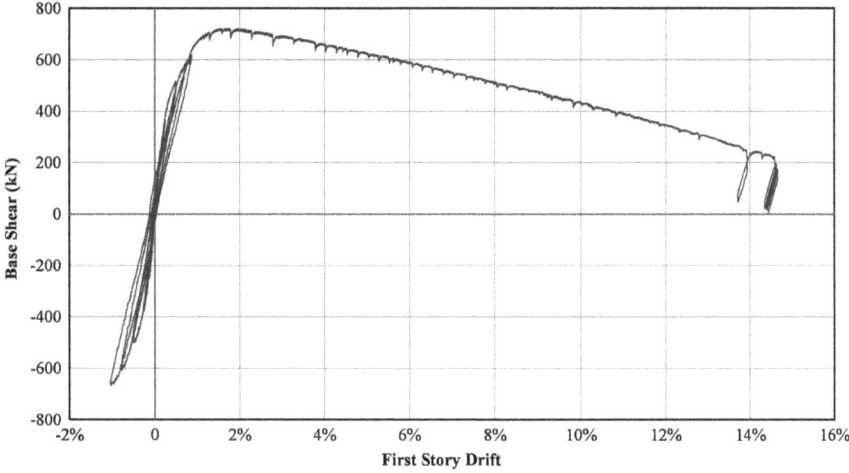

Fig. 4 Experimental base shear versus first story drift ratio relationship of the building

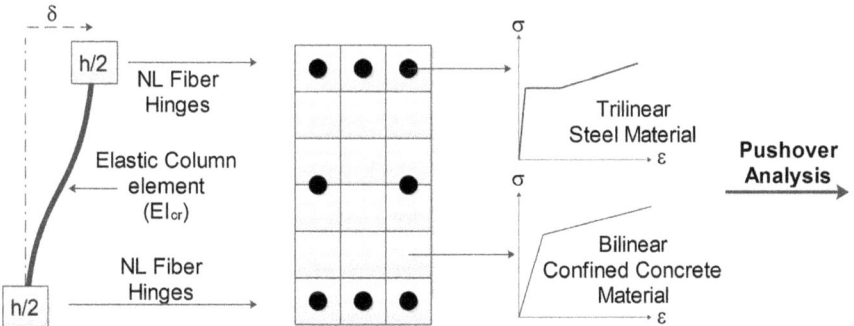

Fig. 5 Nonlinear modeling of columns

Table 2 FRP confined concrete model parameters

Rupture strain ε_f	Shape efficiency factor κ_a	Volumetric FRP ratio ρ_f	Effective confinement stress f_l (MPa)	Confined concrete strength f_{cc} (MPa)	Confined concrete ultimate strain ε_{cc}
0.004	0.494	0.011	2.5	24.6	0.009
0. 5 ε_{fu} = 0.010	0.494	0.011	6.4	33.8	0.015
0.75 ε_{fu} = 0.015	0.494	0.011	9.5	41.4	0.020

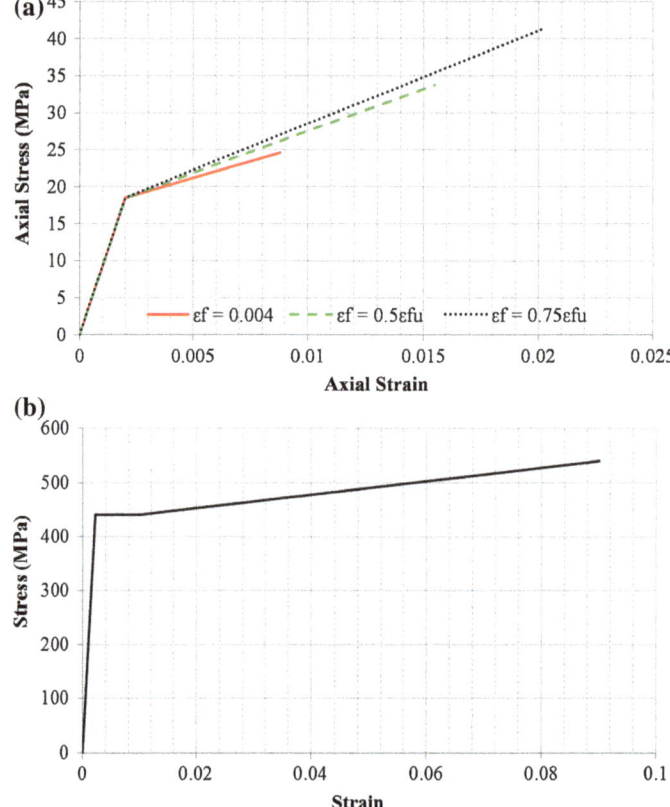

Fig. 6 Considered material models; **a** bilinear confined concrete models for different FRP rupture strain assumptions **b** trilinear reinforcement model

4.2 Nonlinear Analysis of the Test Building

Nonlinear analyses were conducted under gravity loads and incremental lateral (Pushover) loading. Displacement controlled lateral loads were applied to the model with a similar pattern of test loading. Because of expected large displacements, second order effects were also taken into account in nonlinear analysis. In Fig. 7, the predicted base shear-first story drift ratio relationships obtained from the non-linear analyses are compared with the experimental curve. Experimental behaviour is limited to 4% drift ratio, which is a collapse limit according to TSDC (2007) for interstory drift ratio. As seen in Fig. 7, all the predicted behaviours follow the backbone curve of the cyclic part of loading, up to 0.9% drift ratio. Confined concrete model that was calculated by using the 0.004 FRP rupture strain (as recommended by TSDC 2007) seems to be highly conservative (Fig. 7a) which may cause uneconomic retrofit schemes. Increased confined concrete strains

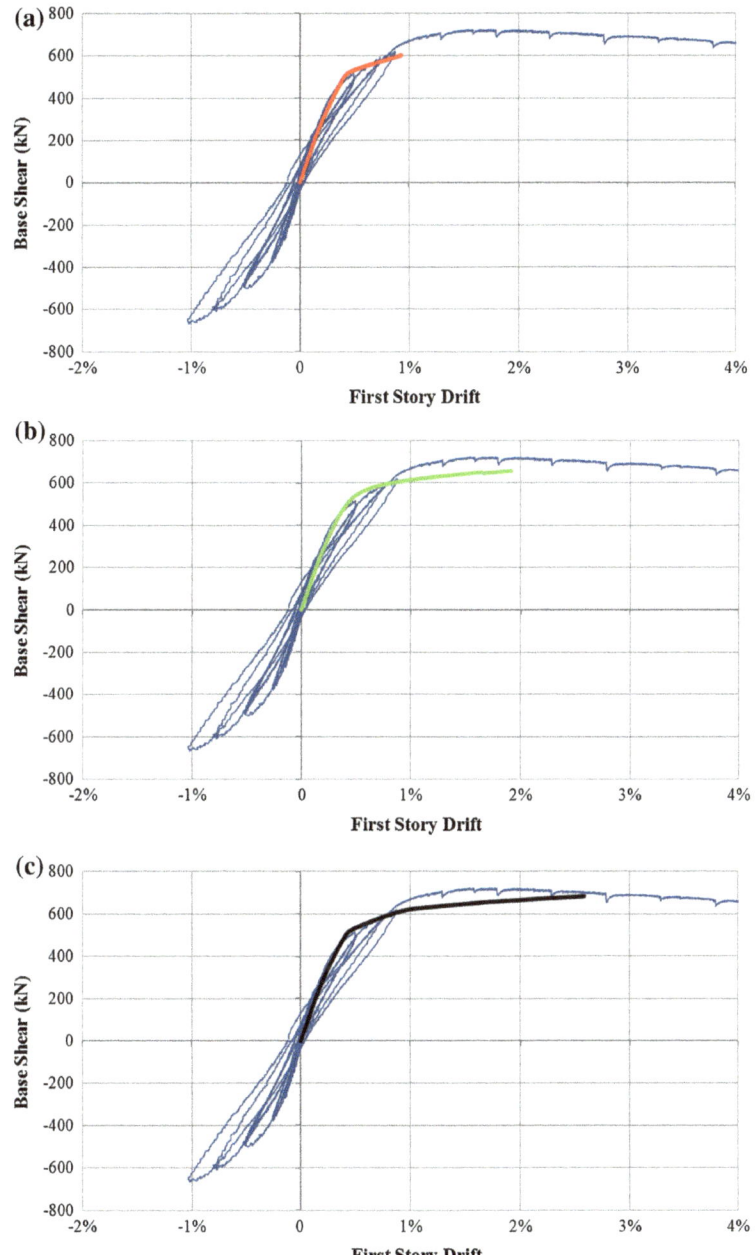

Fig. 7 Comparison of experimental and predicted base shear versus first story drift ratio curves. FRP confined concrete model with **a** $\varepsilon_f = 0.004$ (red curve), **b** $\varepsilon_f = 0.5\ \varepsilon_{fu}$ (green curve) and **c** $\varepsilon_f = 0.75\ \varepsilon_{fu}$ (black curve)

significantly improve the performance of the predicted behaviour, particularly, in terms of ductility (Fig. 7b, c). Ultimate first story drift ratios are calculated approximately as 2% and 2.5% in the analyzed models with 0.5 ε_{fu} and 0.75 ε_{fu} FRP rupture strains, respectively. Additionally, a slight difference can be observed between the experimental and predicted peak lateral load values. This difference can be attributed to mechanical differences between bending and concentric axial loading conditions, slip rotations at top and bottom end of columns and other critical nonlinear analysis parameters such as plastic hinge length of column element. FRP rupture strain assumptions higher than 0.004 for confined concrete model of TSDC (2007) seems to be adequately conservative.

5 Conclusions

In this study, bilinear confined concrete model recommended by TSDC (2007) for nonlinear analysis of FRP retrofitted reinforced concrete buildings is evaluated both experimentally and numerically. FRP rupture strain considered in the model is a critical parameter which is affected by several factors. In order to investigate the code recommended FRP rupture strain values, FRP confined concrete models for three different FRP rupture strain assumptions were used for nonlinear analyses of a full-scale building, tested by the authors. Predicted behaviours are quite compatible with the experimental results, particularly in the linear elastic part and at the beginning of the plastic deformations. Current maximum limit of 0.004 FRP rupture strain according to TSDC (2007) seems to be extremely restrictive, mainly in terms of the displacement capacity predictions. This conservative limit can lead to uneconomical designs and ineffective use of FRP material. FRP rupture strain limit value of 0.5 ε_{fu} gives reasonably acceptable conservatism. Thus, in the new version of Turkish Seismic Design Code, limiting the FRP rupture strain with 0.5 ε_{fu} for enhancing the ductility of a building may ensure economic and efficient FRP retrofit designs, and may make usage of FRP in strengthening applications more attractive.

Acknowledgements Authors are thankful to DowAksa Advanced Composites Holding B.V. and Governorship of Yalova for various supports provided for experimental studies. The contributions of Ilgaz Doğan (DowAksa), Çağlar Göksu Akkaya (Ph.D., ITU), Pınar İnci (Ph.D. Cand., ITU), Ergün Binbir (Ph.D. Cand., ITU), Ali Osman Ateş (Ph.D. Cand., ITU), Ali Naki Şanver (M.Sc. Cand., Rise Eng.), Çağlar Üstün (M.Sc. Cand., RISE Eng.), Duygu Çakır (Civil Eng., Rise Eng.), Gökhan Sarı (Undergrad., Balıkesir University), Emin Amini (Undergrad., Balıkesir University), Tamer Şahna (Undergrad., Balıkesir University), Oğuzhan Sözer (Undergrad., Balıkesir University), Berkay Aldırmaz (M.Sc. Cand., ITU), Ömer Faruk Halıcı (M.Sc., ITU), Mehmet Aksa (Undergrad., Kultur Uni.) are also gratefully acknowledged.

References

ACI Committee 440 (2017) Guide for the design and construction of externally bonded FRP systems for strengthening concrete structures ACI 440.2R-17. American Concrete Institute, Farmington Hills, Mich

Balsamo A, Colombo A, Manfredi G, Negro P, Prota A (2005) Seismic behavior of a full-scale RC frame repaired using CFRP laminates. Eng Struct 27(5):769–780

Di Ludovico M, Prota A, Manfredi G, Cosenza E (2008) Seismic strengthening of an under-designed RC structure with FRP. Earthquake Eng Struct Dynam 37:141–162

fib Bulletin 14 (2001) Externally bonded FRP reinforcement for RC structures. Lausanne, Switzerland

Garcia R, Hajirasouliha I, Guadagnini M, Helal Y, Jemaa Y, Pilakoutas K, Mongabure P, Chrysostomou C, Kyriakides N, Ilki A, Budescu M, Taranu N, Ciupala MA, Torres L, Saiidi M (2014) Full-scale shaking table tests on a substandard RC building repaired and strengthened with post-tensioned metal straps. J Earthquake Eng 18(2):187–213

Ilki A, Kumbasar N (2002) Behavior of damaged and undamaged concrete strengthened by carbon fiber composite sheet. Int J Str Eng Mech 13(1):75–90

Ilki A, Kumbasar N (2003) Compressive behaviour of carbon fiber composite jacketed concrete with circular and non-circular cross-sections. J Earthquake Engineering 7(3):381–406

Ilki A, Kumbasar N, Koc V (2004) Low strength concrete members externally confined with FRP sheets. Struct Eng and Mech 18(2):167–194

Jiang C, Wu Y, Wu G (2014) Plastic hinge length of FRP-Confined square RC columns. ASCE J Compos Constr 18(4):04014003

Tore E, Comert M, Demir C, Ilki A (2017) Collapse testing of full-scale RC buildings with or without seismic retrofit of columns with FRP jackets. In: Paper presented at COST Action TU1207 end of action conference, Budapest, Hungary, 3–5 April 2017

TSDC Turkish Seismic Design Code (2007) Specification for the buildings to be constructed in disaster areas. Ministry of Public Works and Settlement, Ankara, Turkey

Study of Seismic Capacity of Masonry Infilled Reinforced Concrete Frames Considering the Influence of Frame Strength

Hamood Al-Washali, Kiwoong Jin and Masaki Maeda

Abstract Unreinforced masonry infilled-RC frames are widely used in many developing countries. Even though the influence of the masonry walls on the behavior of structure was recognized from the experience of past earthquake disasters, but many practicing engineers still assume that the infill walls are non-structural walls due to incomplete knowledge of the behavior of such structures. In this paper, an experimental study of two ½ scale specimens with different RC frames and masonry infill walls were tested using a static cyclic loading protocol. The main parameter was the influence of changing the RC frame strength on masonry infill seismic capacity. The results showed that shear strength and deformation capacity of masonry infill greatly improved by increasing the strength of boundary frame. The investigation of strength, ductility and initial stiffness based on experimental results and comparative study with existing methods showed large variations between several methods commonly used to assess the seismic capacity of masonry infill.

Keywords Reinforced concrete buildings · Masonry infill · Seismic capacity
Boundary frame strength

1 Introduction

Many of the reinforced concrete buildings in the world and particularly in developing countries use masonry infill as partition walls. The influence of masonry on the structural behavior was recognized from the experience of earthquake disasters

H. Al-Washali (✉) · K. Jin · M. Maeda
Graduate School of Engineering, Tohoku University, Sendai, Japan
e-mail: Hamood@rcl.archi.tohoku.ac.jp

K. Jin
e-mail: Jin@rcl.archi.tohoku.ac.jp

M. Maeda
e-mail: Maeda@rcl.archi.tohoku.ac.jp

© Springer International Publishing AG, part of Springer Nature 2018
R. Vacareanu and C. Ionescu (eds.), *Seismic Hazard and Risk Assessment*,
Springer Natural Hazards, https://doi.org/10.1007/978-3-319-74724-8_32

and experiments by several researchers (i.e., Paulay and Priestley 1992). In general, masonry infill increases the frame strength, which is considered as a beneficial point. On the other hand, masonry infill exerts reaction forces on the RC frame causing additional moment and shear forces, which results in unexpected failure modes FEMA 306 (1998). In addition, the masonry infill greatly increases the frame stiffness that might changes the seismic demand due to significant reduction in natural period of the building. Despite of these distinctive characteristics, many practicing engineers still assume that the walls are non-structural elements due to incomplete knowledge concerning RC frames with masonry infill behavior and complexity in evaluating their failure modes. The seismic performance of masonry infill depends on several parameters such as the confinement effect, masonry type, aspect ratio, mortar strength, etc. Among these factors, the strength of boundary RC frame is a crucial parameter that not only governs the behavior and failure modes of the RC frame but also the masonry infill strength and failure mode, as shown in previous experiments conducted by Mehrabi et al. (1996). However, modelling and estimating strength of masonry infill commonly considers only the material characteristics of masonry and ignoring confining influences of boundary frame.

In brief, the influence of boundary frame strength and its influence on the seismic capacity infill and the overall frame have not been clearly identified. Thus, the objective of this paper is as follows: First, the experimental investigation of RC frame with unreinforced masonry infill is carried out with in-plane cyclic loading tests, where the only varying parameter is the lateral strength of surrounding frames. Second, the investigation of strength, ductility and stiffness based on experimental results and comparative study.

2 Experimental Program

2.1 Test Specimens

Two half-scaled specimens with different RC frames, having same unreinforced masonry infills, are designed. The main variance parameter for test specimens is the ratio of the boundary frame to masonry infill lateral strength defined as β index, as shown in Eq. (1). Specimens are named WF (weak frame) and SF (strong frame) with β of 0.4 and 1.5, respectively.

$$\beta = V_f / V_{\text{inf}} \tag{1}$$

Where V_f is the boundary frame lateral strength which is calculated to be the ultimate flexural capacity of a bare frame with plastic hinges at top and bottom of columns. The V_{inf} is the masonry infill lateral strength calculated based on Eq. (2) which is a simplified empirical equation showing good agreement with experimental database studied by the author (Al-Washali et al. 2017)

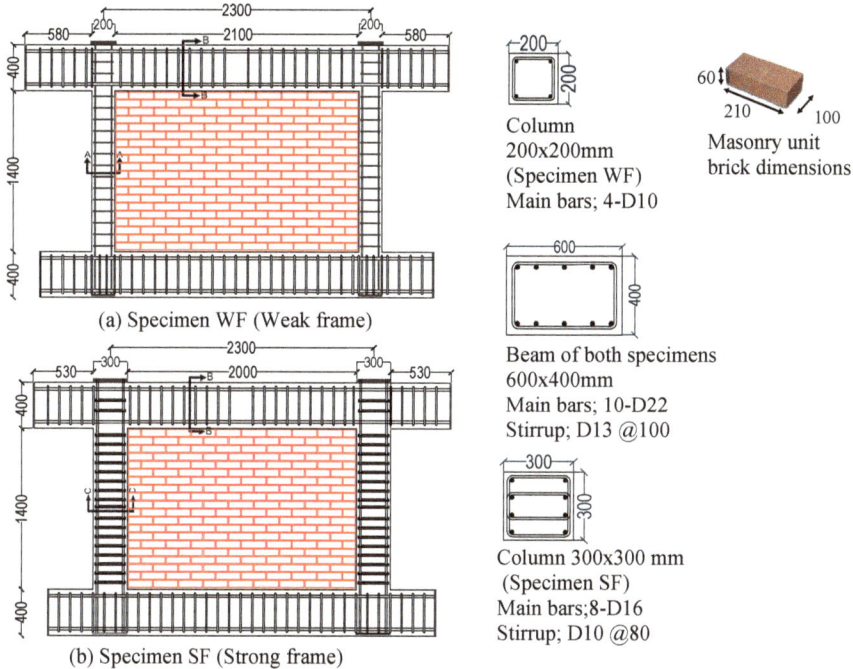

Fig. 1 Dimensions and reinforcement of specimens; units in mm

$$V_{\text{inf}} = 0.05\,fm \cdot t_{\text{inf}} \cdot l_{\text{inf}} \tag{2}$$

where fm is the compressive strength of masonry prism, t_{inf} is the infill thickness, l_{inf} is the infill length.

The specimen dimensions and details are shown in Fig. 1. Both specimens are identical except for the column size and reinforcement, as shown in Fig. 1. The beams were designed to be stronger and stiff enough to simulate a typical case of a weak column and strong beam system observed in existing buildings of old designs.

2.2 Material Properties

The infill panels are constructed using $60 \times 100 \times 210$ mm solid bricks conventionally used in Japan. A professional mason built the infill, after the frame construction, where its thickness is 100 mm and mortar head and bed joint thickness is about 10 mm. Tables 1 and 2 show the material properties based on material tests where the values represent the mean values of three samples. The masonry prism strength is the masonry prism compressive strength tested according to ASTM C1314 (2011). The material tests were conducted at the same time with the experimental loading for each specimen individually.

Table 1 Material properties of concrete and masonry details

Specimen name	Frame concrete			Masonry prism			Mortar cylinders	Brick unit compressive strength (MPa)
	Compressive strength (MPa)	Elastic moduli (MPa)	Split tensile strength (MPa)	Compressive strength (MPa)	Elastic moduli (MPa)	Strain at peak stress	Compressive strength (MPa)	
WF	24.2	2.3×10^4	2.1	17.3	7840	0.0037	20.2	38.1
SF	28.3	2.3×10^4	2.4	18.6	8140	0.0039	29.2	38.1

Table 2 Reinforcement mechanical properties

Bar	Nominal strength	Yield strength (MPa)	Ultimate tensile strength (MPa)
D6	SD345	476	595
D10	SD345	384	547
D13	SD345	356	555
D16	SD345	370	556
D22	SD390	447	619

2.3 Test Setup and Instrumentation

The loading system are shown schematically in Fig. 2. The vertical load was applied on RC columns by two vertical hydraulic jacks and was maintained to be 200 kN on each column. Two pantograph, attached with the vertical jacks, restricted any torsional and out-of-plane displacement. Two horizontal jacks, applying together an incremental cyclic loading, were attached at the beam level and were controlled by a drift angle of R%, defined as the ratio of lateral story deformation to the story height measured at the middle depth of the beam (h = 1600 mm). The lateral loading program consisted of 2 cycles for each peak drift angle of 0.05, 0.1, 0.2, 0.4, 0.6, 0.8, 1, 1.5 and 2%.

2.4 Experimental Results

The lateral load versus story drift angle of both specimens are shown in Figs. 3 and 4. Cracks and failure patterns after final drift cycle of 2.0% are shown in Fig. 5.

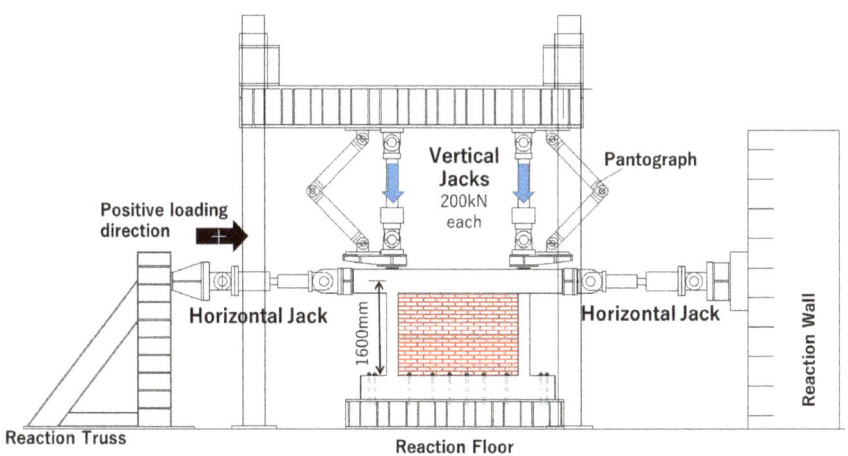

Fig. 2 Test setup

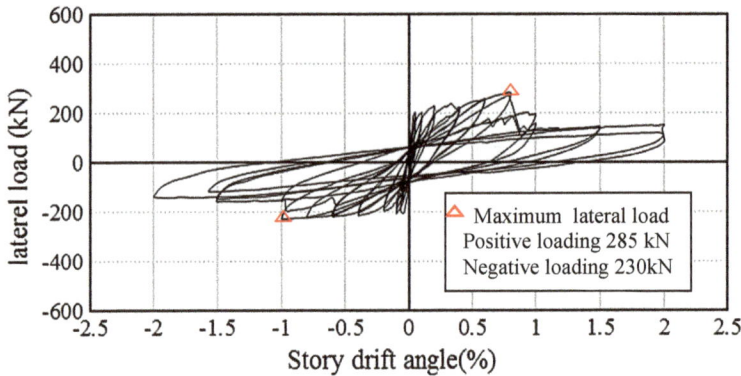

Fig. 3 Lateral strength & story drift angle for specimen WF

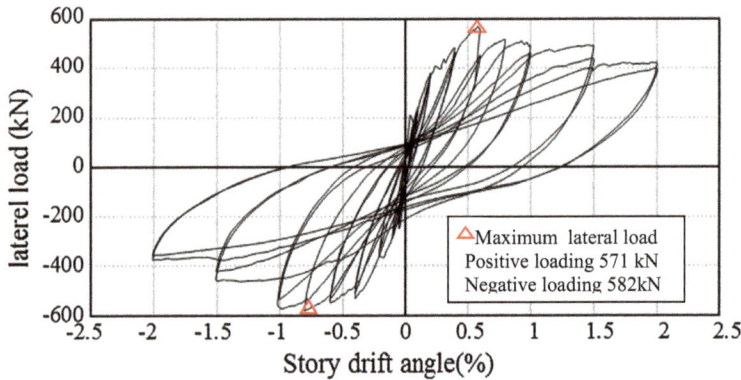

Fig. 4 Lateral strength & story drift angle for specimen SF

For Specimen WF: very small cracks on mortar bed joint and diagonal cracks on bricks near loading corner of infill panel, less than 0.3 mm width, started at early stages of loading just when the drift angle was 0.05%. At drift angles of 0.2 and 0.4%, the longitudinal reinforcement in the tensile column (windward column) yielded at the upper critical section and above its mid-height, respectively, forming failure mechanism similar to a short column, as illustrated in Fig. 6a. Just after reaching the maximum lateral strength, there was a sudden drop of lateral load bearing capacity with extensive cracking and spalling of bricks. After the drift of 1%, the main failure mechanism from diagonal cracks changed to sliding cracks, and clear sliding movement at the mid-height of the infill was noticed. At drift story of 2% in the negative cycle, the concrete around the reinforcement of top compression column spalled-off and main bars buckled.

For Specimen SF: Cracking of infill panel also started at the peak of the first loading cycle, which was relatively similar to the crack width observed in specimen

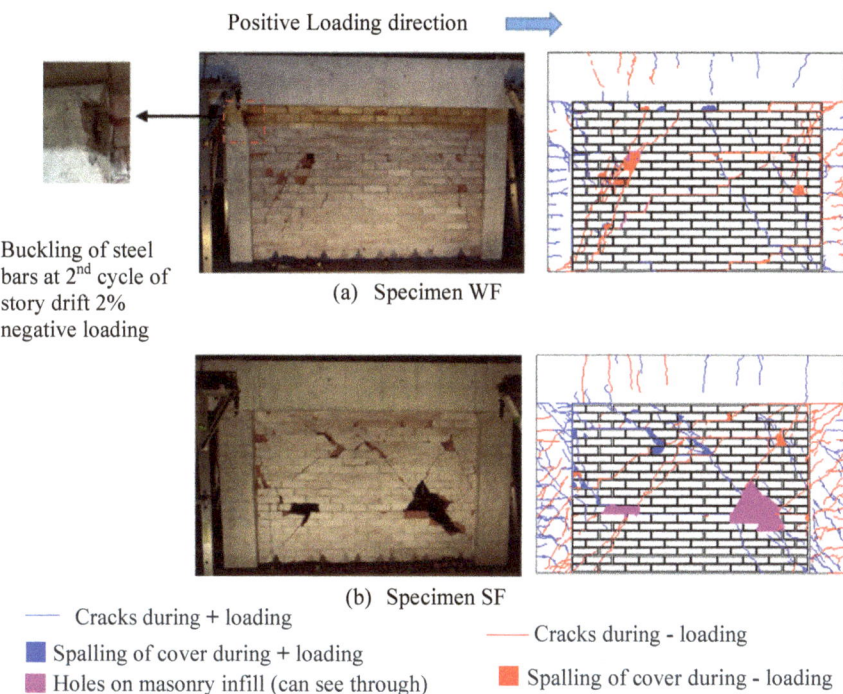

Positive Loading direction

Buckling of steel
bars at 2nd cycle of
story drift 2%
negative loading

(a) Specimen WF

(b) Specimen SF

—— Cracks during + loading

■ Spalling of cover during + loading

■ Holes on masonry infill (can see through)

—— Cracks during - loading

■ Spalling of cover during - loading

Fig. 5 Crack patterns observed at end of the test

WF at this stage. At drift angles between 0.6 and 0.7%, both columns yielded at the locations shown in Fig. 6b. As it reached its maximum strength, the lateral load gradually degraded (contrarily to the sudden degradation of strength in previous specimen WF) with the drift angle increase until the drift angle of 1.5%, where there was a slight drop of the lateral load, after the horizontal sliding between bricks clearly increased. At the drift angle of 2%, the loading stopped as planned, and the masonry infill damage at this point was much greater than observed in the previous specimen WF (see Fig. 5). In spite that columns had many cracks, there was no extensive damage or spalling of concrete cover.

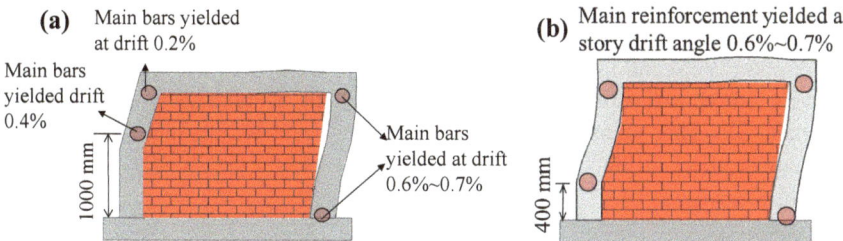

(a) Main bars yielded
at drift 0.2%

Main bars
yielded drift
0.4%

1000 mm

Main bars
yielded at drift
0.6%~0.7%

(b) Main reinforcement yielded at
story drift angle 0.6%~0.7%

400 mm

Fig. 6 Hinge locations formed in RC frame **a** specimen WF **b** specimen SF

3 Discussion of Experimental Results

3.1 Lateral Strength

The maximum lateral load contributed by the masonry infill (V_{inf}) is calculated by deducting the bare frame lateral strength (V_f) from the maximum lateral load of the overall structure (V_{max}), as shown in Eq. (3):

$$V_f = 4Mu/H \tag{3}$$

$$V_{inf} = V_{max} - V_f \tag{4}$$

where Mu is the minimum plastic moment of the column or beam calculated by AIJ provision (2010) and H is the clear height of column (taken here as infill height).

Table 3 shows the experimental shear strength of masonry infills in both specimens, which is the shear force (V_{inf}) divided by the infill cross-sectional area. Even though both infill panels are made by exactly same material and have similar prism compression strength, specimen SF has the shear capacity of 1.48 N/mm^2 which is about 1.5 times the shear strength in specimen WF (0.93 N/mm^2).

3.1.1 Comparison of Lateral Strength with Previous Literature

The in-plane capacity of the masonry infill depends mainly on the type of failure mechanism. The failure mechanism types and identification are different between building standards or researchers. The most recognized failure modes are diagonal compression failure and sliding shear failure modes. The failure mechanism observed in experimental results as mentioned earlier, is a mixture of both: compression and sliding failure. In this study, Table 4 shows the comparison of the infill strength with following methods: Diagonal compression and sliding equation by FEMA 306 (1996), Liauw and Kwan (1985) method, Flanagan and Bennett (1999) method, Paulay and Priestley (1992) method for sliding strength and simplified equation presented by author in previous study (Al-Washali et al. 2017).

As shown in Table 4, Liauw and Kwan (1985) method and simplified method of Flanagan and Bennett (1999) greatly overestimate the strength. The methods proposed to calculate sliding capacity by FEMA 306 (1996), and Paulay and Priestley (1992) greatly underestimate the infill strength. Diagonal compression strength by FEMA 306 (1996) and the simplified equation by AlWashali et al. (2017) showed relatively good estimation for specimen WF, but it underestimated that of Specimen SF. This underestimation is considered due to the ignorance of the confinement effect of the strong boundary frame.

Table 3 Maximum lateral load and shear strength of infill

Test name	Experiment V_{max} (kN)		V_f (kN) Eq. (3)	Experiment V_{inf}(kN) by Eq. (4)		Maximum shear strength τ_{inf} (N/mm^2)		Average V_{inf} of both directions (kN)	Average shear strength τ_{inf} of both directions (N/mm^2)
	+ loading	- loading		+ loading	- loading	+ loading	- loading		
WF	285	230	71	214	159	1.07	0.80	186.5	0.93
SF	571	582	280	291	302	1.46	1.51	296.5	1.48

Table 4 Ratio experimental peak strength/analytical strength

Test specimen	FEMA 306 (1996) (Compre–ssion)	Liauw and Kwan (1985)	Flanagan and Bennett (1999)	FEMA 306 (1996) (Sliding)	Paulay and Priestley (1992) (Sliding)	Simplified method Al-Washali et al. (2017)
Specimen WF	1.21	0.69	0.50	2.32	1.58	1.18
Specimen SF	1.41	0.81	0.66	3.23	2.13	1.62
Average	1.31	0.75	0.58	4.70	1.85	1.40

3.2 Stiffness

The initial stiffness K_0 of infilled frame is taken as the slope between the origin point of the load-displacement curve and the point with the major visible crack in the masonry infill and the RC frame, which was determined as the story drift of 0.1%.

Table 3 shows the comparison between the initial stiffness of overall frames and that of bare frames. Herein, the initial stiffness of bare frame is calculated based on its elastic gross concrete section. The masonry infill greatly increased the initial stiffness up to about 7.1 times that of bare frame in specimen WF. Therefore, in the seismic design, ignoring the contribution of masonry infill to stiffness and natural period of building may cause non-conservative design practice since buildings with lower natural period have greater seismic forces.

The most well recognized method for calculating the infill stiffness is using the equivalent diagonal compression strut, which has the same elasticity and thickness with the infill panel. Paulay and Priestley (1992) recommended using the effective width of strut, where W_{ef} and d_m is the diagonal length of infill panel in Eq. (5).

$$W_{ef} = 0.25d_m \tag{5}$$

Table 5 shows the comparison between experimental and numerical initial stiffness based on the strut width recommended by FEMA 306 (1996) and Paulay and Priestley (1992). The Strut width calculated by FEMA 306 (1996) underestimates the initial stiffness by about 1.9 and 1.08 for specimen WF and specimen SF, respectively. On the other hand, Eq. (3) recommended by Paulay and Priestley (1992) agrees pretty well with specimen WF by the ratio of 0.94, but overestimated specimen SF by the ratio of 1.36. Based on these results, make a proposed assumption of the strut width W_{ef} to be 0.2 d_m (d_m: diagonal length) gives relatively good estimation for the initial stiffness.

Table 5 Comparison of analytical and experimental initial stiffness

Specimen	Exp. stiffness	Calculated stiffness						
	Initial stiffness (kN/mm)	Bare frame			Initial stiffness using diagonal strut model			
		Initial stiffness (kN/mm)	Ratio	FEMA 306 (kN/mm)	Ratio	Paulay et al. (kN/mm) $W_{ef} = 0.25\ d_m$	Ratio
WF	128	18	7.07	68	1.90	136	0.94
SF	150	79	1.91	139	1.08	205	0.73
Average			4.49		1.49		0.84

Note: *Ratio = Experimental/Analytical

3.3 Deformation Capacity

In this study, a backbone curve for RC frames with masonry infills is suggested as shown in Fig. 7. R_{crack}, R_{max} and R_u are the representative drift angles at the cracking, the maximum strength and the strength degradation point, respectively, where the strength degradation point is set to be 80% of the maximum strength.

The simplified backbone curves for specimens WF and SF are shown in Fig. 8. R_{crack} and R_{max} is estimated to be 0.1 and 0.8% for both specimens. For R_u, it was found to be 0.9 and 1.6% for specimen WF and SF, respectively. Therefore, it can be concluded that the influence of surrounding frame strength on R_{crack} and R_{max}

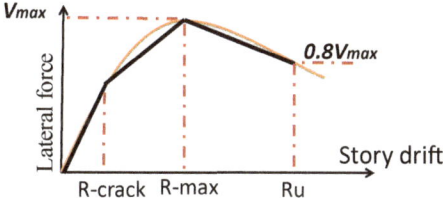

Fig. 7 Idealization of backbone curve

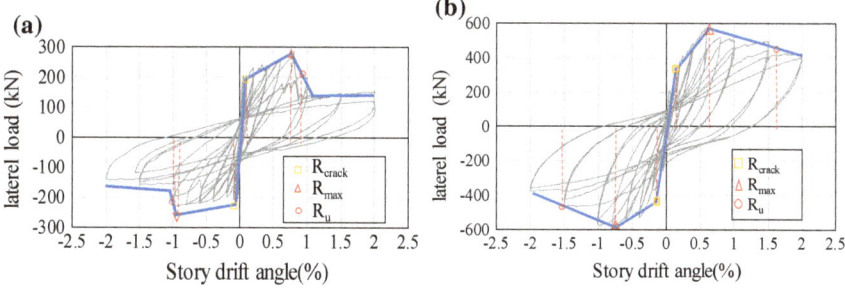

Fig. 8 Backbone curve & deformation limits: **a** specimen WF **b** specimen SF

was slight, but greatly alters the strength degradation slope and R_u. The smooth decrease of strength and improvement of ductility for specimen SF are considered to be due to the confinement by the stronger surrounding frame, which reduces the inelastic deformation of masonry infill.

4   Conclusions

Experimental results to investigate the seismic capacity of unreinforced masonry infilled-RC structure was presented. Based on this study results, the following conclusions can be drawn:

(a) Increasing the ratio of the frame to masonry infill lateral strength named β, greatly improved the infill strength. However, most methods proposed to estimate the infill lateral strength, by previous researchers, underestimate or ignore the influence of β.

(b) The initial stiffness of both frames were almost similar until story drift angle of 0.1%, which is considered the major cracking point. After drift story of 0.1%, the stiffness of both specimens degraded rapidly. Assuming diagonal strut width to be 0.2 times diagonal length gives relatively good estimation of initial stiffness in both specimens.

(c) Drift angles R_{crack} (cracking point) and R_{max} (story drift at maximum strength) was not much influenced by parameter of boundary frame strength. However, R_u drift (degradation point of 80% of maximum strength) was greatly improved by the confinement of stronger columns.

Acknowledgements The study presented in this article was sponsored by the Science and Technology Research Partnership for Sustainable Development (SATREPS) project (Bangladesh, 2015 project headed by Professor Nakano Yoshiaki, University of Tokyo) and it's gratefully acknowledged. Opinions expressed in this article are those of the authors, and do not necessarily represent those of the SATREPS project.

References

Al-Washali H, Suzuki Y, Maeda M (2017) Seismic evaluation of reinforced concrete buildings with masonry infill wall. In: The 16th world conference of earthquake engineering, Chile

American Society of Testing and Materials (ASTM) (2011) Standard test method for compressive strength of masonry prisms. ASTM C1314, West Conshohocken, PA

Architectural Institute of Japan (AIJ) (2010) AIJ standard for structural calculation of reinforced concrete structures

FEMA 306 (1998) Evaluation of earthquake damaged concrete and masonry wall buildings—basic procedures manual. Report No. FEMA 306 (ATC-43 Project)

Flanagan RD, Bennett RM (1999) In-plane behavior of structural clay tile infilled frames. J Struct Eng 125(6):590–599

Liauw TC, Kwan KH (1985) Unified plastic analysis for infilled frames. J Struct Eng 111(7): 1427–1448

Mehrabi AB, Shing PB, Schuller M, Noland J (1996) Experimental evaluation of masonry-infilled RC frames. J Struct Eng ASCE 122(3)

Paulay T, Priestley MJN (1992) Seismic design of reinforced concrete and masonry buildings. Wiley, New Jersey

A Proposal on the Simplified Structural Evaluation Method for Existing Reinforced Concrete Buildings with Infilled Brick Masonry Walls

Matsutaro Seki, Masaki Maeda and Hamood Al-Washali

Abstract The developing countries in the earthquake prone regions in the world are still suffering a lot of casualties as well as building damage. These damages might be caused by inadequate structural design by engineers and/or poor quality control of construction works. In order to contribute to disaster mitigation for existing reinforced concrete (RC) buildings in developing countries, the simplified structural evaluation method based on the philosophy of Japanese evaluation standard; JBDPA (The Japan building disaster prevention association. Standard for seismic evaluation of existing reinforced concrete buildings, 2001) vis-a-vis the international seismic code; IBC (International Code Council, Inc. International Building Code, 2000) was developed by Seki (J Earthq Sci Eng, 2015). However, this evaluation method doesn't consider the infilled brick masonry wall inside the beam and column. The usual RC building has many infilled brick masonry walls but these are not considered in the structural seismic design. They have the benefit in the strength capacity and the disadvantage in the brittle failure mode. For the structural evaluation of existing RC buildings, the consideration of the infilled brick masonry wall is quite important to get the actual behavior during the strong earthquake. The main objective of this study is to take the infilled brick masonry wall into the structural evaluation for the existing RC building in developing countries.

Keywords RC existing buildings · Developing country · Seismic index
Service load index

M. Seki (✉)
Building Research Institute, 1-Tachihara, Tsukuba, Ibaraki, Japan
e-mail: sekimatsutaro@yahoo.co.jp

M. Maeda · H. Al-Washali
Faculty of Architecture and Building Science, Graduate School of Engineering,
Tohoku University, Aobayama 6-6-11, Aoba-Ku, Sendai, Japan
e-mail: maeda@archi.tohoku.ac.jp

H. Al-Washali
e-mail: Hamood@rcl.archi.tohoku.ac.jp

© Springer International Publishing AG, part of Springer Nature 2018   493
R. Vacareanu and C. Ionescu (eds.), *Seismic Hazard and Risk Assessment*,
Springer Natural Hazards, https://doi.org/10.1007/978-3-319-74724-8_33

1 Introduction

The proposed evaluation method is based on JBDPA (2001). The seismic index of structure Is shall be calculated by E_0, S_D and T at each story and in each principal horizontal direction of a building. The irregularity index S_D in the first level screening and the time index T may be used commonly for all stories and directions. The basic and most important structural index E_0 is calculated from C * F formula. The back ground of this formula is based on Blume et al. (1961). This proposed method of this paper is developed for the preliminary screening among buildings, so Is index is basically calculated by the structural and architectural drawings without in situ survey, then it corresponds to the first level screening procedure in JBDPA (2001) (Fig. 1).

2 Proposed Evaluation Method

2.1 Simplified Seismic Index: I_{SS}

$$I_{SS} = E_{SS} * S_{SD} * T_S \qquad (1)$$

where,

E_{SS} Simplified structural index
E_{SS} Maximum values of following three index;

$$\begin{aligned} &\text{(i) } (C_{SSW} + 0.7 * C_{SSB}) * F_W \\ &\text{(ii) } C_{SSB} * F_B \\ &\text{(iii) } \sqrt{(C_{SSW} * F_W)^2 + (C_{SSB} * F_B)^2} \end{aligned} \qquad (2)$$

S_{SD} Simplified Irregularity Index (here assumed to be $S_{SD} = 1.0$)
T_S Simplified Time Index (here assumed to be $T_S = 1.0$).

Fig. 1 Strength index (C) vs. ductility index (F) (JBDPA 2001)

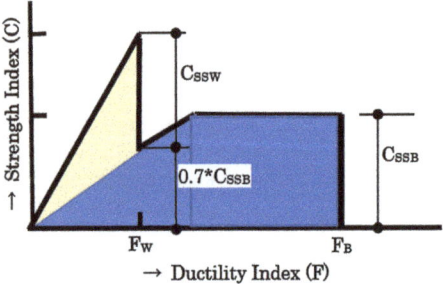

2.1.1 C_{SSB} and C_{SSW} Index; Strength Index

(i) Bare frame

$$C_{SSB} = \tau_c * \Sigma A_C/W \qquad (3)$$

where,

τ_c Average shear strength of column (N/mm^2) (after JBDPA standard)
$h_0/D > 6$ $\tau_c = 0.7$ N/mm^2
$h_0/D \leqq 6$ $\tau_c = 1.0$ N/mm^2
h_0 Clear height of column (mm)
D Depth of column section (mm)
ΣAc Total area of columns (mm^2)
W Total weight of building (N)

(ii) Frame with infilled brick wall

$$C_{SSW} = (2 * \tau_c * \Sigma A_C + \alpha * \tau_w * \Sigma A_w)/W \quad \text{(Commentary A)} \qquad (4)$$

where,

τ_c Average Shear Strength of Column (N/mm^2) (JBDPA 2001)
ΣA_c Total area of culumns (mm^2)
τ_w Averag shear strength of infilled brick wall (mm^2)
τ_w 0.2 N/mm^2
ΣA_w Total area of walls (mm^2)
α Opening reduction factor of infilled brick wall (BSAO 2007)
α $1-\sqrt{\gamma}$ here, $\alpha \geqq 0.6$
γ Opening factor defined in Fig. 2.

2.1.2 F_B and F_W Index; Ductility Index

$$F_B = R_B/\Omega_{0B} \quad \text{(Commentary B)}$$
$$F_W = R_W/\Omega_{0W} \qquad (5)$$

F_B Ductility index of bare frame
F_W Ductility index of frame with infilled brick wall
R_B Response modification factor of frame
R_W Response modification factor of infilled brick wall
       Based on the structural type: Defined in the concerned country's seismic design code.

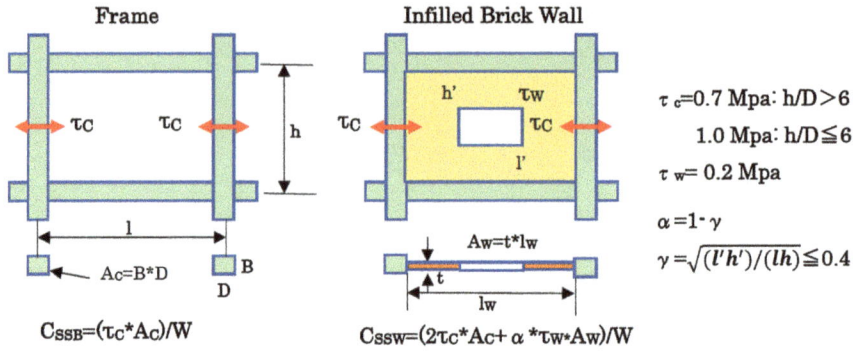

Fig. 2 Definition of C_{SSB} and C_{SSW}

Ω_{0B} Over strength factor of frame.
Ω_{0W} Over strength factor of infilled brick wall
       Based on the structural type: Defined in the concerned country's seismic design code.

2.2 *Simplified Service Load Index: I_{SD} (N/mm²)*

$$I_{SD} = W/\Sigma Ac \qquad (6)$$

where,

W   Total weight of building (N)
ΣA_C Total sectional area of columns (mm²)
       In case of infilled brick wall, Ac is the column's area except the brick wall area.

3 Judgment Index

3.1 *Simplified Seismic Judgment Index: ISS0*

$$I_{SS0} = S_{Da} * I_S \quad \text{(Commentary C)} \qquad (7)$$

where,

I_{SS0} Simplified seismic judgement index
S_{Da} The design spectral response acceleration
I_S The occupancy importance factor.

3.2 Simplified Service Load Judgment Index: ISD0 (N/mm²)

$$I_{SD01} = 0.4 * Fc \quad \text{(Commentary D)}$$
$$I_{SD02} = 0.7 * Fc \tag{8}$$

where,

Fc Designed concrete strength (N/mm²).

4 Judgment Method

4.1 Simplified Seismic Capacity

$$I_{SS} \geqq I_{SS0} : \text{Higher than seismic capacity demand (SA)}$$
$$0.5 I_{SS0} \leqq I_{SS} < I_{SS0} : \text{Lower than seismic capacity demand (SB)} \tag{9}$$
$$I_{SS} < 0.5 I_{SS0} : \text{Remarkably lower than seismic capacity demand (SC)}$$

4.2 Simplified Service Load Capacity

$$I_{SD} < I_{SD01} : \text{Higher than service load capacity demand (DA)}$$
$$I_{SD01} \leqq I_{SD} \leqq I_{SD02} : \text{Lower than service load capacity demand (DB)}$$
$$I_{SD02} < I_{SD} : \text{Remarkably lower than service load capacity demand (DC)} \tag{10}$$

4.3 Final Rank Based on Combination of Seismic Capacity and Service Load Capacity

Final structural rank based on combination of seismic capacity and dead load capacity can be obtained as following Table 1.

Table 1 Final capacity rank of simplified structural evaluation

Final capacity rank	Combination of seismic capacity and service load capacity	Recommendation
A	SA-DA	Safe
B	SA-DB, SB-DA, SB-DB	Detail evaluation recommended
C	SA-DC, SB-DC, SC-DA, SC-DB, SC-DC	Immediately detail evaluation recommended

5   Commentary

5.1   Commentary A

5.1.1   Strength Index of Frame with Infilled Brick Wall

$$C_{SSW} = (2 * \tau_c * \Sigma A_C + \alpha * \tau_w * \Sigma A_w)/W \tag{11}$$

where,

τ_c Average shear strength of column (N/mm^2) (JBDPA 2001)
ΣA_c Total area of culumns (mm^2)
τ_w Averag shear strength of infilled brick wall (mm^2)
τ w = 0.2 N/mm^2
ΣA_w Total area of walls (mm^2)

The shear strength of infill panel (τ_w) was assumed as 0.2 MPa in Fig. 3. As this estimation procedure is basically performed by structural and architectural drawing, in-site material test is not carried out, therefore the minimum and conservative value was decided based on the various research works and seismic codes. The shear strengths in terms of the prism strength (fm) presented by AlWashali et al. (2017) and Sudhir et al. (2014) are shown for the comparison. In case of the strength less than 5 MPa, the proposed 0.2 MPa of shear strength might overestimates the real shear strength of infill wall panel.

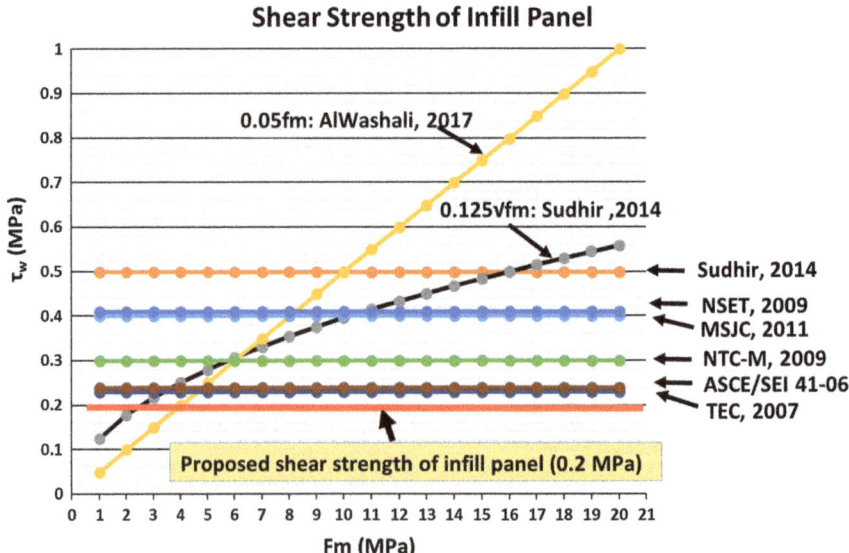

Fig. 3 The shear strength of infill panel (τ_w)

5.1.2 Opening Reduction Factor (α) (BSAO 2007)

$\alpha = 1-\gamma$; (here, $\alpha \geqq 0.6$)
$\gamma = \sqrt{}$(area of opening)/(area of infilled brick masonry wall), (here, $\gamma \leqq 0.4$).

Figure 4 shows the comparison of opening reduction factor between BSAO (2007) and AlWashali et al. (2017). α factor in vertical axis is defined by BSAO (2007) and λ_{OP} factor of horizontal axis is defined by AlWashali et al. (2017). The factor of BSAO (2007) is more conservative than experimental data. According to the experimental data, the effective zone of resisting seismic zone by BSAO (2007) should be less than 0.6. There are many types of infill wall as shown in Fig. 5, therefore this factor's evaluation needs more discussion.

5.2 Commentary B

The relation of lateral seismic force V and lateral deformation (drift) Δ is shown in Fig. 6. Also response modification coefficient R, system over strength factor Ω_0 and deflection amplification factor Cd are shown. These values for the reinforced concrete frame defined in IBC (2000) are shown Table 2. In case of infilled wall frame, ordinary reinforced masonry shear walls (E) may be recommended in Table 2.

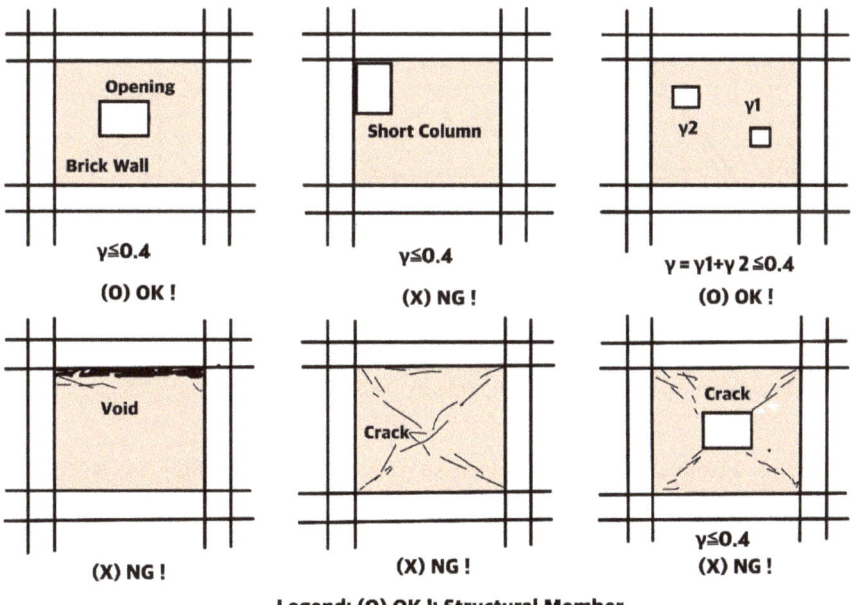

Fig. 4 Reduction factor of shear strength of infilled panel

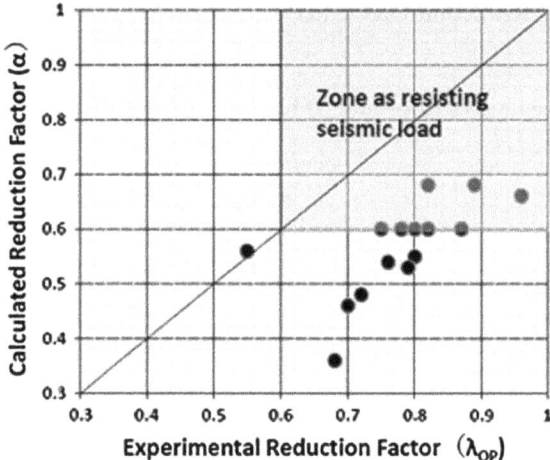

Fig. 5 Various type of the frame with infilled brick wall

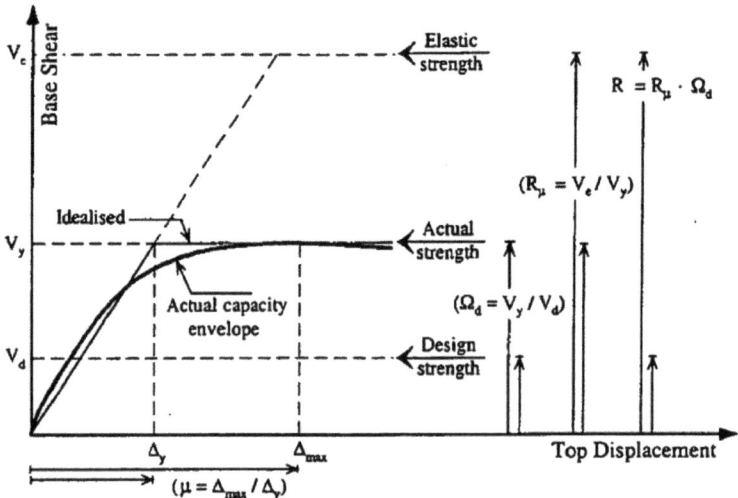

Fig. 6 Relationships between response modification coefficient R and structural over strength factor Ω_d and ductility reduction factor R_μ (Mwafy 2002)

For the seismic evaluation, as Japanese standard JBDPA (2001) is based on the inelastic behavior, ultimate inelastic lateral deformation should be defined. In Fig. 7 the relationships between R factor and Fs factor based on IBC (2000) is shown. R is reduction factor which is the same as ductility factor for elastic design and Fs is the ductility factor for inelastic design. In this proposed simplified seismic evaluation Fs should be used. The relationship between R and Fs can be obtained as Formula 12.

Table 2 Design coefficients and factors for basic seismic-force-resisting system for reinforced concrete moment frames (IBC 2000) (extract)

Basic seismic—force—resisting system	Response modification coefficient (R)	System over strength factor (Ω_0)	Deflection amplification factor (Cd)
Special reinforced concrete moment frames	8	3	5 1/2
Intermediate reinforced concrete moment frames	5	3	4 1/2
Ordinary reinforced concrete moment frames	3	3	2 1/2
Dual system with special moment frames			
E. Special reinforced concrete shear walls	8	2 1/2	6 1/2
L. Special reinforced masonry shear walls	7	3	6 1/2
M. Intermediate reinforced masonry shear walls	6 1/2	3	5 1/2
Dual system with intermediate moment frames			
D. Ordinary reinforced concrete shear walls	5 1/2	2 1/2	4 1/2
E. Ordinary reinforced masonry shear walls	3	3	2 1/2
F. Intermediate reinforced masonry shear walls	5	3	4 1/2

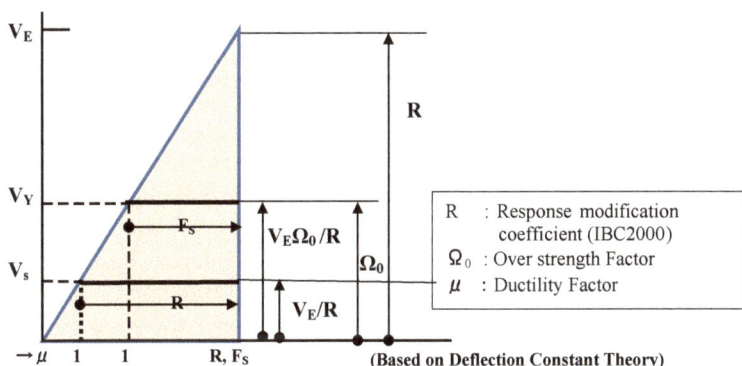

Fig. 7 Response acceleration (v)—ductility index (F_S) relations (IBC 2000)

From Fig. 7, ductility index F_S can be obtained by the following relationships;

$$V_Y/V_E = 1/\mu$$
$$\therefore V_Y = V_S * \Omega_0 = \Omega_0 * V_E/R$$
$$\therefore V_Y/V_E = 1/\mu = \Omega_0/R$$
$$\therefore \mu = Fs = R/\Omega_0$$

$$(12)$$

5.3   Commentary C

The design base shear coefficient S_{Da} in Eq. (7) is usually calculated by the design spectral response acceleration based on the characteristic of building, site soil condition, seismic intensity corresponding to the seismic zone and the occupancy importance factor, etc. These values will be decided considering the concerned country's seismic design code.

5.4   Commentary D

Simplified dead load judgment index I_{SD0} is defined as I_{SD01} is 0.4 * Fc (N/mm^2) based on the JBDPA (2001). From this standard, in the region above 0.4 N_S/(bDF$_C$), the ultimate horizontal deflection angle remarkably decrease and is defined as 0.005 of quite small value. In this evaluation method, as for critical limited value I_{SD01} for the service load judgment of column is assumed as 0.4 N_S/(bDF$_C$) and as for the most critical value I_{SD02} is assumed as 0.7 N_S/(bDF$_C$), respectively.

6   Conclusions

In this paper, a simplified seismic evaluation method based on the structural and architectural drawings was discussed and proposed for utilizing to the preliminary screening stage for the developing countries. The target building is the reinforced concrete moment resisting frame building with infilled brick walls. Seismic evaluation is basically based on the philosophy of The Japan Building Disaster Prevention Association (JBDPA 2001) and International Building Code 2000 (IBC 2000). This evaluation method will be discussed for applying to more advanced in situ simplified evaluation.

Acknowledgements I am thankful for valuable information by Mr. Fumio Kaneko, Mr. Akira Inoue, Dr. Jun Matsuo, OYO International Corporation, Japan and Mr. Yosuke Nakajima, Engineering and Risk Services Corporation (ERS), Japan. This paper is one of the outcome of Bangladesh SATREPS (Japanese side leader; Professor Yoshiaki Nakano, University of Tokyo).

References

AlWashali H, Suzuki Y, Maeda M (2017) Seismic evaluation of reinforced concrete building with masonry infill wall. In: 16th European conference on earthquake engineering, Santiago, Chile, 9–13th Jan 2017
ASCE/SEI 41-06 (2006) Seismic evaluation and retrofit of existing buildings. American Society of Civil Engineers and Structural Engineering Institute

Blume JA, Newmark NM, Corning LH (1961) Design of multistory reinforced concrete building for earthquake motions. Portland Cement Association, Chicago, United States of America

IBC (2000) International Code Council, Inc. International Building Code 2000, March 2000

JBDPA (2001) The Japan building disaster prevention association. Standard for seismic evaluation of existing reinforced concrete buildings (Version 2001) (in Japanese)

Martin-Alarcon DC (2016) Optical monitoring & modeling of masonry behavior under shear load. University of Minho, Portugal, 1st Sept 2016 (Master thesis)

Mwafy AM, Elnashai AS (2002) Calibration of force reduction factors of RC buildings. J Earthq Eng 6(2):239–273

National Society for Earthquake Technology—Nepal (NSET) (2009) Report on development of fragility functions for non-engineered buildings in Bangladesh

Seki M (2015) A proposal on the simplified structural evaluation method for existing reinforced concrete buildings based on the Japanese Seismic evaluation standard vis-à-vis the international seismic code. J Earthq Sci Eng, Publisher ISES, 2015. http://www.joes.org.in

Sudhir KJ, Dhiman B, Indrajit G, Durgesh CR, Svetlana B, Laxmi KB (2014) Application of confined masonry in a major project in India. In: Tenth U.S. national conference on earthquake engineering frontiers of earthquake engineering, Anchorage, Alaska, 21–25 July 2014

The Building Standard Act Enforcement Order (BSAO) (2007, May 8) No. 594, Ministry of Land, Infrastructure, Transport and Tourism, Japan

TMS MSJC-2011 (2011) Masonry standard joint committee's (MSJC) book—building code requirements and specification for masonry structures, containing TMS 402-11/ACI 530-11/ASCE 5-11, TMS 602-11/ACI 530.1-11/ASCE 6-11, and companion commentaries standard by the masonry society

Turkish Code for building in seismic zone (TEC) (2007) Ministry of public works and settlement government of republic of Turkey

A Macro Modelling Blind Prediction of a Cyclic Push-Over Test on a Full Scale Masonry House

Troy Hoogeveen and Joe White

Abstract Gas extraction in the Groningen region of The Netherlands has led to increasing levels of 'induced' seismic activity in the area over recent years. A wide ranging building assessment and strengthening programme is currently underway across the region—primarily focussing on large-scale low-rise residential developments. Delft University of Technology (TU-Delft) recently held a blind prediction test where invited participants attempted to determine the response of a full-scale structure, which is representative of a typical masonry house in the Groningen region. The test structure underwent a prescribed cyclic displacement profile. The authors participated in this blind prediction test using "ANSR", a general purpose analysis program capable of both 2-D and 3-D static, dynamic and time domain analyses. ANSR follows a macro-modelling FEA approach where elements are modelled at a component level. This paper describes how ANSR was used to predict the behaviour of the test structure during the simulated seismic event. It also explores the selection of material properties and their influence on the theoretical performance of the structure. Lastly, the paper describes the benefits of macro-modelling as a viable and cost-effective procedure for undertaking building assessment, particularly where timeframes and accuracy of results are both critical.

Keywords Induced seismicity · ANSR · Netherlands · Calcium silicate

1 Introduction—Current Environment in the Netherlands

The Groningen gas field in the north east of the Netherlands is one of the largest in the world. Via conventional means of extraction, it has been in production since the early 1960s. A known consequence of gas extraction is settlement of soil layers

T. Hoogeveen (✉) · J. White
Holmes Consulting, Groningen, The Netherlands
e-mail: troyh@holmesgroup.com

J. White
e-mail: joew@holmesgroup.com

© Springer International Publishing AG, part of Springer Nature 2018
R. Vacareanu and C. Ionescu (eds.), *Seismic Hazard and Risk Assessment*,
Springer Natural Hazards, https://doi.org/10.1007/978-3-319-74724-8_34

above the gas pocket as the pressure reduces. In the Groningen region this effect has been occurring in sporadic, concentrated and sudden bursts. The results are not dissimilar to tectonic events and are commonly termed 'induced earthquakes'. To date the largest event recorded was approximately 3.6 Mw, with the largest credible event believed to be approximately 5.0 Mw. Whilst these are moderate magnitudes, the 'epicentre' is directly above the gas pocket, at a depth of 2–3 km (Dost and Kraaijpoel 2013). This relatively shallow depth means the associated ground motions have significant magnitudes of acceleration; loosely comparable to what might be expected from a 6.0–7.0 Mw tectonic event at 20–30 km depth. However, the ground motions are also characterised by their short duration and relatively limited horizontal displacement. Vertical displacements are notably similar in magnitude to horizontal displacements (Fig. 1).

The Netherlands is historically not a seismically active zone. A vast majority of the regions current building stock and infrastructure was developed before the onset of the induced events. Design and construction practices in the region have therefore historically not considered seismicity, rendering the region susceptible to significant damage from the expected hazard.

A number of engineering consultancies are in the process of assessing the expected performance of the building stock; a considerable task given the number of buildings, made no easier by the understandable urgency from the public. A coordinated and managed assessment approach is required which balances analysis time, cost and availability of suitably qualified personnel.

Fig. 1 The Netherlands (hatched) with the approximate area affected by induced seismicity (red contour lines) in the Groningen region

2 Typical Dutch Construction Materials and Building Layouts

A significant portion of the building stock across the Netherlands features masonry in some form or another. Clay brick masonry has traditionally been favoured, but in recent decades cheaper alternatives have been sought—hence the emergence of Calcium Silicate (CS) bricks and blocks. CS blocks are particularly utilised in low-rise residential construction, for load bearing walls, both internal and external. The enlarged block dimensions reduce construction times considerably. CS blocks are bonded together by a thin layer of adhesive rather than with traditional lime based mortar (Table 1).

Historically floors and roofs in low-rise residential buildings were formed with timber framing. In recent decades typical floor construction has changed to reinforced concrete, either pre-cast or in-situ.

A significant portion of buildings in the affected region are low-rise terraced or semi-detached houses. There is often a common theme in the layout of such properties; long and solid masonry party walls in the transverse direction, and heavily punctured masonry walls in the longitudinal direction on both the front and rear facades.

Table 1 Typical masonry brick properties in the Netherlands

Symbol	Masonry property	Masonry from			
		Clay bricks <1945	Clay bricks >1945	Calcium silicate >1960	Calcium silicate with thin adhesive joints >1985
f_m	Compressive strength (MPa)	8.5	10.0	7.0	10.0
E_m	Youngs modulus (MPa)	5000	6000	3500	7500
G_m	Shear modulus (MPa)	2000	2500	1450	3000
C	Cohesion (MPa)	0.3	0.4	0.25	0.6
μ	Friction	0.75	0.75	0.6	0.6

Source Nederlandse Praktijkrichtlijn (2015)

3   Assessment Options and Applicability

Over the past five years a number of assessments have been carried out by a variety
of parties. Initially the majority of these assessments were non-linear time history
(NLTH) analyses using full finite element (FE) capabilities and distributed plasticity
models; a micro-modelling approach. Micro-modelling is known to be time consuming; both in construction of the model and analysis run-times. Recently a
number of modal response spectrum (MRS) and non-linear pushover (NLPO)
analyses have been carried out with the aim of increasing assessment efficiencies.
Both of these types of assessments have their benefits and drawbacks.

MRS analyses are not well suited for the assessment of masonry buildings due to
the lack of tension resistance between bricks and the non-linear capabilities of
masonry. Also, a force based assessment tends to generate significant, and often
over-estimated base shears (ElAttar et al. 2014). NLPO analyses on the other hand
offer significant time savings when determining the capacity of the buildings, but do
require more initial model input than that required for an MRS project. It should be
noted that pushover based analyses were not initially developed for buildings of this
construction and hence the methodology needs to be somewhat adapted for low-rise
masonry buildings.

Current consensus in the Netherlands is that NLPO analyses offer a good balance
between analysis time and material complexity whilst utilising drift limits determined from the latest testing. Given that the capacity of the building is determined
separately from demands, NLPO also allows quick re-assessment if the seismic
loading standard is further developed. In the current environment more detailed
knowledge in regards to the seismic hazard is continually evolving and therefore the
ability to separate its influence from that of the buildings capacity is paramount
towards achieving the goal of assessing all affected buildings in a timely manner.

A further possibility aimed toward decreasing analysis time is the use of lumped
plasticity models which significantly reduce model complexity while retaining a
level of computational accuracy which is appropriate for the assessment and
strengthening of existing buildings. Two types of methods that employ this are the
equivalent frame approach and the component (macro) FE modelling approach.

Holmes Consulting's in-house software "ANSR" is built around the ANSR-II
engine (Mondkar and Powell 1979). It has the capability to perform both NLPO and
NLTH analyses utilising a macro modelling approach, where elements are modelled
at a component level. ANSR is a general purpose program capable of 2-D and 3-D
static, dynamic and time domain analyses. The following sections describe how
ANSR was used for predicting the behaviour of a sample masonry structure during
a blind prediction pushover and further discusses the comparison of results following the testing.

4 A Blind Prediction

Delft University of Technology (TU-Delft) recently held a blind prediction competition which considered a sample structure built with a layout and construction materials/methods typical to the Groningen region.

The sample structure was a full scale, two storey CS block and pre-cast concrete floor building. For simplicity, there was no pitched timber framed roof. In the direction perpendicular to loading there were solid masonry walls on both faces of the building. In the direction parallel to loading there were two masonry piers on each face—one longer than the other. This arrangement mimics the solid transverse walls between houses, and the punctured front and rear facades that contain windows. The solid perpendicular walls connected to the pier walls at the building corners to effectively create flanges for the piers. Both the first and second floors were constructed from pre-cast reinforced concrete planks coupled together using fibre reinforced polymer (FRP) to form a rigid diaphragm. All walls could theoretically support the floors, although the pre-cast planks principally spanned between the solid perpendicular walls.

The CS blocks (also known as CASIELs in the Netherlands) were approximately 900 mm long × 650 mm high × 100 mm thick for the piers and 120 mm thick for the solid perpendicular walls. Stretcher bond was typically used, with blocks being placed using a lifting rig. The blocks were glued together with an adhesive rather than traditional mortar (Fig. 2).

The structure was loaded at both the first and second floors simultaneously using displacement controlled hydraulic actuators. There were two actuators at each floor; a left side actuator and a right side actuator; denoted L1 and R1 at the first floor, L2

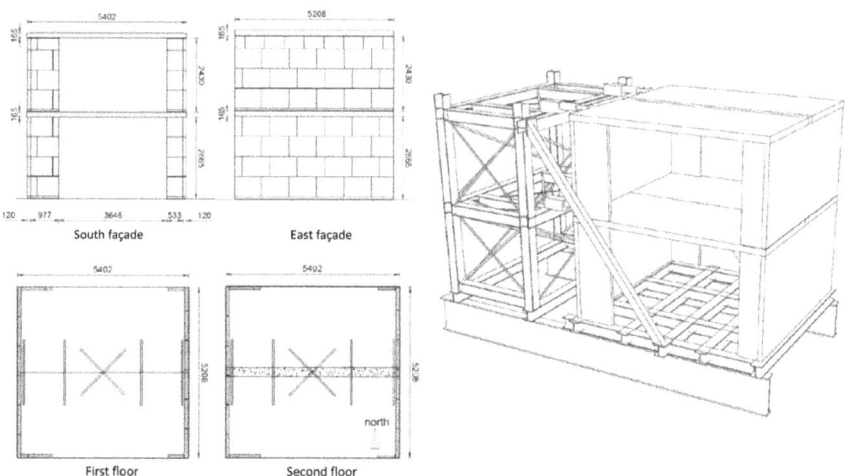

Fig. 2 Left: elevations and plans of the test building, right: schematic of the test structure and loading rig

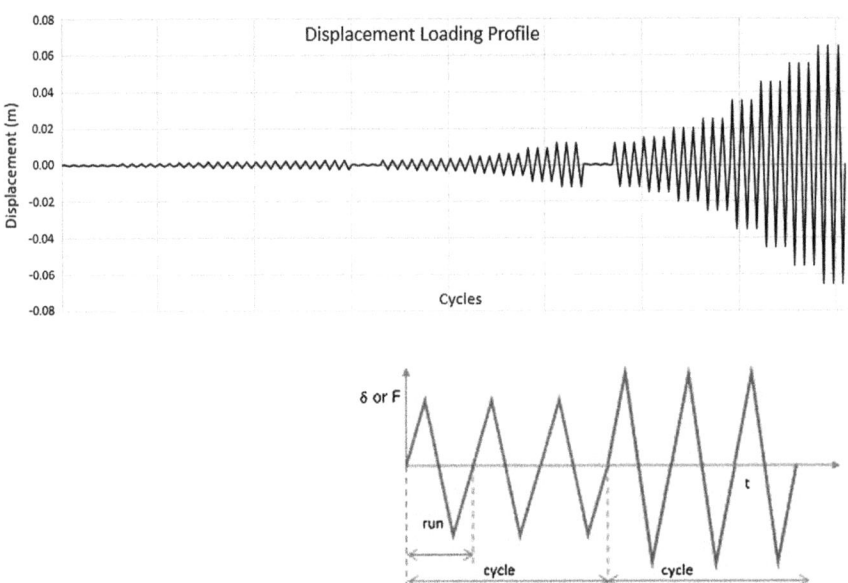

Fig. 3 Top: complete loading profile, bottom: example portion of loading profile

Table 2 Displacement loading procedure to second floor

Cycle number	2nd floor displacement (+/− mm)	Cycle number	2nd floor displacement (+/− mm)
1	0.25	15	4.50
2	0.50	16	6.00
3	0.75	17	9.00
4	1.00	18	12.0
5	1.25	19	0.50
6	1.50	20	12.0
7	1.75	21	15.0
8	2.00	22	20.0
9	2.25	23	25.0
10	2.50	24	35.0
11	0.50	25	45.0
12	2.50	26	55.0
13	3.00	27	65.0
14	3.50	–	–

and R2 at the second floor. The actuators were coupled such that the total force applied at the two floors was the same. However, torsion was not restrained; equilibrium could be gained with varying actuator forces in opposing corners (L1 and R2 for example). The loading profile applied to the structure consisted of 27 cycles of increasing displacement. Each cycle comprising three runs where a run is defined as a push and pull on the structure. The displacement profile started at 0.25 mm with a maximum displacement of 65 mm at cycle 27 (Table 2; Fig. 3).

The cyclic pushover assessment for the blind prediction was undertaken using ANSR. Whilst the process of building the 3D model in ANSR was straightforward, the determination of a number of the inputs was complex (Holmes Consulting Group LP 2014). The main features of the ANSR model shown in Fig. 4 are described under the following points:

- The piers and perpendicular (flange) walls were modelled as plane stress elements with degrading strength and stiffness characteristics.
- Rocking was captured by incorporating gap elements at the corners of all wall panels where separation was expected to occur. These elements were modelled in parallel with brittle truss elements when an adhesive is present at the rocking interface.
- Floors were assumed to form rigid diaphragms for in-plane loading. The floor was modelled as a grillage of beams to include the flexural stiffness of the floor and allow for two-way spanning action. Flexural stiffness was included as it was expected that slab flexure could transfer shear forces between the masonry piers.

A brief description of the key model inputs is summarised below with the aim of providing context for the initial prediction. The implications of values selected for these key inputs are discussed in the section describing the post-test assessment.

- Pier to perpendicular (flange) wall connection: the blocks were not interlocked at the pier to flange intersection, instead they butted up to each other with an adhesive between. Steel 'brick ties' imbedded at the horizontal block joints bridged the vertical joint.
- Kimlaag: a traditional Dutch masonry construction technique where levelling mortar is placed with a short block laid directly on top, to form a flat starting surface for the first full sized block. In this test the base levelling mortar was replaced with an adhesive, bonding the short block down onto the steel base of the test rig, to ensure transfer of the structures base shear. This adhesive in effect worked to suppress a pier rocking mechanism.
- Influence of pre-cast concrete floors: the pre-cast reinforced concrete floors were of sufficient thickness to transfer shear between the opposing corners of rocking walls at the first and second levels. It was important to model its strength and stiffness contribution to the system.
- Application of loading: a prescribed cyclic displacement profile had to be achieved at the second floor level by applying identical but undefined cyclic forces at the first and second floor levels. This was a complex loading regime and required a special purpose dummy frame to be modelled on the side of the structure, to which loading was applied. Even though the stiffness of each storey of the masonry structure differed, with the loading imposed midway between the

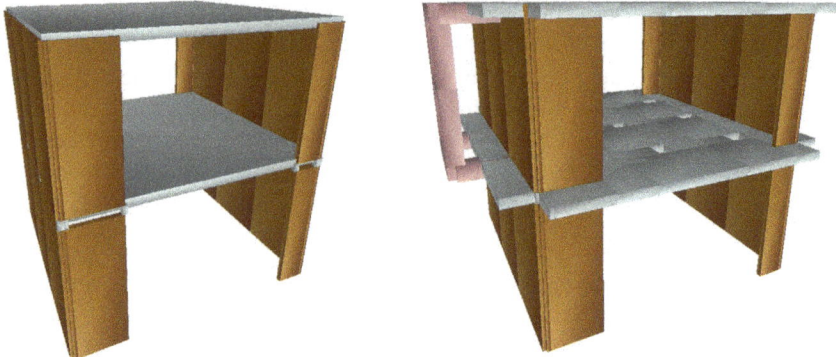

Fig. 4 Left: simplified ANSR model representation, right: ANSR analysis model for the blind prediction

first and second floor levels, the result was equal force at both levels but dif-fering displacements. An iterative approach was then required to match the second floor level displacement to that of the prescribed loading profile.

5 Physical Test Results and Comparison with Initial Blind Prediction

The cyclic displacement response of the second floor was recorded during the actual physical test. The structures backbone curve was derived from this cyclic dis-placement and is illustrated in Fig. 6 for clarity, along with the backbone curve resulting from the ANSR prediction.

From Fig. 6 it is clearly visible that there were notable variations between the experimental results and the initial prediction presented. These are attributable to the following items:

- In the ANSR prediction, the initial spike at the end of the elastic phase corre-sponds to the strength of the adhesive bonding the short block (kimlaag) to the steel base of the test rig. The sharp decline following the peak results from this adhesive fracturing. The prediction used the lowest tensile stress capacity for the adhesive from the range of values provided in the starting documentation. In the physical test this adhesive released at a notably lower level of loading and had no significant impact on the structures performance.
- Following the initial spike, the prediction curve illustrates increasing capacity due to the influence of the flange effect. The flange effect arises when a portion of an out-of-plane wall participates with the in-plane wall in resisting lateral loads (Moon et al. 2006). For this structure the influence of the flange effect is directly related to the strength of the connection modelled between the masonry piers and

the perpendicular flange walls. The prediction used the lowest stress capacity for the adhesive from the range of values provided in the starting documentation. The assumed connection capacity and its effect upon the overall prediction response did not materialise in the physical test—the vertical joint adhesive between pier and flange fractured much earlier than anticipated thus the flanges provided only a negligible influence on the in-plane lateral resistance of the piers.

- The construction of the long piers was such that a thin calcium silicate block was installed adjacent to a full sized block (staggered pattern). This resulted in the thin block being located at the outer edge at the top of the ground level pier and thus it was not confined by the flange wall. It was observed during the testing that under a right to left loading (negative) the thin block was "pinched" by the slab and block below causing it to separate from the full sized block. This eventually led to the loss of block shown in red hatch in Fig. 5. The result of this is two separate local rocking mechanisms in the piers.

Under positive (left to right) loading the "pinch" would be released from the thin block with the restoring weight of the upper level supported on the block edge closest to the flange (denoted as 'P'). Any restoring moment from the thin block would be negligible, therefore the total restoring moment would in effect be derived from the overburden 'P' at a lever arm 'L', which is equivalent to the length of the full sized block (plus the contribution from the short piers).

A local rocking mechanism like this cannot form in the short piers as they are constructed of single stacked blocks. There are no thin blocks present and therefore no points at which high peak stresses need to be resisted nor transferred midway across a block. The issue as a whole can be termed a "size effect".

- The prediction curve was determined using material backbone curves constructed from experimental testing of CS bricks, not CS blocks. The blind prediction testing illustrated the importance of differentiating between brick masonry and block masonry. While this difference at face value appears subtle, it in fact has a significant impact on the performance of the masonry wall. In

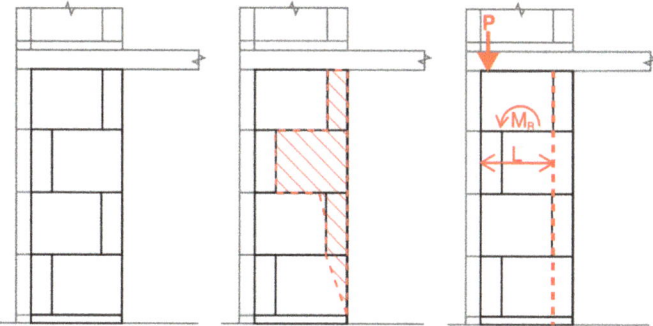

Fig. 5 Long pier at ground level (left) with observed failure (centre) and subsequent rocking lever arm (right)

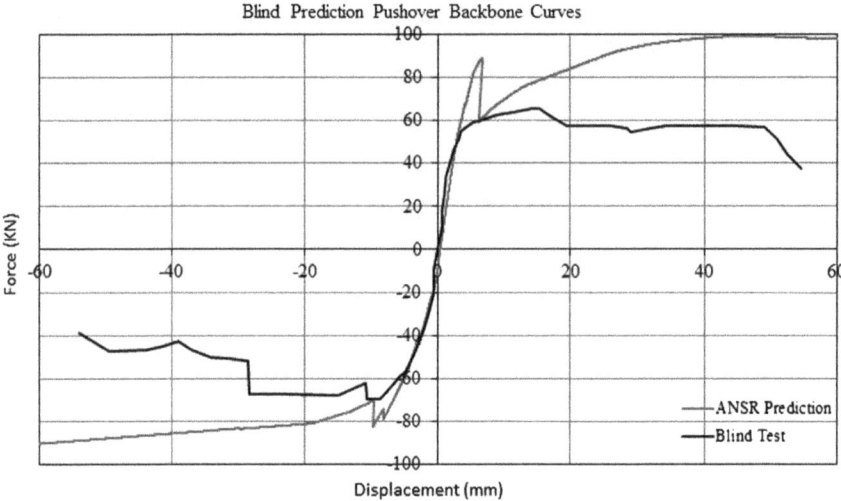

Fig. 6 Comparison of blind test backbone curve with ANSR prediction backbone curve

traditional brick masonry, failure of any one brick in a large wall is inconsequential as it is one of a multitude of bricks. In a block wall, failure of one block can result in the loss of a significant segment of the wall. The result of this is a much more brittle mechanism for the block masonry over the brick masonry (Fig. 6).

6   Post Physical Test Assessment and Discussion

Following the physical test a post-test assessment was carried out by the author. The aim of this was to illustrate the importance of determining the correct inputs and also identifying which of these inputs will have the most influence on the outcome. In this case, simple adjustments were made to the model such as reducing the adhesive shear and tensile properties for the kimlaag and the vertical joint between pier and perpendicular flange walls. The masonry backbone curve, cohesion properties and drift limits were also adjusted to better capture the performance of the glued calcium silicate blocks rather than mortared bricks.

The result of these changes can be seen in Fig. 7 where the post-test backbone curve is plotted against the blind prediction testing backbone curve. The differences are immediately identifiable as the initial peak due to the kimlaag adhesive now only has a negligible influence whilst the contribution from the perpendicular flange walls is also significantly reduced. This results in a structure governed by the capacity of the in-plane calcium silicate block piers.

From the physical testing, rocking of both the first and second level piers characterises the initiation of the failure mechanism. The peak base shear was

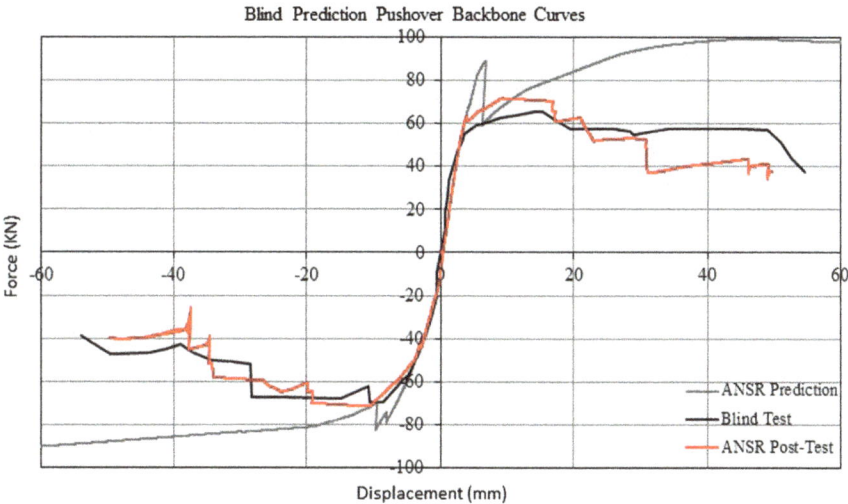

Fig. 7 Comparison of blind test backbone curve with ANSR post-test analysis

attained when this rocking mechanism at both levels was fully activated. During this phase there was some participation from the perpendicular flange walls. Following this peak the rocking mechanism localised in a soft storey at the first level. Collapse was determined to occur when both the long piers at the first level failed at a displacement of 55 mm

In the initial ANSR prediction it was determined that collapse of the structure would not be reached during the 65 mm imposed displacement whereas the post-test assessment indicates that failure would be reached before the full loading profile has been run. As per the physical test, a rocking mechanism formed first with collapse eventually characterised by shear failure of the long in-plane piers leading to flange buckling of the thin, tall blocks.

When undertaking a post-test assessment it is easy to fall into the trap of 'matching' the experimental results by iterating through a number of values until the desired match is achieved. It is important to avoid this by having a clear understanding of the items that are being altered prior to the changes. In this case, it was clear from the testing that the flange effect played only a minor part and that the glued kimlaag layer had a negligible effect on the structures performance as cracking was visible in these particular locations from early in the loading profile. In the case of the calcium silicate blocks, further test data should be used to determine the expected parameters and relevant shear and rocking drift limits.

7 Analysis Approach

Following implementation of the corrected model inputs, ANSR achieved a suitably accurate result and in a reasonable time frame. To undertake an analysis of this size, the ANSR macro model took between 2 and 3 h to build and verify. Following this, a cyclic pushover with full degradation took approximately one minute to analyse, which makes for a highly efficient process and allows for a number of sensitivity analyses to be undertaken within a reasonable time frame. It is understood that other consultants using full FE analysis packages (i.e. micro modelling) spent a number of days building and verifying their analysis models with analysis run times in excess of 24 h.

Holmes Consulting has been using ANSR and a macro modelling approach extensively over the past 30 years. A multitude of projects as well as a number of other blind predictions have been undertaken, achieving successful results and correlations. We have successfully predicted the performance of structural steel frames, reinforced concrete shear walls and reinforced concrete portal (moment) frames (Kelly 2007; Tremayne 2010; Restrepo 2010).

8 Conclusion

In the context of the ongoing assessment process in the Groningen region, striving for analytical accuracy through complex micro-models is not always practical or necessary. This test has demonstrated that macro modelling, in this case using ANSR, achieves sufficient accuracy whilst maintaining feasible and practical analysis build and run times. It has also shown the critical effect of material parameters—the accuracy of such parameters should be of equal focus as finite element size and distribution of plasticity.

Acknowledgements The authors would like to acknowledge TU-Delft for the opportunity to participate in this blind prediction competition and present our entry and post-prediction.

References

Dost B, Kraaijpoel D (2013) The August 16, 2012 earthquake near Huizinge (Groningen). KNMI, De Bilt
ElAttar A, Zaghw A, Elansary A (2014) Comparison between the direct displacement based design and the force based design methods in reinforced concrete framed structures. In: Second European conference on earthquake engineering, Istanbul
Holmes Consulting Group LP (2014) Performance based evaluation of existing buildings: nonlinear pushover and time history analysis: reference manual. Revision 4 (September)
Kelly T (2007) A Blind prediction test of nonlinear analysis procedures for reinforced concrete shear walls. Bull NZ Soc Earthq Eng 40(3)

Mondkar DP, Powell GH (1979) ANSR II analysis of non-linear structural response user's manual, EERC 79/17, University of California, Berkeley (July)

Moon FL, Yi T, Leon RT, Kahn LF (2006) Recommendations for seismic evaluation and retrofit of low-rise URM structures. J Struct Eng 132(5):663–672

Nederlandse Praktijkrichtlijn (2015) Assessing structural safety in the construction of new structures and the renovation/alteration of existing structures. Principles of earthquake loads: induced earthquakes, NPR 9998 (December)

Restrepo JI (2010) Concrete column blind prediction contest 2010. NEES, University of California, San Diego

Tremayne B (2010) Benchmark modelling of a viscous damped steel frame, E-defence blind analysis contest. Hyogo, Japan

Design of Beam Anchorages in Beam-Column Joints in Seismic Structures

Dragos Cotofana, Mihai Pavel and Viorel Popa

Abstract Proper anchorage of beam reinforcement in beam-column joints is necessary for a stable hysteretic behavior of seismically loaded structures subjected to large lateral displacements. Anchorage requirements associated with the use of higher strength steel such as S500 generated the need to increase the anchorage lengths and concrete strengths. Formerly used concrete strength classes such as C20/25 and C25/30 are largely replaced with higher quality concrete such as C40/50 or more. In common design practice, anchorage provisions of the repealed Romanian standard STAS 10107/0-90 are still informally used. This might lead to poor anchorage details and related structural problems. This paper draws attention on the specifications of the current standards for anchorage of beam reinforcement. Practical design rules are presented as well.

Keywords Beam-column joint · Anchorage · Splices · Yielding
Hysteresis · Ductility

1 Introduction

Current design practice of concrete structures in Romania underwent fundamental changes in the last 20 years. These changes are a result of a new set of seismic performance criteria enforced by the new seismic design codes, changes in the

D. Cotofana (✉) · V. Popa
Technical University of Civil Engineering of Bucharest, Bucharest, Romania
e-mail: dragos.cotofana@alliedengineers.ro

V. Popa
e-mail: viorel.popa@utcb.ro

D. Cotofana
SC Allied Engineers Grup SRL, Bucharest, Romania

M. Pavel
SC Altfel Construct SRL Bucharest, Bucharest, Romania
e-mail: mihai.pavel@altfelconstruct.ro

© Springer International Publishing AG, part of Springer Nature 2018
R. Vacareanu and C. Ionescu (eds.), *Seismic Hazard and Risk Assessment*,
Springer Natural Hazards, https://doi.org/10.1007/978-3-319-74724-8_35

519

architectural and functional demands, as well as the improvement of mechanical properties for concrete and steel. All these changes encouraged large span structures with stronger, heavily reinforced, structural elements. The widespread practice to use bars with diameter greater than 20 mm in beams leads to anchorage problems in beam column joints or beam-wall intersections. As most of the Romanian buildings are designed for high ductility class (DCH) using high behavior factors and reduced horizontal seismic loads, high ductility demand is expected. Careful detailing of bars anchorages is essential for a good structural response. This requires special attention from the designers even though the detailing of structural members is often regarded as secondary activity. Rebar pull out should not be considered the only possible anchorage failure mode. Slippage due to concrete crushing in the bent area is the most common failure mode of the anchorage. This common failure mode causes a poor hysteretic behavior and low structural ductility. The use of high behavior factors should always be accompanied by a careful detailing process.

Romanian experience regarding behavior of anchorages in case of earthquakes is limited since the design practice fundamentally changed from the last major seismic event, 1977 Vrancea earthquake. There was a significant change in the architectural request from relatively small span structures with high density of shear walls and frames to large span structures with concentrated strength and stiffness. Moreover, poor quality lightly reinforced concrete elements evolved to high strength concrete heavily reinforced elements. In this paper, anchorage of beams longitudinal reinforcement in beam column and beam wall joints is reviewed. Most important standards provision are presented together with rules for improving anchorages for seismically loaded elements.

2   Common Issues in Anchorage Design

The shear force transfer mechanism within the joint can be described as a combination of the arch (Fig. 1a) and truss mechanisms (Fig. 1b) (Paulay and Priestley 1992). In the arch mechanism, shear forces are directly transferred by the main diagonal concrete struts, requiring no additional reinforcement in the joint except for the confinement reinforcement necessary to increase the concrete strength and deformability. The arch mechanism can be fully engaged if the beam longitudinal rebars are anchored in the joint using hooks or heads (Fig. 2a). The tails of hooks should be oriented towards the midsection of the joint and should be located inside the confined concrete core, as far as possible from the critical section of the rebars. To allow proper installation of the hooked rebars and good concreting conditions a gap of roughly 5 cm is usually maintained between the tail of the hook and the joint horizontal stirrups. It should be reminded that hooks can be reliable used to improve anchorage of tensioned rebars but have negligible effect in case of compressed rebars.

Straight anchorage of beam longitudinal rebars in beam-column joints (Fig. 2b) is easy to build but has limited effectiveness because the arch mechanism cannot be

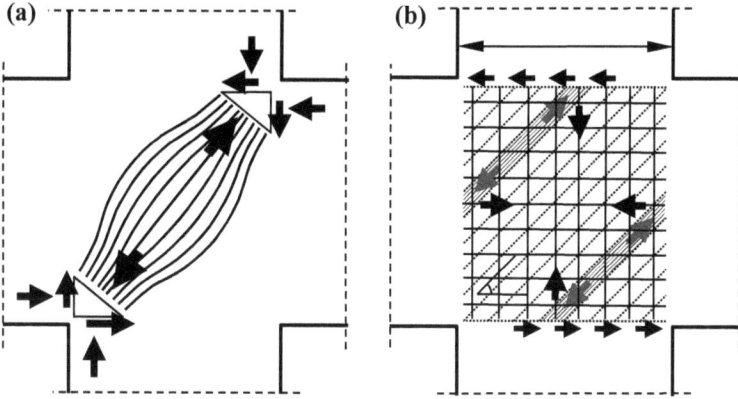

Fig. 1 Arch and truss mechanisms in beam-columns joints

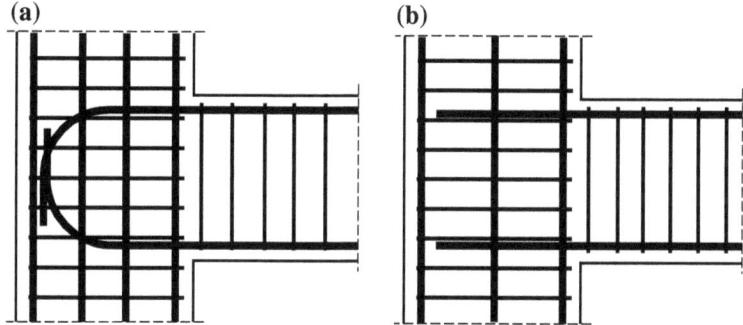

Fig. 2 Anchorage of beam re-bars in beam column joints

fully engaged. In this case, shear is mainly transferred by a truss mechanism which is weaker and softer than the arch mechanism. The arch mechanism cannot be fully engaged because the tensile force in the rebars is not driven behind the joint. Moreover, outside the compressed area of the columns, there is a tensile transversal stress in concrete that impair the anchorage strength of beams rebars. If axial stress in columns is low due to various design criteria, the depth of compressed concrete is relatively narrow and anchorage of beams straight rebars is poor.

Jirsa and Marques (1972) showed that anchorage failure of hooked rebars is caused by the failure of concrete under radial compression stress occurring in the curved section of the rebar (Fig. 3a). The amplitude of this stress is higher as the bend radius is decrease. Therefore, a long tail hook does not necessarily increase the anchorage capacity. As the presence of the hook is necessary to drive the tensile force in the beam rebar behind the joint, it is necessary to increase the horizontal development length to effectively increase the anchorage capacity. To be able to fully benefit from a longer hook tail it is necessary to increase the bending radius as

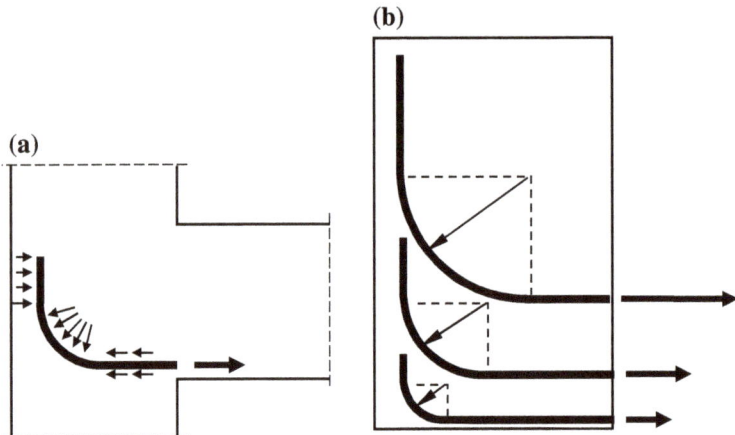

Fig. 3 Force transfer in the bended bars anchorage

well (Fig. 3b). Increasing the tail length without a proper increase of the bending radius is not fully effective in improving the anchorage conditions. Crushing of the concrete in the bending sections generate slip of the reinforcement and subsequent reduction of the beam energy dissipation capacity.

Anchorage can be effectively improved by: increased column cross-section and horizontal development length, increased bending radius and longer tail, sufficient clear space between the anchored rebar, good confinement in the beam column joint. Oversized tails (Fig. 4a) or double bended hooks (Fig. 4b) are not effective because cannot prevent the anchorage failure by concrete crushing in the bending region. This kind of failure does not generate total loss of anchorage, but slip of rebars impairs hysteretic response of the beams.

The radial stress acting of the curved section of the hook tends to straighten the rebar and, generates compression stress acting on the tail of the hook (Fig. 3a).

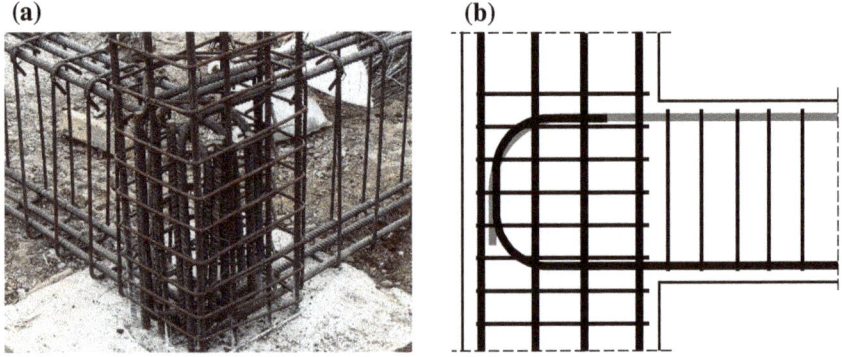

Fig. 4 Beam-column joint with long tail hooked anchorage

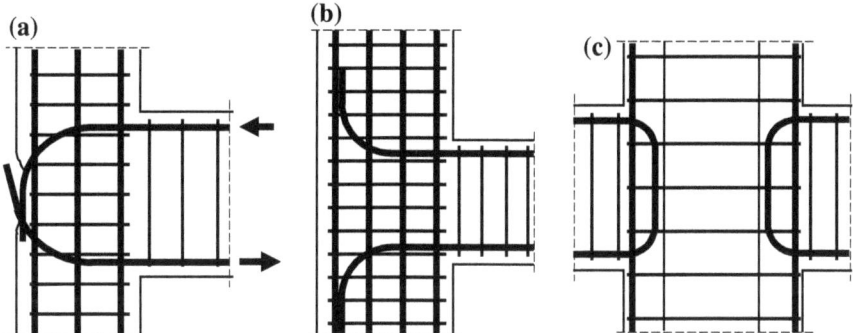

Fig. 5 Unsuitable anchorage details

To prevent loss of anchorage if the concrete cover is spalled (Fig. 5a), hooks should be bended inside the confined core of the joint. Anchorages with hooks oriented outside the joint (Fig. 5b) are not suitable because the arch mechanism inside the joint cannot be directly activated. Anchorage of beam rebars in front of the joint (Fig. 5c) has certain constructability advantages but is extremely weak because none of the resisting mechanisms in the joints can be properly engaged. This technique is mistakenly regarded as effective in case of precast structures or retrofitting works for existing structures.

Headed bars are highly advantageous in case of small sized columns. In this case, the anchors should always be placed inside the column reinforcement cage in the confined concrete core (Fig. 6a). According to Wallace (1996), an anchorage length of 12φ is enough if the head diameter is at least four times the rebar diameter. Heads positioning outside the concrete confined core (Fig. 6b) or in front of the joint (Fig. 6c) is not allowed.

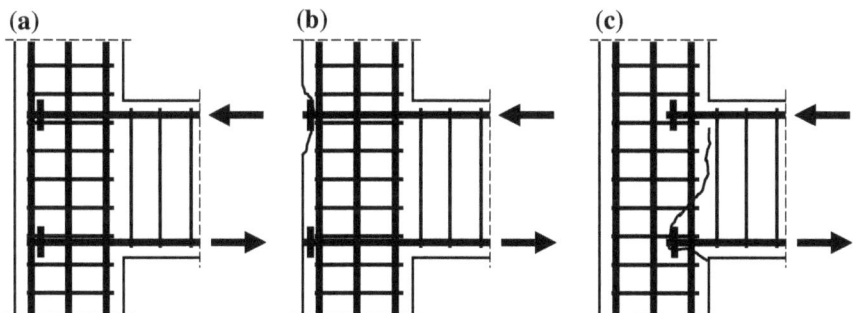

Fig. 6 Mechanical splices anchorage

3   Design Standards Provisions

According to ACI-ASCE-352 (2002), anchorage of bars in structures with expected large nonlinear deformations, can be made only with standard hooks or mechanical anchors. In calculating the anchorage length a maximum 12φ tail length can be considered. The anchorage length is estimated based on the rebar diameter, concrete class, steel grade and steel over strength. Minimal values for development length, l_{dh}, are 15 cm or $8d_{bl}$, d_{bl} being the diameter of the anchored bar. Development length, l_{dh}, is measured from beam-column interface to the tail of the hook (Fig. 7). Table 1 shows values of the development length for $\varphi25$ rebar and S500 grade steel. The 12ϕ length of 90° standard hook is to be added to those values. The bend radius considered in the calculations is 6ϕ.

The European standard EN 1992-1-1 (2004) is referenced by the Romanian Seismic Design Code P100-1/2013 (2013) for the design reinforced concrete structures. Provisions of both standards are to be used in design of bars anchorages for seismic structures. Basically, P100-1 requires measuring of the anchorage length, l_{bd}, from a section located inside the joint at 5ϕ from the critical section. A similar provision is included in EN 1998-1 (2004). The maximum rebar stress for the anchorage design, σ_{sd} should be taken as 1, $2f_{yd}$, where f_{yd} is the design yielding strength of steel. For the common S500 grade steel, the beam rebars stress results as:

$$\sigma_{sd} = 1.2f_{yd} = 522\,\text{MPa} \tag{1}$$

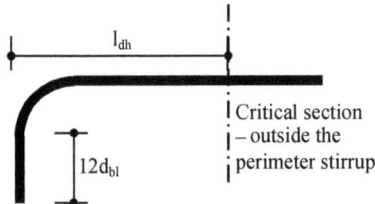

Fig. 7 Hook details in ACI 318

Table 1 Anchorage lengths for $\varphi25$, S500 grade steel, according to ACI-318

Concrete class (as defined in EN 1992-1-1)	Development length
C20/25	22φ
C25/30	21φ
C30/37	19φ
C35/45	18φ
C40/50	17φ

The design value of the ultimate bond stress can be calculated as: $f_{bd} = 2.25 f_{ctd}$, for bottom rebars or $f_{bd} = 1.58 f_{ctd}$, for top rebars.

The difference is mainly caused by the poor anchorage conditions of the top rebars related to the settlement of fresh concrete.

The basic required anchorage length can be calculated as:

$$l_{b, rqd} = \frac{\varphi}{4} \cdot \frac{\sigma_{sd}}{f_{bd}}$$ (2)

resulting in:

$$l_{b, rqd} = \frac{58\varphi \; (\text{mm})}{f_{ctd} \; (\text{Mpa})} \quad \text{for bottom rebars}$$ (3)

$$l_{b, rqd} = \frac{83\varphi \; (\text{mm})}{f_{ctd} \; (\text{Mpa})} \quad \text{for top rebars.}$$ (4)

The design anchorage length can be calculated as:

$$l_{bd} = \alpha_1 \alpha_2 \alpha_3 \alpha_4 \alpha_5 l_{b,rqd} \geq l_{b, \min}$$ (5)

where all α factors are equal to 1.0 for compressed rebars.

Considering that in case of seismic action, beam rebars will yield in tension and compression near the joint, the reduction factor of 0.7 provided by EN 1992-1-1 for rebars under tension cannot be applied.

Values of the design anchorage length for various concrete classes are given in Table 2. It can be observed that concrete strength classes lower than C30/37 are little usable because of the large required anchorage lengths. The anchorage length calculated for a $\varphi 25$ mm diameter rebar yielding in compression anchored in C20/25 concrete is 2.05 m for the upper reinforcement layers and 1.42 m for the bottom reinforcement layers. It is obvious that such long anchorage lengths can hardly be accommodated in practical projects. Moreover, according to EN 1992-1-1, the hook contribution to the anchorage length can be considered only for tensioned bars making the anchorage of compressed rebars an impossible task.

EN 1992-1-1 allows the reduction of the anchorage length with 30% in case of hooked bars under tension, if the concrete cover is larger than 3φ and the distance

Table 2 Anchorage lengths S500 grade, according to EN 1992-1-1 and P100-1/2013

Concrete class	l_{bd}	
	Bottom	Top
C20/25	57φ	82φ
C25/30	48φ	69φ
C30/37	43φ	62φ
C35/45	39φ	56φ
C40/50	34φ	49φ

between bars is more than 6φ. These conditions are hardly met in current seismic structures where large diameter re-bars are spaced at the minimum distance required by the standard.

To accommodate such a long anchorage length, it is necessary to use hooked bars instead of straight bars. The minimum hook tail length of 5φ corresponds to a minimum bending diameter of 7φ for $\varphi \geq 16$mm. If from anchorage conditions a longer tail length is necessary, the mandrel diameter should be increased as well. The equation to determine the minimum mandrel diameter, $\varphi_{m, \ min}$, takes into account the value of the tensile force from ultimate loads at the start of bend, F_{bt}, the half of the center-to-center distance between bars perpendicular to the plane of the bend, a_b, the bar diameter and the concrete design compressive strength, f_{cd}:

$$\varphi_{m, \ min} \geq \frac{F_{bt}}{f_{cd}} \left(\frac{1}{a_b} + \frac{1}{2\varphi} \right) \tag{6}$$

The tensile force in the rebar at the start of the bend can be calculated based on the design value of the ultimate bond stress considering a uniform distribution along the straight part of the anchorage length. It can be observed that straight development lengths ranging from 14φ to 24φ are necessary to reduce by 50% the tensile force in the rebar from the maximum value, F_y, for concrete classes between C20/25 to C40/50. The necessary straight development lengths for different percentages of tensile force reduction and different concrete classes are given in Table 3. Mandrel diameter for different residual stresses at the beginning of the hook, relative to the steel yielding stress, calculated according to EN 1992-1-1, are given in Table 4. If the residual force at the beginning of the hook is equal to 50% from the yielding force the mandrel diameter should be increased up to 13φ, depending on

Table 3 Development lengths according to SR EN 1992-1-1

Concrete class	F_{bt}/F_y				
	0.1	0.3	0.5	0.7	0.9
C20/25	43φ	33φ	24φ	14φ	4φ
C25/30	36φ	28φ	20φ	12φ	4φ
C30/37	32φ	25φ	18φ	10φ	3φ
C35/45	29φ	23φ	16φ	9φ	3φ
C40/50	26φ	20φ	14φ	8φ	2φ

Table 4 Mandrel diameter according to SR EN 1992-1-1

Concrete class	F_{bt}/F_y				
	0.1	0.3	0.5	0.7	0.9
C20/25	2φ	8φ	13φ	18φ	24φ
C25/30	2φ	6φ	10φ	15φ	19φ
C30/37	1φ	5φ	9φ	12φ	16φ
C35/45	1φ	4φ	7φ	10φ	13φ
C40/50	1φ	4φ	6φ	9φ	12φ

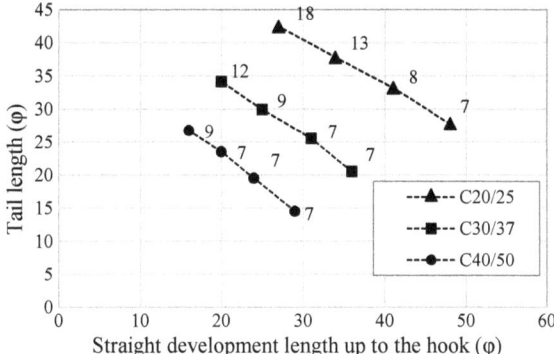

Fig. 8 Mandrel diameter for different values of the tail length

the concrete class. Mandrel diameters calculated for different values of the tail length and the straight development length up to the bending zone are given in Fig. 8.

Large anchorage lengths can be hardly provided in case of compressed reinforcement because the contribution of hook or mechanical anchors cannot be considered. This generates the need for oversized columns. To reduce the anchorage length the contribution of the compressed concrete in reduction of the compression stress in the rebar should be considered. However, the behavior of the compressed rebars in beams under cyclic loading in nonlinear range is not an issue to be dealt by regular designers.

In case of horizontal bars crossing the beam-column joints of concrete frames designed for seismic action, bond failure can be generated by simultaneous action of tension and compression forces in the rebar at both critical sections. To avoid bond failure the embedment length of the rebar and, consequently, the minimum column width should be selected based on the rebar diameter. According to ACI-318, the maximum diameter of the horizontal bars crossing the joint should not exceed 1/20 of the column size. EN1998-1 and Romanian code P100-1/2013 has similar provisions. In many practical situations, designers tend to ignore this limitation resulting in large diameter rebars passing through narrow concrete joints. In Romanian practice, the large cross-section columns resulting from lateral drift limitation requirements are, in the most cases, just enough to accommodate large diameter rebars such as $\varphi22$ mm or $\varphi25$ mm.

4 Basic Rules for Good Practice

Minimum allowed concrete classes for seismic structures, such as C20/25 or C25/30, are not compatible with S500 steel grade in seismically loaded multistory buildings because of the anchorage issues. Beam's reinforcements cannot be anchored in a reasonably sized column for low concrete quality joints. Anchorage issues should be considered when designing the size of columns and concrete strength class. Based on EN1992-1 provisions, as a rule of thumb, the columns cross-sections should be at least 25φ, where φ is beam reinforcement maximum diameter. Anchorage is with hooks bended inside the joint confined concrete core is recommended. Anchorage without hooks, with hooks in front of the joint and hooks placed outside the confined concrete core of the joint is not effective.

Oversized hooks cannot prevent anchorage failure caused by the concrete crushing under radial compression stress in the bending area. If the necessary hook tail is longer than 5ϕ, the dowel diameter should be increased over 7φ, the minimum requirement in EN 1992-1-1. For oversized tails, the resulting mandrel diameter is not usable in practical situations. A feasible way to solve this issue is to increase the column size or to reduce the diameter of the anchored reinforcement. Alternatively, movement of the critical section away from the face of the joint by locally increasing the beam strength can be considered, as presented by Postelnicu and Popa (2009). Oversized hooks and double bended hooks should be avoided. The maximum diameter for beams reinforcement should be decided in accordance with columns size. The diameter should be limited especially in the upper floors where the axial force in columns is small. Headed bars are recommended for corner and perimeter beam-column joints. The heads should be placed, in the confined concrete core of the joint, inside the column stirrups.

5   Conclusions

In case of structures exposed to severe ground motions, anchorage of beam reinforcement should be considered as a design criterion when the columns cross-section are set up.

EN1992-1-1 anchorage lengths requirements are highly restrictive in comparison with other international standards (for example, ACI 318). This is particularly valid in case of compressed rebars for which the contribution of hooks or heads cannot be considered. Contribution of the compressed concrete in reducing the compression stress in the rebar at the critical section should be considered but can hardly be evaluated in regular design cases. Additional research is necessary to improve the standard provisions.

Rather low-quality concrete such as C20/25 is not compatible with the common steel S500, largely used in Europe nowadays. For seismic applications, minimum concrete strength class C30/37 should be used in combination with S500 steel grade. C40/50 should be regarded as average concrete strength class for reinforced concrete multistory buildings in seismic areas.

References

ACI-ASCE 352 (2002) Recommendations for design of slab-column connections in monolithic reinforced concrete structures. ACI Struct J 85(6):675–696

CEN (2004) Eurocode 2: design of concrete structures—part 1-1: general rules and rules for buildings. European Standard EN 1992-1-1, Brussels

CEN (2004) Eurocode 8: design of structures for earthquake resistance—part 1: general rules, seismic actions and rules for buildings. European Standard EN 1998-1, Brussels

Code ACI (2011) Building code requirements for structural concrete and commentary (ACI 318M–2011). American Concrete Institute, Detroit, Michigan

Jirsa JO, Marques JL (1972) A study of hooked bar anchorages in beam-column joints. Department of Civil Engineering, Structures Research Laboratory, University of Texas at Austin

Ministry of Regional Development and Public Administration of Romania (2013) Seismic design code. Part 1—design provisions for buildings, P100-1/2013, Bucharest

Paulay T, Priestley MN (1992) Seismic design of reinforced concrete and masonry buildings. John Wiley & Sons Inc., New York

Postelnicu T, Popa V (2009) Proiectarea nodurilor cadrelor de beton armat in codurile de proiectare actuale, AICPS Review Nr.2-3/2009 (in Romanian)

Wallace JW (1996) Use of mechanically anchored bars in exterior beam-column joints subjected to seismic loads. Department of Civil Engineering, University of California, LA

Annex 1

6th National Conference on Earthquake Engineering
&
2nd National Conference on Earthquake Engineering and Seismology

Program at a glance

Wednesday June 14, 2017	
13:00–14:00	Registration of participants
14:00–15:00	Opening ceremony
15:00–15:45	***Keynote lecture 1**—Kyriazis Pitilakis*
15:45–16:30	***Keynote lecture 2**—Radu Vacareanu*
16:30–17:00	Coffee break
17:00–17:45	***Keynote lecture 3**—Dan Dubina*
17:45–18:30	***Keynote lecture 4**—Mircea Radulian*
18:30–20:30	Welcome cocktail

Location: Romanian Academy, Calea Victoriei 125, Sector 1, Bucharest

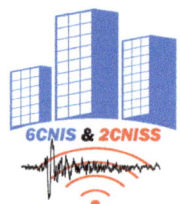

6th National Conference on Earthquake Engineering
&
2nd National Conference on Earthquake Engineering and Seismology

Thursday June 15, 2017		
8:00–9:00	Registration of participants	Hall I.2, FCCIA/UTCB
9:00–9:45	*Keynote lecture 5—Sierd Cloetingh*	Hall I.2, FCCIA/UTCB
9:45–10:30	*Keynote lecture 6—Iunio Iervolino*	Hall I.2, FCCIA/UTCB
10:30–11:00	Coffee break	Room P1B, FCCIA/UTCB
11:00–12:45	Parallel Session 1—Seismicity of Romania *Invited lecture—Gheorghe Marmureanu*	Hall I.1, FCCIA/UTCB
	Parallel Session 2—Seismic risk evaluation and management of emergency situations *Invited lecture—Taiki Saito*	Hall I.2, FCCIA/UTCB
	Round Table—P100-3/2008	Hall I.4, FCCIA/UTCB
12:45–13:40	Lunch break	Room P1B, FCCIA/UTCB
13:40–14:30	A word from sponsors	Hall I.2, FCCIA/UTCB
14:30–15:15	*Keynote lecture 7—Mauro Dolce*	Hall I.2, FCCIA/UTCB
15:15–17:00	Parallel Session 3—Geotechnical earthquake engineering *Invited lecture—Dan Balteanu*	Hall I.1, FCCIA/UTCB
	Parallel Session 4—Workshop "RO-RISK—Disaster risk assessment at national level" *Invited lecture—Alex Barbat*	Hall I.2, FCCIA/UTCB
	Round Table—Resilience-based assessment of structures	Hall I.4, FCCIA/UTCB
17:00–17:30	Coffee break	Room P1B, FCCIA/UTCB
17:30–19:00	Parallel Session 5—Seismic design and assessment *Invited lecture—Alper Ilki*	Hall I.1, FCCIA/UTCB
17:30–19:00	Parallel Session 6—Seismic risk evaluation and management of emergency situations *Invited lecture—Emil-Sever Georgescu*	Hall I.2, FCCIA/UTCB
20:30–23:00	Gala dinner	

Location: FCCIA/UTCB, Bd. Lacul Tei nr. 124, Sector 2, Bucharest

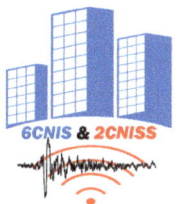

6th National Conference on Earthquake Engineering
&
2nd National Conference on Earthquake Engineering and Seismology

Friday June 16, 2017		
9:00–9:45	*Keynote lecture 8—Alik Ismail-Zadeh*	Hall I.2, FCCIA/UTCB
9:45–10:30	*Keynote lecture 9—Humberto Varum*	Hall I.2, FCCIA/UTCB
10:30–11:00	Coffee break	Room P1B, FCCIA/UTCB
11:00–13:00	Parallel Session 7—Innovative solutions for seismic protection of building structures *Invited lecture—Matsutaro Seki*	Hall I.1, FCCIA/UTCB
	Parallel Session 8—Seismicity of Romania *Invited lecture—John Douglas*	Hall I.2, FCCIA/UTCB
	Round Table—Quick post-earthquake evaluation of buildings	Hall I.4, FCCIA/UTCB
13:00–14:00	Lunch break	Room P1B, FCCIA/UTCB
14:00–14:45	*Keynote lecture 10—Roberto Paolucci*	Hall I.2, FCCIA/UTCB
14:45–15.30	*Keynote lecture 11—Koichi Kusunoki*	Hall I.2, FCCIA/UTCB
15:30–17:00	Parallel Session 9—Seismic design and assessment *Invited lecture—Masaki Maeda*	Hall I.1, FCCIA/UTCB
	Parallel Session 10—Seismic design and assessment *Invited lecture—Viorel Popa*	Hall I.2, FCCIA/UTCB
17:00–17:30	Coffee break	Room P1B, FCCIA/UTCB
17:30–18:15	Parallel Session 11—Seismicity of Romania	Hall I.1, FCCIA/UTCB
	Parallel Session 12—Seismic design and assessment	Hall I.2, FCCIA/UTCB
18:15–18:45	Closing ceremony	Hall I.2, FCCIA/UTCB

Location: FCCIA/UTCB, Bd. Lacul Tei nr. 124, Sector 2, Bucharest

6th National Conference on Earthquake Engineering
&
2nd National Conference on Earthquake Engineering and Seismology

DETAILED PROGRAM

Wednesday June 14, 2017
Chairperson—Dan Balteanu

13:00–14:00	Registration of participants
14:00–15:00	Opening ceremony
15:00–15:40	***Keynote lecture 1**—Kyriazis Pitilakis* Site classification and definition of seismic actions in the revision of EC8
15:45–16:30	***Keynote lecture 2**—Radu Vacareanu* Outcomes of seismic risk assessment for Romanian buildings and a roadmap for resilience
16:30–17:00	Coffee break
17:00–17:45	***Keynote lecture 3**—Dan Dubina* Design assisted by testing of seismic resistant steel structures
17:45–18:30	***Keynote lecture 4**—Mircea Radulian* Vrancea source investigation: milestones since 1977 to the present
18:30–20:30	Welcome cocktail

Location: Romanian Academy, Calea Victoriei 125, Sector 1, Bucharest

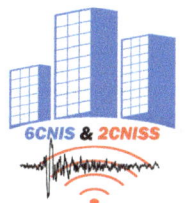

6th National Conference on Earthquake Engineering
&
2nd National Conference on Earthquake Engineering and Seismology

Thursday June 15, 2017		
8:00–9:00	Registration of participants	Hall I.2, FCCIA/UTCB
9:00–9:45	***Keynote lecture 5***—*Sierd Cloetingh* Lithosphere dynamics and sedimentary basins: from the deep Earth to the surface	Hall I.2, FCCIA/UTCB
9:45–10:30	***Keynote lecture 6***—*Iunio Iervolino* Seismic risk of modern structures: what damage to expect after strong earthquakes and where to expect it	Hall I.2, FCCIA/UTCB
10:30–11:00	Coffee break	Room P1B, FCCIA/UTCB
11:00–12:45	**Parallel Session 1**—Seismicity of Romania Chairpersons—S. Cloetingh & M. Radulian	Hall I.1, FCCIA/UTCB
11:00–11:30	*Invited lecture 1—G. Marmureanu* Historical earthquakes: new intensity data points using complementary data from churches and monasteries	
11:30–11:45	*Seismicity analysis using earthquakes energy* V. E. Toader, I. A. Moldovan, C. Ionescu, A. Marmureanu	
11:45–12:00	*The space-time distribution of moderate- and large-magnitude Vrancea earthquakes fits numerically-predicted stress patterns* M. A. Anghelache, H. Mitrofan, F. Chitea, A. Damian, M. Visan, N. Cadicheanu	
12:00–12:15	*Time series analysis of geospatial and field data for earthquake hazard assessment of Vrancea active zone in Romania* M. Zoran, D. Savastru, D. Teleaga, D. Mateciuc	
12:15–12:30	*The seismogenic sources from the west and south-west of Romania* E. Oros, M. Popa, M. Diaconescu	
12:30–12:45	*The seismic network of NIRD URBAN-INCERC for strong motion recording: a valuable tool for 50 Years of seismic monitoring activity* C. S. Dragomir, I. G. Craifaleanu, V. Meita, D. Dobre, E. S. Georgescu, A. Cismelaru	

(continued)

(continued)

Thursday June 15, 2017		
11:00–12:45	**Parallel Session 2**—Using targeted risk in seismic design codes: a summary of the state of the art and outstanding issues Chairpersons—J. Douglas & D. Lungu	Hall I.2, FCCIA/UTCB
11:00–11:30	*Invited lecture 2—T. Saito* *Real-time safety assessment of* *disaster management facilities against earthquakes*	
11:30–11:45	*A plea for people centred perspectives on seismic risk evaluation* I. Nenciu, M. Neagu, B. Suditu, M. Sercaianu	
11:45–12:00	*Risk management of earthquakes in Bucharest central area, the role of urban design* M. Bostenaru Dan	
12:00–12:15	*Evaluation of feasibility of pre-earthquake strengthening of RC buildings in Bucharest, Romania* F. Pavel, R. Vacareanu	
12:15–12:30	*Risk-targeted seismic design maps for Romania* R. Vacareanu, F. Pavel, V. Coliba, I. Craciun	
12:30–12:45	*Uniform risk-targeted seismic design maps for Romania* V. Coliba, R. Vacareanu, F. Pavel, I. Craciun	
11:00–12:45	Round Table—P100-3/2008 Convener—Viorel Popa	Hall I.4, FCCIA/UTCB
12:45–13:40	Lunch break	Room P1B, FCCIA/UTCB
13:40–14:30	A word from sponsors	Hall I.2, FCCIA/UTCB
14:30–15:15	**Keynote lecture 7**—*Mauro Dolce* The 2016–17 seismic sequence of central Italy: main scientific features and technical emergency activities	Hall I.2, FCCIA/UTCB
15:15–16:45	**Parallel Session 3**—Geotechnical earthquake engineering Chairpersons—K. Pitilakis & A. Aldea	Hall I.1, FCCIA/UTCB
15:15–15:45	*Invited lecture 3—D. Balteanu, M. Jurchescu, M. Sima, M. Micu, G. Kucsicsa* *Interdisciplinary research on seismically and* *rainfall-induced landslides in Romania*	
15:45–16:00	*Site studies for seismic risk mitigation in urban areas* S. F. Balan, B. F. Apostol	
16:00–16:15	*Investigation of local site responses at the Bodrum Peninsula, Turkey* H. Alcik, G. Tanircan	
16:15–16:30	*Ground types for seismic design in Romania* C. Neagu, C. Arion, A. Aldea, R. Vacareanu, F. Pavel	

(continued)

(continued)

Thursday June 15, 2017		
16:30–16:45	*Advanced geo-data modelling to assess earthquake-induced liquefaction deposits for Bucharest city* E. Calarasu, C. Arion, C. Neagu	
15:15–17:00	**Parallel Session 4**—Workshop "RO-RISK—Disaster risk assessment at national level" Chairpersons—A. Barbat & B. Duduc	Hall I.2, FCCIA/UTCB
15:15–15:45	*Invited lecture 4—A. H. Barbat, L. G. Barbu, S. Oller, A. Cornejo, C. Escudero, X. Matínez* *Seismic analysis of the containment building of an existing nuclear power plant*	
15:45–16:00	*National risk assessment—challenges, results & lessons learnt* B. G. Duduc, F. Senzaconi, I. Radu	
16:00–16:15	*Fragility and vulnerability curves for existing reinforced concrete buildings in Romania* I. Damian, C. Rusanu, V. Oprisoreanu, A. Papurcu	
16:15–16:30	*Seismic exposure analysis in RO-RISK project* C. Arion, C. Neagu, F. Pavel	
16:30–16:45	*Seismic risk assessment for residential buildings in Romania* F. Pavel, R. Vacareanu, C. Arion, C. Neagu, M. Iancovici, V. Popa	
16:45 –17:00	*The contribution of GIS to seismology. Case study: the assessment of seismic hazard and risk in Romania* D. Toma-Danila, C. O. Cioflan, I. Armas, E. F. Manea	
15:15–17:00	Round Table—Resilience-based assessment of structures Convener—Florin Pavel	Hall I.4, FCCIA/UTCB
17:00–17:30	Coffee break	Room P1B, FCCIA/UTCB
17:30–19:00	**Parallel Session 5**—Seismic design and assessment Chairpersons—A. Ilki & M. Iancovici	Hall I.1, FCCIA/UTCB
17:30–18:00	*Invited lecture 5—A. Ilki, E. Tore, C. Demir, M. Comert* *Code based performance prediction for a full-scale FRP retrofitted building test*	
18:00–18:15	*Toward the seismic evaluation of "Carol I" Royal Mosque in Constanta—ambient vibration measurements* A. Aldea, C. Neagu, E. Lozinca, S. Demetriu, S. M. Bourdim, F. Turano	
18:15–18:30	*Nonlinear design optimization of reinforced concrete structures using genetic algorithms* A. Pricopie, I.-V. Cimbru	
18:30–18:45	*A macro modelling blind prediction of a cyclic push-over test on a full scale masonry house* T. Hoogeveen, J. White	

(continued)

(continued)

Thursday June 15, 2017		
18:45–19:00	*Static and dynamic approaches on the low-rise rc frames seismic capacity evaluation and damage quantification* V.-A. Paunescu, M. Iancovici	
17:30–19:00	**Parallel Session 6**—Seismic risk evaluation and management of emergency situations Chairpersons—Emil-Sever Georgescu & Cristian Arion	Hall I.2, FCCIA/UTCB
17:30–18:00	*Invited lecture 6—E.S. Georgescu, K. Steinbrueck, A. Pomonis* *New archival evidence on the 1977 Vrancea, Romania earthquake and its impact on disaster management and seismic risk*	
18:00–18:15	*Presentation of structural systems and characteristic parameters of Romanian buildings for application of the RVS method* C. Stere, C. Ursu	
18:15–18:30	*Application of advanced tools for the evaluation of natural hazards preparedness: the Romanian participation in the E-PreS project* I.-G. Craifaleanu, E.-S. Georgescu, V. Meita, C.-S. Dragomir, D. Dobre, A. Cismelaru	
18:30–18:45	*Seismic preparedness of the population of Bucharest, Romania: questionnaire results* I. Calotescu, F. Pavel	
18:45–19:00	*Seismic risk assessment through the lenses of a civil engineer: the need of structural health monitoring* A. Tiganescu	
20:30–23:00	Gala dinner	

Location: FCCIA/UTCB, Bd. Lacul Tei nr. 124, Sector 2, Bucharest

6th National Conference on Earthquake Engineering
&
2nd National Conference on Earthquake Engineering and Seismology

Friday June 16, 2017		
9:00–9:45	***Keynote lecture 8**—Alik Ismail-Zadeh* From lithosphere dynamics and earthquake modelling through seismic hazard and risk assessments to disaster risk reduction: A long way towards seismic safety and sustainability	Hall I.2, FCCIA/UTCB
9:45–10:30	***Keynote lecture 9**—Humberto Varum* Influence of infill masonry walls in the seismic response of buildings: from field observations to laboratory research	Hall I.2, FCCIA/UTCB
10:30–11:00	Coffee break	Room P1B, FCCIA/UTCB
11:00–13:00	**Parallel Session 7**—Innovative solutions for seismic protection of building structures Chairpersons—M. Seki & D. Dubina	Hall I.1, FCCIA/UTCB
11:00–11:30	*Invited lecture 7—M. Seki* *A proposal on the simplified structural evaluation method for existing reinforced concrete buildings with infilled brick masonry walls*	
11:30–11:45	*Development of two types of buckling restrained braces* C. I. Zub, A. Stratan, A. Dogariu, C. Vulcu, D. Dubina	
11:45–12:00	*Seismic protection of structures equipped with magneto-rheological dampers* C. Vulcu, D. Dubina, N. Popa, L. Vekas, G. Ghita, T. Sireteanu, I. Borbath, R. Oprescu	
12:00–12:15	*Re-centring capacity of dual steel building frames with replaceable thin-walled shear panels* C. Neagu, F. Dinu, D. Dubina	
12:15–12:30	*Seismic performance of eccentrically braced frames with bolted links under different types of seismic motions* A. Stratan, A. Chesoan, D. Dubina	
12:30–12:45	*Robustness performance of seismic resistant steel moment connections* I. Marginean, F. Dinu, D. Dubina	
12:45–13:00	*Towards European prequalification criteria for steel moment joints in multistory buildings frames* C. Maris, A. Stratan, D. Dubina	
11:00–13:00	**Parallel Session 8**—Seismicity of Romania Chairpersons—A. Ismail-Zadeh & C. Ionescu	Hall I.2, FCCIA/UTCB
11:00–11:30	*Invited lecture 8—J. Douglas, A. Gkimprixis* *Using targeted risk in seismic design codes: A summary of the state of the art and some recent work*	

(continued)

(continued)

Friday June 16, 2017	
11:30–11:45	*Seismic Intensity estimation using macroseismic questionnaires and instrumental data—Case Study Bârlad, Vaslui County* I. A. Moldovan, B. Grecu, A. P. Constantin, A. Anghel, E. F. Manea, L. Manea, R. Partheniu
11:45–12:00	*Earthquake mechanism and correlation with seismogenic zones in the southern and eastern part of Romania* A. Bala, M. Radulian. E. Popescu, D. Toma-Danila
12:00–12:15	*Testing the macroseismic intensity attenuation laws for Vrancea intermediate depth earthquakes* M. M. Rogozea, I. A. Moldovan, A. P. Constantin, E. F. Manea, C.-O. Cioflan, L. M. Manea
12:15–12:30	*Evidence for earthquake damage on St. Michael church in Cluj-Napoca, Romania* M. Kázmér
12:30–12:45	*On the ground motions' spatial correlation for Vrancea intermediate-depth earthquakes* I. Craciun, R. Vacareanu, F. Pavel
12:45–13:00	*Abnormal animal behavior prior to the Vrancea (Romania) major subcrustal earthquakes* A. P. Constantin, I. A. Moldovan, R. Partheniu

11:00–13:00	**Round Table**—Quick post-earthquake evaluation of buildings Convener—Eugen Lozinca	Hall I.4, FCCIA/UTCB
13:00–14:00	Lunch break	Room P1B, FCCIA/UTCB
14:00–14:45	***Keynote lecture 10***—*Roberto Paolucci* 3D physics-based numerical simulations: advantages and limitations of a new frontier to earthquake ground motion prediction	Hall I.2, FCCIA/UTCB
14:45–15.30	***Keynote lecture 11***—*Koichi Kusunoki* A new structural health monitoring system for real-time building damage evaluation	Hall I.2, FCCIA/UTCB
15:30–17:00	**Parallel Session 9**—Seismic design and assessment Chairpersons—M. Maeda & D. Marcu	Hall I.1, FCCIA/UTCB
15:30–16:00	*Invited lecture 9—M. Maeda, H. Al-Washali* *Damage due to earthquakes and improvement of seismic performance of reinforced concrete buildings in Japan*	
16:00–16:15	*Study of seismic capacity of masonry infilled reinforced concrete frames considering the influence of frame strength* H. Al-Washali, K. Jin, M. Maeda	
16:15–16:30	*Specific aspects in seismic analysis of embedded structure* C. Ilinca, A. Popovici, C. Anghel	
16:30–16:45	*How to improve the seismic behaviour of steel structures by using "dog-bone" configurations* H. Köber, M. Stoian	
16:45–17:00	*Comments about the seismic design of concentrically braced frames* H. Köber, R. Marcu	
15:30–17:00	**Parallel Session 10**—Seismic design and assessment Chairpersons—T. Saito & V. Popa	Hall I.2, FCCIA/UTCB
15:30–16:00	*Invited lecture 10—V. Popa* *Challenges of earthquake resistant design for buildings in Bucharest*	

(continued)

(continued)

Friday June 16, 2017		
16:00–16:15	*The use of artificial neural networks in assessing the state of effort in three-dimensional models with solid elements* M. Budescu, A.-E. Pandelea, L. Soveja	
16:15–16:30	*Virtual lifelong learning platform in the field of seismic design* R. Pascu, I.-G. Craifaleanu, O. Anicai, L. Stefan, V. Popa, V. V. Oprisoreanu, I. Damian, A. Papurcu, C. Rusanu	
16:30–16:45	*Anchorage of beam reinforcement in beam-column joints: overview of the Romanian standards* D. Cotofana, M. Pavel, V. Popa	
16:45–17:00	*On the seismic jerk* R. Sofronie	
17:00–17:30	Coffee break	Room P1B, FCCIA/UTCB
17:30–18:15	**Parallel Session 11**—Seismicity of Romania Chairpersons—A. Bala & C. Cioflan	Hall I.1, FCCIA/UTCB
17:30–17:45	*Presignal signature of radon (Rn222) for seismic events* M. Zoran, R. Savastru, D. Savastru, D. Mateciuc	
17:45–18:00	*Ground motion intensity versus ground motion kinematics. Some alternative intensity measures* H. Sandi	
18:00–18:15	*Combined solutions for an integrated GNSS study over NW Galati seismogenic area issued from GPS continuous and campaign measurements* E. I. Nastase, A. Muntean, C. Ionescu, V. Mocanu, B. Ambrosius	
17:30–18:15	**Parallel Session 12**—Seismic behaviour of engineering structures Chairpersons—K. Kusunoki & M. Pavel	Hall I.2, FCCIA/UTCB
17:30–17:45	*Experimental study on eccentrically braced frames with a new type of bolted replaceable active link* A. Ashikov, G. C. Clifton, B. Belev	
17:45–18:00	*Assessment of multi-storey steel structures behaviour concentrically braced with or without BRBs* S. C. Ionescu-Lupeanu, A.-S. Dima	
18:00–18:15	*Blind prediction of structural behavior for a reinforced concrete structure subject to torsion and nonlinear effects* C. A. Ursu, C. C. Stere, V. V. Oprisoreanu	
18:30–19:00	Closing ceremony	Hall I.2, FCCIA/UTCB

Location: FCCIA/UTCB, Bd. Lacul Tei nr. 124, Sector 2, Bucharest

Annex 2

Alex Horia Barbat	Masaki Maeda	**International advisory committee**
Sierd Cloetingh	Roberto Paolucci	
Mauro Dolce	Kyriazis Pitilakis	
John Douglas	Rajesh Rupakhety	
Iunio Iervolino	Taiki Saito	
Alper Ilki	Matsutaro Seki	
Alik Ismail-Zadeh	Humberto Varum	
Koichi Kusunoki	Akira Wada	
Alexandru Aldea	Constantin Ionescu	**Scientific committee**
Emil Albotă	Dan Lungu	
Andrei Bălă	Dragoș Marcu	
Dan Bălteanu	Vasile Meiță	
Virgil Breabăn	Iren Adelina	
Mihai Budescu	Moldovan	
Sorin Demetriu	Mihaela Popa	
Dan Dubină	Viorel Popa	
Benone Gabriel	Mircea Radulian	
Duduc	Francisc Senzaconi	
Emil Sever	Daniel Stoica	
Georgescu	Radu Vacareanu	
Mihail Iancovici		
Paul Ioan		
Cristian Arion	Ancuța Neagu	**Organizing committee**
Dan Bîtcă	Cristian Neagu	
Ileana Calotescu	Paul Olteanu	
Carmen Cioflan	Cristian Onofrei	

(continued)

(continued)

Veronica Colibă	Vasile Oprişoreanu	
Dragoş Coţofană	Andrei Papurcu	
Ionuţ Crăciun	Raluca Partheniu	
Mihai Dragomir	Daniel Paulescu	
Bogdan Ghinea	Florin Pavel	
Helmuth Kober	Mihai Pavel	
Eugen Lozincă	Ionuţ Radu	
Eduard Năstase		

Lightning Source UK Ltd.
Milton Keynes UK
UKHW02n0609180418
321195UK00002B/12/P